INVITATION TO
Contemporary Physics

2nd edition

INVITATION TO
Contemporary Physics

2nd edition

Q. Ho-Kim
Université Laval, Canada

N. Kumar
Raman Research Institute, Bangalore, India

C. S. Lam
McGill University, Canada

 **World Scientific**

NEW JERSEY • LONDON • SINGAPORE • SHANGHAI • HONG KONG • TAIPEI • BANGALORE

Published by

World Scientific Publishing Co. Pte. Ltd.

5 Toh Tuck Link, Singapore 596224

USA office: Suite 202, 1060 Main Street, River Edge, NJ 07661

UK office: 57 Shelton Street, Covent Garden, London WC2H 9HE

British Library Cataloguing-in-Publication Data
A catalogue record for this book is available from the British Library.

INVITATION TO CONTEMPORARY PHYSICS (2nd Edition)

ISBN 981-238-302-6
ISBN 981-238-303-4 (pbk)

Printed in Singapore by Mainland Press

Preface

This book, first published some ten years ago, is an attempt to communicate to a non-specialist readership the main results in selected areas of modern-day physics. The last decade, which finally brought to a close an eventful century, was marked by significant advances in science and technology, achieved against the background of a shifting global political landscape.

In 1990, the Hubble Space Telescope, a 2.4 meter reflecting telescope, was deployed in earth orbit by the crew of the space shuttle Discovery; together with other orbiting observatories that followed close behind, it vastly expanded the horizons of our observable cosmos and worked to better define our universe and our place in it. That same year, the Human Genome Project was launched with the goal to identify the genes in human DNA and to determine the sequences of the base pairs that make it up; when this project is completed, not only will we know more about ourselves than ever before, but we will also have learned, in the process, the ways nature works. At about the same time, the HyperText Transfer Protocol (HTTP) was created that went on to become the standardized means of information transfer over computer networks, thereby inaugurating the information age and changing forever the way we live and work.

To reflect these important developments and a number of others, we have rewritten several chapters of the first edition and revised or updated all of the others. To these we have added three completely new chapters, on Bose–Einstein condensation, nanoscience, and quantum computation, three emerging areas with great potential for impact and applications in physics and beyond.

It is the objective of this book to present the essential concepts and observations of contemporary physics in language as simple as possible, without much mathematics but not without rigor. We have tried to write at a level that corresponds to a lower-undergraduate course, although, occasionally, the nature of the topic being discussed makes a more advanced treatment unavoidable.

As a textbook, it may be regarded as our modest contribution to a renewed approach to teaching introductory physics, in which concrete real-life examples happily cohabit with the usual elements of a traditional course. Whether taking a logical bottom-up or a thematic top-down approach, the physics teacher would want above all to keep her students interested and motivated: she will find here a source of fascinating topics at the research frontiers for open-classroom discussions or essay assignments.

This work also addresses the general reader who has a keen interest in physics. Physics, just as science in general, is not only about nature; it is also about people: it is a human pursuit, as old as civilization, as ingrained in our nature as our search for happiness. As a human activity, it shapes our intellect, molds our view of the world and of ourselves; but, for good or ill, it also affects our everyday life. It behoves us all, as ordinary citizens, to keep ourselves constantly informed of its progress and be alert to its issues and implications. Given the way science is built up and the pace at which advances are being made, those who stay behind are bound to fall farther and farther behind.

The reader should regard our book as an invitation to deeper meditation or further studies; you will find at the end of each chapter suggestions of possible avenues to more extended explorations. Mathematics is the natural language of physics, and, how ever hard we try, we cannot fully appreciate physics without equations. You may wish to check your understanding of the subject at the quantitative level by attempting to solve some of the end-of-chapter problems with the help, if necessary, of the hints scattered throughout the chapter or in Appendixes A–C and a peek or two at the solutions given in Appendix D.

C. S. Lam would like to thank Hoi-Kwong Lo and Patrick Hayden for their help on Chapter 6.

March 2003

Contents

Contents

Symmetry of Nature and
Nature of Symmetry

1

1.1   What Is Symmetry That We Should Be Mindful of It?

Our immediate sense of symmetry comes from looking at objects around us. It may well be that the idea of symmetry is very primitive and comes naturally to the human mind. Perhaps the human mind can grasp it internally all by itself. But we shall leave these questions to the philosopher and to the artist. Instead, let us for a moment turn experimentalist and consider a sphere. Then we will be left in no doubt that we are in the presence of a perfect symmetry. We may view the sphere actively by turning it around every which way we like and find that it looks the same. We may view it passively by keeping the sphere fixed but shifting ourselves around it and find again that it looks just the same. It is this unchanging aspect of sameness against a changing viewpoint that symmetry is all about. But then we have to get sophisticated. We have to abstract the general idea of symmetry and make it free from this static and rather limited visual setting. This we must do and in doing so we will see more, and not less than the artist can, for all his sensitivity and imagination, ever hope to see. There is much more subtlety familiar in the world of physics than meets the eye. However, we will continue to use the same word for it: symmetry.

Symmetry suggests a sense of balance and proportion, of pattern and regularity, of harmony and beauty, and finally of purity and perfection. These synonyms just about sum up all our subjective reactions to the symmetries that abound in Nature, with her myriads of inanimate objects and life forms — the celestial spheres of the sun, the moon and the planets, the hexagonal snowflake with its six-fold symmetry, the five-fold symmetry of the starfish and of many a wild flower, the bilateral symmetry of the butterfly with its outstretched wings and of the man in his poise (Fig. 1.1). One even speaks of the fearful symmetry of the tiger. Examples will fill volumes. And as life imitates Art and Nature, we find something of it reflected in the art forms created by man — be it sculpture, architecture, painting, poetry or music. It is true though that in most of these cases the symmetry is only approximate. As a matter of fact the ancient Greeks used to intentionally

1

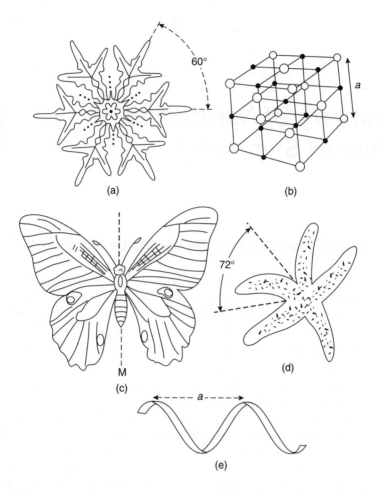

Figure 1.1:   (a) Snowflake with six-fold axis; (b) crystal of common salt; (c) butterfly with bilateral symmetry; (d) starfish with five-fold axis; (e) right-handed helix.

and secretly introduce some degree of asymmetry in their otherwise symmetric designs. (After all, there is no perfect beauty that has not in it a certain *strangeness of proportion*). The fact remains, however, that the human mind is absolutely fascinated by symmetry. In physics, the term symmetry takes on an objective meaning which is much deeper and far more precise, almost more austere than our vague feelings of it can command. Let us get acquainted with it.

Now, we can hardly do better than just repeat the definition of symmetry given by the great German mathematician Hermann Weyl — a thing is symmetrical if there is something you can do to it so that after you have finished doing it, it looks the same as it did before. This is an operational definition — it can decide. The 'thing' here is the *object* of interest. What you do to it is called the *Symmetry operation* or *transformation*. And 'looks the same' is yet another name for *invariance*. The 'look' itself is some discernible property of the object that remains invariant. Thus, there has to be an object with a discernible property that remains invariant under the action of the '*group*' of symmetry transformations. Now, the point of all this is that the object itself can be just about anything. It depends on our interest and on the level or the depth of our enquiry. At its simplest, the object may be a mere geometrical figure (a hexagon, a helix or a lattice), or the geometrical shape of a material body (a snowflake, a screw or a crystal of common salt) (Fig. 1.1). The symmetry operations involved here are purely geometric in nature — rotation by $360/6 = 60$ degrees or multiples of it about the six-fold axis of rotation, mirror reflection in the plane of the bilateral symmetry, translation in space by a repeat distance, or combinations of these (Fig. 1.1). The object and its transform must be superposable if the symmetry is true. (This is obviously not so for a screw, or a helix. Although the screw is intrinsically identical with its mirror image, the two are not superposable. We will return to this interesting case later). But at its subtlest the object can be a mathematical entity, a (differential) equation expressing a physical law. Now, how do you rotate, reflect or translate an equation anyway? Well, we really do not do so literally. We perform these transformations passively on the independent variables, *i.e.*, the space-time coordinates occurring in the equation accompanied then by suitable transformations on the dependent variables. The invariance then is the invariance of the *form* of the equation under these symmetry transformations. More properly, it is called covariance. Thus, for instance, an expression $x^2 + y^2 + z^2$ is invariant under any rotation of the Cartesian coordinate system (x, y, z) with its origin fixed at $x = 0$, $y = 0$ and $z = 0$. It just becomes $x'^2 + y'^2 + z'^2$, where the primed quantities are the coordinates of the same point, but with respect to the rotated (primed) coordinate system (x', y', z'). Similarly, the wave equation

$$\frac{\partial^2 \phi}{\partial x^2} + \frac{\partial^2 \phi}{\partial y^2} + \frac{\partial^2 \phi}{\partial z^2} - \frac{1}{c^2} \cdot \frac{\partial^2 \phi}{\partial t^2} = 0$$

keeps its form under the above symmetry transformation, and additionally under translation in space and in time. Just replace the unprimed quantities by the primed quantities. In particular $\phi(x, y, z, t)$ becomes $\phi'(x', y', z', t')$ and is numerically equal to it. If you take ϕ to be the pressure or the density, then the wave equation begins to describe a sound wave propagating in a medium such as air or water which is homogeneous (translationally invariant in space), isotropic (rotationally invariant) and unchanging in time (translationally invariant in time). In fact this equation has a much higher symmetry and it can describe the widest range of wave phenomena that occur in Nature. These symmetries of the medium (and the medium may well be vacuum as in the case of light) *almost uniquely* fix the form of this equation. Such is the restrictive power of symmetry.

One speaks of the symmetry of a particular law. Thus, we have the spherical symmetry of the Coulomb law of electrostatic attraction between a negatively charged electron and a positively charged nucleus of an atom. The Coulomb potential energy varies as the inverse of their distance apart, independent of the direction. The force on the electron, being the gradient of potential, of course, varies as the inverse of the square of this distance and is directed radially inward. But for an atom embedded in a molecule or a solid, the potential law governing the motion of the electron has the symmetry of its environment which is necessarily lower than the spherical symmetry of the free atom. Much of the chemistry of molecules and the physics of solids depend on these environmental symmetries.

As we probe matter deeper, we uncover special laws that govern the goings-on at the nuclear and the subnuclear level — the domain of the elementary particles and the fundamental interactions between them. Here we encounter yet another kind of symmetry different from the space-time symmetries described above. These are the so-called *local gauge symmetries* that seem to be at the very heart of the nature of things. It is already present in the interaction of light with charged particles, where it was discovered first. But, of this more later.

A note of caution at this stage is in order. The symmetry of a law as expressed by a symmetric equation does not necessarily lead to symmetric phenomena resulting from it. The states of a system, the processes or the events represent the allowed solutions of the governing (differential) equation. But a particular solution gets selected by the initial conditions that can be imposed at will. These conditions need not have the symmetry of the system. And so it happens that the law of gravitational attraction between the earth and the sun is spherically symmetric, and yet the orbit of the earth round the sun is an ellipse — a foreshortened circle, with the sun at one of its foci. The same is true of a man-made satellite orbiting the earth. Its orbit depends on its height and the velocity at the time of its injection into orbit. A symmetry operation will not leave the particular orbit invariant but carry it into another, albeit allowed orbit. Thus, in general the particular solutions (realizations) or the events or physical conditions themselves are not invariant. What is indeed invariant is the governing equation that fixes only the correlations between the successive events. The idea that the state of a system can have a symmetry lower than that of the governing law takes on a deep physical significance as we will see when we discuss the phenomenon of *spontaneous symmetry breaking*, which is the most symmetrical way of breaking the symmetry. In this we may catch a glimpse of the act of creation whereby Nature seems to have generated the observed diversity of fundamental laws as a result of a descent from the most symmetric, possibly a '*grand-unified*' law of interactions.

The all pervasive nature of symmetry is in itself a sufficiently strong reason for us to be mindful of it. But the most compelling reason of all is that symmetry is a great ordering principle and we can make it work for us. We will now demonstrate this power with the help of some simple and some not-so-simple examples.

To start with, symmetry simplifies things. Suppose you are asked to draw a butterfly with its out-stretched wings. Now, all that you really have to do is to draw only the left, or the right half of the butterfly, preferably on a tracing sheet. The other half is related to it by mirror reflection. It is more of the same. You can simply fold the sheet along the median line of bilateral symmetry and re-trace over your half-drawing. That is all. The reflection symmetry has halved your work, or very nearly so. In general, an n-fold symmetry divides your work by n. This is really a common trick and we should imagine that the makers of patterns use it all the time. This is, however, a trivial example.

A highly non-trivial example of reduction of a problem by symmetry is provided by the case of a hydrogen atom. Here we have an electron bound to the nucleus (proton) by the attractive Coulomb potential which is spherically symmetric. In order to appreciate reasonably well the promised reduction of the problem, we have to describe the atom properly. It is now well known that in the domain of the very small, and that is where the atoms belong, the proper theoretical framework is that of *Quantum Mechanics*, and not the *classical (Newtonian) mechanics* that describes our sensible world of middle dimensions so well (see Appendices A and B). Thus, we have to abandon the classical view of the hydrogen atom as a miniature solar system with sharply defined orbits for the planetary electron. We have, instead, an all pervasive waviness associated with the motion of the electron. We can picture the state of the electron as a fuzzy cloud around the nucleus, with the proviso that the density of the cloud at a point gives the probability (density) of finding the point-like electron at that point. (This replacement of the classical certainty of sharply determined orbits by the quantum uncertainty of dicey probabilities of being found somewhere is most disturbing. It was so to Einstein himself who was, ironically, one of the founders of this 'plutonic' republic of Quantum Mechanics, but never quite belonged there as a citizen. Quantum Mechanics is today the established *framework* theory for everything in the physical universe. Its predictions differ from those of classical mechanics and the difference gets more and more pronounced as we go deeper into the domain of the small). To get these probabilities one has to solve a certain wave equation, the *Schrödinger equation*, for the wave function ψ, which is complex in general. The probability is then simply $|\psi|^2$, the square of its absolute magnitude. All that is important for our discussion is to note that ψ has both radial as well as angular dependence. The spherical symmetry of the Coulomb potential now helps us factor out the angular dependence and determine it completely without having to solve the Schrödinger equation. The spherical symmetry by itself determines the allowed values of the angular momentum $\ell\,(=0,1,2,\ldots)$ and its component $m\,(=-\ell,-\ell+1,\ldots,\ell-1,\ell)$ along a chosen direction in units of Planck's constant h divided by 2π. These are the labels, called *quantum numbers* that symmetry provides to specify completely the angular aspect of the state of the system. This is no mean reduction of the problem. In fact one can do better than this. In addition to the spherical symmetry, the Coulomb law has yet another 'dynamical' symmetry following from a certain special value of a parameter in its form, namely that the force involves the square of the reciprocal of the distance, and not any other power such as the cube or the fourth power, and so on. (This is, of course, a rather hidden dynamical symmetry and is for the preoccupied eyes of the mathematical physicists only). Properly treated, this symmetry solves the remaining radial problem too and provides yet another label, the principal quantum number $n \, (= 1, 2, \ldots)$ that fixes the allowed electronic energies.

That there is something special about the inverse square law which singles it out from among all possible central forces, can be seen from the following fact. Consider the motion of the earth around the sun, or better still the motion of a man-made satellite around the earth. The orbits are elliptical as we know. But the real point, which hardly ever gets emphasized, is that the orbit closes upon itself! This will not be the case if you deviate ever so slightly from the inverse square law. For small deviations the orbit will still be close to being an ellipse but the ellipse will slowly precess or turn around the focus. The motion of perihelion (the point of closest approach to the sun) of the orbit of the planet Mercury around the sun may be viewed as due to small deviation from this dynamical symmetry of the inverse-square law caused by Einstein's general relativistic corrections to Newton's law of gravitation.

Now we turn to another aspect of this great ordering principle, namely, that symmetry classifies things. All classification is based on identification of a set of common characteristics. Thus we have the classification of the animal kingdom into vertebrates and invertebrates depending on the presence or the absence of the vertebral column. The *periodic table* of elements prepared by the great Russian chemist Mendeleyev is a classic example of classification. The most striking and rigorous example of classification by symmetry is the grouping of crystalline forms of solids. A crystal is a periodic arrangement of atoms in space. It can have spatial symmetries of discrete translation, discrete rotation and reflection and, of course, combinations of these. Symmetry considerations have led to the remarkable result that only a finite number of distinct groupings of these symmetry elements are possible. These are the celebrated 230 space groups of crystallography! Any of the nearly countless varieties of crystals, no matter how complex, must belong to one of these groups. We must hasten to add, however, that the crystals belonging to a given space group are certainly not identical, no more than all the vertebrates in the animal kingdom are identical. Finding the space group of a crystal is the first step towards understanding its molecular structure.

A much more profound example of symmetry-based classification in physics is the classification of identical particles as *fermions* (after the great Italian physicist Enrico Fermi) and as *bosons* (after the great Indian physicist S.N. Bose). Here the symmetry is with respect to permutation, or more simply, reshuffling. Let us understand this. Consider a set of particles located arbitrarily in space. Let the particles be identical in all respects, *i.e.*, the same mass, the same charge, and so on. You may think of a pack of cards, somewhat unusual in the sense that all the cards are alike — only queens of diamond, say. Now it is clear that any permutation of these identical particles (the same as reshuffling of the identical cards) will leave our system unchanged. After all, a permutation involves just pairwise interchanges, and interchanging identical objects changes nothing. But not quite. The different permuted configurations are undoubtedly identical but they are distinguishable all the same. The reason for this is that nothing prevents you from keeping track of these identical particles as these are being moved around to their new permuted locations. This knowledge is sufficient to distinguish between the different permuted configurations even though the objects being permuted are identical. Thus the identical particles are distinguishable even if only by virtue of being initially located differently. You may wonder if this knowledge is of any consequence and if this distinction between identity and indistinguishability is not mere nitpicking. Classically, you are right. But as we have noted earlier, the correct framework for dealing with microscopic particles is quantum mechanics. And, most importantly, quantum mechanics does not allow sharply defined trajectories. It replaces them with an irreducible fuzziness. Therefore, even in principle, we really cannot keep track of our identical particles in the process of permuting them as we did before. This idea of indistinguishability is brought home rather forcefully if you consider, *e.g.*, a pair of algebraic equations $x^2 + y^2 = 13$ and $x + y = 5$. These two equations are left invariant if we interchange x and y and hence permutation symmetric. Now, you can readily solve these two equations. You get either $x = 3$ and $y = 2$, or $x = 2$ and $y = 3$. Thus all you can say is that one of them equals 2 and the other equals 3, but which one is which you cannot say even in principle. So is the case with our identical particles. We can only say how many are there at a given point of space (*i.e.*, the occupancy) but it is meaningless to ask which ones. This *indistinguishability* when treated properly leads to the great divide of identical particles into two classes — the fermions (*e.g.*, electrons, protons, neutrons, neutrinos, etc.) and bosons (photons, mesons, etc.). Identical fermions, electrons, say, exclude each other in that not more than one can occupy the same state. This is the Fermi statistics — kind of negative feedback at work. In contrast to this, any number of identical bosons, photons say, are allowed to occupy the same state. This is the Bose Statistics. In fact, bosons tend to clump together, a kind of positive feedback. What determines whether a given set of identical particles will be fermions or bosons requires deeper analysis of relativistic invariance. It is beyond our scope to go into that. But the result is simple. It turns out that a particle can have an intrinsic angular momentum called spin. You may roughly picture it as a spinning top much the same way as the earth spins about its own axis in addition to orbiting around the sun. The spin angular momentum is immutable (you cannot stop it spinning). It is quantized in multiples of $h/2\pi$, denote by slashed $\hbar$. Now the rule is that particles with integral spin $(0, \hbar, 2\hbar, \ldots)$ are bosons and those with half odd-integral spin $(\hbar/2, 3\hbar/2, 5\hbar/2, \ldots)$ are fermions. This connection between spin and statistics has been one of the marvels of the symmetry principles in physics. The fact that two electrons (fermions with spin half) cannot simultaneously occupy the same point of space with their spins pointing in the same direction (*i.e.*, cannot be in the same state) is responsible for the stability of all matter, and for the fortunate circumstance that your hands do not go through the table which they might be resting on. For, in doing so the electrons in your hand must go through the electrons in the table which is clearly forbidden. The clumping tendency of photons (bosons with spin unity), on the other hand, makes it possible for any number of them to condense into a given state — *Bose condensation* (See Chapter 4). This is what makes laser beams so coherent (See Chapter 2 on Lasers). Similarly, superfluidity of ^{4}He (the isotope of helium with total spin zero), namely that it can flow through the finest capillaries without any viscosity, is due to the same Bose condensation of these atoms in the lowest energy state at low temperatures close to absolute zero. ^{3}He, the fermionic isotope, on the other hand, behaves differently even though chemically the two isotopes are identical.

Elementary particle physics abounds in examples of order brought about by classification of the zoo of particles based on certain postulated, rather abstract and well concealed symmetries without knowledge of the details of the underlying laws (see Chapter 9).

Symmetry is also highly restrictive. It limits the possibilities allowed without detailed knowledge of the system. The classic example is the forbidden five-fold axis of rotational symmetry in a crystal. The only allowed ones are the two-fold, three-fold, four-fold and the six-fold axes. The compatibility of the rotational and the translational symmetries rules out the five-fold axis as also the higher order axes of rotation. The five-fold axis is also conspicuous by its absence on the floor designs, or the tiling of a plane called tessellation. You see the square, the equilateral triangular and the regular hexagonal motifs, but never a regular repeating pentagonal pattern with the five-fold symmetry. However, individual molecules and other objects can and in fact do have the five-fold axis. Just think of a pentagram or the starfish. It is an interesting thought that living organisms like the starfish may adopt the five-fold symmetry as a natural defence against the deadly 'capture' by the rigid crystalline formation.

The really restrictive power of symmetry in physics derives from the overriding conservation laws that it imposes — the conservation of energy, momentum, angular momentum and charge. We will return to this when we discuss this connection between invariance and conservation laws. Processes violating these are simply forbidden.

Symmetry is at its most powerful when it predicts. Let us illustrate this with an example from solid geometry. Suppose you are interested in regular convex polyhedra (poly = many, hedra = faces). A regular polyhedron is a volume bounded by plane faces which are identical regular polygons. A simple cube (the common dice, for earth) is one such polyhedron. It has six faces that are square (you can call it a regular hexahedron). There are other regular polyhedra, namely the tetrahedron (for fire) with four equilateral triangular faces, the octahedron (for air) with eight

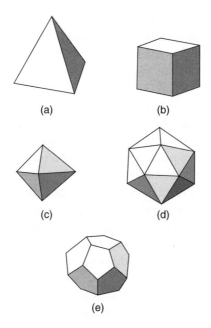

Figure 1.2: Five platonic solids: (a) tetrahedron; (b) cube; (c) octahedron; (d) dodecahedron. (e) icosahedron;

equilateral triangular faces, the dodecahedron (for quintessence) with twelve regular pentagonal faces and finally the mysterious icosahedron (for water) with twenty equilateral triangular faces (Fig. 1.2). These are the so-called Platonic Solids contemplated by the Greek Pythagoreans. The question is if there are more. Well, the answer is a definite no. Symmetry forbids any other occurrence. This is a restrictive aspect of symmetry. The predictive aspect is just the flip side of the coin. If there are intelligent inhabitants in some distant galaxy interested in these exotic dice-forms, we can predict that they will find just these five and no more.

But the real predictive power of symmetry is seen in particle physics. The basic idea is just this. Having identified or guessed the symmetry of the governing law, the processes, or the states or the particles related by the symmetry operations are all treated *at par*, *i.e.*, equally allowed and intrinsically the same. Thus, if you find one, the others, the missing ones are predicted. This is, for example, how the short-lived particle called Ω^- was predicted by Gell–Mann in 1962, and later confirmed in 1964 as the missing member of the family of ten objects (resonances) predicted on the basis of a postulated symmetry $SU(3)$. This was a historic triumph of symmetry in physics.

There are two other aspects of symmetry with far-reaching consequences. These are its unifying and creative powers. We will return to this point later.

There is an ingenious way crystallographers use the power of symmetry constructively. Suppose you need to know the structure of a complex molecule. It may be a protein with some hundred thousand atoms, or a fragment of DNA. These are very important but complex molecules. Proteins are the building blocks of cells and enzymes, while DNA (Deoxyribonucleic acid) carries the genetic information for making these proteins. Now, you cannot use ordinary light to probe these. Its wavelength of several thousand Angstroms (1 Å $= 10^{-8}$ cm) is much too large to reveal the finer molecular details on the scale of a few Angstroms. We must use X-rays with wavelengths of about an Angstrom or so. If you shine X-rays on a sample containing these molecules, placed and oriented randomly, the scattered waves of X-rays will interfere randomly to produce a mere smudge on a photographic plate. If, however, you could somehow arrange the molecules periodically in space, that is to say if you could crystallize the substance, the waves scattered from the molecules would interfere constructively in certain well-defined directions and thus produce a systematic pattern of bright sharp spots (the diffraction pattern) on the plate. This is like making Fourier series analysis of a periodic function. One can invert this to get at not only the periodic structure of the crystal lattice but also the structure of the molecules making it up! (One only hopes that the imposed crystalline arrangement has not done too much violence to the molecule whose structure we were interested in). This is why crystallographers-turned-molecular biologists round the world are preoccupied with crystallizing these substances. At this point, we should note that the crystalline order as a necessary condition for getting sharp X-ray spots has been called into question recently with the discovery of the so-called *quasicrystals* by D. Shechtman, I. Blech, D. Gratias and J. W. Cahn (1984). The first quasicrystal was an alloy, $Al_{14}Mn_{86}$, *i.e.*, 14 atomic per cent aluminum and 86 atomic per cent manganese. Since then many more have been found. These materials show sharp diffraction spots like any other good crystal but the arrangement of spots has a five-fold symmetry which is, of course, forbidden in the real space crystal lattice. The conclusion is that the conventional crystalline order is not necessary for sharp spots in X-ray diffraction. A two-dimensional quasicrystal is exemplified by the so-called Penrose aperiodic tiling of a plane with motifs of two rhombuses fitted as pieces of a jigsaw puzzle (Fig. 1.3). The smaller and the larger rhombuses have angles 72 degrees and 108 degrees, and 36 degrees and 144 degrees, respectively, and their areas and numbers are in the golden ratio $= (1 + \sqrt{5})/2$. This is the intellectual property of the Oxford mathematician Roger Penrose, who constructed it for play. The tiling has no translational symmetry of the conventional crystals and yet would give a sharp diffraction pattern. It is now known that quasicrystals may be viewed as a projection of ordinary-crystalline order from hypothetical higher dimensional spaces.

Our discussion of symmetry so far has been rather discursive. But, as we have remarked repeatedly, symmetry is a very precise concept. The proper language for a systematic study of symmetry is that of *group theory*, which is a highly developed branch of mathematics. The basic idea is simplicity itself. Identify *all* the symmetry operations that leave a given object invariant. Call them $A, B, C, \ldots$. This is then

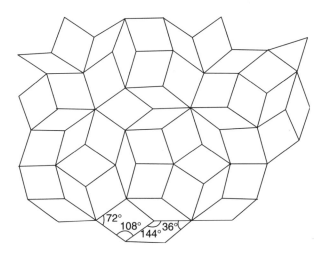

Figure 1.3: Penrose aperiodic tiling of a plane.

an exhaustive list. It is clear from the very definition of symmetry that the successive applications of any two operations, first A and then B, say, will also leave the object invariant. Therefore, the combined operation must also be one of the symmetry elements we have listed exhaustively above. Let it be C. Then we can write $C = BA$. Mark the order of A and B in BA. It means A operates first, followed by B and the result is the same as C. This is a kind of multiplication, composition or successive operation, that gives the interlocking of the various symmetry operations. We say that the symmetry operations are closed under this multiplication. Next, we note that doing nothing at all to the object is also a symmetry operation because it trivially leaves it invariant. In fact it leaves it alone! We denote this trivial symmetry operation of 'doing nothing' by E (This is a fairly standard notation). Finally, we note that reversing a symmetry operation is also a symmetry operation — it restores *status quo ante*. Remember that the reverse of a clockwise rotation by an angle θ is an anticlockwise rotation by the same angle θ about the same axis. In obvious notation, we denote the reverse (more properly called inverse) of A by A^{-1}. It is now clear that $E = A^{-1}A$. (That is applying a symmetry operation followed by its inverse amounts to doing nothing). We are all set now. A set of elements having a law of multiplication (successive operations) under which the set is closed, with an identity (doing nothing) and where each element has a unique inverse is called a *group*. The symmetry operations then form a group. We are now compelled by the sheer logic of it. The inner structure of the symmetry group is given completely by enumerating the results of all pairwise multiplications, *e.g.*, $C = BA$. Constructing a multiplication table is like finger-printing the symmetry. Identical multiplication tables imply identical symmetry structures no matter how physically different the objects themselves may be. It is all very nice, but what can we do with all this, you may ask. Well, you can do a lot. An example will help

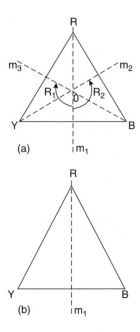

Figure 1.4: Symmetry elements of (a) equilateral triangle; (b) isosceles triangle.

illustrate the point. Suppose the symmetry of the physical law in question turns out to be the symmetry of an equilateral triangle living on a plane (Fig. 1.4).

The symmetry operations are then E (identity), R_1 and R_2 (clockwise and anti-clockwise rotations by 120 degrees, respectively, about the three-fold symmetry axis) and the three reflections m_1, m_2 and m_3 in the three mirror-lines (medians). One can readily construct the multiplication table. Thus verify, for example, $R_1 R_2 = E$, $m_1 m_2 = R_2$, $m_2 m_1 = R_1$, and so on. Remember that we have completely specified the symmetry structure of the physical law that governs our physical system. The latter may be a molecule with an atom literally at the center of an equilateral triangular environment formed by three other identical atoms. (Situations analogous to but more complicated than this are very common in chemistry, *e.g.*, an atom, or rather a doubly charged ion of copper at the center of a regular octahedron formed by the negatively charged atoms of oxygen in copper sulphate). Let us assume now that this system can exist in one of the three states, or *linear combinations of these*, which are permuted among themselves under the symmetry operations. Of course, we are assuming here that it is meaningful to speak of such linear superpositions. This is indeed the basic structure underlying quantum mechanics (see Appendix B). Thus we can identify the three vertices of our equilateral triangle with these three states. We provocatively label them R (for red), Y (for yellow) and B (for blue). It is easily verified that our symmetry operations indeed permute them in all possible ways. (There are six ways in which three objects can be permuted and the number of elements in our symmetry group is also six). While the symmetry operations do permute the three states among themselves, they do not mix them indiscriminately. Indeed, they split the possible linear combinations into two sets (called multiplets more properly) such that only members of the same multiplet mix among themselves. One multiplet, call it $W = R + Y + B$ has only one member (a singlet). The other has two members, $Z_1 = R - 2Y + B$ and $Z_2 = R - B$. Now Z_1 and Z_2 mix freely under our symmetry operations and, therefore, they are intrinsically the same — they differ according to our viewpoint only. Thus, for instance, they should have the same energy (or mass). We say that the multiplet is two-fold degenerate. Their energy, however, must be in general different from that of W with which they do not mix under symmetry. In this simple case we could write down this multiplet structure by mere inspection. In general one has to use the multiplication table in a systematic way. It is called the representation theory of groups. In our example W and $\{Z_1, Z_2\}$ provide, respectively, one- and two-dimensional representations. We can go further and lower the symmetry to that of an isosceles triangle by pulling one of the vertices out (Fig. 1.4b). Our symmetry group now will consist of only two elements $\{E, m_1\}$. It is a sub-group of the earlier larger group. The result is that the doublet is further split into two singlets. We now have three non-degenerate (unequal) levels.

This splitting or reduction of degenerate multiplets with the progressive lowering, or descent, of symmetry is well known and well studied in chemistry and solid state physics, where the symmetry is mostly geometrical and known from structure. The situation is quite different in elementary particle physics where the symmetry is rather abstract and not directly accessible. Here symmetry takes on a creative role. This is made possible by the fact that a given group uniquely specifies the possible multiplet structures it can support. Thus one can postulate a symmetry and then work out the multiplet structures it implies and compare with the observed families of closely related particles. This is the idea underlying the unending quest for symmetries, *e.g.*, $SU(2)$, $SU(3)$ and so on. One is limited only by his ingenuity and insight. Thus $SU(3)$ (special unitary group of rotations in a three-dimensional complex space) has a multiplet with eight members (eight-dimensional representation) and one with ten members (ten-dimensional representation) that fitted so well the observed families of eight baryons and ten hyperons — behold the 'unreasonable' effectiveness of symmetry in physics!

Finally, a remark on the group multiplication. Note, that in our example we had $m_1 m_2 = R_2$ and $m_2 m_1 = R_1$. Thus unlike ordinary multiplication of numbers, the order in the applications of symmetry operations is important. We say that m_1 and m_2 do not commute. Such a symmetry group is said to be non-abelian. The important group of rotations in three-dimensional space $SO(3)$ is non-abelian. The corresponding group of rotations in a plane is Abelian. An amusing demonstration of this is the following. You fly out of the North Pole down the zero-degree longitude through Greenwich to the equator. You will be over the Atlantic, south of Ghana. This is a rotation by 90 degrees about the east-west axis. Now you turn and follow the equator to longitude 90 degrees east. You should be over the Indian ocean east of Sumatra. This amounts to a rotation by 90 degrees about the north-south axis. Now, you perform these operations in the reverse order. Start out at the North Pole and turn by 90 degrees eastwards. But since you are right on the axis of rotation, you just stay put. Next fly down the zero degree longitude till you reach equator, and thus you end up over the Atlantic, south of Ghana, thousands of kilometers away from your earlier destination in the Indian ocean. It turns out that most of the symmetries in Physics are non-Abelian, and that makes it richer. Abelian symmetry gives only non-degenerate one-dimensional or single-state multiplets.

1.2 Space-Time Symmetries: Invariance and the Great Conservation Laws

Objects are located in space. They endure in time. This is true of all events and processes, of beings and becomings, that ultimately involve the elementary particles and their interactions that make up the world of physics. Admittedly, this is a highly reductionist viewpoint but you can hardly fault it. It seems reasonable, therefore, that the study of symmetries of objects and phenomena must be preceded by a proper study of the symmetries that this background space-time continuum may have. For obvious reasons we will call these the framework symmetries. These symmetries must be established as facts of experience, no matter how compelling *a priori* they may appear to be. To the best of our knowledge, then, the following symmetries are true.

Space is homogeneous. That is to say that the absolute position of an object is irrelevant. What it operationally means is that if we perform an experiment at a location and then repeat the same experiment somewhere else, in outer space, say, the results will be identical — translationally invariant in space. By the 'same experiment' we mean that all conditions relevant to the experiment must be reproduced exactly. Thus, if the change in earth's gravity in going out there is relevant, then the earth must be transported along with the apparatus. One may argue that this claim is then vacuous inasmuch as any discrepancy between the results of the two experiments can always be blamed on something that may have escaped our attention, to wit our altered position with respect to the distant stars! Now, this is perverse because it is possible to isolate our experiment far enough to any desired degree of accuracy by including larger and larger regions of space as part of our experimental set-up and, because one can assume reasonably that all effects are essentially local in nature. Ghosts are not admitted! In any case, there is nothing to suggest violation of this translational symmetry.

Next comes isotropy of space, or the irrelevance of absolute direction. Operationally, it means that if we perform a certain experiment and then rotate our entire setup to a new orientation and repeat the same experiment, the results will be identical. We can re-word all our earlier provisions and arguments in support of this. So far there is no empirical evidence in support of a preferred direction in space. Thus isotropy of space is a good symmetry.

There is an interesting connection between these two symmetries. Isotropy (relative to every point of space) implies homogeneity but not *vice versa*. This is readily proved. Let P_1 and P_2 be two points in space. Draw a sphere passing through P_1 and P_2, with center, say, at O. You can draw any number of such spheres. Now, viewed from O, P_1 and P_2 are related by isotropy and, therefore, are equivalent. You can repeat this process till you cover the entire space and thus establish homogeneity of space.

Next comes homogeneity of time, or time-translation symmetry. There is no absolute origin of time. If you perform an experiment now and repeat the same experiment at a later date, the results will be identical. Indeed, without these symmetries the universe will hardly be comprehensible. We should perhaps mention here that there is evidence that the universe is finite, though unbounded, and that it had a beginning some 15 billion years ago — the *Big Bang*. We hope that we are at a sufficient remove from this boundary (though there is actually none) and initial conditions to ignore these symmetry breaking effects here and now.

Finally, to these irrelevancies, namely those of absolute position, absolute direction and absolute time, we add the irrelevance of absolute rest, or of absolute uniform motion. Consider two unaccelerated platforms in uniform relative motion, that is to say that one platform is moving with a constant velocity as seen by an observer who is stationary on the other platform. Now, if we perform an experiment on one of these platforms and then repeat the same experiment on the other, the results should be identical. Thus no *local* experiment, *i.e.*, without reference to the other platform, will detect any effect that can distinguish between these two unaccelerated platforms — there is no absolute uniform motion. This equivalence of unaccelerated platforms is the great symmetry expressed by the *principle of relativity* and was a wonderful achievement of Galileo. Acceleration is, on the other hand, absolute and can be detected locally by an accelerometer — a mass attached to one end of a spring, the other end of which is fixed to the platform. (In all these discussions, we will ignore the presence of gravitation). This is quite consistent with our every day experience. We are hardly aware of the velocity with which the lift, by which we may be traveling, is moving except at the times of start and stop, *i.e.*, when there is acceleration or deceleration.

A platform is, more formally, a set of points at rest relative to one another. It is convenient to introduce a rectangular coordinate system (x, y, z) at rest with respect to these points. One may also assume a clock, an atomic clock, say, attached to every point of this set. The identical clocks may be synchronized by exchanging light signals. Thus, if A and B are two points and if t_1 is the time at which a light signal is sent out from point A, and if t_2 is the time at which the signal is received at and reflected by the point at B, and finally if t_3 is the time at which the signal is received back at the point A, then the clocks at A and B are synchronized if $t_2 - t_1 = t_3 - t_2$. Note that this is purely by symmetry and does *not* require knowledge of the speed of light. Thus an elementary event is completely located by giving spatial coordinates (x, y, z) and the time of its occurrence t, read out by the clock at the point (x, y, z). Such an unaccelerated platform equipped with the markers $(x, y, z; t)$ is called a Galilean frame of reference S, say. Another Galilean frame S', say, will have a primed space-time coordinate system $(x', y', z'; t')$. Now the relativistic invariance asserts that the laws expressed in terms of the primed and the unprimed space-time coordinates should have the same form. The question now is how the primed and the unprimed space-time coordinates of the same event are related. The relativity of motion encountered in everyday life, also called Galilean relativity, would suggest the following answer. Time intervals are absolute. So are the space intervals. This means that, if $(x_1, y_1, z_1; t_1)$ and $(x_2, y_2, z_2; t_2)$ are the space-time coordinates of two events observed in a Galilean frame S, and $(x'_1, y'_1, z'_1; t'_1)$ and $(x'_2, y'_2, z'_2; t'_2)$ are those for the same two events but in another Galilean frame S', then the time interval $t_{12} = (t_1 - t_2) = t'_{12} = (t'_1 - t'_2)$ and the space interval squared $r^2_{12} = (x_1 - x_2)^2 + (y_1 - y_2)^2 + (z_1 - z_2)^2 = (r'_{12})^2 = (x'_1 - x'_2)^2 + (y'_1 - y'_2)^2 + (z'_1 - z'_2)^2$.

This leads to the rules of vector addition of velocities and displacements well known from our high-school days. The symmetry operations here are the familiar translation and rotation (re-orientation) in space, 'boosting' to a relatively uniformly moving frame, and time translation. Galilean relativity is, however, based on our common experience with slow objects moving at small velocities, *e.g.*, the speed limit of about 100 kilometers per hour on national highways. Compare this with the speed of light, 1080 million kilometers per hour in vacuum. Can we extrapolate our tardy experience to such high velocities? Let us see. In Galilean relativity the speed of light in vacuum would depend on the relative velocity of the source of light and the observer. One can then, in principle, chase light and even outrun it. Or one can run just fast enough to keep pace, bringing light to a relative standstill. This is true, for instance, in the case of sound. But sound propagates only in a medium, *e.g.*, air. Light, however, can propagate in vacuum. Is vacuum too filled with an all pervasive medium — the 'aether' as was indeed thought for quite some time? This hypothetical medium, the aether, could then provide the preferred frame of reference at absolute rest, making thus the different Galilean frames moving relative to it in principle non-equivalent. Even the most careful laboratory measurements and astronomical observations have, however, failed to detect this aethereal medium. Einstein did not like this loss of symmetry anyway. The point is that light is an electromagnetic wave whose propagation in vacuum relative to a Galilean frame is described by a wave equation of the type we wrote down in the last section. Notice that the speed of light 'c' occurs explicitly in this equation. The invariance (or rather covariance) of this equation with change from one Galilean frame to another then demands the invariance of the speed of light. Thus, we have the fundamental postulate of the absolute constancy of the speed of light for all

Galilean frames of reference — the basis of *Einstein's special theory of relativity*. The changes from one frame of reference to the other are the symmetry operations that leave the speed of light unchanged. It is clear that for this to be so, the notion of *absolute* time interval t_{12} as separate from that of the absolute space interval r_{12} between the two events labeled 1 and 2 inherent in the Galilean relativity must be abandoned. Einstein's relativity replaces these two with a single *absolute* invariant interval s_{12} between the two events, given by $s_{12}^2 = r_{12}^2 - c^2 t_{12}^2$. Three-dimensional Euclidean space and time $(x, y, z; t)$ are replaced by a four-dimensional space-time (the *Minkowski world*) that treats time t as just another co-ordinate to label the events, *at par* with space coordinates (x, y, z) (Fig. 1.5).

An event is now located at a world-point (x, y, z, t) — the semicolon that set time apart from space has now been replaced by a common comma. The transformation from (x, y, z, t) to (x', y', z', t') is now the symmetry operation of displacement and rotation (Lorentz transformation) in this four dimensional world, keeping in mind,

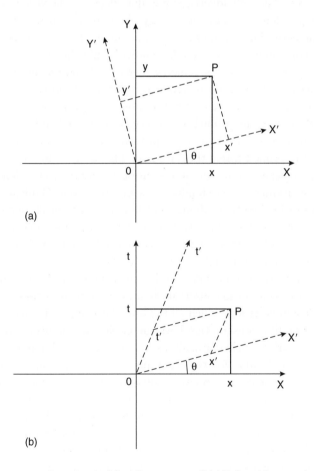

Figure 1.5:   Rotation in (a) ordinary space; (b) Minkowski space-time.

however, the technical point about the minus sign that occurs in $s_{12}^2 = (x_1 - x_2)^2 + (y_1 - y_2)^2 + (z_1 - z_2)^2 - c^2(t_1 - t_2)^2$. Einstein's special relativistic space-time symmetry now demands that the laws of physics be invariant under the Lorentz transformation from one Galilean frame (unaccelerated or *inertial frame*) to another. This replaces the old Galilean invariance with its absolute time intervals. Indeed, the Maxwell wave equation for the propagation of an electromagnetic disturbance in vacuum has the Lorentz invariance, but *not* the Galilean invariance. One may mathematically absorb the minus sign by defining an imaginary time $\tau = it$, with $i = \sqrt{-1}$ and treat time formally completely *at par* with space. But it is actually better to leave it as such, as a gentle reminder that time is, after all, qualitatively different from space. The negative sign implies that the interval s_{12} can vanish without the two events coinciding in space-time, *i.e.*, $s_{12} = 0$ but $r_{12} \neq 0$, $t_{12} \neq 0$. We can even have s_{12} negative. (We say that the Minkowski world has an indefinite metric).

The geometry of this four-dimensional world has important and interesting physical consequences. The well advertised popular effects — the variation of mass with velocity, the equivalence of mass and energy, the Lorentz contraction and time dilation, all belong here. The speed of light in vacuum is the limiting speed that cannot be exceeded. Our main concern here is, however, only the symmetry aspect of relativity — the great framework symmetry. Let us note one highly counter-intuitive aspect of it because it has a deep significance for our discussion of invariance and conservation law later. Since it is only the interval s_{12} that remains invariant from the unprimed frame to the primed one, it is clear that we can have t_{12} zero but t'_{12} non-zero. That is to say that in the unprimed frame the two events are simultaneous, but in the primed frame they are not. This is the relativity of simultaneity that totally demolishes the notion of absolute time interval. (Incidentally, one may have an uneasy feeling, when simultaneous events in one Galilean frame appear non-simultaneous in the other, about what happens to their chronological order of occurrence — which is older of the two. Well, relativity does allow a certain amount of play in this game of courtesy, but there is an absolute past and an absolute future even here consistent with notions of cause and effect).

We now turn to the deep connection between these relativistic space-time symmetries (invariances) and the conservation laws. Inasmuch as these symmetries are the framework symmetries to which all the basic laws of physics are subject, we will call the corresponding conservation laws the Great Conservation laws. Consider a physical process written schematically as $x + y \rightarrow z + w$. A quantity is said to be conserved if its total value for the reactants $x + y$ is the same as its total value for the products $z + w$ of the process as observed in a given Galilean frame. Thus we speak of conservation of energy, linear momentum and of angular momentum. It turns out that the conservation of energy follows from the invariance with respect to translation in time. The conservation of linear momentum follows from the invariance with respect to translation in space (homogeneity of space). The conservation of angular momentum follows from the invariance with respect to rotation in space (isotropy of space). A proper discussion of conservation of these quantities (and even their definition in general) as a consequence of the invariances requires the introduction of 'action' and 'action principle.' This is beyond our scope. The important point to note is that this connection between invariance and the conservation law is not restricted to any specific dynamical laws such as Newton's laws of motion. The connection is purely kinematic. For the specific case of mechanical systems where, for instance, momentum is mass times velocity, one may derive conservation of linear momentum by applying Newton's three laws of motion. And so on for energy and angular momentum. But the connection is really much more general. After all, there are non-mechanical objects, light for instance, that also carry energy, momentum and angular momentum. We should note in passing that just as isotropy of space implies (but is not implied by) homogeneity, conservation of angular momentum implies conservation of linear momentum, but not *vice versa*.

Much of the restrictive and predictive power of these symmetries comes from the associated conservation laws. The striking example is radioactivity (β-*decay*) in which a neutron was thought to decay into an electron, a proton and something else. The electric charge is conserved as required by another invariance called *global* gauge invariance to be discussed later. However, a careful reckoning of energy and momentum of the system before and after the reaction led to an imbalance. Thus a new particle was suspected as a decay product that carries the missing energy and momentum. It was predicted to be neutral and to have zero rest mass. Also, recalling that the neutron, the electron and the proton all carry spin half (angular momentum $\hbar/2$), conservation of angular momentum required the then unknown particle to carry spin half. All this was confirmed happily later. This is the now well known but elusive elementary particle, the electronic anti-neutrino denoted by $\bar{\nu}_e$, and the corrected process reads $n \rightarrow e^- + p + \bar{\nu}_e$. These particles are now routinely and abundantly produced in laboratories, in nuclear reactors as well as accelerators.

The full power of these great framework symmetries is realized only when these are combined with the great framework theory — Quantum Mechanics (Appendix B). But this will take us very far afield. We will be content with just mentioning it. In addition to these continuous symmetries, there are discrete spacetime symmetries too. One of them is the symmetry under space reflection, also called the mirror symmetry or parity. This produces enigmatic effects in ordinary laboratory physics and chemistry as also in the extraordinary processes involving elementary particles. We will take this up next.

1.3 Reflection Symmetry

We have spoken of objects having bilateral symmetry, also called the left-right symmetry. A butterfly with outstretched wings or a maple leaf for example. When an object is reflected in a mirror, the left and the right sides of it get interchanged.

Thus, an object having bilateral symmetry is by definition superposable on its mirror image. The mirror is just an optical device that enables us to visualize the result of reflection of objects in space through a plane. For these reasons the terms bilateral symmetry, left-right symmetry, mirror symmetry and the symmetry under space reflection are all used interchangeably. In physics, handedness is often referred to as *chirality*.

There is something that sets this symmetry apart from the rest that we have discussed so far. As noted above it is a discrete symmetry unlike the continuous symmetry of rotation or translation, say. Changes caused by continuous symmetry operations can be made arbitrarily small. Not so with discrete ones. You reflect or you don't: The excluded middle — there is nothing in between. Also, unlike these, it is a *non-performable* symmetry operation. Space reflection involves turning the object inside out laterally, an operation we can hardly perform continuously. But we can and we do visualize it by the optical trick of reflecting it in a mirror. Having visualized it so, nothing prevents us from making a physical copy of the image, using silly putty, say, which can then be tested for superposability on our object. This is the operational meaning of reflection symmetry as applied to shapes of material objects or geometrical figures. The non-performable nature of reflection symmetry conceals an important aspect of it that we will try to uncover now. Consider an arbitrarily shaped object and its reflection in a mirror. A rather handy example would be, well, your right hand itself. Its mirror image is constructed by translating every point on the hand to a point on the other side of the mirror, along the line perpendicular to the plane of the mirror and equidistant from it (Fig. 1.6a).

Now, it is clear that this image (ideally your left hand) is not superposable on your right hand. Such an asymmetric object is called 'handed,' and with very good reason. Have you ever tried your left-hand glove on your right hand? And yet nothing else is more like my right hand than its mirror image, that is my left hand. The reason that the two cannot be superposed is an inconvenient circumstance of life, namely, that the hand is a three-dimensional object and so is our physical world (space) in which cooped up we live. In a world of higher dimensions, the right hand could have been turned around by a temporary excursion into the extra dimensions, and thus superposed on the left hand. This can be demonstrated quite easily with an example taken from the world of lower dimensions — of a two-dimensional object, a flatlander living on a plane which is embedded in our familiar three-dimensional space. Thus, the symbol "Om" in Fig. 1.6b can be superposed on its mirror image by simply folding the paper along the mirror line M. The act of folding involves a temporary lift or escape into the third dimension coming out of the plane of the paper. Science fiction is full of such excursions into the extra dimension — the "tesseract" in "A Wrinkle in Time" by Madeleine L'Engle is a fascinating case in point. These extra dimensions, somewhat curled up and rather inaccessible, are also the subject of serious thought by the physicists of our times. But we are digressing. All this suggests that an object and its mirror image are intrinsically the same. To

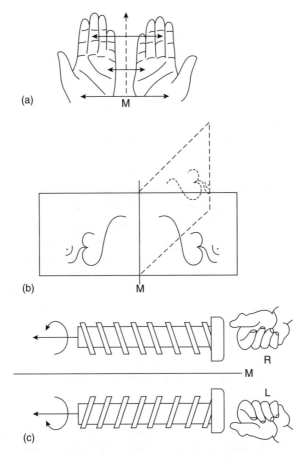

Figure 1.6: (a) Mirror reflection; (b) continuous reflection via third dimension; (c) right-handed screw reflected as left-handed screw.

emphasize this we call them a pair of *enantiomorphs* or antipodes — an impressive name for ordinary mirror images.

What is in a hand that makes it so 'handed'? To understand handedness more thoroughly we can think of a screw — the common screw that we use to fasten things together, and without which much of our civilized world would simply come apart. The common screw is nothing but a helical ridge, called thread, cut into the surface of a cylinder, or a cone if it is a tapered screw. When the screw is turned as indicated by the circular arrow (Fig. 1.6c), it advances (or recedes) along its axis as indicated by the linear arrow. This makes the screw a machine that converts a rotary motion about its axis into a translational motion along that axis. You may think of the Archimedes Screw that was used by the Egyptians to raise the waters of the Nile, and is still in use for similar purposes. It should be clear that there are two and only two classes of screws possible. These correspond to the two possible relations between the circular and the linear arrows. Let us give them names. Suppose that you clasp the screw in your right hand with your fingers pointing in the direction of the circular arrow while your thumb stays parallel to the axis of the screw. Now, if your thumb points in the direction of the linear arrow, then the screw is said to be right-handed. If, on the other hand, it points in the opposite direction, then the screw is said to be left-handed (Fig. 1.6c). It is easy to see that the mirror image of a right-handed screw is a left-handed screw. The two form a pair of enantiomorphs. That there are two classes of screws, *i.e.*, the two-ness of it, is an absolute fact. But defining them as the left- and the right-handed screw is, of course, a matter of convention — a very useful convention though, which is followed uniformly all over the civilized world. This has been made possible by the intimate contacts we have had over centuries of togetherness, and not a little by our admirable practice of shaking hands. But a convention all the same. The non-triviality of this is brought home by the following thought provoking circumstance. Suppose we establish radio-contact with some advanced civilization in a galaxy far away. Such an eventuality can not be ruled out, thanks to the project SETI (Search for Extra-Terrestrial Intelligence) mounted by some serious-minded people. Now we should have no difficulty convincing our distant correspondents that these are the two classes of screws possible. But try hard as we may, we will not be able to explain to them what we mean by the right-handed screw. This is the famous problem of Ozma (named after the mythical prince Ozma in Lyman Frank Baum's classic "Wonderful Wizard of OZ"). The *Ozma problem*, suggested by Martin Gardner, is a deep problem of communication theory, and its solution involves deeper understanding of symmetry in physics. We will return to it briefly later.

From the reflection symmetry of geometrical shapes let us now pass to the real question. Are the various laws of physics symmetric with respect to space reflection? To fix ideas consider a simple molecule CH_4, the molecule of methane (marsh gas found commonly in marshy lands). The molecule consists of a carbon atom surrounded by four equidistant hydrogen atoms arranged at the vertices of a regular tetrahedron (Fig. 1.7).

The molecule is clearly reflection symmetric, *i.e.*, it is superposable on its image. We can break this reflection symmetry by replacing the four hydrogen atoms by four different atoms (or groups of atoms) X, Y, Z and W, say. Thus, for example if $X = H$, $Y = CH_3$, $Z = C_2H_5$ and $W = OH$, we get a molecule of butyl alcohol. Numerous other examples are possible. Chemists refer to such a molecule as having an *asymmetric carbon atom*, and the pair of enantiomers are called *stereoisomers*. Such a molecule is handed because it is no longer superposable on its mirror image. Thus for example, if you look down the XO direction, the atoms Y, Z and W will be seen as arranged either clockwise or anticlockwise. Now suppose we synthesize this molecule in the laboratory starting from the elements C, H and O. The question is which one of the pair of stereoisomers we will get. The answer is simply this. If the laws governing the chemical reaction are symmetric with respect to space reflection,

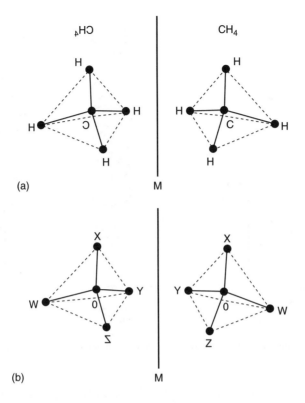

Figure 1.7: Mirror reflection of molecules: (a) CH_4 with symmetric carbon atom; (b) OXYZW with asymmetric carbon atom, hence optically active.

then the probabilities of getting the two stereoisomers are strictly equal. Therefore, at the end of the reaction we will get a mixture of the two in equal proportions. Chemists call this a racemic mixture. The mixture will have no *net* handedness. A law is said to be reflection symmetric, if a process or phenomenon and its mirror image are *equally* allowed by that law. Experimental evidence strongly suggests that the laws of physics that govern processes at low energies, like chemical reactions, are indeed reflection symmetric. Thus the molecules of butyl alcohol in our example and the molecule of its mirror image will have the same physical and chemical properties, *e.g.*, the same boiling point, the same freezing point, the same density and, of course, the same molecular weight. Next we will demonstrate the predictive power of this symmetry of the physical law — we will predict optical activity. Consider again our handed molecule with the asymmetric carbon atom. A molecule of sugar is perhaps a more pleasing example. Sugar molecules are also handed but a bit more complex. Now, we can hardly experiment on a single sugar molecule. So consider trillions of these identical sugar molecules — a solution of the sugar molecules in water, for example. The water molecules (H_2O) are mirror symmetric and, therefore, any handedness at all will be due only to the sugar molecules. We can and we will ignore

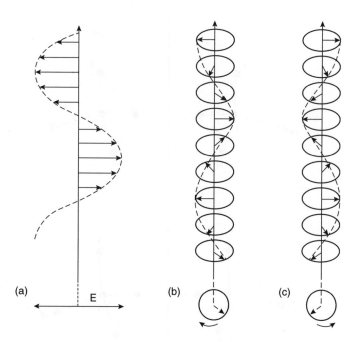

Figure 1.8: (a) Plane-polarized light; (b) left-circularly polarized light; (c) right-circularly polarized light.

the solvent (water) completely in what follows. In the aqueous solution the sugar molecules are located and oriented randomly. This, however, does not neutralize or average out their handedness. After all, a chest full of left gloves can hardly be confused with a chest full of right gloves, no matter how randomly the gloves are placed. We will now send a beam of light from a laser, say, through our sugar solution. But first let us remind ourselves of some elementary facts about light.

Light is a transverse electromagnetic wave. The electric and the magnetic fields oscillate sinusoidally in time and space with a given frequency and wavelength. They are perpendicular to each other and to the direction of propagation of the wave (hence transverse). Light can be circularly polarized. Here the tip of the electric vector describes a helix with its axis along the direction of propagation. It may be left- or right-circularly polarized according as to whether the helix is left- or right-handed (Fig. 1.8).

Light can also be linearly (or plane) polarized if the oscillating electric vector lies in a plane containing the direction of propagation. Finally, we note that a linearly (plane) polarized light may be viewed as a vector addition of the two oppositely circularly polarized light waves of the same frequency and wavelength. We are all set now. Let the beam of light passing through our sample of handed sugar solution be circularly polarized. The mirror image of this process will be an oppositely circularly polarized light passing through a sugar solution of opposite handedness. Given the reflection symmetry of the governing law, the two enantiomorphic processes must

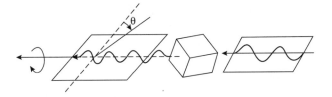

Figure 1.9:   Rotation of plane of polarization of light by optically active medium.

be equally and identically allowed. In particular the speed of light in the two cases must be exactly the same. But what if we keep our sugar solution the same and only reverse the sense of circular polarization of light. Well, this is not symmetry related to the earlier situation, and there is no sufficient reason to expect the speed of light to remain the same — it will in general be different. Thus the speed of light in a handed medium depends on the sense of circular polarization of the light! This effect can be made more spectacular by taking our light to be plane polarized. Recall that it may be viewed as a superposition of two oppositely circularly polarized light waves. Now that these two components must travel with different speeds, they will get out of phase as they traverse the handed medium. This results in the twisting of the plane of polarization of the light relative to that of the incident light (Fig. 1.9). This twisting or rotation of the plane of polarization is called optical activity. It is perhaps the most dramatic manifestation of handedness of the medium. The substance (sugar in our case) is said to be dextrorotary (dextro = right) if the plane of polarization twists as a right handed screw. It is said to be levorotary (levo = left) if it twists as a left-handed screw. A racemic mixture of the two will leave the plane of polarization unchanged. (Somewhat confusingly, the opposite convention is also in use).

Let us re-emphasize that optical activity is due to the handedness of the substance. The law itself is even-handed, *i.e.*, reflection symmetric. This is expressed perhaps most forcefully by the famous example of a milk drinking kitten of the Looking Glass world. Milk contains asymmetric molecules of sugar, proteins and fats. So does, of course, the body of the cat. And conventional cats love conventional milk. Reflection symmetry now demands that the reflected kitten love the reflected milk just as much, and fare just as well in all respects.

The living world is, however, far from being racemic. Thus, practically all the 20 odd amino acids that make up the proteins of the living cells are left-handed. The proteins in the living cells, in turn, have a helical backbone which is almost always right-handed. Each of the sugar phosphate chains in the double helix of the information bearing molecule DNA is a right-handed (double) helix, and a man has typically 10^{11} kilometers of it. Left amino acids are common and are assimilated by our body, but the right amino acids are rare and filtered out by our kidneys. The nicotine commonly found in tobacco is known to be harmful but its reflected stereoisomer is rare and much less offensive. Limone, in perfumes, has the pleasant orange scent — its enantiomer smells like turpentine. The same is true of other biochemicals such as the lactic acid found in milk, or the table sugar (sucrose) found in sugarcane, etc. The sugar D-glucose is found throughout the animal kingdom but its mirror image L-sugar is unknown except in laboratory synthesis. (Handedness of drug molecules poses a serious problem — one has only to recall the tragedy of the thalidomide babies with birth defects caused by the wrong handedness of the drug molecules involved. Preparing *optically pure* compounds, *i.e.*, those of a given handedness, is, however, difficult and expensive).

There are indeed few exceptions to the rule that anything from lactic acid to the double helical DNA, having handedness, will occur biologically in only one form. Indeed if we let a colony of bacteria feed on a *racemic* (optically inactive) mixture of (L) and right (R) sugars, the bacteria would feed preferentially on L-sugars, and then leave the mixture right-handed and optically active. The question now is, do we understand this dominance of handedness, or shall we say high-handedness of the living world when the governing laws themselves are so just and even-handed? Well, not quite. But a highly plausible answer is something like this. The observed handedness of the living matter may be the result of a fantastic amplification of an initial chance asymmetry, ever so slight. This is made possible by the positive feedback inherent in the process of multiplication (reproduction) by self-replication that is all pervasive in the animate world. To see this clearly let us simplify things to the absurd limit and consider the first (single) helical strand of the DNA molecule ever formed. We know that it is potentially equally likely to be right- or left-handed. But once formed it has got to be just one of them. So let it be right-handed. Now this single right-handed strand proliferates or multiplies by self-replication. It acts as a template and makes a copy of itself which is now necessarily right-handed. The process gets repeated over and over again. This is the positive feedback at work that may lead to the necessary amplification of an initial chance event over the aeons of chemical and biological evolution, and thus produce the handed life as we know it today. This is made all the more plausible by the observation that the inorganic world, by contrast, seems to be quite racemic. Consider the mineral quartz for example. It is one of the crystalline forms of silicon dioxide (SiO_2), the common silica sand. The basic unit here is SiO_2 which by itself is mirror symmetric, making quartz optically inactive when dissolved. But in a quartz crystal the units are arranged in the form of parallel helices which can be either left- or right-handed, making quartz optically active. In Nature both the forms occur with equal frequency.

Does this not go against the laws of thermodynamics, the entropy principle, that makes states of equal energy equally probable? The left- and the right-handed strands are, of course, energetically equivalent, being related by reflection symmetry. Well, the point is that the thermodynamic statement is about a system in thermal equilibrium. But the living state is far from equilibrium. It is self-organized and maintained at the cost of 'freely' available energy that comes eventually from the sun. Once the cell is dead, the right-handed helices of the DNA molecule will begin to flip their handedness, and gradually tend to the racemic state as dictated by thermodynamics. Indeed, the rate of racemization can be, and has been, used for the dating of dead cells older than 40,000 years or so, much better than the conventional dating based on the decay of ^{14}C, a radioactive isotope of carbon. Louis Pasteur regarded handedness as a sign of life. Racemization signaled death.

Are the fundamental laws of physics all strictly symmetric under space reflection? Is the antipodal world of the *Looking Glass* just as legal as our conventional world? We now know that the answer to this question is a definite *no*. There are fundamental processes such as the β-decay (radioactivity) controlled by the so-called *weak interaction* that break this reflection symmetry. There is a screw at the very heart of Nature. To see this we have to get more sophisticated. We have seen how to reflect geometrical figures and shapes of material objects. But how do we reflect magnetism? Take a bar magnet with the poles N and S marked on its ends. Its mirror reflection will be just another bar magnet with the letters N and S laterally inverted (Fig. 1.10).

But this is a naive reflection of the body of the magnet. It hardly addresses the real question of how to reflect the magnetism of it. The magnetic field of the magnet may be regarded as due to an electric current circulating in a loop around the body of the magnet as indicated by the circular arrow (Fig. 1.10). Remember Ampère's Law! (Incidentally, when Ernst Mach learnt of the sideways deflection

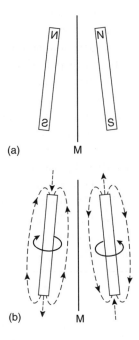

Figure 1.10: Mirror (M) reflection of (a) body of a magnet; (b) magnetism.

of a compass needle when placed below and parallel to a current carrying wire, he was shocked out of his wits as he thought it to be violating the left-right symmetry. With our picture of the magnet, now we see that Mach's shock was a false alarm as there is no such symmetry in this situation to start with). It should be clear now that when reflected, the sense of the circular arrow will reverse. And so will the polarity of the magnet. With this we are all set to describe the phenomenon that shook the world of physics — the fall of parity. Parity is yet another name for reflection symmetry. (Note that space reflection actually means inversion through the origin — that is letting (x, y, z) go to $(-x, -y, -z)$. In three-dimensional space, however, the inversion through the origin and the reflection in a mirror are related through a mere rotation by 180 degrees about an axis perpendicular to the mirror plane. And, of course, the rotation symmetry is not in doubt).

Take an atom of cobalt, the isotope ^{60}Co to be precise. The nucleus of ^{60}Co has a spin. It is like a spinning top. This makes it a tiny magnet with the magnetic poles on the spin axis. As before, this is equivalent to having a circulating current loop indicated by the circular arrow. We now apply a magnetic field. The nuclear magnet will align parallel to this field just as a compass needle aligns parallel to earth's magnetic field. The ^{60}Co is a radioactive nucleus. It decays by emitting, among other things, electrons, the β-rays, in all directions. The question is whether they come out equally in all directions, or there are some preferred directions. We can know this by placing detectors all around our sample and counting the number of electrons coming out in a given direction in a given interval of time. In particular let us compare the number of electrons shot out of the north pole (N) parallel to the field with the number shot out of the south pole (S) antiparallel to the field. If we perform this experiment as Chien-Shiung Wu did in 1957, we will find that more electrons are shot out of the south pole than out of the north pole. Is this consistent with the reflection symmetry of the underlying law? To answer this all we need to do is to look at the process reflected in a mirror (Fig. 1.11).

Everything looks the same except that the sense of the circular arrow and, therefore, the polarity, is reversed. Thus in the mirror world, more electrons would come out of the north pole antiparallel to the field than out of the south pole. The reflection symmetry is violated! (This violation of parity was predicted by the two Chinese-American physicists T. D. Lee and C. N. Yang in 1956 on theoretical grounds. It was confirmed by C. S. Wu in 1957. The same year Lee and Yang won the Nobel Prize in Physics). In fact the nuclear spin of the ^{60}Co nucleus (the circular arrow) and the preferred direction of electron emission define a left-handed screw. Nature is weakly left-handed after all! This is more than a mere convention. One could perhaps use the ^{60}Co decay to communicate to our distant correspondent the meaning of the left and the right, and thus solve the Ozma problem. But not quite, as we will presently see.

We can go further and, as it were, pinpoint the screw by looking at the full β-decay reaction: ^{60}Co $\rightarrow$ ^{60}Ni $+ e^- + \bar{\nu}_e$. The neutrino (or rather anti-neutrino $\bar{\nu}_e$), as we have noted earlier, is an elusive particle with zero rest mass. This particle has

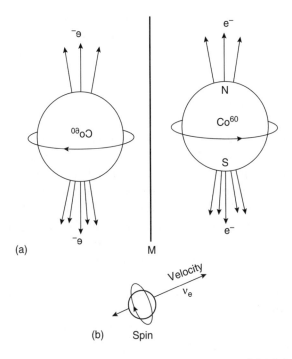

Figure 1.11:   (a) Parity violation in β-decay of cobalt-60 nucleus; (b) left-handed neutrino.

a zero charge and relativity requires it to move with the speed of light. The neutrino has spin one-half and this is important for us, since relativity demands that the spin of this massless particle be either parallel or antiparallel to its velocity! Now the spin (the circular arrow) and the velocity (the linear arrow) form a screw or a helix that can be either right-handed or left-handed. In Nature we find only left-handed neutrinos (and right-handed antineutrinos) (Fig. 1.11). Here lies the screw at the heart of Nature! All reactions involving these handed objects violate parity.

Besides *parity* ($\mathcal{P}$), there are two other discrete symmetries — *charge conjugation* ($\mathcal{C}$) and *time reversal* ($\mathcal{T}$). The symmetry operation of *charge conjugation* ($\mathcal{C}$) replaces a particle with its antiparticle, denoted by an overhead bar. A particle and its antiparticle have the same mass but equal and opposite electric charges, among other things. Thus we speak of the anti-electron e^+ (commonly called positron), antiproton ($\bar{p}$), antineutron ($\bar{n}$), antineutrino ($\bar{\nu}$), and so on. The photon (γ) is its own antiparticle. Charge conjugation symmetry demands invariance of physical laws under the operation $\mathcal{C}$. Thus a reaction $X + Y \rightarrow Z + W$ and its conjugate $\overline{X} + \overline{Y} \rightarrow \overline{Z} + \overline{W}$ should proceed at the same rate. The antiworld is as allowed as our conventional world. And yet we see more electrons around than positrons, more protons than antiprotons and so on. The asymmetry seems to be of a cosmological origin, not fully understood at present. One thing should, however, be clear. We can hardly expect particles and antiparticles to co-exist in close proximity. They would annihilate immediately producing a flash of radiation, *e.g.*, $e^- + e^+ \rightarrow \gamma + \gamma$. This positron annihilation is used in solid state physics to study electrons in metals.

Time-reversal symmetry demands invariance of the law under the operation of time reversal $(\mathcal{T})$. Thus, if we take a movie of a process and then re-run the reel backwards, what we observe will be an equally allowed process. The reaction $X + Y \rightarrow Z + W$ is as legal as the time-reversed reaction $Z + W \rightarrow X + Y$. The time-reversal operation $(\mathcal{T})$ requires reversing all velocities and spins in detail and interchanging past and future. Thus, at the level of elementary processes, there is no arrow of time. Microscopically, every process is reversible. (But how do we reconcile this microscopic reversibility with the all too common irreversibility of processes in complex systems — the irreversibility at the macroscopic level? What about ageing for instance? There is a thermodynamic arrow of time no doubt. The connection between the time-reversal symmetry of the microscopic laws and the observed asymmetry of complex processes has been and continues to be a subject of much debate. We will not pursue this matter here any further).

Like parity $(\mathcal{P})$, the time-reversal $(\mathcal{T})$ and the charge-conjugation $(\mathcal{C})$ symmetries are also approximate. There are subnuclear reactions in which $\mathcal{C}$ and $\mathcal{T}$ are individually violated. But amazingly, the combined action of these approximate symmetry operations (in any order) is an exact symmetry of nature with no violation known. Thus, if in any process we replace all particles by their respective antiparticles, reflect the resulting process in a mirror, and then reverse all velocities and interchange past and future, we will get an equally allowed process. This celebrated $\mathcal{CPT}$ theorem expresses a deep symmetry of Nature. "All Hell will break loose" if $\mathcal{CPT}$ invariance is ever found to be violated.

Finally, what about the Ozma problem? We now know why it is not sufficient just to ask our otherworldly correspondent to repeat the ^{60}Co experiment, as he (or she) may belong to the antiworld (of antimatter). There will always be an ambiguity inasmuch as both a right-handed helix of matter and a left-handed helix of antimatter will interpret the results of the experiment equally well. We must somehow ascertain before-hand whether they are made of matter or antimatter. It turns out that this is in fact possible. There are subnuclear reactions that violate time-reversal symmetry and eventually provide us with a method of ascertaining the material versus antimaterial nature of the distant world. The details are much too complicated, but the happy ending is that the Ozma problem is solved in principle.

1.4   Gauge Symmetry

We have been talking mostly about the geometric symmetries of space-time. These are the general framework symmetries without which the physical world will hardly be comprehensible. They seem so natural, almost *a priori*, that we take them for granted. Thus, the failure of symmetry under space reflection, even though a discrete and non-performable one, came as a great shock. Now we are approaching a symmetry of an entirely different kind — the *gauge symmetry*. It is special, it is abstract and it appeals only to a preoccupied mind. Here we are requesting invariance of the law that there be, with respect to transformations that are simply outrageous. And yet the experience of the last five decades points to these gauge symmetries as the basic dynamical principles on which the fundamental interactions (forces) of Nature are designed. The familiar *electromagnetic* interaction that controls much of the low-energy physics and all of chemistry, the *strong* interactions that hold neutrons and protons together in the nucleus, the *weak* interactions responsible for the radioactive decay of unstable nuclei, and possibly even the universal *gravitation* that holds the planets and the stars together, all seem to fit in with this general scheme as gauge fields.

A proper understanding of gauge symmetry in physics requires a background knowledge of the framework theory, *quantum mechanics*, which is frankly outside the scope of this discussion (see, however, Appendix B). It is possible to get acquainted with the basic idea of gauge symmetry from an example that we know from our first year in college — the example of a simple harmonic oscillator (SHO). It will be a caricature, but real enough for our purpose. Let us get down to it without further apology.

Consider a particle performing a simple harmonic motion in a plane, *i.e.*, a two-dimensional SHO. What it means is that both its x and y coordinates oscillate sinusoidally with the same frequency. Thus the particle will in general describe an elliptical trajectory in the x, y-plane. It is convenient to combine the two motions along the x and the y axes into the motion of a single complex variable $z = x + iy$, where $i = \sqrt{-1}$ is the imaginary unity that keeps the real and the imaginary parts of z from getting scrambled up. The position of the particle in the x, y-plane is now labeled by a single complex variable z. This is the familiar Argand diagram, or the Gauss plane, for complex numbers (Fig. 1.12).

The magnitude of z is $r = \sqrt{x^2 + y^2}$, which is the distance of the particle from the origin O, and the polar angle θ is its angular position, where $\tan\theta = y/x$. The SHO is described by the equation $d^2z/dt^2 + \omega^2 z = 0$. Here $2\pi/\omega$ is the time

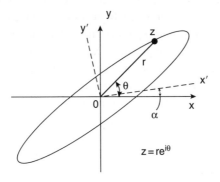

Figure 1.12: Elliptical trajectory of a two-dimensional harmonic oscillator in the complex plane representation.

period of oscillation. (We can even make ω time dependent and have parametric oscillations). As remarked before, the real and the imaginary parts of this equation do indeed describe the simple harmonic motions along the x and y axes.

Now comes the crucial observation. We have reckoned the angle θ from the x-axis. But this is just a matter of convenience. The absolute origin of the angle is irrelevant. And with this irrelevancy comes the freedom of choice. We could, for example, rotate our x, y axes anticlockwise by an angle α to new axes x', y' and reckon θ from the new x'-axis. This trivially amounts to replacing θ by $\theta - \alpha$. We say that we have re-gauged θ. All we have to do is to multiply our equation by $e^{-i\alpha}$ and absorb this phase factor by redefining $z' = ze^{-i\alpha}$, and our equation reads the same in terms of z'. Nothing really has changed. We could do this, of course, because α was a constant, *i.e.*, time independent. This irrelevance of absolute θ and the associated invariance of our equation is what we call the *global gauge freedom* and invariance. Global because it was an overall shift of θ, fixed and the same for all time. Encouraged by this, we now become more demanding. We demand freedom of choosing α differently at different times. That is to say we demand invariance under time-dependent shift $\alpha(t)$. This is the *local gauge invariance*, *i.e.*, local in time. But with $\alpha(t)$ varying with time, the factor $e^{i\alpha}$ can no longer be absorbed by the re-definition of z because of the time-derivative occurring in our equation. It will generate additional terms involving time-derivatives of $\alpha(t)$. Our earlier invariance of the equation is obviously lost. The question is if we can regain it with as little and as reasonable, or natural, a modification as possible of our original equation. In other words, can we introduce something that will *compensate* for these additional terms? It comes as a pleasant surprise that the answer is *yes*. All we have to do is to replace the time-derivative d/dt occurring in our equation by $d/dt - iA(t)$ with the proviso that re-gauging θ locally as $\theta - \alpha(t)$ should be accompanied by a re-gauging of $A(t)$ as $A(t) - d\alpha/dt$. Here $A(t)$ is the compensatory, or the 'gauge' field. That is all! But what have we gained after all this, you may well ask. Let us see. The time-dependent shift $\alpha(t)$ amounts to rotating our reference frame with an angular velocity $d\alpha/dt$. Now we may recall from our high-school mechanics that such a rotation gives rise to 'fictitious forces,' namely the centrifugal force and the Coriolis force acting on our particle. The centrifugal force is the radially outward directed force you feel while riding a merry-go-round. This is the force that makes the rotating earth bulge out at the equator. The Coriolis force is the force that makes you swerve sideways when you try to walk on a rotating platform. This is the force that deflects the winds and the ocean currents to the right (left) in the Northern (Southern) hemisphere due to Earth's rotation. After a little calculus our equation will show that the 'gauge field' $A(t)$ generates precisely these forces automatically. Thus, the requirement of local gauge invariance has created the right kind of forces acting on the particle in accord with experience. Is this not wonderful? This is the essence of local gauge symmetry.

It is now believed that all the fundamental forces of nature, the *electromagnetic*, the *weak*, the *strong*, and even the *gravitational*, are generated just this way. One has to simply identify the correct global symmetry (the irrelevancy) that is to be gauged locally. This is where all the ingenuity and the insight of the theorist lie. We have spoken of the irrelevance of the absolute origin of space and time, and the irrelevance of the absolute orientation in the Minkowski space-time. When these global symmetries are gauged locally, we get Einstein's *general theory of relativity* that replaces the old-fashioned Newtonian gravitation acting in the old-fashioned Euclidean space. Thus gravitation appears as a gauge field. The idea of local gauge invariance really comes into its own only when it is combined with the framework of quantum mechanics, with all its built-in redundancies, irrelevancies and unobservables. For instance, as we have remarked earlier, the absolute phase of the wave function ψ of an electron is irrelevant. It can be changed globally by an arbitrary constant. But when we gauge it locally, the compensating force turns out to be just the electromagnetic force that we know so well from our experience. It couples to (acts on) the charges and the currents as it should. In point of fact, should we replace the single independent variable t in our oscillator equation by the three space co-ordinates (x, y, z), generalize the gradients appropriately, and let $z(t)$ become $\psi(x, y, z)$, our equation will become the Schrödinger equation for a charged particle moving in a magnetic field represented by the gauge field $A(x, y, z)$, the so-called 'vector potential.'

When this gauge principle is applied to relativity-plus-quantum mechanics, it becomes the formidable gauge-field theory of physics today. The principle of local gauge invariance has become the guiding principle in our quest of fundamental understanding in the domain of the very small as well as the very large. Let us hasten to add that the same general principle appears again and again in our world of middle dimensions — the physics of *condensed matter*. So, next time you hear of gauge invariance, it may well be the gauge theory of ordinary glass, or its magnetic cousin, the 'spin glass.'

1.5 Spontaneous Symmetry Breaking (SSB)

Finally, we come to discussing an idea which is as deep as the idea of symmetry itself, or perhaps even deeper. Its time came much later. But now it is seen as a physical principle that holds the key to unifying all the fundamental forces of Nature, the *electromagnetic*, the *weak*, the *strong* and possibly even the *gravitational*. This has been in one form or another, the all-time dream of physicists. It is already partially realized now, and some say that the end is in sight. But, first, what is *spontaneous symmetry breaking*? Let us define it. A symmetry is said to be broken spontaneously if the symmetry of the state of the system is lower than (is a subgroup of) the symmetry of the force law governing the system. Mark you, we do not break the symmetry of the law itself. We have already hinted at such a possibility — remember the elliptical orbit of the earth around the sun in spite of the spherical

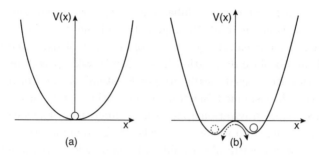

Figure 1.13:   Particle in a symmetrical potential well: (a) symmetric state; (b) spontaneously-broken-symmetry state.

symmetry of the gravitational force of the sun! To fix the idea, let us consider two examples, the first a trivial one, taken from mechanics, and the second a highly non-trivial one taken from statistical mechanics where it all began.

Take a piece of wire. Bend it in a U-like shape and hold it vertically. Now slip a bead on the wire and let it slide freely on it. It is common knowledge that the bead will oscillate for a while and eventually settle down (due to friction) at the bottom of the U-wire (Fig. 1.13), this being the state of lowest potential energy (equilibrium).

The gravitational potential energy measured from the bottom is proportional to the height. Thus, the U-wire is really a 'potential well' with a single potential minimum at the bottom. Notice that the potential is symmetrical about the vertical through this minimum. Thus the state of the system has the same symmetry as the potential (the force law). Now, let us flatten the bottom part of our U-wire and finally make it convex upwards. We will now have two local minima of the potential located symmetrically about the midpoint which now becomes a local maximum. What should we expect now? The potential is still symmetrical about the vertical through the midpoint, but this is now a state of unstable equilibrium. A disturbance, however small, will tilt the balance in favor of one or the other of the two minima and the bead will roll down accordingly. Let it roll down to the right-side minimum. Now, the symmetry of this lopsided state is definitely lower than the symmetry of the potential which is still symmetrical about the vertical axis. This is *spontaneous symmetry breaking*. Broken symmetry agreed, but what is spontaneous about it, you may ask. After all we did need some disturbance to break it. Well, the point is this. The disturbance needed to break the symmetry of the state can be made arbitrarily small. Even the tiny thermal jiggling of molecules in the wire will do. The effect produced, namely the rolling down to one of the minima, is totally out of proportion to this tiny disturbance which could in principle be made almost zero — we have here a *critically poised atom*! This is why it is called spontaneous. (One is reminded of *Buridan's ass*. The hapless ass was placed symmetrically between two identical bales of hay. The ass was hungry but the very symmetry (equidistance) of the two options forbade him from making up his mind and, as the parable goes, he starved to death. But, of course, we know that the ass will eat — the slightest bias, even if an autosuggestion, or merely thinking about it, may make him turn to one or the other of the two stacks of hay!).

Now, we turn to the physically interesting example of a system of many interacting particles in thermal equilibrium. The branch of physics that deals with such systems is called *statistical mechanics* (see Appendix C). Most inanimate systems are of this kind. A good example is that of a ferromagnet. Take a piece of iron. For our purpose, we may regard the atoms of iron as tiny magnets, compass needles if you like. The origin of these tiny magnets, or the magnetic moments as they are called, lies in the spinning electrons. But this detail is not relevant for our discussion. These tiny magnets, shown as arrows in Fig. 1.14, interact with each other. The interaction is due to the quantum-mechanical 'exchange' of electrons because of their indistinguishability. But this is again a detail not important for our discussion. What is really important is that the interaction energy depends on the *angle between* these magnetic moments. Thus if we turn all the atomic magnetic moments around by the same angle about the same axis, the energy of the system will remain the same. We say that the law governing the system is spherically symmetric. Furthermore, for a ferromagnet the energy is minimum when the magnetic moments are all parallel to each other. It is clear, therefore, that these atomic moments will tend to align parallel to each other. At high temperatures, however, the thermal agitation will make these moments point in different directions at random so that there is no net magnetization. The state of the system will be spherically symmetric — it has the same symmetry as the law of interaction. We call this disordered, high-temperature symmetric phase the paramagnetic phase. As the sample is sufficiently cooled, however, the interaction energy favoring parallel alignment of the magnetic moments wins over the disrupting tendency of the thermal agitation. When this happens the magnetic moments align parallel to each other on average and thus the

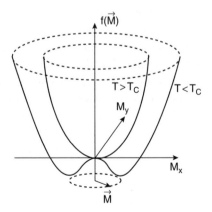

Figure 1.14: Spontaneous symmetry breaking in a ferromagnet at T_c. $f(\mathbf{M})$ denotes thermodynamic potential.

system develops a net magnetization **M**, which grows in magnitude with decreasing temperature. This low-temperature ordered state is called the ferromagnetic phase. The temperature T_c at which the system makes the continuous transition from the high-temperature disordered phase to the low-temperature ordered phase is called the critical temperature, or the Curie temperature. The magnetization which is a measure of order is referred to as the *order parameter*. (The physics of continuous phase transition, often called the second-order phase transition, at and about the critical temperature has been an extremely active area of research of our times. It is determined almost entirely by the symmetry and the dimensionality of the order parameter and is quite independent of the microscopic details of chemical composition, etc. This '*universality*' of the 'critical behavior' is most fascinating but we must let it pass). The question now is what should be the direction of this net magnetization **M**. Inasmuch as the energy depends only on the relative orientation of the atomic moments, all directions of **M** are equally probable statistically. And yet in a given realization some direction of **M** must get selected. This state is then symmetric only for rotations about this direction of **M** — it has only an axial symmetry which is a subgroup of the full spherical symmetry of the interaction law. The symmetry is thus *spontaneously broken*! For a large, in principle infinite, system an arbitrarily small magnetic field or anisotropy will fix the direction of **M**. The connection with our mechanical example should be obvious. The order parameter (magnetization **M**) plays the role of displacement x. Instead of mechanical potential energy $V(x)$ which was to be minimized, here we have a thermodynamic potential (free energy) $f(\mathbf{M})$, which is to be minimized. The only detail that differs is that whereas the position x in our mechanical case was a scalar (one-dimensional), the order parameter **M** is a vector. Thus, in the mechanical case the two equivalent minima were separated by a potential barrier in the broken symmetry phase, while in the ferromagnetic case all the equivalent minima (differing only by the direction of **M**) are degenerate (*i.e.*, have the same free energy) and **M** can in principle *freely* gyrate among them. Perhaps a better mechanical analogue would have been the marble in a punted wine bottle. (Incidentally, this freedom leads to the possibility of certain waves propagating in the broken symmetry phase whose frequency tends to zero as the wavelength tends to infinity (*i.e.*, they are massless). We call these *Goldstone modes*. For the (antiferro-)magnetic case, these are the spinwaves). All this plays an important role in the physics of phase transition in *condensed matter*. We emphasize that this is a highly cooperative phenomenon resulting from interaction among large number, infinite in principle, of particles, their spins in this case.

Now, how can all this possibly bring about unification of the fundamental forces of Nature? This is a magnificent and highly technical obsession of contemporary physics. We will try to give just the flavor of it in plain words. Any symmetry can be broken spontaneously. In particular and most importantly, it can be the *gauge symmetry*. It is the combination of gauge symmetry and *spontaneous symmetry breaking* that is central to unification. Consider the simplest case when the matter consists of charged particles (electrons). As we have already seen, the global symmetry (namely, the irrelevance of absolute phase) when gauged locally, generates the electromagnetic field automatically, which in the quantum version is the photon. (It is inherent in this mechanism that the photon — the gauge field — be massless). Now in quantum theory, the interactions between particles are mediated by the exchange of quanta of some field (much the same way as the exchange of a handball between two players will exert an effective force of repulsion between them. For attraction, let them exchange boomerangs!). Thus, the photon mediates interaction between charges. It also follows from general quantum principles that the range of interaction be inversely proportional to the rest mass of the quanta exchanged. This is why the range of the electromagnetic interaction is infinite. This intimate 'genetic' connection between matter (electrons) and the gauge field (photons) leads us to expect an induced change in the character of the gauge field when the matter undergoes a phase transition in which the very global symmetry, whose local gauging generated the gauge field, undergoes spontaneous breakdown. In point of fact, it would be very surprising if it were otherwise. We already see this effect in a superconductor in the laboratory (see Chapter 3 on Superconductivity). Here electrons undergo a transition in which the phase of this collective (macroscopic) wave function takes on a definite value, breaking thus the global symmetry spontaneously. This induces the photon to acquire a non-zero mass, making it impossible for it to propagate very far into the superconductor. This explains the famous Meissner effect, namely that the magnetic field is expelled from the bulk of a superconductor.

This was the simplest, but a most striking demonstration, of the change of character of a gauge field induced by the SSB of the matter (field). All we have to do now is to generalize to more complicated internal symmetries that can be imagined and indeed have been postulated. Thus, there may be several gauge fields generated by local gauging. They may be all symmetry related and thus of the same character. Now, if the matter undergoes SSB, the group of these gauge fields may be split into subgroups, and different subgroups may acquire different masses. Successive phase transitions (and the associated SSB's) may generate thus a gamut of fields with different characters. This generation of different gauge fields (fundamental forces) by the descent of symmetry due to SSB, from the single most symmetric initial entity is the dream of the *grand unified theory* (GUT). It has already been partially realized in the *unification* of the *electromagnetic* and the *weak* interaction by Glashow, Salam and Weinberg for which they won the 1979 Nobel Prize in Physics.

One has a plausible scenario in mind that the universe began totally symmetric with a Big Bang some 15 billion years ago. As it expanded it cooled and underwent successive phase transitions. The associated SSB's led to the diversity of fields that survive at the present epoch. And what a diversity — if the strong interaction measures unity on a certain scale, the electromagnetic interaction will measure 10^{-2}, the weak interaction 10^{-5} and the gravitational interaction 10^{-34}!

The strong interaction acts on hadrons (protons, neutrons, pions, etc., or their postulated building blocks, the quarks) but not on leptons (electrons, neutrinos, etc.)
and is short-ranged. The weak interaction involves neutrinos and has a still shorter-
range. The electromagnetic interaction acts on all charged particles and has infinite
range. The gravitational interaction is the weakest of all, but acts universally on
everything and has infinite range. And yet all may have a common origin. This re-
minds one of the concept of the Nirguna Brahma of the ancient Hindus — formless,
featureless, totally symmetric pure existence, from which all diversity originated,
shall we say, by spontaneous symmetry breaking!

This brings us to the end of our exploration of symmetry. We have seen its
power. Obviously, symmetry cannot answer all the *why*'s and *how*'s, but it does
reduce them to fewer *why*'s and *how*'s. To the philosophical question of why Nature
is so symmetric, we can perhaps answer thus. Symmetry is, in the ultimate analysis,
absence of bias. It is an expression of justice. There is a principle of insufficient
reason against asymmetry. A sphere is admitted. But a deviation from sphericity
must bide our question.

Galileo had spoken of the great Book of Nature. We should perhaps add that
the first and the last Chapters of this Book are on *symmetry* and its *spontaneous
breakdown*, respectively.

1.6 Summary

Symmetry means invariance of an object with respect to a set of operations called
symmetry operations to be performed on it. The object may be the geometrical
form of body such as a crystal of common salt, and the set of operations may the
geometrical operations of translation along a direction, rotation about an axis, or
reflection in a plane. The symmetry operations may be continuous or discrete,
physically performable or non-performable. More importantly, the object may be
a law of nature itself expressed mathematically by a certain equation. Symmetry
then means the invariance, or rather covariance, of the form of the equation under
the mathematical transformations corresponding to the symmetry operations, that
may not be geometrical in nature. Symmetry is a powerful physical principle that
helps us not only simplify calculations and classify and unify diverse objects, it also
restricts the possibilities in the absence of complete knowledge of the physical world.
It creates new physics when a symmetry is requested on intuitive grounds. There is
a branch of mathematics called *group theory* that provides the proper and powerful
language for dealing with symmetry. Symmetry has played a fundamental role in
quantum physics, particularly in the domain of high-energy physics, where its predictive power has been fully vindicated. The rather abstract idea of *gauge symmetry*
is one of the profoundest concepts produced by the human mind. Any symmetry,
however compelling aesthetically it may be, must be established experimentally.

Thus, *parity* signifying the left-right symmetry between an object or a process and its mirror image turned out to be false in certain fundamental processes involving neutrinos. Symmetry can also be broken spontaneously when there is a phase transition. The idea of spontaneous symmetry breaking has played a decisive role in our understanding of phase transitions in general, and in the context of grand unified theories and the early universe in particular. The search for deeper, hidden symmetries of Nature continues.

1.7 Further Reading

Books

- M. Gardner, *The New Ambidextrous Universe*, 3rd Ed. (W. H. Freeman and Company, New York, 1995).
- H. Weyl, *Symmetry* (Princeton University Press, Princeton, 1952).
- I. Hargittai, *Symmetry 2: Unifying Human Understanding* (Pergamon Press, Oxford, 1989).
- A. Zee, *Fearful Symmetry: The Search for Beauty in Modern Physics* (Macmillan Publishing Company, New York, 1986).

Lasers and Physics

2

2.1 Invitation to the New Optics

The laser may turn out to be one of the most significant inventions of our times. A product of modern quantum mechanics, it generates a light endowed with many remarkable properties and qualitatively very different from the light hitherto available to us from conventional sources. This form of light, which gives us a completely new tool for probing nature, already transforms and broadens to an extraordinary extent the ancient science of optics. It gives us a radically new power of control of light that opens up seemingly limitless applications in arts and sciences, in medicine and technology. Physicists have used lasers to study minute details of the structure of atoms and molecules, to catch atoms in flight, and to perform delicate experiments to test the very foundations of quantum mechanics. Biologists have used lasers to study the structure and the degree of aggregation of various biomolecules, to probe their dynamic behavior, or even to detect constituents of cells. Mathematicians actively involved with nonlinear complex systems have been intrigued by the possibility that their ideas could be tested by observing the dynamical instabilities exhibited by some lasers. And not only scientists or engineers — artists and dentists, soldiers and spies have also been touched by this invention.

The term *laser*, which is an acronym for *light amplification by stimulated emission of radiation*, is an apt description of the device. The principle on which the laser is based can be traced back to a work by Albert Einstein who showed in 1917 that 'stimulated,' or induced, radiation could be obtained from atoms under certain conditions. But the actual invention of the laser did not come until 1958 when Arthur Schawlow and Charles Townes and, independently, Alexander Prokhorov demonstrated that it was possible to amplify this kind of radiation in the optical and infrared regions of the spectrum. Soon thereafter, the first laser beam was obtained.

The laser is a device for producing a very tight beam of extremely intense and highly coherent light. To appreciate these remarkable properties of laser light that make the laser the unique tool it has come to be for research and applications, we will begin by considering light as it is normally found and discuss its characteristic features, so as to contrast them with the distinctive properties of laser light. There exist now many types of lasers, which use different substances as active media, achieve atomic excitations through different techniques and generate light at different wavelengths. They all share, however, the same basic principles. We will describe many applications of lasers, not only in technology, but also in basic research in physics where the use of this tool has led us to new directions, taken us to new frontiers.

2.2 Conventional Light Sources

Light is ordinarily produced in hot matter. In 1704, in his second great work, *Opticks*, Isaac Newton wrote:

> "Do not all fixed Bodies, when heated beyond a certain degree, emit Light and shine; and is not this Emission performed by the vibrating motions of their parts?"

2.2.1 *Light and Electromagnetic Radiation*

Light can be described as a perturbation of space of the kind one may observe in the vicinity of an electric conductor or near the path of a rapidly moving charged particle. Once we have observed how iron filings on a glass plate through which a current-carrying wire passes point to form closed circles about the central wire, or watched a small probe charge attached to the end of a very thin thread respond to the particle's motion, we are left in no doubt that space nearby is pervaded with a certain distribution of force that is called *electromagnetic field*. The precise way in which this field varies in space and time is described concisely by a set of differential equations, due to James Clerk Maxwell (1873), which replaced and generalized all previous empirical laws of electricity and magnetism. According to the theory encapsulated in these equations and later confirmed by experiments, the electromagnetic field is a disturbance that propagates from point to point in all accessible directions and behaves at large distances from the source as a wave. For this reason we may refer to it as an *electromagnetic wave*. One of its basic properties is that it can convey energy through empty space without transferring matter, always moving at the same high speed, the speed of light ($c = 300\,000$ km/s in empty space), and when it stops moving, it ceases to exist. We implicitly refer to this kind of energy transfer whenever we use the term '*electromagnetic radiation.*' An electromagnetic wave has all the characteristics, except visibility, of light: it can be reflected, refracted or diffracted. Light, in effect, is the *visible form* of electromagnetic radiation.

A light wave, like any other electromagnetic wave, is described by the variations in space and time of two vectorial quantities, namely, an electric field $\boldsymbol{E}$ and a

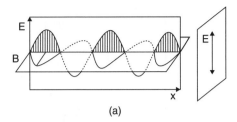
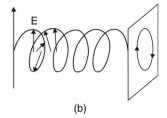

(a)   (b)

Figure 2.1:   Representation of an electromagnetic wave as space and time variations of electric field $\boldsymbol{E}$ and magnetic field $\boldsymbol{B}$. In (a) $\boldsymbol{E}$ stays at all times in a plane passing by the propagation line, and the polarization is said to be planar. In (b) $\boldsymbol{E}$ changes direction as it evolves, and the polarization is nonplanar.

magnetic field $\boldsymbol{B}$. These vectors remain at all times perpendicular to each other and perpendicular to the direction of propagation. So the wave in question is a traveling transverse wave, much like the ripples on a disturbed water surface. We refer to the direction of the electric field vector as the wave *polarization*. Over a period of time, the electric field defines a vibration pattern which may be projected in the $\boldsymbol{E}$–$\boldsymbol{B}$ plane as a line segment, a circle or an ellipse (Fig. 2.1). You can easily discover in which way a light wave is polarized: let it pass through a polarizer — a tourmaline crystal or a Polaroid filter — and observe the output as you slowly rotate the polarizer. A typical polarizer is a substance composed of long straight molecules aligned perfectly parallel to one another which strongly absorb the electric component parallel to the molecules, but let the perpendicular component pass on through with almost no absorption.

Electromagnetic waves differ from one another in their characteristic *wavelengths*, λ, the distances from one peak to the next. The whole range of electromagnetic wavelengths, called the *electromagnetic spectrum* (Fig. 2.2), covers values from the very small (for high-energy gamma rays) to the very large (for low-energy radio waves). A small portion of it forms the *optical*, or *visible spectrum*, extending from about 400 nm (violet) to about 700 nm (deep red). (1 nm is 1 nanometer, or one billionth of a meter.) Instead of wavelength,[1] we may equivalently speak of wave *frequency*, ν, which refers to the rate of vibration and is measured in cycles per second, or Hertz (Hz); or also *angular frequency*, ω, which is just the ordinary frequency multiplied by 2π, and so is given in radians per second (rad/s). In practice, we can characterize a given radiation by its wavelength, λ, or by its frequency, ν or ω. For example, orange light has wavelength $\lambda = 600$ nm, or frequency $\nu = 5 \times 10^{14}$ Hz, or $\omega = 3 \times 10^{15}$ rad/s. It is the same information said in different ways (Problem 2.1).

In many respects, light behaves as a stream of quantum particles, which we call *photons*. The production of an electric current by a sheet of copper when irradiated by intense light (a process known as the *photoelectric effect*) and the absorption

---

[1]$\lambda = c/\nu$, where λ and ν are the wavelength and ordinary frequency of the wave; the angular frequency ω is defined as $\omega = 2\pi\nu$. A brief discussion of the concepts of waves and fields can be found in Appendix A.

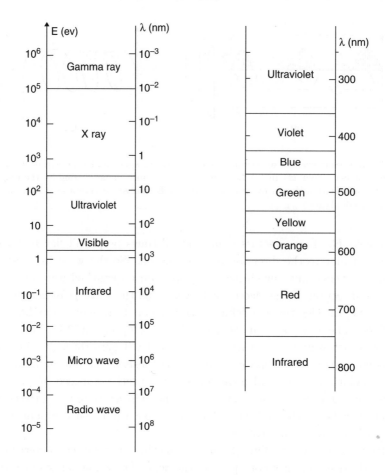

Figure 2.2: Electromagnetic spectrum shown in the left diagram, as a function of energies in electron volts (eV) and wavelengths in nanometers (nm). Right diagram exhibits the optical part of the spectrum.

or scattering of light by atoms are just two examples of phenomena that are more naturally explained by the *corpuscular model* than the *wave model* of light. The photon has all the attributes of a particle: it has a mass and charge, both of which are exactly vanishing; it has an intrinsic angular momentum (or spin) equal to 1 (in suitable units), meaning it can be described by a vector quantity, such as E. The photon, of course, travels with the speed of light, and has two other attributes of a particle, energy and momentum. The simple proportionality relations between the energy (or momentum) of a photon and the frequency (or wavelength) of the wave associated with it, provide the links between two apparently diverging pictures.[2]

[2] $E = h\nu = \hbar\omega$, where E is the energy of the photon, ν (or ω) the frequency of the corresponding wave, h a numerical constant called the Planck constant, and $\hbar = h/2\pi$. Aspects of the wave–particle duality are discussed in Section 6 of this Chapter and Appendix B. Momentum p is related to wavelength λ by $p = h/\lambda$ (de Broglie's relation).

2.2.2 Spontaneous Radiation

An atom, like any other stable quantum system, is characterized by a set of discrete energy states (or energy levels); the atom may exist as a stable system in any one of these allowed states but in no others. The state of lowest energy, or the ground state, is the one in which the atom is normally found, and to which it ultimately returns after being excited. When an atom is placed in an electromagnetic field, the field acts on the atomic electrons, transferring energy to them. If an electron can convert the energy gained into potential energy for itself, then it, and therefore the atom to which it is bound, may be lifted to an allowed higher-energy, or excited, state. This transition (*photon absorption*) can occur only if the energy added will just raise the energy to one of the allowed values; otherwise there is no absorption.

Once an atom has reached an excited state, it stays there only for a short lapse of time — close to the (*radiative*) *lifetime of the state*, or some tens of nanoseconds (1 ns = 10^{-9} s) — then gives up its excess energy by dropping spontaneously to a lower energy level, emitting electromagnetic radiation (*i.e.*, photons) in the process. This transition is called a *spontaneous emission*. The word 'spontaneous' refers to the fact that the transition is not provoked by the action of any external force, but is rather the result of an interaction of the atom with the all-pervasive electromagnetic field present in the medium itself. The frequency of the radiation emitted, ω, and the energies E_2 and E_1 of the initial and final levels are related by *Planck's law*: $\omega = (E_2 - E_1)/\hbar$. Since the allowed energies of an *isolated* atom are discrete, the corresponding emission frequencies are also discrete, in other words, non-continuous (Problem 2.2).

But a *continuous*, broad range of frequencies exists in other situations: atomic transitions to unbound states (whose energies are not restricted to discrete values), radiation of an incandescent liquid or solid (where atoms are packed closely), and radiation of a hot body (where frequent collisions cause loss of energy to the medium).

The light that emerges from any such source is non-directional, non-monochromatic, and incoherent.

It is *non-directional* because each point of the source radiates isotropically, with equal probability in any accessible directions in space. One may attempt to obtain radiation in a selected direction by placing a screen with a small hole in it some distance from the source, or by focusing the light output into a narrow beam with a mirror or a lens. But, obviously, part of the light will fall outside the collecting angle and will be lost. Even with the best available point sources, such as arc lamps, the resulting beam will nevertheless spread.

Perhaps you would think that light waves behave as perfectly sinusoidal curves, oscillating rhythmically and indefinitely over long distances. They do not. Ordinary light actually comes in a jumble of very short wave trains. The light vibration may change in shape and orientation over the duration of the wave train, depending on a large number of perturbations that might have affected the radiating atom during the emission process. What the eye perceives in the short lapse of time needed to make an observation is an average of the effects produced by an enormous number of unrelated, tiny wave packets it receives. Thus, even if the beam may have some dominant pattern of vibration at some given instant, the vibrations of its field components keep continually changing, favoring now one pattern, now another. Natural light emitted by ordinary light sources, the sun or any other star, behaves in this way; it exhibits no long-term preference as to vibration pattern: it is *unpolarized*.

The lack of *coherence* in ordinary light results from both the finite size of the wave packets and the spatial spread of the radiating points in the light source. Suppose that two such points, separated by a small distance, emit, at fixed time intervals, short identical wave packets along two intersecting paths. These wave packets can meet and interfere with each other only if they overlap at the inter- section. Also, only if the wave trains are sufficiently long, does the interference pattern remain stable long enough to be seen. Now, consider several such point sources radiating identical wave packets at random, and observe the optical inter- ferences on a screen some distance away. If the point sources are closely spaced, the interferences are almost identical for all sources, and a stable pattern emerges. But if the emitting points are distributed over a relatively large volume — as they are in conventional sources — the wave packets follow paths of very different lengths, and produce at any point on the screen successive interferences which fluctuate in brightness one instant to the next. Such a pattern is very unstable and can hardly be visible to the eye: it is *incoherent*. (This is not necessarily a bad thing: you wouldn't want to read by a coherent light.)

An idealized source that can emit infinitely long sinusoidal waves at a fixed frequency — *e.g.*, in radiative transitions between two infinitely stable and ex- tremely sharp levels — is said to emit a *monochromatic* radiation at that frequency. Such is *not* the case with sources of *continuous* radiation; neither is it with ordinary sources of *discrete* radiation, as we now discuss.

Suppose that a certain number of atoms (or ions or molecules) have been raised somehow to an upper atomic level E_2. They will spontaneously decay, or relax, to lower energy levels, giving up their excess energy in the process. And this, in two ways: One by *radiative* relaxation (in which the radiation energy materializes as photons, directly measurable with photodetectors); the other by *nonradiative* relaxation, which occurs mainly in solid-state materials (here, the energy goes into setting up mechanical vibrations of the surrounding crystal lattice and not into producing radiation).

The lifetime of an atomic state E_2 is the average length of time for finding the atom in that state before it relaxes, regardless by radiative or nonradiative processes. Its reciprocal is related to the *decay rate*, denoted by γ_2 or, equivalently, the *probability* of atomic transition of state E_2. It tells us how fast atoms belonging to a collection of identical members spontaneously relax from level E_2 according to an exponential decay law (Fig. 2.3). This law simply says that after each time interval $\tau_2 = 1/\gamma_2$ the atomic population in level E_2 decreases by 63%. A discrete

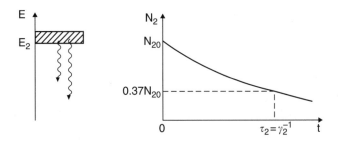

Figure 2.3: Spontaneous relaxation of a population of N_2 atoms from the single-atom energy level E_2.

transition $E_2 \rightarrow E_1$ gives rise to an exponentially decaying signal oscillating with the transition frequency, $\omega_a = (E_2 - E_1)/\hbar$. The radiation carried by such a wave, although most intense at frequency ω_a, exists at other adjacent frequencies as well. The distribution of its intensity over all possible frequencies, called the *frequency spectrum* of the signal, is a bell-shaped curve centered at ω_a and has a linewidth $\Delta\omega_a = \gamma_2$. When γ_2 includes the contributions from the decay to all lower energy levels, it represents the *lifetime line-broadening* of level E_2 (Problem 2.3).

Other mechanisms may accelerate the relaxation of atoms from a given level and contribute to further broadening that level. In gases, the most important of all such mechanisms is collisions between the radiating atoms and other atoms in the medium. In solid-state materials, the major contributor is the modulation of the atomic-transition frequency by lattice vibrations. All these line-broadening mechanisms act on all of the constituent members in the same way, so that the response of each member is equally and *homogeneously broadened.*

In other situations, however, different atoms in a system of identical atoms may have their resonance frequencies unequally shifted, such that the resulting values of the resonance frequencies for individual atoms, ω_a, are randomly distributed about some central value ω_{a0}. When a signal is sent through the medium, it cannot pick out distinct responses from individual atoms, but it will receive a cacophony of overlapping responses from all the atoms present. This gives the effect of a broadening of the transition, an effect generically referred to as *inhomogeneous broadening* (Fig. 2.4).

The prime example of broadening of this type is the Doppler broadening: the resonance frequencies change randomly, following *Doppler shifts* on each gas atom, an effect akin in nature to the apparent change in pitch of the whistling from a passing train. Besides internal motion, atoms in a gas have a random kinetic motion, or *thermal motion*. When an atom moving with a velocity v_x along the x direction interacts with an electromagnetic wave of frequency ω traveling with velocity c along the same direction, the frequency of the wave as seen by the atom is shifted to a new value given by $\omega' = (1 - v_x/c)\omega$. This means that the applied signal can resonate with the atomic transition only when the transition frequency, ω_{a0}, coincides with the Doppler-shifted signal frequency, ω'. Alternatively, we may

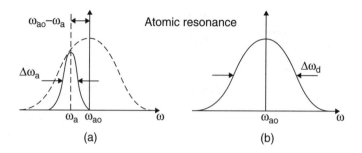

Figure 2.4: (a) Individual atomic response with homogeneous broadening $\Delta\omega_a$ is Doppler-shifted from ω_{a0} to ω_a. (b) Inhomogeously broadened atomic transition.

say that, seen in the laboratory frame, the atomic resonance frequency ω_{a0} appears to change to a new value, $\omega_a = (1 + v_x/c)\omega_{a0}$. Typically, the Doppler broadening, v_x/c, amounts to a few parts per million at room temperatures (Problem 2.4).

In short, for reasons rooted in atomic and molecular physics, ordinary light sources are bound to generate light over a *broad range of frequencies* without supplying much power at any particular frequency.

2.2.3 Summary

Light is the visible form of electromagnetic radiation; its wavelengths range from 400 nm to 700 nm, a small part of the whole electromagnetic spectrum. Light from conventional sources is produced in radiative collisions and also when excited atoms in heated matter spontaneously relax to lower energy levels, releasing their excess energy as radiation. The light that emerges from those sources runs in all accessible directions of space, and is a mixture of very short unrelated wave trains: it is non-directional, non-monochromatic, and incoherent.

2.3 What is a Laser?

In matter at normal temperatures, atoms are never at rest. Their nervous, random motion produces in the system a kind of pressure we call heat. It also takes them into a collision course with one another, constantly shifting about their energies and changing their states. But when the system reaches *thermal equilibrium* with its surroundings, there are just as many particles coming into each atomic state as there are leaving it, and the atoms will then be statistically distributed among all the allowed quantum states with a profile uniquely determined by the ambient temperature. At a typical finite temperature, the *Maxwell–Boltzmann distribution*, as this statistical law is called, will show the atomic populations of different energy states to decrease smoothly and rapidly as the energy increases (Fig. 2.5). At room temperatures, practically all the atoms sit in the ground state (Problem 2.5).

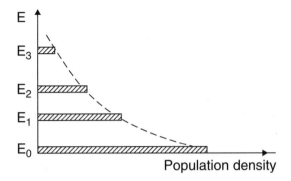

Figure 2.5: Maxwell–Boltzmann distribution of the population densities of atomic levels in an ensemble of atoms at some finite temperature T.

2.3.1 Stimulated Radiation

As we have seen in the previous section, emission of light by ordinary sources, all of which are atomic systems in thermal equilibrium, occurs when an atom absorbs energy, jumps to an excited state, and then spontaneously decays to a lower level, radiating its excess energy as a photon. But there is another mechanism, known as *stimulated emission*, in which atoms in an upper level are triggered, or stimulated, by an incoming photon of a specific energy to drop to a lower level, such that the energy of the photon exactly matches the energy difference of the two atomic levels. The incoming photon is *not* absorbed by the atom; rather, it vibrates the pair of levels in question, so that the atom de-excites and emits a photon identical to the incident one. In this process, not only do we gain two photons in return for one in each step, but also obtain photons with unique properties: they are in phase, have the same wavelength and travel in the same direction (Fig. 2.6).

Stimulated emission makes negligibly small contributions to light emission by thermal sources, but because it can produce energy that has exceptional qualities and increases exponentially with each successive step, it has a great potential for applications, in particular in lasers.

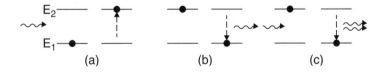

Figure 2.6: Spontaneous and stimulated transitions in atoms. (a) Atomic excitation upon absorption of a photon. (b) Emission of a photon in spontaneous downward relaxation of atom. (c) Stimulated emission occurs when the excited atom is induced to decay by a photon of frequency equal to the atomic transition frequency.

2.3.2 *Laser Action*

Let us consider a large collection of identical atoms, and concentrate first on two energy levels that we call *upper* (or *active*) and *lower* (or *terminal*) levels. To achieve laser action between them, they must satisfy two basic conditions:

(1) De-excitation of the upper level must proceed by stimulated emission rather than by spontaneous emission. For stimulated emission to occur, an excited atom must not decay spontaneously before a photon of the right frequency has found its way into the system. To promote this possibility, we must choose a pair of levels such that the upper level has a very long lifetime for radiative decay to the lower level — millions of times longer than an ordinary excited state, or up to a few milliseconds. Such a long-lived state is called a *metastable state*. An atom excited to such a state will stay there long enough for the 'right' photon to arrive and trigger it to de-excite and emit a second photon. We refer to this step as the *lasing transition*.

(2) To sustain stimulated emission, there must be more atoms arriving in the upper level than leaving it. This can be achieved by continuously pumping (exciting) atoms into that level at a rate greater than the rate at which they leave. And because it is a metastable state, atoms can stay there without de-exciting while the population is being built up. Eventually, when there are far more atoms in the upper level than in the one below, these two levels reach a thermally unstable situation known as *population inversion*.

However, it is not possible to break thermodynamic equilibrium with only two levels: for any pair of levels, it is equally likely that the upper level is populated by absorption as it is depopulated by emission, and no matter how hard pumping works, the Boltzmann factor will prevent the number of excited atoms exceeding the number of atoms in the lower state in thermal equilibrium. At best the two populations are equalized. We get around this restriction by employing more than two atomic states in the dynamics.

In a three-level lasing system, we have a third level (called the *pump level*) located above the metastable level. Because it is chosen to have a short lifetime, atoms excited into it drop rapidly into the metastable state, where they will stay for a while. The lasing transition then proceeds from here on just as in a two-level system. Once returned to the lower level, atoms are pumped back up rapidly into the pump level, and the chain of events repeats itself (Fig. 2.7).

The lasing action works even better with four levels: to three levels that play the same roles as described before, we now add another level below all three in energy, so that the terminal level just above it can be rapidly emptied by fast relaxation. The general idea, of course, is to have a huge number of atoms in the upper laser level ready for lasing duty, and almost none in the lower to prevent losses of the emitted photons through absorption.

A suitable lasing medium can certainly amplify light, but, by itself, it cannot sustain an energy production large enough to make a useful beam. To transform an

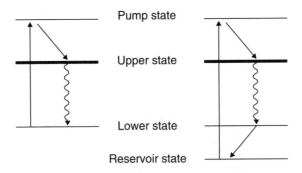

Figure 2.7:   Laser action in three- and four-level systems.

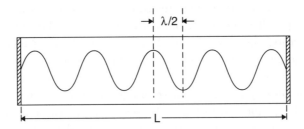

Figure 2.8:   Resonance condition in a standing-wave laser cavity of length L.

active medium into a generator of light, one encloses it in a *resonant cavity*, which is essentially a narrow cylinder closed at both ends by small, slightly curved mirrors facing each other. One of the mirrors is totally reflecting and the other partially reflecting. The qualitative '*resonant*' implies that only specific longitudinal (or axial) modes of oscillation can be supported, namely those leading to standing waves pinned at both ends (Fig. 2.8). Although many modes are theoretically possible, only those with frequencies close to the laser frequency remain stable for a long time.

The mirrors at the cavity ends create conditions favorable to *resonant feed-back* such that the stimulated photons are forced to flow back and forth across the length of the cavity, stimulating further emission as they go. This fosters amplification of a highly directional beam. Any emission in directions other than that of the cavity axis will not be redirected back into the medium, and will just be lost (Problem 2.7).

Let us now see how a three-level cavity-enclosed laser works (Fig. 2.7). At the beginning, the system is in thermal equilibrium, with the atomic populations distributed among the three (lower, upper, and pump) states according to statistical laws. Injecting a suitable energy to the system raises atoms to the pump state, from which they must rapidly decay (by radiative or nonradiative processes) to the upper laser state. The atoms arriving here will eventually de-excite either by spontaneous or stimulated transitions. If they arrive much more quickly than they leave, the

upper level will in the end have more inhabitants than the lower level, reversing
the statistical situation. Initially, as the population inversion is being built up,
spontaneous transitions from the upper laser level may occur, and photons emerge
in all directions. Most will be lost to the system. However, a few might travel along
the axis of the cavity, encounter further excited atoms and stimulate them to decay,
emitting more photons in the same direction which *lock on to the phase* that the
first spontaneously generated emission happens to have. As the stimulated photons
make repeated passes in the medium along the cavity, the herd will multiply into a
rapidly amplified beam of light down the axis of the system.

To keep the medium radiating, we maintain the population of the active level
above that of the terminal level. So the atoms that have radiated and fallen to the
ground state are continuously pumped back to the active level where they are again
available to further stimulation. The steady, coherent wave that now travels in a direction precisely parallel to the axis of the cavity is reflected backward and forward
between the mirrors, and grows in amplitude with each new passage. The energy
output by individual atoms — which are distributed over a relatively large volume
and yet radiate with the same phase — adds up coherently to yield a powerful radiation. When the *laser gain* (the increase in light intensity in the medium) exceeds
the *loss* of the beam intensity (caused by scattering at the mirrors, spontaneous
emission and off-axis propagation), the system has reached the *threshold* for laser
action, and a cascade of photons will break out through the output half-reflective
mirror in a short, sharp burst: we have a *pulsed laser beam*. If we keep re-enforcing
the population inversion in the medium, then we can produce a *continuous-wave
laser beam* (Problem 2.8).

2.3.3 Laser Light

In contrast to ordinary light, *laser light is highly intense, directional, monochromatic, and coherent.*

Laser light has well-coordinated waves that can keep a constant phase relationship with each other over *time* and through *space*. Its *temporal coherence* is
measured by the lengths of its wave trains or, equivalently, the widths of its spectral bands. Radiation emitted by a laser, just as ordinary light, could take place
throughout the Doppler-broadened distribution of frequencies (Fig. 2.9). However,
two factors contribute to make the linewidth of a laser gain considerably narrower.
First, stimulated emission is more likely to occur at the lasing-transition frequency,
at the center of the spectrum, than at any other frequencies. Secondly, when the
laser operates with a resonant cavity, only radiation corresponding to standing
waves, such that a whole number of their half-wavelengths fit perfectly between
the mirrors, can be supported as a cavity axial mode, and subsequently amplified
(Fig. 2.8). Within the range of the allowed frequencies, a large number of such
modes may be supported by long cavities (since the mirror spacing is usually much

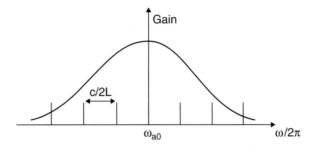

Figure 2.9: Gain profile of a typical laser system. The available axial-mode frequencies ν are equally spaced, separated by $c/2L$, where L is the cavity length. The centrally located frequency coincides with the natural atomic transition frequency, $\nu_0 = \omega_{a0}/2\pi$.

greater than the radiation wavelength); the emission is then said to be *multimode*. But for a short cavity, only one mode may lie within the gain bandwidth, and the laser emission is *single-mode*, producing an *extremely monochromatic* light output. For example, a helium–neon laser — a type of gas laser commonly used for reading bar-codes at supermarkets — can be so designed that the emitted light emerges in wave packets a hundred kilometers long with frequencies within a spectral band only a kilohertz wide. The monochromaticity quality of laser light, like its coherence, arises primarily from the resonant-cavity properties of the laser resonator rather than from the quantum properties of the lasing medium (Problems 2.9, 2.10).

We can demonstrate the temporal coherence of laser light (Fig. 2.10) by directing two laser beams on a photomultiplier (which generates a current by the photoelectric effect) connected to an oscilloscope (which displays visually the changes in the varying current). We assume that the two lasers are programmed to emit simultaneous pulses with two slightly different frequencies, ω_1 and ω_2. The two beams are then combined to give signals with frequencies ranging from $\omega_1 - \omega_2$ to $\omega_1 + \omega_2$.

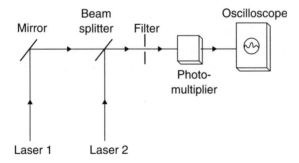

Figure 2.10: Experimental set-up to observe optical beats. The beams from two lasers of nearly identical frequencies are directed by a mirror arrangement to a photomultiplier connected to an oscilloscope. The resulting signals have many compound frequencies. If all but those having a frequency equal to the difference between the input frequencies are filtered out, beats can be observed.

A filter is used to remove all but the lowest frequencies so that the signal actually received by the photomultiplier is a single modulated sinusoid of frequency $\omega_1 - \omega_2$. The intensity of the signal is large when the superposed waves interfere constructively; it decreases and vanishes when they interfere destructively. Then, again, it grows larger. With the passing of time, the recurring pattern of enhanced and reduced interferences moves along with the wave velocity. This phenomenon is known as *beats*. It arises from a stable interference of two long waves of nearly identical frequencies, and cannot be observed with an untreated incoherent light. Musicians often use acoustical beats to bring their instruments in tune at the beginning of a concert. They listen to the beats, and adjust one instrument against another to reduce the beat frequency until it disappears.

Spatial coherence, on the other hand, is sensitive to the size of the light source. Light emerging from the aperture of a laser system diverges slightly. But when focused through a suitable lens, it always gives a point image as if emitted by a point source. To demonstrate spatial coherence, we can repeat the two-slit interference experiment, first performed by Thomas Young in the early 1800s. In this experiment, light passes through two parallel slits and falls on a screen placed some distance away. When the light waves emerging from the two openings fall *in phase* (crest on crest) at some point on the screen, they produce a bright spot. When they arrive *out of phase* (crest on valley), they cancel out, producing a darker spot. This superposition of waves gives rise to a pattern of alternating bright and dark lines on the screen, each indicating a half-wavelength difference in lengths between the two intersecting optical paths. The spacings between the fringes depend on the relative obliquity angle of the interfering incoming waves: the greater this angle is, the closer the fringes are.

When the experiment is performed with ordinary light, the source must be small or made to appear small, and must be placed some distance from the slits so that the wave fronts when reaching them are as nearly plane as possible. If neither condition is met, no pattern of useful contrast will form. But when a laser is used as the source of light, it can be placed directly in front of the slits and a clear, stable pattern can be seen. More remarkably still, observations of interference of waves from different lasers that emit long wave trains of well-defined frequencies become practical, a feat evidently not possible with ordinary light sources (Problems 2.11–2.13).

2.3.4 Summary

A laser device has three essential components:

(1) a laser medium — a collection of atoms, molecules or ions in gaseous, liquid or solid state — which allows population inversion in two levels and produces coherent light by stimulated emission;

(2) a pumping mechanism, which injects energy into the medium and thereby initiates and sustains the population inversion;

(3) a resonant structure, which confines radiation emitted by stimulated transitions and promotes its amplification in selected modes by repeated passages through the medium.

The light produced by a laser, though basically of the same nature as the light generated by any thermal source, also has significant differences. Laser light is more intense, directional, monochromatic, and coherent than conventional light.

2.4 Types of Lasers

The first working laser model was announced by Theodore Maiman in 1960. Soon after, lasers of different types were built. They all incorporate the three basic components described earlier: an *active medium*, a *pumping mechanism*, and a *resonant structure*. Lasers use a variety of substances as active media, and can produce intense light at frequencies that range over the whole visible spectrum and beyond.

2.4.1 Solid-state Lasers

In Maiman's laser, ruby was used as the active medium. Here, as in many other solid-state lasers now in operation, the ruby crystal is machine-tooled into a cylindrical rod about five centimeters long and half a centimeter wide. Its ends are polished flat, parallel, and are partially silvered. It is placed at the center of a coil of xenon-filled flashtube that can produce intense light. Ruby is aluminum oxide (Al_2O_3) in which a small fraction of aluminum has been replaced by chromium. The chromium ion absorbs green and yellow light, and lets blue and red pass through, which gives ruby its characteristic color. Upon absorbing energy in the blue-green spectral region, it is excited to a broad band of levels from which it quickly falls to a relatively long-lived level lying immediately above the lowest energy level. As more and more ions throughout the crystal reach the metastable state, the population of this state rapidly exceeds that of the ground state, and we have an inversion of population. Soon, a few excited ions spontaneously de-excite, releasing photons which go on to strike other still-excited chromium ions, triggering off a cascade of photons, mostly with a wavelength of about 700 nm (Fig. 2.11).

To keep the ruby laser in operation, it is necessary to pump at least half of the chromium atoms in the crystal to the active level. Such an effort consumes a good deal of energy. It can be significantly reduced in substances that allow a lasing transition to end not in the ground state itself, which is always densely populated, but at a level that lies at some energy above it. The population of such a level is normally sparse and, provided a suitable metastable state exists, population inversion can be achieved with only a small energy expenditure by placing a relatively small number of atoms in the metastable state. Neodymium is such a substance. When

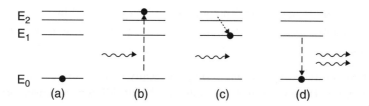

Figure 2.11:　Principle of operation of the three-level ruby laser. Chromium ions are raised from the ground state (a) to a band of levels upon absorbing a photon (b). The atoms relax downward to a metastable state (c), from which they decay and radiate by stimulation (d).

energy is injected into a Nd:YAG (neodymium-doped yttrium-aluminum garnet) crystal, the neodymium atoms are excited to a band of levels, from which they drop to the ground state in three steps. The first and the last transition are fast and spontaneous, and do not contribute to the laser beam. The intermediate transition proceeds by stimulation and produces coherent light at 1 064 nm.

2.4.2　Gas Lasers

Whereas most solid-state lasers operate in the *pulsed* mode, gas lasers are capable of *continuous-wave* operation. A gas laser consists of a tube filled with atomic or molecular gas and placed in a resonant cavity. The pumping energy is provided by a high-voltage electric current. Energetic electrons are injected into the tube by electric discharge and, through collisions, boost the gas atoms or molecules to excited quantum states. What makes gas lasers highly efficient is that there always exist many transitions capable of laser emission that terminate on levels above the ground state, making feasible a four-level laser operation. De-excitation of the terminal levels can be accelerated by adding to the active medium other gases that provoke more frequent collisions and convert the surplus internal energy more quickly to kinetic energy. The high rate of depopulation of the terminal levels and a continuous repopulation of the active levels are the two factors that contribute to make gas lasers highly efficient continuous sources of light.

The first gas laser used *atomic* neon (Fig. 2.12). It generated a continuous beam with excellent spectral purity, but with a power output low compared with that of solid-state lasers. The advent of *molecular*-gas lasers changed all that. The carbon-dioxide (CO_2) laser, the prime example of this type of laser, is capable of producing beams several kilowatts strong. The energy spectrum of the CO_2 molecule is far richer than that of each of its atomic components. Besides excitations of the individual electrons, the molecule can change its internal energy through oscillations of its component atoms about their mean positions, or through rotations of the system as a whole. So each electronic state is associated with a set of vibrational levels, which are in turn accompanied by rotational levels. The spacings of the electronic levels in the molecule are comparable to those found in atoms — a

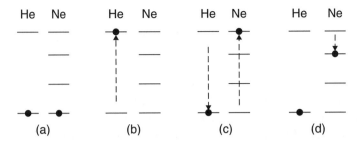

Figure 2.12: Principle of the four-level helium–neon laser. Both atoms are initially in the ground state (a). Helium atoms are excited by electron bombardment (b) and subsequently transfer their surplus energy, via collisions, to neon atoms which are excited to high-energy levels (c). When stimulated by an incoming photon, the neon atom fluoresces, contributing a photon to the laser beam; it then falls back to the ground state in steps (d).

few electron volts — but the spacings of vibrational and rotational levels are smaller by factors of ten and one hundred, respectively.

In CO_2 the lasing transition occurs between rotational levels belonging to different vibrational bands, and emits a stimulated photon in the infrared. Their output can be controlled by a *Q-switch*, which is simply a rotating mirror replacing one of the usual cavity mirrors. Normally, it is oriented so as to interrupt the photon flow in the cavity, but when it lines up with the opposite stationary mirror, the path is restored and the beam of accumulated energy passing through the gas column touches off a massive avalanche of photons. Operating with this scheme, a CO_2 laser can produce sharp, nanosecond (ns) pulses of energy which can reach peaks a thousand times greater than the average power it normally produces in a continuous-wave operation.

Even shorter pulses can be achieved by '*mode-locking*.' As we have seen in the previous section, lasers usually allow within their bandwidth many axial-mode frequencies separated by spacings that depend on the cavity length. These modes are uncorrelated (independent) in phase, which leads to a randomly fluctuating laser field. However, if they could be correlated (made to oscillate with comparable amplitudes and constant phase relationships), they would interfere regularly in step to produce periodic strong pulses. The oscillating radiation thus produced could be visualized as a stable pulse that propagates back and forth between the cavity mirrors. The duration of the resulting laser pulse will vary in inverse proportion to the oscillation bandwidth: the larger the gain bandwidth is, the shorter the pulse produced. More precisely, the spectral bandwidth $\Delta\nu$ controls the shortest duration of the optical pulse T by the relation $T \geq 1/\Delta\nu$. So gas lasers with bandwidths of about 10^{10} Hz are limited to pulses no shorter than 0.1 ns. To obtain picosecond (1 ps is 10^{-12} s) pulses, the bandwidth must be greater than 10^{12} Hz. For this purpose, *dye lasers* (in which the active medium is a complex organic dye in liquid solution) are a good choice. Invented in the 1980s, they operated at about 620 nm, and generated pulses lasting 100 fs (1 fs or femtosecond is 10^{-15} s), and 30 fs as technology was perfected. Later, in the 1990s, the invention of titanium-sapphire-based lasers brought in a new revolution: Ti:sapphire oscillators now produce 10–20 fs pulses routinely, and 4–5 fs in optimized configurations (Problem 2.14). The frontier of the ultrafast world was pushed back further in early 2003 when researchers showed that they could produce isolated bursts of coherent extreme ultraviolet photons in controlled and reproducible shape that lasted a few hundred attoseconds (1 as is 10^{-18} s), breaching for the first time the femtosecond barrier.

To have an idea of the time scale we are dealing with, consider this: in 1 second light goes around the earth 7 times; in 50 fs it travels a distance of 15 μm, smaller than the thickness of a hair; 3 as is to the second what 1 s is to ten billion years, or the present age of the universe. Used as an observational tool, a femtosecond laser detects ultrafast events in a way analogous to a powerful electron microscope resolving atom-sized details.

2.4.3 Semiconductor Lasers

Solid-state lasers usually operate at a single frequency or, at best, a few frequencies. Gas lasers, especially molecular-gas lasers, are more versatile; they can generate power at a very large number of discrete wavelengths lying within a narrow band. Still, their radiation frequencies cannot be continuously varied and controlled, or, as one says, *tuned*. This may be a drawback for some applications. Fortunately, full tunability is available in lasers based on semiconductors.

In crystalline solids, atoms arrange themselves in a regular pattern, or *lattice*. The strongly bound inner atomic orbits for electrons are unaffected by outside forces, and remain essentially unchanged as if in isolated atoms. But the energy level of each of the outer orbits, perturbed by interactions with other electrons and atoms in the solid, now broadens into a *band* of some 10^{21} closely packed individual levels. The bands become wider with increasing energy, and are usually separated from one another by no-electron's intervals, or *gaps*, although in some materials adjacent bands may overlap. An electron may have an energy lying in one of the bands but not in any of the gaps.

In general, the bands are fully occupied from the lowest energy all the way up to a certain limit (the *Fermi level*), above which the electron population of the bands drops off abruptly to zero. Such an energy spectrum indicates that all electrons remain firmly in their atomic orbits. In most *metals*, however, electrons in the external orbits are more loosely bound, and some may even escape their orbits and move about freely[3] in interatomic space. Normally occupied states may then become empty — equivalently, '*holes*' are created — and normally empty states may become occupied. The presence in metals of free (or delocalized) electrons, unattached to any specific atomic sites, makes such materials good *conductors* of electricity and

---

[3]Of course, the conduction electrons are not free since they are influenced by other electrons and atoms. What is meant is that they are not attached, or bound, to any particular atom. Later, we use the same terminology in 'free-electron laser,' even when the electrons propagate in magnetic fields.

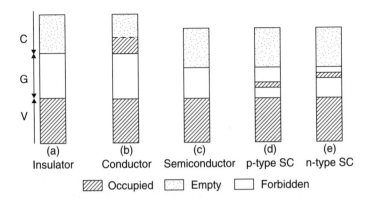

Figure 2.13: Valence bands (V), conduction bands (C), and forbidden band gaps (G) for (a) an insulator, (b) a metal, (c) a pure semiconductor, (d) a type-p semiconductor, and (e) a type-n semiconductor.

heat. In electrical *insulators*, there exist no free electrons, and the lowest empty band (called the *conduction band*) is separated from the highest occupied band (the *valence band*) by a wide *band gap* (Fig. 2.13).

Semiconductors have an energy spectrum similar to that of insulators, only with a smaller gap between the valence and conduction band. They exhibit electronic properties halfway between those found in conductors and insulators. In practice, researchers usually spike (dope) them with traces of foreign atoms, so as to modify their energy spectra and so also their conduction properties. Interactions with the embedded impurities may remove some electrons from the valence band and introduce others in the conduction band, creating conditions more favorable to conduction. So the situation normally observed in metals can be recreated in some solids in a way that may better suit our purpose. In a *type-p* (positive) semiconductor, the donor atoms have empty levels just above the top of the valence band of the receptor material; these empty levels can be readily reached via thermal excitations by valence electrons, leaving holes behind, which then act as *carriers of positive charges*. In a *type-n* (negative) semiconductor, the donor atoms have electrons on a level just below the conduction band of the receptor. These electrons can be easily excited to the conduction band, where they act as *carriers of negative charges*.

A *semiconductor* (or *diode*) *laser* in its simplest form is a *junction diode* formed by the juxtaposition of a type-n and a type-p semiconducting crystal (Fig. 2.14). The opposing faces of the two crystals are polished flat and parallel. They are separated by a thin, undoped semiconducting layer. An electric current can be fed into the heterostructure by connecting its p-component to the positive pole of an external electric source and its n-component to the negative pole. Electrons injected into the system move from the n-layer to the p-layer, whereas the holes present in the crystals move in the opposite direction, quickly filling up the junction region — electrons in the conduction band and holes in the valence band. In other words,

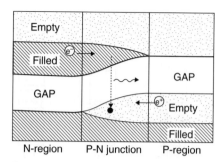

Figure 2.14: Principle of operation of a semiconductor laser. Layers of p- and n-semiconducting materials are separated by a thin layer of the same, but undoped material (p–n junction). Electrons are pumped into the n-region. At the junction, the electrons drop into empty states of the p-region, emitting photons. This recombination radiation is amplified by the geometry of the junction plane, the mirrors at both ends of the junction and a proper choice of the refractive index such that the junction can act as a wave-guide.

levels lying at the bottom of the conduction band hold many new electrons, while the top levels of the valence band lose many. There exists then a population inversion between the valence and the conduction band, exactly the situation required to produce a sustained light emission when the conduction electrons, which exist in abundance all along the plane of the junction layer, are stimulated to drop into the empty states of the valence band: electrons and holes recombine to radiate at a frequency determined by the band gap. This radiation steadily grows as it propagates between the two reflecting inner faces of the crystals, which act as wave-guides. The amplified waves finally emerge as a laser beam at one end of the junction region.

Semiconductor lasers are highly efficient sources of energy because every electron fed into the system contributes a useful photon (nonradiative decay being negligible), and no radiation is wasted in non-coherent transitions. Their output is also tunable. For example, gallium-arsenide lasers emit light at a wavelength around 900 nm at room temperatures. As we gradually lower the ambient temperature, the emission wavelength continuously decreases to 840 nm. We can further widen the range of available wavelengths by inserting impurities in the semiconducting medium. For instance, lasers based on heavily doped gallium-indium arsenide can operate between 840 and 3100 nm. There now exist many lasers of this kind, based on different materials and designs, and operating over a wide range of wavelengths, in pulse or continuous-wave mode.

2.4.4 *All Those Other Lasers*

These descriptions give just an idea of the various media capable of sustaining laser action. We can find lasers with a very wide range of wavelengths, from microwaves through infrared and visible to the ultraviolet and X-ray region of the spectrum. If we include the *maser* (a device historically preceding the laser and generating not visible but microwave radiation), then we have a choice of wavelengths anywhere from centimeters to nanometers, two extremes separated by seven orders of magnitude. Power outputs range from a few milliwatts to hundreds of kilowatts in continuous-wave operations, and hundreds of terawatts in pulsed lasers. There is also an enormous choice of pulse durations, anywhere from the millisecond to the femtosecond level. The physical dimensions of different types of lasers also vary widely: At one end of the scale, we find the low-power semiconductor lasers in sub-millimetric sizes and, more and more commonly, the even smaller quantum-well lasers (to be studied in Chapter 5). At the other extreme, we have enormous facilities, like the high-power neodymium-glass laser called NOVA and even the more powerful National Ignition Facility, both at the Lawrence Livermore National Laboratory in the USA.

2.4.5 Summary

Since the first ruby laser was presented to the public in 1960, many other types have become available. Coherent light is now obtainable in wavelengths extending from the far infrared to the near ultraviolet and beyond, in the microwave and X-ray regions. Lasers are classified according to the media in which laser action takes place. The main types are:

(1) Solid-state lasers, where the active elements are impurity ions (*e.g.*, transition metal ions, notably Cr^{3+}; or rare-earth ions, notably Nd^{3+} and Er^{3+}) embedded in a solid matrix; optical pumping is the commonly used excitation technique.

(2) Gas lasers, where the active elements are atoms, molecules or ions in gas or vapor phase, and pumping is achieved by passing a strong electric current through the gas.

(3) Semiconductor (or diode) lasers, using a variety of semiconductors as active media (notably GaAs), which operate on the principle of electron–hole recombination radiation.

(4) Dye lasers, in which the active medium is a complex organic dye in liquid solution or suspension; the most common dye laser is rhodamine 6G, which is tunable over 200 nm in the red portion of the spectrum.

2.5 Applications of Lasers

Nowadays we can find lasers performing tasks in many fields: from measurement to detection, cutting to etching, metallurgy to microelectronics, surgery to armament. Of course, most of these tasks can also be done with other tools, but the unique properties of laser beams give them clear advantages. More significantly, lasers have played a key role in opening new and often unexpected directions of research: optoelectronics, femtochemistry, laser-based biological technology, and so on. We will describe some of these applications and developments in the present and the following section.

2.5.1   Optoelectronics

Many of the most familiar applications of lasers derive from the *extreme brightness at specific wavelengths* of the light beam, which can be many orders of magnitude greater than the output obtained from the best conventional light sources. The laser beam can be captured with a suitable lens and focused to an extraordinary density into a *very small spot* whose size, perhaps a micron across, depends only on the resolution of the lens, not on the laser aperture itself. And this can be achieved for *durations* varying from femtoseconds through picoseconds to the continuum, depending on the laser type and operational mode.

Engineers now routinely use this tremendous source of energy to heat, weld, melt or cut small areas of any materials in operations that require a high precision and good control of the amount of power to be applied.

They also use tightly collimated laser beams to make microcapacitors by cutting meander paths through a conducting film vapor-deposited on some substrate, or to fabricate highly specialized microcircuits by performing discretionary wirings on general purpose circuits. They can synthesize semiconducting compounds (*e.g.*, CdTe, CuInSe) and oxides (CuO, SnO) by bringing thin films of the deposited substances rapidly to high temperatures, or modify chemical reactivity of the irradiated spots by changing the characteristics of the adsorbed molecules.

And, of course, in that area familiar to the public — data storage — the laser has been instrumental in launching it on an exponential growth curve that it has followed in recent years. Optical compact disks (CDs) have become one of the cheapest and most convenient ways of storing digital information; they have all but wiped out the vinyl record and are replacing bulky reference books. The digital information on a read-only CD is encoded in microscopic bumps of varying lengths along a single, continuous, extremely long, circular track. The pattern is stamped on a clear polycarbonate plastic layer coated with a thin aluminum film, covered in turn by a protective acrylic lacquer. A bump is 0.5 microns wide, 0.8 microns long, and 0.12 microns high. To read the data, a laser beam passes through the plastic substrate and reflects off the metallic layer to hit an optoelectronic sensor that detects changes in light as the beam scans over successive hills and flatlands. The electronics in the disk drive interpret the changes in reflectivity to read the stored data (bits). The focused spot produced by the objective lens has a diameter limited by the light wavelength and the numerical aperture of the lens. With a wavelength of 780 nm (infrared), the laser produces a spot less than 1 micron across, sufficient to resolve the bits on CDs.

A single-layer DVD (digital video disk) can store 4.7 Gbytes of data (1 byte is 8 bits), seven times more than a standard CD. This gain in capacity is due mainly to smaller pit length, tighter tracks and more efficient error correction. To read data, one uses a laser emitting at a shorter wavelength (650 nm) and a larger lens aperture to make the light spot smaller. Since their introduction in 1995, DVDs have known a tremendous success and now exist in several formats, readable and recordable.

The next-generation optical disks will be the high-density DVDs, of which there are already several candidates (Advanced Optical Disk, Blu-ray Disk). They use blue or violet (405 nm) lasers to read smaller pits, increasing data capacity to around 15 to 30 Gbytes per layer.

2.5.2 Optosurgery

The unique properties of lasers have also attracted, early on, the interest of practitioners in the medical field. A well-focused laser image, flashing intense pulses less than a millisecond in duration, makes it an ideal surgical tool: it can make precise small cuts and cauterizes as it cuts; it can stop blood circulation in a small volume of tissues; it can melt away the constricting plaque inside blood vessels that could lead to a heart stroke. A laser beam can be carried inside the human body on optical-fiber light guides, and used to attack ulcers and tumors in internal organs. Because of the absence of any contact, it ensures perfect asepsis and excellent cicatrization. Specially designed compact lasers have now widely replaced the more traditional tools in ophthalmology to weld torn retinas to their support by coagulation and to remove the degenerative blood vessels that cause diabetic retinopathy, in dermatology to treat angiomas or tumors of blood vessels, and in urology to pulverize kidney stones via a laser-initiated shock wave.

Biologists are exploiting similar techniques on the microscale: they now use laser tweezers and laser scissors to perform minimally invasive manipulations on living cells. *Laser tweezers* make use of continuous, low-irradiance beams to trap and grip individual molecules or bacteria. Let us try to understand this process: when a laser light shines on a small transparent object, its rays are refracted and bent, transferring momentum to the target. When the geometry of the arrangement of the beam and the object is correct, the transferred momentum pulls the target in the direction of the laser beam. So, with the beam grasping the target, the laser operator can drag the object from place to place by moving the beam. *Laser scissors*, in contrast, employ microsecond or femtosecond pulses of high irradiance focused in a very small effective spot. Biologists may use them to inactivate a selected part of a chromosome in dividing cells, when they want to modify its properties without totally destroying it. They also may drill a micron-size hole in a cell membrane and insert molecules into the cell without permanently damaging the membrane (the hole seals within a fraction of a second). We don't understand yet exactly how the process works, but only that photon absorption heats the target and initiates in it

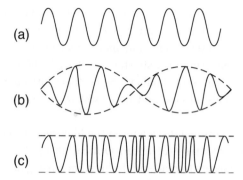

Figure 2.15: Two common methods of wave modulation. In amplitude modulation (b), the amplitude of a perfectly sinusoidal carrier wave (a) is modulated according to some lower-frequency wave. In frequency modulation (c), the frequency of the carrier is modified according to some definite pattern.

chemical reactions. With laser scissors and laser tweezers, biologists have powerful tools to probe and manipulate cells and organelles, to cut DNA molecules into fragments for analysis.

2.5.3 Communications

The advent of the laser marked a new era in long-distance communications. Light, it is true, had been used for signaling since time immemorial. But it had been handicapped in its development as a practical means of communication by the noisiness and feebleness of existing light sources. So it was replaced at the turn of this century by the more efficient and more versatile electrical techniques. With the invention of the laser, light has become, once again, the focus of interest as a vehicle for long-distance, high-volume communications. This renewed interest is justified both by the inherent superior capacity of light for transmitting information and by the special properties of the laser light itself.

The fundamentals of digital optical communication are straightforward. Let us say we want to send something, like a message. First, we digitize it by breaking it up into binary bits. Then we use a laser to produce the light and encode the bits as light pulses, add information to them with a modulator (Fig. 2.15), transmit the modulated wave through a medium (*e.g.*, optical fibers), receive it with a photodetector at the other end, and, finally, use a demodulator to recreate the message.

The capacity of a communication system is measured by the maximal information it can transmit per unit time (*e.g.*, 10^9 bits per second). This capacity depends crucially on the *frequency of the carrier wave*: the higher the frequency is, the more rapidly the oscillations follow one another, and so the more numerous bits of information can be imprinted on the carrier wave. In the lower frequencies, around 1 MHz (megahertz, or 10^6 Hz), used by ordinary radio broadcasts, only voice and music can be transmitted; but at 50 MHz, electromagnetic waves can carry the complex details of television pictures. Imagine the amount of information that visible light, at frequencies some ten thousand times higher, could transmit! Over the past decades, radio engineers have extended the usable radio region of the spectrum to all frequencies between 10^4 Hz and 10^{11} Hz — from the navigation bands to the bands reserved for use by microwave relays and radar stations. But with the insatiable needs of modern society, the available capacity for transmission of the electromagnetic spectrum has almost reached its point of saturation. The visible region of the spectrum, which ranges in frequencies from 0.4 PHz to 0.7 PHz (1 PHz or petahertz is 10^{15} Hz), could support ten thousand times more transmission channels than all the present radiowave and microwave portions combined, and could satisfy our communication needs for many years to come. In addition, the infrared and far-infrared light, which is available in many types of gas or semiconductor lasers, holds another attraction: it suffers little loss in transmission on earth and out into space, because the earth's atmosphere is partially transparent to that waveband.

Carrier waves used in long-distance transmission of information — much like sheets of paper used in written communications — play a key role in determining the quality of the transmission. Laser light waves are ideal for use as carrier waves in intercity communications for at least two reasons.

First, the *spatial coherence* of the laser light makes it possible to have a highly directional, well-focused beam over very large distances. The narrowness of the beam means that a large fraction of the radiation output can be coupled to the transmitting medium. The power output, already substantial to begin with, suffers little loss, and so can provide perfect conditions for transporting broadband information over great distances.

Secondly, the *monochromaticity* of laser light is a decisive advantage because it helps to preserve the integrity of the information transmitted. When a light signal containing a mixture of colors travels through a dispersive medium — an optical fiber, for example — its spectral shape is inevitably distorted. Waves of different colors travel at different speeds; the higher the frequency, the lower the speed. Blue light falls behind, red light gets ahead, and the pulse spreads out unevenly. The distortions, which are appreciable in pulses with broad bandwidths, could pose severe limits on transmission. Consider for example the high-intensity light-emitting diodes commonly used in optical communications. Their output has a spectral bandwidth of about 35 nm, centered in the infrared part of the spectrum. A pulse from these sources would spread over 65 cm/km, which would limit the signal rate to about 1.5×10^8 pulses per second. In contrast, a pulse from an infrared diode laser, which has a bandwidth of about 2 nm, would suffer a wavelength dispersion of only 4 cm/km, allowing a transmission rate twenty times better.

Researchers have been working hard to develop advanced light sources based on good tunable lasers that provide the needed bright, multi-wavelength light, and modulate it rapidly. State-of-the-art light sources are tunable over about 10 nm, switch on and off faster than 20 billion times a second, and can send 100 channels down a single glass fiber.

2.5.4 Holography

The laser has also been instrumental in the rapid development of another field, *holography*. Invented in 1947 by Dennis Gabor, holography is a photographic process that does not capture an image of the object being photographed, as is the case with the conventional technique, but rather records the *phases* and *amplitudes* of light waves reflected from the object. The wave amplitudes are readily encoded on an ordinary photographic film, which converts variations in intensity of the incident light into corresponding variations in opacity of the photographic emulsion. Recording the phases is another matter since the emulsion is completely insensitive to phase variations. But here comes Dennis Gabor with a truly ingenious idea: why not let the light reflected from the subject interfere with a reference coherent light on the photographic plate so as to produce *interference patterns*, which are then visible to the film emulsion?

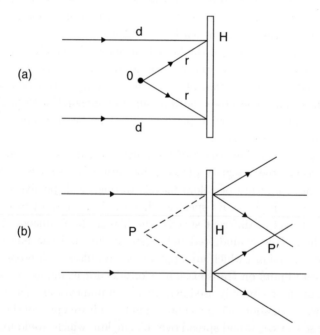

Figure 2.16: (a) Making of a hologram. Interference patterns are produced and recorded on a photographic plate H when the direct (d) and the reflected (r) beams interact on the plate. (b) Viewing of the hologram. When the developed hologram H is illuminated by the same coherent light as used in the recording, two images are formed: a virtual image *P* and a real image *P'* at points symmetric with respect to H.

To understand Gabor's idea, let's first consider a simple arrangement in which the object being photographed is just a point (Fig. 2.16). A beam of perfectly coherent light splits into two parts: one goes on to illuminate the *object* before being reflected away, the other — the *reference wave* — goes directly to a high resolution emulsion plate. The divided beams recombine on the plate and produce a clear pattern of alternating dark and bright circular fringes. The spacings of the fringes depend on the angle between the reflected and direct waves as they together strike the plate. The greater this angle is, the more closely spaced the fringes are. The pattern tends to be coarser at the center of the plate, directly facing the illuminated object, than near the edges, where the reflected beam makes a greater angle with the direct beam. So the variations in the spacings of the fringes give an exact measure of the phase variations of the reflected waves. Similarly, local variations in the amplitude, or intensity, of the reflected waves translate into local variations in the contrast of the fringes. In other words, the perfectly coherent waves of the reference beam act as carrier waves on which is impressed the information transmitted by the light reflected from the object. The waves thus modulated are then recorded by the photographic emulsion. Once the plate is developed in the traditional way, it has the 'whole picture' of the object photographed: it is a *hologram.*

When a hologram is illuminated by a collimated beam of coherent light, it shows the same properties as a grating surface: the transparent slits on the negative let the light rays pass on through — some undisturbed, other bent — and effectively act as sources of radiating cylindrical waves. These waves reinforce each other in certain directions, and produce *diffractions* of varying degrees of intensity. For example, the two directions of strongest reinforcement can be constructed — as suggested by Christiaan Huygens, one of the first proponents of the wave theory of light in the seventeenth century — by drawing lines tangent both to a wave front emerging from each slit and to the wave fronts emerging a period earlier from the two adjacent slits. We can draw in this way two series of parallel lines, representing moving wave fronts, going away from the hologram in two diverging directions (Fig. 2.17). Each direction defines a diffracted wave. Its obliquity depends on the separation of the slits; the finer the grating spacings, the greater the diffraction angles. Since the fringes become more closely spaced as one moves away from the center of the pattern, it is evident that the two diffracted waves have reverse curvatures. One wave *diverges* away from the hologram and seems to emanate from the point where the real object was placed; it produces a *virtual image* visible to an observer placed on the opposite side, in front of the hologram. The other *converges* as it moves away from the hologram to form a *real image* at the point symmetric to the position of the virtual image with respect to the hologram. This image can be seen with the eye or recorded with a camera.

Now, if we take as the subject for our experiment a realistic three-dimensional object, we still can obtain an interference pattern by making the split beam of coherent light recombine on a photographic plate, just as before. Of course, the

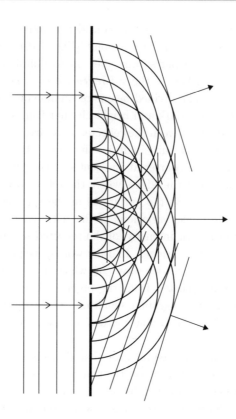

Figure 2.17: Diffraction pattern as could be formed in a hologram reconstruction. The figure shows an unscattered wave parallel to the direction of the incident light, and two diverging diffracted waves which could be prolonged backward to meet at the virtual image point.

data recorded are much more complex. Each point on the surface of the object reflects light to the entire photographic plate; conversely, each speck of emulsion receives light from all reflecting parts of the object. So the local variations in opacity and spacings of the interference fringes on the plate are directly related to the irregularities in the impinging waves and, ultimately, to the complexity of the reflecting surface. When the hologram obtained from the development of a film exposed in this way is placed in a beam of coherent light, two sets of strong diffracted waves are produced — each an exact replica of the original signal-bearing waves that impinged on the plate when the hologram was made. One set of diffracted waves produces a virtual image which can be seen by looking through the hologram; it appears in a complete three-dimensional form with highly realistic perspective effects. The reconstructed picture has all the visual properties of the original object and, for optical purposes, can be as useful as the original object.

Several recent developments have contributed to make this photography-by-reconstruction-of-light-waves an exciting field of research. One of them is the introduction of *three-dimensional holography*. An ordinary hologram is a two-dimensional

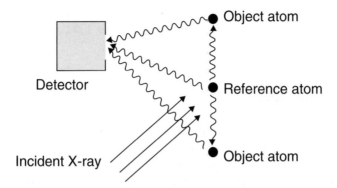

Figure 2.18: X-ray holography. Incident radiation triggers an atom to emit photons, which reach the detector, either directly (forming the holographic reference wave), or after scattering off other atoms (forming the object wave).

recording of wave fronts as variations in opacity of a thin photographic film. A three-dimensional hologram is a thick plate of high resolution emulsion that can record interference fringes throughout its thickness. This is possible when the reference beam and the reflected beam make a large angle between them as together they strike the plate; the interference fringes will then be much finer than the thickness of the emulsion layer. The plate, when developed, acts much like a three-dimensional grating or a crystalline lattice. When exposed to a beam of light, it diffracts light in the same way as a crystal would diffract X-rays.

In optical holography, the resolution of the image is limited by the wavelength of the light to several hundred nanometers. Using hard X-rays and γ-rays offers the potential for obtaining atomic resolution. The challenge here is to obtain a source with sufficient coherence; researchers have solved this difficulty by using atoms within the sample as the source.

The technique works as follows (Fig. 2.18). Atoms within the sample are stimulated by an external source of, say, X-rays. As the atoms relax, they emit fluorescent photons with a wavelength of about 0.1 nm. The radiation can reach the detector directly, forming the reference wave, or after scattering off nearby atoms, forming the object wave. The two waves meet and form an interference pattern that can be mapped out by varying the angular position of the detector around the sample. This produces a hologram.

This technique and its extensions, using X-rays and γ-rays, have been applied to image light atoms, noncrystalline, and doped samples. Researchers believe that biological and other large molecules, which are not amenable to study by X-ray diffraction methods, may be good candidates for holographic investigations.

2.5.5 Summary

We have discussed some of the applications that take advantage of the exceptional qualities of the laser beam. The high concentration of energy in a tightly collimated beam of light provides a particularly useful tool in microelectronics, medicine and biology. On the other hand, the coherence of laser light is a crucial factor in long-distance optical communications and holography.

2.6 Quantum Optics

In the previous sections we have related the invention of the laser and the subsequent developments of one of the most important technologies of our time. The laser technology in turn has dramatically stimulated not only the field that has spawned and nurtured it but also many other scientific endeavors, and even has initiated many new, unexpected lines of research. Using the laser as a tool, physicists, chemists, biologists and medical researchers have pushed their respective fields to new frontiers. In what follows, we will focus on physics, and discuss how physicists make use of the light newly available to them to scrutinize matter from various angles — its structure, bonding and interactions with light — and to explore many fundamental aspects of physics — the wave–particle duality and the reality of quantum-mechanical entities.

2.6.1 Atomic and Molecular Spectroscopy

Atoms and molecules are quantum-mechanical systems which can exist, unlike classical objects, only in a certain number of discrete states, but no others. These states are directly determined by the composition and the dynamics particular to each system and the general laws of quantum mechanics. Each quantum state (or energy level) is defined uniquely by a set of physical characteristics, called *quantum numbers*, two examples of which are the energy and the angular momentum (or spin). The full set of such states is unique to the system (atom, molecule, etc.), and so can serve as its signature. It is essential to know exactly the physical properties of the atom in its various states, because by comparing them with calculations, the physicist can identify the system, learn about its structure and the forces shaping it.

One of their favorite approaches is to measure, whenever possible, the radiative transitions between levels and, from information on levels already identified and knowledge of the electromagnetic force governing such processes, extract bits of facts from unknown levels. Unfortunately, the signals they detect for a given quantum transition are, as a rule, spread around the expected transition energy (or frequency) rather than located precisely at this energy itself, as one would anticipate from conservation of energy. In Section 2, we mentioned that this kind of radiation distribution, or line broadening, arises mainly from two effects. First, quantum levels in real physical systems are fuzzy rather than sharply defined (because of their finite lifetimes, various perturbations they are subjected to), so that any transitions may take place within these uncertainties. Second, the frequencies of all these transitions are further altered by the Doppler shift arising from the thermal motions of atoms or molecules; this effect is most pronounced in high frequency transitions in gaseous samples involving low-mass particles at high temperatures. What one observes then is not a single-emission mode but many transitions, closely related, yet differing in frequencies. The upshot is that the object of the experimenter's quest, the energy level, is hidden somewhere beneath this background of noise.

One of the most significant contributions of the laser to atomic and molecular spectroscopy is to reduce this noise by eliminating the (first-order) Doppler broadening, which is (except in transitions involving very short-lived excited states) the major factor limiting the resolution attainable in conventional spectroscopy. Another notable success of the laser is that, with its high intensity energy output, it makes accessible to the experimenter's scrutiny transitions that can only be reached by processes involving more than one photon at a time.

2.6.1.1 Single-Photon Transitions

In a typical spectroscopic experiment making use of a laser, the light output from a tunable dye laser with very stable lines and very small widths is beamed into an atomic (or molecular) gas sample, traverses it once, then is reflected back into the gas from the other end. The intensity of the reflected beam is measured for different laser frequencies and, at the peak intensity, the laser is tuned to the frequency of an atomic transition unaltered by any Doppler effects.

In this experiment, atoms moving in the gas generally see the direct and reflected beam at Doppler-shifted frequencies, one up, the other down. If the laser light is sufficiently intense to excite a large number of atoms, the velocity distribution of gas atoms at ground states will be depleted at some definite velocity and we will see two dips in the distribution, one produced by each beam, at *different* atomic velocities (Fig. 2.19). The separation of the two dips gives a measure of the atoms longitudinal

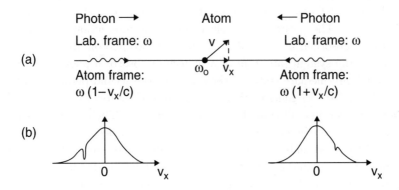

Figure 2.19: (a) Doppler effect for an atom moving obliquely to two counter-propagating laser beams; photon frequency ω is shifted to $\omega(1 \pm v_x/c)$. (b) Velocity-distribution of atoms. Dips in the distribution are produced by the two beams tuned to resonate with obliquely moving atoms, and occur at $v_x = \pm c(1 - \omega_0/\omega)$, where ω_0 is the atomic transition frequency.

velocities. Whenever the two dips merge together, it means that the *same* sets of atoms interact with the two beams. This can happen only when their longitudinal velocities vanish, *i.e.*, when the Doppler effect is absent. Then, provided the laser light is strong enough to excite a sufficiently large number of atoms, the direct beam will bleach a path through the sample, completely depleting it of a set of atoms moving at right angles to the beams over a range of velocities comparable to the radiative transition linewidth. It follows that less of the reflected beam will be absorbed, resulting in a greater intensity than at any other frequencies, and its strong signal will display the spectrum of the transversely moving atoms *free of Doppler broadening*.

With this technique, experimenters were able to study the fine structure of the famous red line of atomic hydrogen at wavelength 656.5 nm, and determine the value of the Rydberg constant, one of the fundamental physical constants, with an unprecedented accuracy.

2.6.1.2 Multiphoton Transitions

So far we have discussed electromagnetic transitions involving only one photon at a time. They are practically the only ones possible in weak fields. Generally, the probability for a radiative transition to occur is sensitive to two factors which vary in opposite directions with the number of photons involved. First, it is proportional to the response of the system to the electromagnetic field, a factor that decreases a hundred times for each additional participating photon; thus multiphoton excitations are normally weaker by many orders of magnitude than allowed one-photon transitions. Secondly, it depends strongly on the intensity of the incident radiation, so that multiphoton transition probabilities may become significant when a sufficiently strong source of light is used. In other words, in the strong field created by an intense laser light, multiphoton processes will become observable.

To illustrate, let us consider the excitation of state i to state f by absorption of *two photons*. The nature of the electromagnetic interaction is such that only one photon is absorbed or emitted at a time. So two-photon absorption is a two-step process: the system (an atom or a molecule) absorbs a photon and passes to some allowed intermediate state m, then absorbs the second photon and jumps to state f (Fig. 2.20). In general, there are many such states m accessible through one-photon absorption, all contributing to the process. Their contributions to the total transition probability interfere destructively with deep minima when waves following different available routes come up badly out of step, and constructively with high maxima when the radiation frequency matches the frequency of a resonant intermediate state. If there exists an accessible *resonant* intermediate state in the system, that is, if there is an observable state with a well-defined lifetime in the range 10^{-6}–10^{-9} s, then the second photon needs to arrive only within this lapse of time after the first absorption for the process to complete successfully. The transition is being carried out sequentially, an absorption completed before the next starts. But

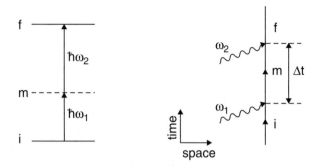

Figure 2.20: Two-photon absorption process via an intermediate state. (a) Energy level diagram; $\hbar\omega_1 = E_m - E_i$, $\hbar\omega_2 = E_f - E_m$. (b) Space-time representation of the sequential two-photon absorption. For a transition via virtual state m, $\Delta t \approx 10^{-15}$ s.

if there are no accessible intermediate resonant states, the transition still can go through, provided the second photon arrives within the flyby-time, say, 10^{-15} s, of the first. The state m is not observable, it is a *virtual* state, a very fleeting situation created by a temporary distortion of the system under the applied force. It is in this particular circumstance that the laser can play a uniquely useful role.

Of course, if one-photon transitions between the two levels can occur at all, they will dominate over any other transition modes. But such transitions are not always physically allowed; whether they are or not depends on the nature of the interaction and the conservation of energy, spin or other symmetries. Those considerations are conveniently summarized in a set of rules called *selection rules*. Consider, for instance, the possibility of exciting the hydrogen atom from its ground state, 1s, to an excited state, 2s. Assume no observable states exist between the two. When the atom is in either of these states, it has a spherically symmetric configuration. But the interaction primarily responsible for one-photon transitions, the electric dipole mode, is represented by a mathematical object that has the symmetry of the spatial vector; in particular, it changes sign under inversion (an operation that flips the signs of all position vectors). When acting on a symmetric state it changes the state to one having the same symmetry as itself, and so cannot lead to a symmetric state: the transition 1s $\rightarrow$ 2s cannot occur via one-photon absorption, it is said to be *forbidden*. But if a second photon with the right attributes (*e.g.*, angular momentum) happens to be there, while the system is still distorting after a dipole absorption, it can act in the reverse direction to bring the system back to a symmetric configuration: the transition 1s $\rightarrow$ 2s is *allowed* in the two-photon mode.

In fact, this transition, which is of great importance in physics and astrophysics, has been observed in a recent experiment using a tunable laser as the driving force for atomic excitations. Photons at half the transition frequency are sent toward the gas sample in two parallel beams from opposite directions with opposite circular polarizations. When a moving atom absorbs two photons in rapid succession, one from each beam, it gains an energy exactly double of the photon energy and devoid of any Doppler effects.[4] Thus, the transition frequency can be measured with enough accuracy to determine the Lamb shift of the hydrogen ground state, a purely quantum effect, which plays a crucial role in verifying the validity of relativistic quantum theory.

Ultrafast lasers, even with modest energy, can produce huge peak power, making them very suitable for inducing multiphoton absorption. And their femtosecond pulse durations allow the researcher to create, detect and study in real time very short-lived events, such as the electron–hole recombination in a semiconductor or the initial steps in a chemical or biological reaction.

2.6.2 Nonlinear Optics

So, under intense light, atoms and molecules undergo complicated mutations in their private little worlds. How can this affect our world? What does it do to the medium as a whole? And how does light behave through all that? A short answer to these questions is, in effect, intense radiation elicits from the medium a collective response quite unlike that observed with weaker light, which in turn alters its own properties, and even produces an output with novel features.

2.6.2.1 Harmonic Generation

As we have seen, when an intense electromagnetic field is applied on an atomic or a molecular medium, photons are absorbed one by one, either singly or multiply, and individual atoms or molecules are carried to excited states. This picture of the microscopic world translates into the appearance of *nonlinearities* in the optical bulk properties (*e.g.*, refractive index and susceptibility) of the medium.

Normally the presence of a weak field does not affect the medium; the substance acts as an inert background through which the field propagates. Its response will vary linearly with the field intensity. But in a strong field, the medium itself is modified in a field-sensitive way. Its response depends on both the applied field and the modified medium, so that the overall dependence on the applied field is rather complicated. To illustrate, let's consider how a playground swing is brought back to its vertical position. For small deviations, the restoring force is proportional to the deviation angle; for larger deviations, it can be expressed as a finite combination of successive powers of the angle. But for still larger angles, it will vary in an even more complicated manner. In the same way, atoms always respond in a complicated manner to external forces. But this *nonlinearity* is normally overshadowed by the dominant linear effects; it becomes significant, leading to detectable effects, only when the applied field is not much smaller than the interatomic field. A typical

---

[4]The two photons have energies $(1 - v_x/c)\hbar\omega$ and $(1 + v_x/c)\hbar\omega$ in the atom frame, so together they contribute an energy of $2\hbar\omega$ to the atom. Here, ω is the light frequency in the laboratory frame, v_x the atom velocity in the direction parallel to the beams.

laser at 1 MW cm^{-2} irradiance produces fields of 10^6 volts/m. Though appreciably smaller than the typical electric field in an atom (about 10^{11} volts/m), they are already sufficient to induce nonlinear effects discernible to sensitive detectors.

To be specific, let us consider the effects of the electric component of an intense electromagnetic field on a dielectric (non-conducting) material. We assume the field to be monochromatic at frequency ω. As it propagates through the medium, the applied field separates the positive and negative charge distributions in each atom, inducing a time-dependent electric field, called the *induced dipole moment*. This effect on the whole collection of atoms can be expressed in terms of a macroscopic quantity, the electric dipole moment per unit volume or, simply, the *electric polarization*. It is a space-time dependent vector field, and can be expanded in powers of the applied field. In contrast to the applied field, the induced polarization is anharmonic; it oscillates with more rounded peaks and deeper valleys than the field itself, and can be decomposed into components of various frequencies, or *harmonics*. These include, besides the frequency of the applied field, all of its whole multiples. The simplest term, linear in the field, is responsible for classical weak-field phenomena, such as ordinary refraction or absorption; its oscillations faithfully retrace the field vibrations with the same frequency but a smaller amplitude. The higher-order terms, in higher powers of the field, arise from processes involving several photons, and can be visualized as anharmonic oscillations of less intense higher harmonics.

The polarization vector has the same symmetry as the electric field vector which, let us recall, changes sign under inversion. If the medium also has that symmetry, even-order terms must vanish because they have the wrong symmetry, they do not change signs under inversion. Examples are isotropic systems (crystals, liquids or gases); they support only odd-order processes. However, for molecules and lattice sites in crystals without inversion symmetry, both odd and even powers of the field may appear. Since for a moderate field, contributions decrease in importance with increasing orders, the dominant nonlinear term in a nonisotropic medium is generally the second-power term. Quartz, calcite and anisotropic crystals are materials of this kind.

The induced polarization in a *linear medium* has a very simple frequency dependence. If the applied field has a single frequency, this is precisely the frequency of the induced oscillations; if the field has components at several different frequencies, their contributions to the polarization add up, and the response of the medium is simply proportional to the sum of all different components. Nothing new so far. But if you have an *anisotropic nonlinear medium* (*e.g.*, quartz), then a laser operating at frequency ω induces a polarization that radiates at both ω and 2ω. In other words, you send red light in through a piece of quartz, you get both red and violet light out (of course, at lower intensities). Similarly, if the crystal is shined on by two beams of frequencies ω_1 and ω_2, it will generate waves of frequencies $2\omega_1$, $2\omega_2$, $\omega_1 + \omega_2$, and $\omega_1 - \omega_2$. Finally, for an example of polarization in an *isotropic medium*, when the applied field contains components with three distinct frequencies, ω_1, ω_2, and ω_3, the dominant nonlinear components will have 22 different harmonics with frequencies of the types $3\omega_1$, $2\omega_1 + \omega_2$, $\omega_1 + \omega_2 + \omega_3$, $2\omega_1 - \omega_2$, $\omega_1 + \omega_2 - \omega_3$, and ω_1. If the input frequencies are equal, with common value ω, the polarization radiates only at ω and 3ω.

The practical implication of this discussion is clear: you can use a suitable nonlinear medium to boost the frequency of a strong field to higher values. In fact, this idea has been applied to produce tunable VUV radiation by generating third harmonics from a dye laser beam. The nonlinear medium is chosen so that the third harmonic of the field (or any other combination of input frequencies you wish, for that matter) falls close to a resonance of the medium. This harmonic will then dominate over all others. Let us take an example. The output of a pulsed dye laser, tunable over the 361–371 nm wavelengths, is focused onto a sample of krypton gas. Now the krypton atom has a known allowed three-photon transition to a state labeled 5s, at wavelength 124 nm. Thus, three light quanta will be absorbed almost simultaneously by the atom, which on decay emits a single photon at a frequency triple the input frequency. Other harmonics are suppressed. Third-harmonic generations in krypton and other noble gases are a proven method of producing VUV radiation of narrow bandwidths, high intensity and good tunability. The technique has even been applied to molecular gases (CO, N_2, C_2H_2 and HCl, just to name a few) to produce VUV and extreme UV radiation.

2.6.2.2   Phase Conjugation

Let us return now to the general example of harmonic generation in a nonlinear isotropic medium under the action of a three-component strong field, which we briefly mentioned earlier. We will refer to this process as *four-wave mixing*. We want to study a particular radiation produced in this process by three incident beams (of frequencies ω_1, ω_2, ω_3 and momenta $\boldsymbol{k}_1$, $\boldsymbol{k}_2$, $\boldsymbol{k}_3$) through nonlinear frequency mixing. Of all the possible harmonics generated, we focus on the radiation component at frequency $\omega_1 + \omega_2 - \omega_3$ and momentum $\boldsymbol{k} = \boldsymbol{k}_1 + \boldsymbol{k}_2 - \boldsymbol{k}_3$. If now we let the three incoming beams have a common frequency, ω, then the emitted wave will also have this frequency ($2\omega - \omega = \omega$). Its momentum is determined by the geometry of the incoming beams: we let the first two (*pump beams*) counter-propagate, $\boldsymbol{k}_1 + \boldsymbol{k}_2 = 0$, so that the radiated beam has momentum opposite to that of the third (the *probe beam*), $\boldsymbol{k} = -\boldsymbol{k}_3$. This means the emitted beam is identical to the third, except it travels in the reverse direction, retracing the steps of the incident wave all the way to its source (Fig. 2.21). For this reason, it is also called a *time-reversed* wave, being the time-reversed replica of the probe wave or, more correctly, a *phase-conjugate* wave because of its dependence on the complex conjugate of the probe wave.

The nonlinear medium, excited by the pump beams to states at twice the pumping energy, acts as a perfect *reflector* for the probe beam, which is turned back precisely along the retrodirection. When you gaze into such a mirror you will not

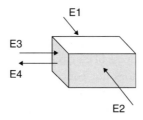

Figure 2.21: Phase conjugation by nonlinear four-wave mixing. Two strong pump beams (E_1 and E_2) counterpropagate into a nonlinear medium; at the same time, a probe beam (E_3) travels into the medium and generates a fourth wave (E_4) having the same frequency but moving in the opposite direction.

see your face but just the dots of light scattered by the corneas of your eyes. If the probe light is distorted or scrambled along its way, say, by a frosted glass plate, the distortions will be gradually but completely undone as the light wave retraces its path through the glass to its source. The medium may also act as an *amplifier* of the incident light if the pumping fields are sufficiently strong; the reflective index of the medium depends directly on their intensities. So the light reflected on a phase-conjugate mirror may even be brighter than the light beamed in.

The reader may have noticed a certain resemblance between the production of phase-conjugate waves and conventional holography. In holography the reference beam and the object beam interfere on an emulsion plate to produce a hologram. After the film is developed, the static hologram can be recreated with the same reference beam, and a realistic image of the object obtained. In four-wave mixing, the nonlinear medium acts as the photographic emulsion, the probe beam as the object beam, and the pumping beams as reference beams. The probe beam and each of the pumping beams interact to produce a wave pattern in the medium — a sort of dynamic, real-time hologram. The phase-conjugate beam is radiated when the other probe beam is reflected from the hologram.

The remarkable optical properties of phase-conjugate wave systems — *e.g.*, perfect retro-reflectivity, perfect homing ability, cancelation of aberrations in wavefronts, and amplification — all point to applications in both practical and fundamental domains. For example, a high quality beam can be transmitted through a turbulent atmosphere, collected and possibly amplified by a phase-conjugate laser system, and sent back to its point of origin free of degradation (provided the intervening atmosphere does not change appreciably in the interval). Such beams can be used in tracking satellites, self-targeting of radiation in fusion, processing images (comparing fingerprints, identifying cells and their mutations, and so on) or in realizing novel classes of ultra sensitive detectors, sensitive enough to respond to gravitational waves. Not to be neglected are the potential benefits for spectroscopic studies: the properties of the observed conjugate-phase beam could be used to probe the nonlinear medium that produces it, giving us further insight into matter and its photonic properties.

2.6.3 *Is Quantum Physics Real?*

It is very significant that the laser, a product of quantum theory, has played an essential role in the ongoing process of clarifying some deep questions still remaining in that theory. The problem does not concern any technical aspects of the theory nor its overall validity, but rather the interpretation of quantum mechanics, how to reconcile its strange character with our intuitive, common sense perception of nature.[5]

Light has characteristic wave-like properties: it can be made to produce diffraction and interference patterns. It can also knock electrons out of metals to produce a photoelectric current, an effect representative and indicative of a particle-like behavior. The co-existence of the wave and particle properties in the photon — its *wave–particle duality* — is one of the first basic realizations of the theory, with implications that defy easy interpretations.

Consider again the now familiar two-slit interference experiment. The interference pattern that one observes in this experiment is a clear proof of the wave-like nature of light. But one can also argue just as convincingly in terms of quanta: the corpuscular photons, each arriving at a definite point on the screen and each leaving its own speck, all cooperate to build up a mosaic of spots which gradually takes the form of a regular interference pattern by the law of averages of large numbers. This is the first surprising conclusion: before the advent of quantum theory, the world was completely predictable; now it looks as if events in the quantum world are only known in a probabilistic sense. In Young's experiment, when one of the two slits is plugged, only a bright spot on the screen marks the image of the open slit; no traces of interference fringes. We would certainly fail if we attempt to reconstruct the interference pattern by superimposing the patterns obtained separately with each individual aperture acting alone. The photons behave quite differently from the way they did before; they 'know' that this time only one hole is open and pass on through the aperture. The argument is in no way based on a perceived collective wave-like behavior of the whole group of quanta but only on the inherent character of each individual. How do they know, if they are independent, indivisible particles? In classical physics, a particle moves along well-defined paths. Not so in quantum physics. Suppose a photon is in a certain state at a certain time, and you want to calculate the probability to find it in some other state at a later time. You simply allow the photon to go wherever it wants to go in space and time, provided only that it starts and ends in the two given fixed states. You obtain the required probability by adding together the contributions from all possible paths from a large number of identically prepared photons.

Wave–particle duality is not confined to the photon alone. Electrons, atoms, particles of matter and quanta of energy, all have both wave and particle behavior.

---

[5]See Appendix B for a review of quantum theory.

Wave–particle duality, probability of events, and undetermined paths, all this implies an inescapable degree of *indeterminacy* in the quantum world that is not due to experimental limitations, but belongs integrally to quantum mechanics, and is perfectly compatible with the best accuracy obtainable in measurements. Suppose we want to observe an electron under a 'microscope' by illuminating it with a strong radiation. At the instant when the electron's position is measured, *i.e.*, just when the probing light is diffracted by the electron, the latter makes a jump, changing its momentum discontinuously. So, just when the position is determined, the momentum of the electron is known only up to a certain degree which corresponds to the discontinuous change. The greater this change is, the smaller the wavelength, and hence the more precise the position measurement. Conversely, if we want to measure the momentum of the particle accurately, its location becomes unavoidably uncertain.

Let Δx denote the latitude within which a coordinate x is determined in a large number of similarly prepared systems, and Δp the latitude within which the x-component of the conjugate momentum p is determined, independently, in an identical experimental arrangement. The indeterminacy in the values of position and momentum is then given by the inequality $\Delta x \Delta p \geq \hbar/2$. This equation means that no matter how hard we try, we *cannot* know the values of x and p to a better precision than indicated: $\hbar/2$ (a very small number) is the best we can do for the combined deviations. There exist similar inequalities involving other pairs of conjugate variables, such as different cartesian components of the angular momentum. All these formulas, referred to as the *Heisenberg uncertainty relations*, describe the irreducible level of uncertainty in our knowledge of those pairs of variables when they are measured in identical conditions.

In quantum mechanics, a particle (or system) is described, not by its trajectory as in classical mechanics, but by its wave function, or *state function*, which contains complete information about the particle. Another basic tenet of quantum mechanics is the *superposition of states*, which asserts that from any two independent quantum states of a system, other states can be formed.

The principle of superposition leads to many disconcerting conclusions, not only at the invisible, quantum level but also at the macroscopic level as well. To dramatize the kind of philosophical problem one might encounter with the superposition principle, Erwin Schrödinger, one of the founders of quantum mechanics, devised the following thought experiment. Suppose a cat is penned up in a steel chamber along with a poisoning device that has equal probability of releasing or not releasing a deadly poison within one hour. As long as the box remains sealed and as far as we know, the poor animal is neither live nor dead — it is equally likely to be live and dead! And it will remain in this uncertain state until we open the box and have a look, at which time it either jumps out fully alive or remains quite dead, either way with 100% probability.

2.6.3.1   Delayed-Choice Experiment

An experiment that could sharpen the concept of wave–particle duality was proposed some time ago by John Archibald Wheeler. It is basically a modern version
of the classic Young interference experiment in which the two slits are replaced by
the two arms of an interferometer (Fig. 2.22). A pulse of laser light, so severely
attenuated that at any time it carries only one photon into the apparatus, is split by
a beam splitter (BS1) into two beams (A and B). These beams are later deflected
toward the lower right of the set-up by two mirrors (M). A detector is placed at
the end of each of the two light paths. Two situations could be envisaged. In one,
a second beam splitter (BS2) is placed at the crossing of paths A and B. With a
proper adjustment of the lengths of the two arms of the interferometer, interference
signals can be recorded by the two detectors. This result would be evidence that
the photon came by both routes, thereby showing its wave-like property. In the
other, the second beam splitter is removed, and the detectors will indicate whether
the photon came along one of the two possible paths, A or B, thereby revealing its
particle-like property.

Wheeler then asks whether the result of the experiment would change if the
experimenter's decision for the mode of observation — with or without the second
beam splitter, *i.e.*, detecting wave-like or particle-like properties — is made *after*
the photon has passed the first beam splitter. Theoretically, one would decide to
put in or take out the second beam splitter at the very last moment. A photon

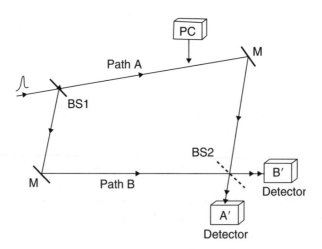

Figure 2.22:   Modern interference experiment. A single-photon pulse enters an interferometer via
a beam splitter (BS1) and the two beams may follow paths A and B. In the absence of the second
beam splitter (BS2), detectors A' and B' will reveal the route taken by the photon, either path
A or path B. With BS2 in place, this particle-like information is lost, and the two detectors will
record a wave interference signature. In the delayed-choice version of the experiment, the second
beam splitter is put in place, and one of the light paths may be interrupted by actuating a Pockels
cell switch (PC) installed for this purpose.

will take 15 ns to travel a distance of 4.5 m, which is the length of each route in a typical experimental arrangement. This would not give enough time for an ordinary mechanical device to switch between the two modes of measurement, but it would be feasible with the use of a Pockels cell, which can respond in six nanoseconds or less. The switch is installed on one of the light paths, which can be interrupted by applying a voltage to the Pockels cell. In an experiment recently performed, measurements are made in two ways. In the *normal* mode of operation, the Pockels cell is open when the light pulse reaches the first beam splitter and remains open during the whole transit time of light through the apparatus. In the *delayed-choice* mode, the cell is normally closed, and is flipped open a few nanoseconds *after* the pulse has passed through the beam splitter, and thus, has been well on its way to the detectors. For data collection, the operation is switched back and forth between the two modes, normal or delayed-choice, with each successive light pulse, and the photon counts are stored in different multichannel analyzers. Note that in either mode, the data come from many single-particle events, and the information obtained results from a time average, not an ensemble average as in many-particle experiments. The results of the experiment show that there is no observable difference whatsoever between the interference patterns obtained in the normal and delayed-choice modes.

Thus, the observations are completely consistent with the mainstream understanding of quantum physics: the photon behaves like a wave when undulatory properties are observed, and like a particle when corpuscular properties are measured. In other words, the photon in the interferometer resides in an ambiguous state that leaves many of its properties indefinite until a measurement is made. An indefinite property becomes definite only when it is observed, the transition from indefiniteness to definiteness is performed by some 'irreversible act of amplification.' Or, as Wheeler puts it: "In the real world of quantum physics, no elementary phenomenon is a phenomenon until it is recorded as a phenomenon."

2.6.3.2 To Catch an Atom

Perhaps the most convincing proof of the reality of the quantum world would be to capture some of its creatures and hold them in place for all to see. This has become feasible with *laser cooling and trapping.*

When light interacts with matter, it transfers to the medium some of its own momentum. This momentum transfer manifests itself as a mechanical force acting on the atoms. This force is not constant; it fluctuates over time because the photons scatter at random times and because the recoil of the atoms following excitation and spontaneous emission is also random. However, an average force can be defined: it is parallel to the direction of propagation of light and equal in size to the product of the photon momentum and the photon scattering rate. It is largest when light resonates with an atomic transition. Let us consider a jet composed of atoms having a strongly allowed transition close to the ground state. A laser beam with a frequency slightly below the atomic resonance frequency is directed against the atomic jet. As an atom is moving against the beam, it sees light Doppler-shifted toward resonance, and so becomes subject to a maximum scattering force which can effectively brake the motion of the atomic jet. Once the atoms are sufficiently slowed down and cooled, perhaps down to near one degree Kelvin, they can be captured and confined in a small region of space by an appropriate configuration of applied electromagnetic fields. For ions, trapping poses no problem. But for neutral atoms, it is more delicate, but still feasible. The technique used is based on the fact that a neutral atom, with its weak magnetic dipole, interacts with a magnetic field and so can be controlled by this field.

Laser cooling is the basis for many interesting experiments. It can be used to study collisions between very cold atoms and ions. Such interactions can be observed with excellent energy resolutions.

Now that small thermal motions are largely eliminated by cooling, quantum effects can emerge free of interference. Observations can be made in single-particle systems rather than in collections of many particles. An interesting example of the latter class of experiments is the observation of *individual quantum jumps* in single ions. In this experiment, atoms are cooled, trapped and confined in a region of space; then excited by a laser field. Two transitions are possible, driven by two lasers, one to a dominant, strongly fluorescing level with a normal, short lifetime; the other to a metastable state lasting several seconds. Normally, the strong emission is easily detected; but whenever the atom makes a transition to the metastable state, this strong fluorescence ceases, and a period of darkness follows. This period of darkness ends when the atom decays from the metastable state to the ground state, at which point the strong transitions resume, accompanied by spontaneous radiation. The atom will flash on and off like a tiny lighthouse, signaling each time the absence or occurrence of the weak transition.

Last but not least, cooling atoms may allow certain collective aspects of the particle dynamics to emerge: when cooled to and below a certain low critical temperature, a gas of a certain type of atom enters a new phase of matter in which a large fraction of ultracold atom fall to the lowest-energy single-particle state, creating what is known as the *Bose–Einstein condensate*, a new state of matter in which the atoms move in unison in a way similar to the photons in a laser beam. The reader will find a discussion of this subject in Chapter 4.

2.6.4 Summary

The laser has played a significant role in recent advances in atomic and molecular spectroscopy and in nonlinear optics, and also in clarifying some fundamental aspects of quantum mechanics.

A major contribution of the laser to spectroscopic studies is to eliminate the Doppler broadening of transition line widths, particularly in atomic and molecular gases where this effect has severely limited the resolution attainable in experiments using conventional electromagnetic sources for excitations. The intense monochromatic laser light makes accessible states that are normally closed to one-photon excitations — the ones usually available by conventional means — but are allowed to transitions via multiple-photon absorptions. Multiphoton excitations have been used together with Doppler-free techniques to deepen our understanding of the structure and dynamics of gaseous atoms and molecules.

Under the intense radiation of a laser output, atoms and molecules are massively excited to high-energy states. These strong interactions lead to the appearance of nonlinearities in the response of the medium, such that the optical bulk properties of the medium depend on the applied field strength in a more complicated way than just the first power. The irradiated medium will in turn generate higher harmonics, which can give us further insight into the medium itself, or can be put to other uses. One such use is the generation of a phase-conjugate beam by a nonlinear medium under the combined action of three appropriate laser beams.

Recent advances in detection techniques and the availability of high quality laser light have made many experiments involving single-particle events possible. The key point is that information gathered in these experiments results from a *time average* of successive single-particle events rather than from an ensemble *statistical average*. We stressed the importance of the delayed-choice interference experiment and various works on the cooling and trapping of atoms.

2.7 Looking Beyond

Ever since the invention of the laser, researchers have been continually extending its range of usefulness and improving its performance by exploring new lasing media and testing novel system designs. They have also been probing the restrictions that are common to all working lasers and considered indispensable to their functioning.

A helium–neon laser emitting one milliwatt of power contains several quadrillion active neon atoms and many more background helium atoms. The laser gain is reached when about one billion stimulated photons are bouncing between the resonator mirrors. In other words, several million active atoms are needed to maintain each photon in the resonator. Even in the most efficient conventional lasers, only one photon in 100 000 is useful. Because of this enormous waste, the system requires a high threshold energy to ensure the presence of a sufficiently large number of atoms in the upper laser state to sustain stimulated emission. This is why lasers require relatively high currents to work. Or do they?

From the economical viewpoint, an ideal laser should have zero threshold: only the tiniest amount of energy would be needed to bring it to self-sustained light amplification. For this to be possible, the active component must be very small (of the order of the wavelength of the light emitted) and designed such that every photon produced, even by spontaneous radiation, contributes to lasing action.

Imagine growing a layer of the lasing medium with a thickness equal to one half the wavelength of the emitted light: photons would then be confined within the layer and allowed just one state, the one that corresponds to the fundamental optical mode. So they would have no other choice than to contribute to this one possible wave, which would amplify to an intense beam. No photons would go to waste. The laser would be thresholdless. At present, we have not yet reached this limit, but researchers devoted to nanometric-sized lasers are making great strides in this direction, as we will see in Chapter 5.

Michael Feld and Kyungwon An took another tack: they built a laser with a single atom. They based their idea on an elementary process called *quantized Rabi oscillation*, which works as follows. Suppose you have a sample of two-level atoms enclosed in an electromagnetic cavity. First, the atoms absorb photons with energy matching the difference in energy between the two atomic levels. Once all the atoms have reached the upper level, they cannot absorb any more, and so the process must reverse itself, and the atoms begin to de-excite and return energy to the system until the upper level completely empties itself. Then the cycle starts all over again, and you have a system oscillating between two states.

The resonator of a single-atom laser consists of two precisely aligned mirrors with ultrahigh reflectivity, about one millimeter apart; its dimensions are carefully adjusted so that quantum-mechanical coupling occurs (*i.e.*, emitted photons can build up inside). Excited two-level (*e.g.*, barium) atoms stream one by one into the resonator. The incoming atoms emit the first photons, and this light is further amplified by the Rabi oscillations. As the number of photons in the cavity goes up, it becomes increasingly more likely that an atom passing through the resonator emits another photon. All the photons thus produced share the same direction, wavelength and phase, producing a weak (picowatt) beam of coherent light.

In this type of laser, each transition could in principle produce one useful photon; in the prototype built by Feld and An, only half the energy absorbed by the barium atoms was converted to laser light. Not too bad compared with the efficiencies of conventional lasers, which range from 1 to 30 percent. Note that even in this case — one atom, two levels — the requirement of an inverted 'population' was satisfied.

But, is population inversion really indispensable? This question has more than an academic interest for many researchers, who hope to produce coherent radiation at ultraviolet or higher frequencies by inner-shell excitations: as the pump power increases rapidly with the excitation energy, it becomes harder to achieve the required population inversion. We recall that in lasers this process has the dual purpose of lifting more atoms to the upper laser level so that they can be drafted into lasing duty, and keeping fewer in the lower level so that they absorb less of the emitted light and thereby sabotage the lasing action. But, what if we manage to block or drastically reduce this absorption? Can we then *lase without inversion* (LWI)?

Consider a gas of atoms that have three levels interacting with an electromagnetic field. Suppose that the two lower levels, g and g', may be excited to a higher level, u. Quantum mechanics tells us how to calculate the transition probability from the two lower levels to the upper level: square the sum of the two probability amplitudes. When the conditions are right (when there is coherence between g and g'), the various terms will mix so as to cancel each other, and the transition probability for photon absorption by u vanishes. On the other hand, excited atoms residing in u may relax to g and g' by independent transitions, with non-vanishing probabilities. It appears then, with u blocked to transitions by photon absorption from g and g', it would be possible to attain laser gain even if there are many more atoms sitting in the lower levels than in the upper level. That this new scheme of making laser light is feasible has been demonstrated by several groups of researchers, who have produced coherent light beams at 480–600 nm by LWI with Rb, Cd, and Ne gases.

Finally, the *free-electron laser* (FEL) is an entirely different way of producing coherent radiation, advocated since 1971 by J.M.J. Madey. A beam of relativistic electrons produced by an electron accelerator passes through a transverse, periodic magnetic field (technically an *undulator*), itself enclosed in a resonator, and exchanges energy with an electric radiation field. As the electrons travel through the undulator they accelerate from side to side and spontaneously radiate in the forward direction. Some of this radiation remains in the resonator. As more electrons arrive and are forced to move from side to side, they emit photons, but now in the presence of the stored radiation field; this is the classical analog of quantum stimulated radiation. An intricate energy exchange takes place between the electrons, the undulator and the radiation field, the outcome of which being the presence of separate groups of slow and fast electrons; the beam becomes bunched on the scale of the radiation wavelength. The bunched electrons then radiate coherently, and go on to amplify existing radiation.

The FEL has several advantages. Because a single medium (the electrons) provides gain in all spectral regions, and because the conditions in the resonator can be adjusted at will, the device is broadly and easily tuned. Because waste energy leaves the system as kinetic energy of the electrons at nearly the speed of light, and because the lasing medium cannot be damaged by high fields, it can generate very high (gigawatt) peak powers. FELs can produce, in principle, radiation at any wavelengths, but they are most needed in the far-infrared and the X-ray region, where no conventional laboratory lasers operate. At present, the shortest wavelength achieved in an FEL is 240 nm. But numerous plans to build VUV and X-ray FELs are being hatched in several countries.

2.8 Further Reading

History

- Arthur Schawlow, *Lasers and Physics: A Pretty Good Hint*, Physics Today, December 1982, pp. 46–51.
- Charles Townes, *How the Laser Happened* (Oxford U.P., Oxford, 1999).
- http://www.bell-labs.com/history/laser (*Invention of the laser*).

Applications

- G. Thomas *et al.*, *Physics in the Whirlwind of Optical Communications*, Physics Today, September 2000, pp. 30–36.
- M. Tegze *et al.*, *Imaging of Light Atoms by X-ray Holography*, Nature **407** (2000), p. 38.
- J.A. Wheeler, W.H. Zurek, ed., *Quantum Theory and Measurement* (Princeton U.P., Princeton, 1983).
- R.B. Griffiths and R. Omnès, *Consistent Histories and Quantum Measurements*, Physics Today, August 1999, p. 26.
- M.W. Berns, *Laser Scissors and Tweezers*, Scientific American, April 1998, pp. 62–67.
- H. Metcalf and P. van der Straten, *Laser Cooling and Trapping* (Springer, Berlin, 1999).

Laser Types

- P. Mandel, *Lasing without Inversion*, Contemporary Physics **34** (1994) 235–246.
- P.L. Gourley, *Nanolasers*, Scientific American, November 1998, pp. 56–61.
- M.S. Feld and K. An, *The Single-Atom Laser*, Scientific American, July 1998, pp. 57–63.
- William Colson *et al.*, *Putting Free Electron Lasers to Work*, Physics Today, January 2002, pp. 35–41.
- http://www.llnl.gov/science_on_lasers (*High-Energy Lasers at LLNL*).

2.9 Problems

2.1 For a photon, an energy E of 1 eV is equivalent to a frequency ν of 2.4×10^{14} Hz or a wavelength λ of 1240 nm. Use these equivalences and the relations among E, ν and λ to obtain the energies and frequencies corresponding to the wavelengths of 500 nm (green), 1 nm (X-ray) and 10^4 nm (infrared).

2.2 The three lowest levels in the hydrogen atom have the energies $E_1 = -13.6$ eV, $E_2 = -3.4$ eV and $E_3 = -1.5$ eV. (a) How much energy is needed to raise the atom from the ground state to the first excited state E_2? (b) What happens when the atom de-excites from E_3 to E_2? (c) Can the atom in the ground state absorb a photon of 10.0 eV?

2.3 Two atomic levels have relative line widths $\gamma_2/\gamma_3 = 100$. What are their relative lifetimes?

2.4 Make a rough estimate of the relative Doppler broadening, $\Delta\nu/\nu$, of the resonance transitions in atomic hydrogen gas at room temperature. The average kinetic energy is related to the thermal energy by $Mv_x^2/2 = kT/2$. The mass M of the atom is given by $Mc^2 = 10^9$ eV.

2.5 Bodies in thermal equilibrium have characteristic temperatures T, measured in degrees Kelvin, or equivalently, thermal energies E_T related by $E_T = kT$, where k is the Boltzmann constant. An energy of 1 eV is equivalent to a temperature of 12 000 K. Find the energies corresponding to temperatures 300 K (room temperature) and 6 000 K (at the solar surface).

2.6 Explain why 'stimulated emission makes very small contributions to light emission by atomic systems in thermal equilibrium.'

2.7 Which qualities of laser light depend most on the presence of a cavity? Explain what happens when a laser has no resonant cavity.

2.8 The irradiance (power per unit area incident on a surface) of sunlight on earth is 1 400 W/m^2. (a) A laser produces a beam of 1 mW in power, 1 mm in diameter at wavelength of 700 nm. Calculate its irradiance, the photon flux (number of photons per second), and the number of photons that a 1 m cavity contains at any time during lasing. (b) NOVA produces 100 kJ of infrared light in 3 ns pulse lengths. Calculate its power and irradiance assuming a beam diameter of 10 mm.

2.9 The length L of the cavity and the wavelengths λ_n of the resonant radiation modes are related by $L = n\lambda_n/2$, where $n = 1, 2, 3, \ldots$ (a) Suppose the laser wavelength is $\lambda_R = 500$ nm. If you want to have monochromatic laser light, what value of L would you take? If $L = 5$ cm, how many modes will the cavity support? Which are the most likely? (b) Calculate the wavelength spacing $\Delta\lambda_n = |\lambda_{n+1} - \lambda_n|$ and the frequency spacing $\Delta\nu_n = |\nu_{n+1} - \nu_n|$. Note how they depend on n. Show that for large n, $\Delta\nu_n = c(\Delta\lambda_n)/\lambda_n^2$.

2.10 We may characterize a wave train by its coherence time τ_0 (its average duration) or its coherence length $L_0 = c\tau_0$, and the spectral width of the emitted radiation by $2\Delta\nu$ (the intensity falls by half over an interval of $\Delta\nu$). We have the relation $L_0 \approx \lambda^2/\Delta\lambda = c/\Delta\nu$, where λ is the radiation wavelength (from Problem 2.9, $\Delta\nu = c\Delta\lambda/\lambda^2$). Calculate the coherence length L_0 of the following sources: (a) Conventional lamp: $\lambda = 546$ nm, $\Delta\lambda = 10$ nm. (b) He–Ne laser: $\lambda = 0.628$ μm, $\Delta\nu = 1$ MHz. (c) Nd:YAG laser: $\lambda = 1.06$ μm, $\Delta\nu = 12$ GHz.

2.11 The divergence of a spatially coherent beam of diameter d is measured by the divergence angle $\theta = \lambda/d$. Calculate θ for (a) He–Ne laser: $\lambda = 0.628$ μm, $d = 2$ mm; and (b) Ruby laser: $\lambda = 0.694$ μm, $d = 0.2$ mm.

2.12 The beam from a ruby laser ($\lambda = 0.69$ μm) is sent to the moon after passing through a telescope of 1 m diameter. Calculate the beam diameter on the moon assuming that the beam has perfect coherence (the distance to the moon is 400 000 km).

2.13 Consider the following version of Young's two-slit experiment. Take a plain window shade, and make two parallel, closely-spaced thin slits on it. Place on one side of it a lamp that can be dimmed such that it emits only a few photons per second, and on the other side, a blank screen. Darken the room, then do the following experiments: (a) Cover one slit so that light can pass only through the other slit. What do you observe? Repeat, inverting the roles of the slits. (b) Now uncover both slits, and place behind each a photomultiplier designed to make a clicking sound each time a photon passes through it. Does either of the detectors click, or can both click at the same time? (c) Now remove the detectors so that light can pass freely through the slits. Describe what you observe.

2.14 In mode locking, the bandwidth of the laser radiation $\Delta\nu$ determines the lower limit of the duration T of the pulse by the condition $T \geqslant 1/\Delta\nu$. (a) If you want to have a 100 fs pulse, what bandwidth should you have? (b) For a 1 m long cavity, how many modes should be locked in to have a 100 fs pulse?

Superconductivity

3

Disordered states are all alike. But every ordered state is ordered in its own strange way. And the most strangely ordered of these is the state of superconductivity — so much so that we may not even make a mental image of it. For superconductivity, as we will see, is a truly quantum effect coherent over a large, macroscopic scale. It has manifest in it the wave nature of matter that is normally spoken of atoms and molecules, that is, in the domain of the very small. For these reasons superconductivity is somewhat hard to understand except for a preoccupied mind. But we can easily get acquainted with the superconductor by watching its behavior, which is quite robust and readily amenable to ordinary experiments. And when we do this we will be left in no doubt that we are in the presence of something very extraordinary, almost bizarre.

3.1 Zero Electrical Resistance

Superconductivity is the total disappearance of electrical resistance of a material at and below a sharply defined temperature (T_c) which is characteristic of that material (Fig. 3.1).

This *critical temperature* (T_c) can, however, be very low, typically close to the absolute zero of temperature. A superconductor is thus a perfect conductor. And so an electric current once set up in a superconducting ring will go on circulating undiminished forever, or very nearly so. A simple-minded calculation would give the time of decay of this current much longer than the age of the universe, which is some 15 billion years! Very precise laboratory measurements of the decay of the supercurrent in a superconducting coil estimate the decay time to be about a hundred thousand years, which is long enough. This persistent flow of electricity, the supercurrent, is just about as close as we can get to man's recurrent dream of a *perpetual motion* — of the second kind (no frictional loss). Today, a superconducting magnet with the supercurrents circulating in its winding coils is a common sight in the low temperature physics laboratories around the world as a quiet source of constant magnetic field as long, of course, as it is kept cold enough.

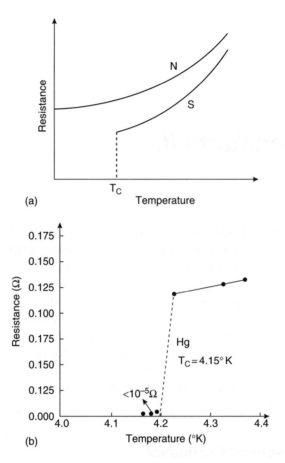

Figure 3.1: (a). Temperature dependence of resistance for a superconductor (S) and a normal metal (N); (b) superconducting transition in mercury after Kamerlingh Onnes.

This amazing phenomenon of superconductivity was discovered by the great Dutch physicist Heike Kamerlingh Onnes back in 1911. Kamerlingh was studying the low temperature behavior of electrical resistance of metals in his world famous low-temperature laboratory at Leiden of which he was the director. Just three years earlier, Kamerlingh had liquefied the last and the noblest of the permanent gases, helium. Helium boils at the incredibly low temperature of 4.2 K (Kelvin), that is just 4.2 degrees centigrade above the absolute zero of temperature — the lowest temperature possible as ordained by the laws of physics. This circumstance made it possible for Kamerlingh to observe things very close to absolute zero for the first time. Kamerlingh found to his great surprise that the resistance of his sample of frozen mercury (chemical symbol, Hg) dropped almost abruptly to zero, within experimental limits, when it was cooled below about the boiling point of liquid helium, 4.2 K (Fig. 3.1). Kamerlingh was quick to realize that the resistance was not just low — it was essentially zero! He could observe the persistent current that flowed without an external source such as a battery. He called it *superconductivity* and the name stuck. For this discovery Kamerlingh won the Nobel Prize in Physics for the year 1913. His life-long preoccupation with low temperatures earned him the informal title of the 'gentleman of absolute zero.'

Since its discovery almost ninety years ago now, superconductivity has remained one of the greatest surprises of physics. Why should zero resistance be so surprising? Let us understand this first.

3.1.1 *Metallic Resistance*

Metals are, by definition, good conductors of electricity. Now, even the best and the noblest of them all such as copper, silver and gold do offer some resistance to the flow of electricity. This is what causes the wasteful heating of the wire, the *copper loss*, in transmitting electrical power from one point to another. And for this, of course, we have to pay as an invisible part of our electricity bill. Remember Ohm's Law: Amperes (I) = Volts (V)/Ohms (R) and Watts $(W) = I^2R$. This ohmic dissipation can be reduced by going to lower operating temperatures. This is readily understandable. Much of the electrical resistance is due to the incessant thermal jiggling of the atoms or the ions (atoms that have lost one or more of their loosely attached outermost electrons) that perturbs the otherwise free flow of electricity. More precisely, the ultimate carriers of electricity are the freely moving electrons that abound in a metal. There are roughly 10^{23} of them in each cubic centimeter of the metal. In equilibrium, that is, in the absence of an applied electric field or potential difference, there are, on the average, as many electrons moving in any given direction as in the opposite one, and so there is no net current. In the presence of an electric field, however, there are relatively more electrons moving in the direction opposite to the applied electric force and this excess makes up the directed electric current. Now, the thermal vibrations of the background ions 'scatter' these electrons randomly in all directions and tend to neutralize this excess current, causing electrical resistance. Indeed, the electron would accelerate indefinitely but for this continual scattering that offers the necessary friction forcing the electron to settle down to a steady thermal drift — much the same way as the mechanical friction or, better still, the fluid viscosity (treacliness) limits the flow of a liquid through a metal pipe or a glass capillary in spite of a head of pressure. Now, the lower the temperature, the lesser the intensity of heat motion, and hence the smaller is the resistance. Thus, for example, the resistance of a specimen of copper will go down by a factor of about a thousand or more as it is cooled from room temperature of 300 K (26.85°C) to liquid helium temperature, 4.2 K. (The Kelvin (K) and the Celsius (C) scales of temperature are simply related: Degrees Kelvin = Degrees Celsius + 273.15. Thus, absolute zero, written 0 K, corresponds to −273.15°C). But the fall is a smooth one. Ideally, then, the resistance should vanish at the absolute zero of temperature, where all thermal agitation ceases. We say ideally because this would be true only if the specimen was a perfect crystal — a perfectly periodic array of atoms or ions. (It is a profound result of quantum mechanics that the electrons are not scattered by such a perfectly periodic arrangement of 'scatterers' no matter how strong the individual scatterer may be. This has to do ultimately with the wave nature of the electron). But, of course, real materials are far from possessing this perfect crystalline symmetry. There are the ubiquitous defects — impurities, misplaced atoms, or missing atoms (vacancies). These deviations from perfect crystalline symmetry can scatter electrons and, therefore, offer resistance. Thus, we have to live with this *residual* resistance even at the absolute zero of temperature, which is inaccessible anyway. (Incidentally, the lowest recorded temperature achieved so far in the laboratory is about a nanokelvin, that is, a billionth of a degree Kelvin. But even this is *not* quite zero. And, of course, never mind the cost of refrigeration). Against this normal behavior consider Kamerlingh's sample of frozen mercury that had lost all its resistance at and below about 4.2 K. Now the point is that the thermal agitation of the atoms at this low temperature, while admittedly small, is still far from being zero. Also, the randomly placed defects that were present above 4.2 K are still very much around and look just as obstructive. In fact nothing much has changed in the material by way of its chemistry or crystal structure — but the resistance has vanished completely. The scatterer has somehow lost the 'will' to scatter — it lets the electrons pass by uninterrogated. It is as if the cloud of electrons flows past these obstacles, ever adjusting, ever adapting but never quite getting perturbed — ghostlier than a ghost! The scatterer does not scatter. The dog does not bark. And that is the strange thing!

3.1.2 *Superconductivity is Common*

One may get the impression that such a bizarre phenomenon as superconductivity must be a rare occurrence. But this is simply not true. A quick look at a modern *periodic table of elements* will convince you of this. Of the 92 elements prominently displayed, 68 are metals, and of these at least 26 are superconductors. Then there are others that become superconducting when *pressed* hard enough. Thus, silicon, which is not only not a metal but a semiconductor (of which transistors and computer chips are made), begins to superconduct under the pressure of a few tens of kilobars (kilo = thousand, bar = 1 atmospheric pressure). But there are exceptions. The magnetic metals, iron (Fe), nickel (Ni) and cobalt (Co) refuse to ever superconduct — ferromagnetism seems inimical to superconductivity. The same is true of the light alkali metals such as sodium (Na), potassium (K), etc. But the most notable exceptions are the noble metals like copper (Cu), gold (Au) and silver (Ag), which are normally the best conductors of electricity. In fact, ironic as it may seem, it turns out that good superconductors (the ones with high T_c such as niobium (Nb) for example) are 'bad' normal metals. We will soon begin to see why this has to be so. Besides these elemental superconductors, we have thousands of superconducting alloys (metallic mixtures), organic compounds, and now even earthy ceramics, that are found to superconduct — may their tribe increase! In

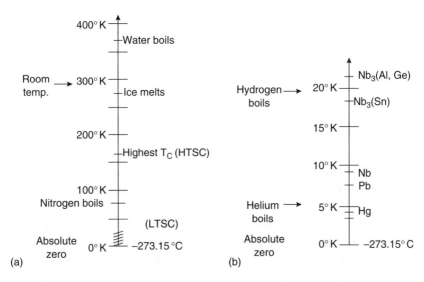

Figure 3.2: (a) Absolute temperature scale marking well known transitions; (b) expanded scale showing T_c's of some superconductors.

Fig. 3.2 we have listed some of the common superconductors along with their critical temperatures marked on the absolute (Kelvin) scale. Note the crowding at the lower end of the scale — the low-temperature superconductors (LTSC). Until about five years ago the highest critical temperature known was 23.2 K, for a compound of niobium and germanium (Nb_3Ge). Now the record is about 165 K, held by a mercury (Hg)-based high-temperature superconductor (HTSC).

Thus, superconductivity is indeed very common. It is just that the critical temperatures are abysmally low. If, however, we are willing to leave our terrestrial laboratories and look elsewhere, there is high temperature superconductivity in abundance. For instance, there are strong theoretical reasons to believe that the interior of the neutron star is a neutronic superfluid and a protonic superconductor, with a T_c of about a hundred million degrees (see Chapter 8). Nearer home we have the case of the planet Jupiter. It is again suspected that hydrogen, the major constituent of the giant planet, is crushed to a metallic density under its gravitational pressure of about a million atmospheres (megabar), and the metal so formed is a superconductor with a T_c of several thousand degrees Kelvin.

Here on earth, however, until very recently, superconductors lived only in the liquid helium cryostat (*i.e.*, *dewar*, a sophisticated thermos or vacuum bottle that keeps cold things cold and, of course, hot things hot). It is precisely this coldness that has kept superconductors confined to the low-temperature laboratories of the world, away from the gaze of the public eye. We may call these the *liquid helium (LHe) superconductors*. More than half a century of uninterrupted worldwide research could barely push the critical temperature to little over 20 K. So much so that one began to doubt seriously if higher T_c's were possible at all. And then came the breakthrough in 1986, when J. G. Bednorz and K. A. Müller of International Business Machines (IBM) at Zurich announced in the September issue of the German journal *Zeitschrift für Physik* their discovery of an earthy, ceramic superconductor with a T_c of more than 30 K. For this they won the Nobel Prize in Physics for the year 1987. This led in quick succession to superconductors with still higher critical temperatures ranging from 90 K to 125 K, thus bringing the age of the LHe-superconductors to a sudden end. It also initiated the era of the *liquid nitrogen (LN_2) superconductors.* Liquid nitrogen boils at a comfortable 77 K. The highest recorded T_c stands at about 165 K, which is really not very far from the lowest temperature recorded on Earth (183 K). And now there is already some responsible talk of room temperature superconductors — the *holy grail* of solid-state physicists.

These events of the last fifteen years have altered our view of superconductivity. It is now a serious belief that in the coming decades, superconductors may revolutionize human conditions more decisively than the laser, or nuclear power, or even the transistor ever could. We ought to get more than just acquainted with superconductivity.

3.2   Infinite Magnetic Reluctance

Zero electrical resistance is the defining property of a superconductor. It is also by far the most striking properly of a superconductor. But the deciding property of a superconductor is really its infinite reluctance to admit magnetic fields in its interior. A superconductor is a perfect diamagnet, which as we shall see is more than being just a perfect conductor. Let us understand what all this means. Take a piece of a superconducting metal like tin (Sn) and hold it at a temperature above its T_c so that it is in the normal resistive state. Now, place it in a static magnetic field which may be conveniently produced by a permanent magnet, or by a solenoid carrying electric current. Now, a normal non-magnetic metal like tin is indifferent to magnetic fields. It is almost as good as vacuum, and the magnetic lines of force run right through it undisturbed. (Of course, the act of placing the sample in the magnetic field involves initially some motion through the magnetic field which, by the *Faraday law of induction*, induces an electric field in the metallic sample. This, in turn, generates *eddy currents* in the metal and the associated stray magnetic fields. But these transient effects die down rather quickly, and we assume that we have waited long enough for this to happen). Now, let us cool the sample sufficiently and, lo and behold, at a certain temperature around T_c the sample turns superconducting and the magnetic lines of force (the flux) are expelled totally from the bulk of the superconductor. This happens unless, of course, the external field is much too strong, in which case superconductivity is suppressed and the sample remains normal down to 0 K (Fig. 3.3).

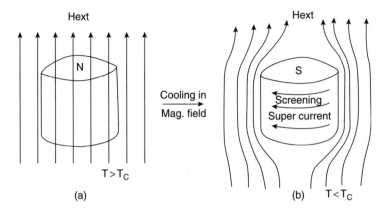

Figure 3.3: Meissner effect: (a) normal (N) state; (b) superconducting (S) state.

This dramatic phenomenon of flux expulsion or exclusion, is the famous *Meissner–Ochsenfeld* effect named after the discoverers W. Meissner and R. Ochsenfeld (1933). The process is reversible, that is, the flux lines re-enter the sample if it is re-heated through the same temperature, T_c. What really happens is that in the presence of the external magnetic field, persistent supercurrents are generated in the superconducting sample of a magnitude, sense and detail which is just right so as to produce a field that cancels the external field throughout the interior of the sample (see Fig. 3.3). We call these *screening currents*. They flow mostly on the surface of the sample, almost skimming it. Note that we are talking here about static magnetic fields and these *screening currents* are not to be confused with the *eddy currents* that are induced even in normal metals but by a time-dependent field (Faraday's law of induction). In a perfect conductor (infinite conductivity) these inductively induced eddy currents will be infinitely large so as to totally screen out any time-dependent magnetic field. Even in a metal such as copper, which is merely a good conductor, because of this screening the alternating currents flow only on the surface up to skin depth which is about a centimeter at 60 Hz and one-twentieth of a millimeter at 1 MHz. Thus, the central core of a thick copper wire hardly carries any 'AC' current. A superconductor is not merely a perfect conductor (zero resistance) — a perfect conductor will not exclude a *static* magnetic field. On the other hand we can readily reason why perfect diamagnetism (flux expulsion) must imply zero resistance. Assume to the contrary that our perfect diamagnet had a finite resistance. But then the persistent screening currents must continually dissipate energy — the $I^2 R$ loss, remember! The question now is where could this energy possibly come from. Perhaps from the energy stored in the magnetic field. But the magnetic field is given to be static (constant in time) and, therefore, it cannot supply the necessary energy. We seem to have a problem here. There is clearly no source of energy available to our system. Having thus eliminated the obvious, what remains, no matter how improbable, must be the true explanation — in this

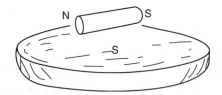

Figure 3.4:   Magnetic levitation of a bar magnet above a superconductor (S).

case, namely that our perfect diamagnet had no resistance to start with and hence there was no energy loss to be accounted for. We have just proved that a perfect diamagnet is also a perfect conductor. The converse is not true. It is for this reason that perfect diamagnetism is regarded as being a more fundamental property, in fact the deciding property, of a superconductor.

Perfect diamagnetism (flux expulsion) implies that the superconductor is repelled away from a strong magnetic field. Thus, for example, we can have a bar magnet floating above a superconducting surface (Fig. 3.4). This *magnetic levitation* has led to the possibility of having ultrafast trains gliding on a frictionless magnetic cushion.

3.3 Flux Trapping

There is an interesting corollary to the Meissner effect, which is the trapping of magnetic flux by a superconductor. Let us repeat our experiment demonstrating flux expulsion, but this time with a sample in the shape of a *hollow* cylinder. As the temperature is lowered through its critical value in the presence of the magnetic field, the flux lines are again expelled from the *bulk* of the material as expected. Nothing, however, comes in the way of the flux lines threading the *hollow* of the cylinder. Persistent screening currents will flow near the inner and the outer surfaces

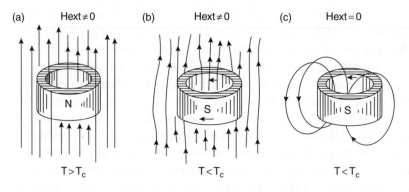

Figure 3.5:   Flux trapping by a hollow superconducting cylinder on field cooling.

of the hollow cylinder as shown in Fig. 3.5. And now let us gradually remove the externally applied magnetic field leaving our sample all by itself. But what about the flux lines passing through the hollow of the cylinder? Surely they cannot escape sideways because in doing so they must traverse the surrounding superconducting material and this is forbidden by our perfect diamagnet — the flux is trapped. The persistent screening current now circulating near the inner surface of the hollow cylinder will sustain this trapped flux. What we have really got here is a permanent bar magnet. It is robust. You could carry it around in your pocket except for the inconvenience of having to keep it cold enough. Viewed differently, you have created a non-polluting device for storing energy — the trapped magnetic flux and the circulating screening currents form a kind of flywheel, if you like, that stores (magnetic) energy *at almost zero entropy.*

3.4 Wholeness of Trapped Flux

The curious case of the *trapped flux* becomes all the more curious if we enquire further. What is the amount of magnetic flux trapped in the hollow of the cylinder? It turns out, and we will shortly know why, that the flux thus trapped cannot have an arbitrary value. It has to be an integral multiple of a certain basic unit of flux, denoted by ϕ_0. That is to say that it must be a whole number when measured in lots of ϕ_0. Fractions, or half-measures, are not allowed! This wholeness is the celebrated *flux quantization*, and ϕ_0 the quantum of flux. This unit of flux is extremely small but still macroscopic enough. To have an idea of how small it is, imagine a circular wire loop of diameter 0.1 millimeter facing the earth's magnetic field. Then the loop will intercept about 100 of these flux quanta! Small as it is, the trapped flux quanta can be counted by jiggling our cylinder in and out of a coil and then measuring the voltage (electromotive force) induced in the coil due to the changing flux linkages (Faraday's law of induction). This was indeed done by B.S. Deaver and W.M. Fairbank in their classic experiment in 1961 that confirmed this quantization of trapped flux. As we shall see later, $\phi_0 = hc/2e = 2 \times 10^{-7}$ gauss centimeter-squared ($= 2 \times 10^{-15}$ tesla meter-squared) where h is Planck's constant, c the speed of light and e the magnitude of the electric charge on the electron. Planck's constant gives away the hidden quantum nature of superconductivity. It is a remarkable fact that the quantum nature of superconductivity was anticipated by Fritz London in 1935, long before the fully microscopic theory of superconductivity was given by John Bardeen, Leon N. Cooper and J. Robert Schrieffer in 1957, the celebrated *BCS theory* for which the trio was awarded the Nobel Prize for Physics in 1972. In fact, London had predicted flux quantization, but being far ahead of his time, he missed the all important factor 2 in the denominator of $\phi_0 = hc/2e$. We now know after BCS that it is '$2e$' and not 'e,' and thereby hangs the tale of 'two electricities' — the *electron pairing* theory of superconductivity.

3.5 Temperature and Phase Transition

The irresistible zero electrical resistance and the irrepressible infinite magnetic reluctance that set in at the critical temperature should leave us in no doubt that a qualitative change of state has taken place. There is a branch of physics that describes these changes of states of matter in general terms — *thermodynamics*, or *statistical mechanics* if we are interested in a microscopic treatment (see Appendix C). Temperature plays the central role here. This section is a brief digression intended to acquaint ourselves with some simple but powerful ideas that make the change of state understandable.

But first some quick remarks on the absolute (Kelvin) scale of temperature that we have already spoken of several times. Temperature is the *intensity of heat*. It measures the energy of the random jiggling of atoms, molecules, electrons, spins, or more generally, of the *dynamical degrees of freedom* that our system may have. Absolute temperature measures it absolutely. Thus, at the absolute zero of temperature the thermal energy is zero — all motion comes to a standstill. (There is, of course an irreducible *zero-point* motion even at the absolute zero of temperature which is of a purely quantum nature, and is appreciable for the so-called quantum liquids of which we will speak later. In fact, the superconductor is one such *quantum liquid*.) It is clear that any property that at all depends on temperature can be used to detect changes in temperature. Thus, the common household thermometer uses the thermal expansion of mercury for this purpose. One can also use, for example, the change of electrical resistance of metals, alloys or semiconductors to measure temperature charges. Thus, the Platinum (Pt) resistance thermometer is a prime standard for measuring temperatures down to $-260°C$. But how can we meaningfully specify *equal* intervals of temperature? To say that equal changes in the length of the column of mercury in our thermometer give equal intervals of temperature is nothing more than an assertion that mercury expands equally for equal changes in temperature — clearly a circular statement empty of any objective content. For example, the equal intervals so defined may not be equal on a thermometer using alcohol instead of mercury. The question is if we can define equal intervals of temperature independently of the property of the material. The answer is Yes. As an act of almost pure reason, Lord Kelvin of Britain, one of the greatest of the classical physicists, proved in 1860 that such an absolute scale does exist and is defined in terms of the efficiency of an ideal heat engine. The absolute (or Kelvin) temperature scale (K) so defined is then conveniently graduated so that the boiling and the freezing points of water differ by 100 degrees on this scale just as on the commonly used centigrade scale (C) of Celsius. Then absolute zero (0 K) measures $-273.15°C$ and is the lowest temperature possible. Water freezes at 273.15 K ($0°C$), and 'room' temperature is 300 K ($26.85°C$). It is a profound result of classical statistical mechanics that *every degree of freedom of a system such as a classical gas in equilibrium carries an equal amount of kinetic energy, $k_B T/2$, where k_B is the Boltzmann constant (the law of equipartition of energy)*.

Temperature is the single most important control parameter that determines the states of matter. A solid (ice) melts to a liquid (water) and the liquid (water) boils to a gas (steam) as the temperature is raised through the well defined melting and the boiling points. These are the commonest and perhaps the most important changes of states that have shaped our biological lives and indeed the universe itself (see Chapter 10). Yet another interesting example of change of state is the loss of magnetization when a bar magnet is heated above its *Curie temperature* — the change from the ferromagnetic to the paramagnetic state. The change from the non-magnetic resistive normal state at high temperatures to the diamagnetic superconducting state at low temperatures is also a change of state, in fact closely related to the paramagnetic-to-ferromagnetic change of state. We call these different states different *phases* of the substance. The change of state is called phase transition, and the corresponding temperature the transition temperature.

3.5.1 Order Parameter

There is a feature which is common to all phase transitions. The higher temperature phase is disordered (or less ordered at any rate) while the lower temperature phase is ordered (or more ordered). There is indeed a competition between order and disorder, and temperature decides the winner. Thus, for example, the liquid state is disordered — a snap shot of the liquid state will show atoms positioned more or less at random, while the solid state formed upon freezing displays a periodic arrangement of atoms, which we call a crystal. Similarly, for the magnetic case, the spins (the tiny atomic magnets) point in different directions at random in the paramagnetic phase above the Curie temperature T_c, while they align parallel on average in the ferromagnetic phase below T_c. Indeed, one can define an *order parameter* that vanishes in the disordered phase but assumes a nonzero value in the ordered phase. For a magnet, the choice of the order parameter is obviously the magnetization. In the case of the superconducting transition, however, the nature of the order is too subtle as we will see later. The order parameter is one of the most powerful intermediate concepts in the physics of phase transition. It is an *emergent* quality. It was introduced by the great Russian physicist Leo Davidovich Landau in 1960, who gave a general theory of phase transition based on this crucial concept.

3.5.2 Free Energy and Entropy

What is the basic principle that determines which one of the possible phases our system in equilibrium will be found to be in? For mechanical systems with friction the answer is well known from our high school physics — the system will settle down to a state of *minimum* potential energy. Thus, a marble thrown in a bowl will eventually come to rest at the bottom-most point. This is a one-body problem. A somewhat similar *minimum* principle exists even for our many-body systems with a large, almost infinite number of particles (or degrees of freedom) interacting with one another. Left to itself, our system too will settle down to a final state which will change no more in time — a state of equilibrium. This state will, however, correspond to a minimum of what is called the *free energy* (see Appendix C). Let us see what this means. Consider a microscopic state of our system of energy E. A microscopic state means specifying in detail the momenta (roughly, velocities) and the positions of all the particles (degrees of freedom). Then, the energy E is the sum total of their kinetic and potential energies. Now, the fundamental principle of statistical mechanics is that all such microscopic states, which are possible at all, will occur, but with a probability proportional to $\exp(-E/k_B T)$. Next, strange as it may seem, almost all of these microscopic states of the same energy (E) *look alike* from the macroscopic (average) point of view. And it is the macroscopic viewpoint that matters for all practical purposes. (Indeed, even if we knew the finer microscopic details, we wouldn't know what to do with them. The fact of the matter is that the microscopic description is too fine-grained while our usual probes are too coarse). Thus, for a given macroscopic state of energy E, there will be a large number of the microscopic states corresponding to the number of ways in which the energy E can be partitioned among the many degrees of freedom. Let this number be $g(E)$. Hence, the probability of occurrence of the physically identifiable macroscopic state must be proportional to $g(E)$ times $\exp(-E/k_B T)$. We may re-write this as $\exp(-F/k_B T)$ with the exponent $F = E - k_B T \ln[g(E)]$. Here $\ln[g]$ denotes the 'natural' logarithm of g with respect to base $e = 2.71828\ldots$. So the most probable state is the one that corresponds to the minimum of F, the Free Energy and not of E. The quantity $k_B \ln[g(E)]$ is the mysterious *entropy* and is usually denoted by S. Thus, we must minimize $F = E - TS$, and not just E. At $T = 0$ K, this, of course, reduces to minimizing the energy itself. This lowest energy state, called the ground state, is essentially unique for the system. For this state, g is unity and hence $S = 0$ (remember, the logarithm of unity $= 0$). The *ground state*, the most ordered state, has zero entropy! It is clear that at a sufficiently high temperature the entropy term in F may dominate the free-energy and a different state may be preferred. In general, $g(E)$ and, therefore, entropy is expected to increase with energy — there are then obviously more ways of partitioning it among the various degrees of freedom. The corresponding macroscopic state will also be more disordered. Ordered microscopic states are fewer due to the constraints of order, and hence the corresponding ordered macroscopic state has lower entropy. (Just compare the disorderly crowd and the disciplined military and you will have the general drift of the idea.) And so it happens that high temperature favors disorder. The relationship between the many microscopic states and the single macroscopic state corresponding to them is illustrated best by an analogy with the game of dice. Consider casting two dice simultaneously. Each can come face-up with a number from 1 to 6. Thus there are $6 \times 6 = 36$ possibilities. These are all the possible 36 microscopic states of our system of the two dice. Let the dice be true for simplicity. Then all 36 microscopic states are equally probable. But suppose now that we are interested only in the *sum* of the numbers that the two dice come up with. The sum can vary from $1 + 1 = 2$ to $6 + 6 = 12$. These are then the 11 macroscopic states. Let us label them by the sums 2 to 12. Now you see that the macroscopic state $2 = 1 + 1$ is realized in only one way, and the macroscopic state $3 = 1 + 2 = 2 + 1$ in two ways, and so on. You can easily verify that the macroscopic state 7 is realized in six ways, which is the maximum and hence the most probable state. If you want to push this analogy further then all you have to do is to imagine a large, almost infinite number of dice, and let the dice *not* be true. Then the most probable macrostate is all that will occur overwhelmingly. The others may be regarded as mere fluctuations about this. But a discussion of these fluctuations will take us far afield.

In the case of our superconducting material the fact that for $T < T_c$, the material is in the superconducting (S) state implies that it has a free-energy less than the normal (N) state. The difference $F_N - F_S$ is called the condensation free energy and is denoted by ΔF. (The corresponding ΔE is the condensation energy.) It is clear that ΔF is positive and a maximum at 0 K, falls off to zero at $T = T_c$ and then turns negative for $T > T_c$ when the normal state takes over.

Calculating the free energy is a horrendous task of statistical mechanics. But the principle of phase transition is now clear. The behavior of free energy determines the nature of the phase transition. It may be a discontinuous one, where the energy E changes by a finite amount even though F is (as it must be) continuous. This discontinuous change of E is the latent heat that is given out (absorbed) during freezing (melting) or condensation (boiling). We have all experienced it some time or the other, rather regretfully though — the scalding of the hand exposed to condensing steam from a boiling pot. We call these *first-order phase transitions*. The superconducting transition, on the other hand, is a continuous transition with no latent heat associated with it. The same is true of the magnetic transition. These are called *second-order phase transitions*. Unlike first-order transitions, the changes that take place at and near the second-order phase transition are very subtle. Several physical quantities, such as the specific heat, show singular behavior which is remarkably universal.

3.6 Type I Superconductors

While superconductors are all alike electrically, namely that they all transport electricity without loss once they are below the critical temperature and at very low currents, their magnetic behavior can be really very different. The perfect diamagnetism that we have spoken of typifies a superconductor of *Type I*. Their behavior is understood quite simply. Expulsion of the magnetic field from the bulk of the superconductor requires doing some work against the magnetic pressure of the field thus expelled. It is like blowing up a balloon. You may picture the magnetic

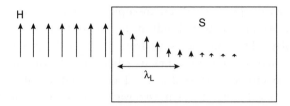

Figure 3.6:   Flux penetration in a superconductor (S). The London penetration length λ_L.

lines of force as elastic strings under tension. They get stretched as they are pushed out sideways (Fig. 3.3). The amount of work done on the system is proportional to the square of the field H. This raises the energy (and, therefore, the free energy) of the superconductor by the same amount. It is clear now that when this exceeds the condensation energy, the superconducting state will no longer be favorable energetically and the sample will turn normal. This defines a *critical field* H_c such that superconductivity prevails only for $H < H_c$. Inasmuch as the condensation energy decreases from a maximum at $T = 0$ K to zero at $T = T_c$, H_c too will behave likewise. Typical examples of a Type I superconductor are mercury, aluminum and tin. The critical field H_c is typically 0.1 tesla (1 kilogauss).

Even below H_c, the flux expulsion is really only partial. Indeed, it is energetically favorable to allow the field to penetrate some distance into the interior of the sample. This is a kind of energy minimization through an optimal compromise as we shall see later. In fact the field diminishes exponentially as $\exp(-x/\lambda_L)$ with the depth x below the surface of the sample (Fig. 3.6). The characteristic length λ_L is called the *London penetration depth*. The screening currents flow mainly within this depth. For the Type I superconductors, λ_L is typically 10^2–10^3 Å (1 Å = 10^{-8} cm). It is smallest at 0 K and grows to infinity (*i.e.*, the size of the sample) as we approach T_c, where the sample turns normal and is filled with the flux lines uniformly.

3.7   Type II Superconductors

Type II superconductors are different, and much more interesting. Discovered in 1937 by the Russian physicist L. V. Shubnikov, they are also the more important of the two types for most practical applications. They show the Meissner effect just like Type I superconductors up to a *lower critical field* H_{c1}. As the magnetic field exceeds H_{c1}, something catastrophic happens: the flux rushes into the bulk of the superconductor and permeates the whole sample. But it does so in the form of filaments, or flux tubes, rather than uniformly (Fig. 3.7). Each flux-tube carries exactly one quantum of flux. As the external field is increased, more and more flux tubes are formed in the sample to accommodate the increased total flux. This goes on until the flux tubes begin to almost touch each other at $H = H_{c2}$, the *upper critical field*, beyond which superconductivity is destroyed. For Type II supercon-ductors, the upper critical field H_{c2} is typically 1 to 10 tesla (10–100 kilogauss).

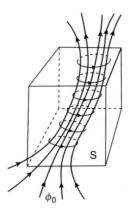

Figure 3.7:   Flux tube in a superconductor (S) carrying single flux quantum ϕ_0.

The flux tubes are very real. You can make them *visible* by sprinkling finely divided iron filings on the surface of the superconductor. The iron filings naturally cling to the foot points where the flux-tubes emerge from the superconductor and thus give them away. Such *flux decoration* experiments demonstrate not just the existence of the flux tubes, but also that these flux tubes are ordered in space as a triangular lattice at low temperatures. This is the so-called *Abrikosov flux lattice*, named after the Russian physicist A.A. Abrikosov who predicted it theoretically in 1957. The flux lattice is very real. It can vibrate elastically and even melt at higher temperatures and form a flux liquid, where the flux tubes can get entangled like the strands of melted polymers and hinder their mobility.

An individual flux tube has an interesting structure. It has a core which is in the normal state in that the superconducting order-parameter (or the condensation energy) is locally depressed to zero. The effect of this depression extends out to a distance ξ from the axis of the tube. This is the so-called *coherence length* and measures the distance over which the superconducting order is correlated — it is the minimum distance over which the superconducting order can change appreciably. Such a coherence is characteristic of all systems ordered one way or another. Thus, the superconducting order-parameter (roughly the condensation energy), which is zero in the normal core, will rise to its full value only beyond ξ, in the superconducting regions between the flux tubes. The magnetic field, on the other hand, will have its maximum value along the axis in the core region, and will fall off to almost zero beyond a distance λ_L away from the axis. Surrounding the core we will have the circulating supercurrents that do this screening of the field. Thus, the flux tube looks like a vortex and this state for $H_{c1} < H < H_{c2}$ is called the vortex state. Here the normal core region co-exists with superconducting regions intervening between the cores. For this reason the vortex state is also called the mixed state.

Let us see now what determines the *type* of a superconductor. The basic principle is the same — the free energy must be minimized. For a superconductor this requires compromise between two competing tendencies that operate at the interface (or the boundary) between the superconducting region of the sample and the region driven normal by the magnetic field. On the one hand it is favorable energetically to let the magnetic field penetrate the superconducting region and thereby reduce the energy cost of flux expulsion. This gain in energy is proportional to the London penetration depth λ_L for a given area of the interface. On the other hand, the superconducting order (or the condensation energy), which is depressed to zero in the core, remains more or less depressed up to a distance ξ from the axis of the flux tube. This costs energy proportional to ξ for the given area of the interface. Thus, for ξ much greater than λ_L, it is energetically favorable to reduce the total area of the interface. It is as if there is a positive interfacial surface energy per unit area (a surface tension). This will correspond to a complete Meissner effect — total expulsion of flux as in a Type I superconductor. For the opposite case of ξ much less than λ_L, it is energetically favorable to increase the interface area as if the surface tension is negative. This is realized by flux tubes filling the sample as in a Type II superconductor. Detailed calculation shows that $\xi \simeq \lambda_L$ is the dividing line. The basic physics here is the same as that of wetting — water wets glass while mercury does not wet it. A more closely related situation is that of mixing of oil and water in the presence of some surfactant that controls surface tension. The mixture may phase-separate into water and oil (Type I) or, globules of oil may be interspersed in water (Type II). The Type II superconductor is indeed a *laboratory* for doing interesting physics.

3.8   The Critical Current

The critical field (H_c for Type I or H_{c2} for Type II) is an important parameter that limits the current carrying capacity of a superconductor. This is because a superconducting wire carrying current generates its own magnetic field (Ampere's Law) and if this self-field exceeds H_c or H_{c2}, superconductivity will be quenched. This defines a *critical current density*, J_c. Clearly, a Type II superconductor with large H_{c2} up to 10 teslas, is far superior to Type I superconductors with H_c of about 0.1 tesla. There is, however, a snag here that involves some pretty physics. Consider a flux tube threading a Type II superconductor, and let an electric current I flow perpendicular to the flux tube. Now, by the Faraday principle of the electric motor, there will be a force acting on the flux tube, forcing it to move sideways perpendicular to the current and proportional to it. Physicists call it the Lorentz force. Once the flux tube starts moving with some velocity, the Faraday principle of the electric generator (the dynamo principle of *flux cutting*) begins to operate — an electromotive force (potential drop, V) is generated perpendicular to both the flux tube as well as its velocity so as to oppose the impressed *current*, I. This leads to dissipation of energy. Remember W (watts) = V (voltage drop) $\times I$ (amperes). We have the paradoxical situation of having a lossy superconductor! Where is the energy dissipated, you may ask. Well, the core of the flux tube is in the normal state and, therefore resistive. The motional electromotive force really acts on this normal region, dissipating the energy. Now, this would make Type II superconductors practically useless. The flux tubes must be *clamped* somehow. There is an ingenious way of doing it. Let us introduce some defects into the material — by adding impurities, alloying, or mechanically by cold working. This can locally depress the order parameter (or even make it zero altogether). Now, it is clear that it will be energetically favorable for the flux tube to position itself such that its normal core overlaps maximally with these defects, where the order parameter (condensation energy) is small anyway. Thus, the flux tube is *pinned* at these pinning centers. Of course, beyond a *critical current*, the Lorentz force will exceed the pinning force and the flux tubes will be released, leading to a snap-jiggle kind of motion and hence to dissipation. (This is a kind of self-organized critical state, much like that of a sand-pile that self-organizes through avalanches as its local slope exceeds some *critical value*.) Proper pinning is the secret of superconductors with high critical currents.

3.9 Understanding Superconductivity

It should be clear at the very outset that superconductivity has to do with the state of the free electrons that make up our metal. The electrons repel one other and are attracted towards the oppositely charged ions that form the background lattice. The ions can collectively oscillate about their mean positions — the lattice vibrations or the sound waves called phonons. This is our many-body system. Thus, an electron moves under the influence of all other electrons and that of the ions. Individually, it can easily be scattered and this is what happens in the normal resistive state. But at temperatures below T_c, the interacting electrons enter into an ordered state that somehow has a collective rigidity against such scattering. What is the nature of this order? It is certainly not that the electrons have crystallized. With such a long-range rigid order in space they could hardly conduct, much less superconduct. No, the electrons remain a liquid, but this liquid has an order of which we may not form a simple mental picture. Here finally we are confronted with their all pervasive quantum waviness, amplified infinitely by their indistinguishability, and reified in the stillness of absolute zero, where the scattering ions have, so to speak, all but gone to sleep — the single, whole *macroscopic wave function* of the many electrons. That such may be the case was anticipated by Fritz London back in 1933, almost 25 years before the fully microscopic theory of BCS was completed. Let us try to see how this may have come about without getting technical.

3.9.1 Fermions

The electron carries an internal angular momentum (spin) which is one-half in units of $\hbar$. The spin can point either parallel or antiparallel to any direction that can be chosen arbitrarily. The quantum-mechanical state of a single free electron in a metal is then specified, or labelled, by its energy, its momentum and the direction of its spin. The latter can be conveniently taken to be either up or down. As we saw in Chapter 1 on Symmetry, spin-half particles (electrons) are fermions, and no more than one electron can occupy the same state. Thus, at a given point in space you can have at most two electrons with opposite spins — the Pauli exclusion principle. This social (or rather asocial) behavior, called Fermi statistics, is the direct consequence of indistinguishability of identical particles in quantum mechanics and their half spin. Thus, the ground state of a system of free electrons, that is the state at the absolute zero of temperature, is obtained by placing two electrons with opposite spin directions in the lowest one-electron orbital state, two electrons with opposite spin directions in the next higher one-electron orbital state and so on until we have accommodated all the electrons in the system. Thus, there will be a highest occupied one-electron state, defining an energy level, called the *Fermi level* of energy usually denoted by E_F. (see Fig. 3.8).

This distribution of occupation numbers among the allowed energy levels is called *Fermi statistics*. This is precisely how we build up atoms. A solid is like a large extended atom. The only difference is that in an atom there is a single attractive center, the nucleus, and then the electrons are confined or localized around it like the planets around the sun. In a solid, on the contrary, there are many equivalent nuclei, and a given electron moving under their influence (potential) is as likely to be on one of them as on any other. Thus, the electronic wavefunction is extended over the whole sample like a plane wave. This corresponds to freely moving electrons. Now, recall that the kinetic energy of a free electron is proportional to the square of its momentum. Thus, the Fermi energy E_F will define a *Fermi sphere* in the space of momenta such that all states within the sphere are occupied while the states outside are empty. Such a Fermi sphere is referred to as the *Fermi sea* and the surface as the *Fermi surface*. In a crystal the Fermi surface can have a complicated shape reflecting the symmetry of the crystal lattice.

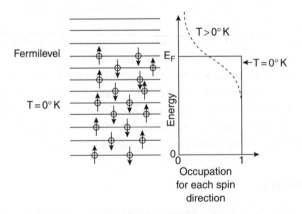

Figure 3.8: Fermi-distribution at zero and finite temperatures.

For a macroscopic system size, the allowed energy levels are very closely spaced — they nearly form a continuum, or a band. In a metal there are empty states available just above the Fermi level into which an electron can be accelerated by an external electric field, no matter how small. This is what makes a metal a good conductor — there is room at the top! (Incidentally, in an insulator, by contrast, there are no allowed higher-energy levels, arbitrarily close to the Fermi level into which an electron may be promoted by a small electric field, and hence no conduction. We say that the Fermi level lies in a forbidden gap — between a lower filled (valence) band and an upper empty (conduction) band of allowed states. A semiconductor is merely an insulator with a small band gap.) In a metal, the electrons at the Fermi level (E_F) move with large speeds, the Fermi speed v_F — typically 10^6 metres per second (*i.e.*, a third of one hundredth of the speed of light). Because of the exclusion principle, however, it is clear that only electrons lying close to the Fermi level can participate in any low-energy phenomena, which is much of the solid-state physics — including superconductivity.

The above picture at 0 K is only slightly modified at finite temperatures. All that happens is that some of the electrons lying below the Fermi level get thermally promoted to the empty states just above it. Thus, the sharp step in the Fermi-distribution at 0 K gets smeared out over an energy interval of about $k_\mathrm{B}T$ which is much smaller than E_F. Typically, $k_\mathrm{B}T_\mathrm{c}/E_\mathrm{F}$ is 10^{-4}.

One final remark about the electrons in a metal. The electrons, of course, repel each other by strong long-range Coulomb forces. One may wonder how we can possibly talk ourselves out of this and treat them as a Fermi gas of particles moving independently of each other, in the given potential of the background ions. Surely a moving electron creates a disturbance around it by pushing other electrons out of its way, for example. This is a complicated many-body system — we speak of a Fermi liquid. It turns out, however, that these effects can be by and large absorbed in a re-definition of our particles: a *bare* electron is dressed with a cloud of disturbance around it. The *quasiparticles* so defined now move more or less independently of one another. Quasiparticles carry the same charge and spin as the bare electron. But they have a different effective mass, and interact with a relatively weak, short-ranged (screened) Coulomb repulsion. All we have to do then is to read quasi-particle (quasi-electron) whenever we say particle (electron). This tremendous *reduction* of strongly interacting *bare* particles to weakly interacting quasiparticles is due to the great Russian physicist Landau. It is called the Landau quasi-particle picture.

3.9.2 Bosons

Now, we turn to the other species of particles, the *bosons*, that have a completely opposite social behavior — the *Bose statistics*. Bosons are particles with spin equal to zero, or an integer. An example of direct interest to us is that of ^{4}He, an isotope of helium having two protons and two neutrons in the nucleus and two electrons outside of it, with the total spin adding up to zero, *i.e.*, it is a *boson*. (Remember that for all phenomena involving low energies, as indeed is the case in *condensed matter physics*, we do not probe the internal fine structure of a composite particle like ^{4}He. It acts just like any other elementary particle with spin zero.) Bosons, unlike fermions, tend to flock together. That is to say that any given one-particle state can be occupied by any number of Bose particles. And this leads to a remarkable phenomena called Bose–Einstein (B–E) condensation. To see this, consider an ideal gas of N Bose particles (*i.e.*, non-interacting Bose particles) with N very large, almost infinite. The ground state of the system, that is, the state at 0 K, can be readily constructed by simply putting all the N particles in the lowest one-particle state, which is the state of zero momentum and zero energy. What we have is a macroscopic occupation of a single one-particle state. The occupation number is proportional to the size of the system — it is *extensive*. We call this phenomenon *Bose–Einstein condensation* (see Chapter 4.) As we raise the temperature, we expect a finite fraction of the particles to be excited, or promoted to higher energy levels (Fig. 3.9), thus depleting the *condensate* partially. The gas of excited particles in equilibrium with the condensate forms a kind of interpenetrating two-fluid system. Finally, at and above a characteristic temperature $T_{\text{B–E}}$, the condensate is depleted totally. The temperature $T_{\text{B–E}}$ is called the *Bose–Einstein temperature*, and ideally, *i.e.*, for non-interacting Bose particles, it depends only on the number density and the mass of the Bose particles. It increases with increasing density and decreasing particle mass. Thus, something drastic must happens at $T_{\text{B–E}}$. At and below this temperature, a finite fraction of atoms condenses into a single one-particle state of zero momentum, and the fraction ideally grows to a maximum (unity) as the temperature is lowered to absolute zero. For ^{4}He, regarded as an ideal Bose system, the calculated $T_{\text{B–E}}$ is about 3 K. Now, ^{4}He undergoes a phase transition at 2.17 K, below which its viscosity drops abruptly by a factor of at least a hundred million. It can flow through the finest capillaries without any viscous drag. It becomes a *superfluid*! This phase is called He-II to distinguish it from the normal

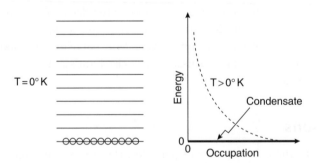

Figure 3.9: Bose distribution at zero and finite temperatures. Thick horizontal peak at zero energy signifies Bose–Einstein condensate below the lambda point.

phase of liquid helium called He-I. The transition point is called the *lambda point* (T_λ) because the temperature dependence of the specific heat near T_λ has the shape of the Greek letter lambda (λ). The proximity of the λ point (2.17 K) to the Bose–Einstein temperature (3 K) is no coincidence. It is now believed that superfluidity is due to Bose–Einstein condensation. The difference between T_λ and the ideal T_{B-E} is attributed to the fact that helium is after all not an ideal Bose gas. The atoms of ^{4}He repel each other strongly at short distances due to their hard core, and attract weakly at long distances due to *van der Waals forces* (the latter act even between neutral atoms and molecules and make small particles stick together).

3.9.3   Bose Condensation and Superfluidity

But why should Bose–Einstein condensation give superfluidity? The argument runs something like this. It is not just that we are allowed to put any number of Bose particles in a given single-particle state, it is rather that they tend to flock together. Thus, if a Bose particle is scattered from an initial state to a final state that is already preoccupied by N particles of its kind, then the probability of this process is enhanced by a factor $(N+1)$ — it is like the rich getting richer. Consider now the situation where the condensate is moving with a certain velocity relative to the walls of a capillary. Let a particle be scattered out of the condensate due to its interaction with the wall. This is the kind of process that would give rise to viscous drag. Now the probability of this particle being scattered *back into* the condensate is proportional to $(N^* + 1)$, where N^* is the number of bosons in the condensate, which is a macroscopic number, almost infinite. It is clear, therefore, that the particle will relapse almost immediately into the condensate with probability unity. Thus, the condensate has a collective rigidity against scattering — hence the superfluidity.

3.9.4   Phonon Mediated Attraction

But what has all this got to do with the superconductivity of metals, where we have instead electrons obeying Fermi statistics? If only the spin-1/2 electrons could somehow form bound pairs, with necessarily integral spin, they would then behave like bosons and undergo Bose–Einstein condensation! But in order to form these pairs the electrons must attract each other, and not repel as they normally do. This then is the big question. It was, however, shown by H. Fröhlich that the electrons can indeed effectively attract one another in the presence of a deformable (polarizable) lattice of ions that is, of course, always present as the background. The Fröhlich mechanism is roughly the following. An electron attracts the ions in its immediate neighborhood. The ions respond by moving, ever so slightly, towards it creating thus an excess of positive charge around it. We say that the electron has *polarized* the lattice. Another electron is now attracted towards this polarization localized around the first electron, and in doing so it is effectively attracted towards the first electron. This is very much like the water-bed effect. Imagine two people lying on a water-bed: each one tends to fall into the depression in the bed created by the other. You can easily demonstrate this effect by putting two marbles on stretched linen held taught and watch them roll towards each other.

One is still left with the uneasy feeling that this rather indirect attraction may not be strong enough to overcome the direct repulsion between the two electrons. The argument gets somewhat subtle here and involves some interesting physics. As we have already remarked, the electrons, that is our quasiparticles, repel via a short-ranged, screened Coulomb potential of a range of the order of the mean spacing between electrons. This is about an Ångstrom or two. But what is most important is the fact that this potential acts instantaneously, that is to say that it depends only on the present positions of the two electrons under consideration. The indirect attractive interaction, on the other hand, involves the tardy movement of the ions. The ions are sluggish because of their relatively large mass, which is at least a few thousand times the mass of the electron. The response time of the ions may be taken to be the period τ_D of their harmonic oscillations, and is typically 10^{-12} s. This corresponds to their typical oscillation frequency, 10^{12} Hz, the so-called *Debye frequency*. This means that the local polarization induced by the electron at a point will persist for a time τ_D even after the electron has moved away from that point. Now, the electron moves at the Fermi speed v_F, typically 10^8 cm s^{-1}. It would thus have moved a distance $\tau_D v_F$ which is about 10^{-4} cm (10^4 Å) during this time τ_D. This is much larger than the mean spacing between the electrons, or the range of the screened Coulomb repulsive interaction. One may, therefore, expect a second electron to come around and feel the attraction of the persistent polarization left behind by the first electron, and still be too far away from it to feel its direct repulsion. This is the essence of *dynamical screening* — a rather subtle effect. Thus, the direct repulsion is strong but instantaneous, while the indirect attraction is weak but retarded, and it is this difference of the time scales that makes the weak attraction prevail over the stronger repulsion.

In the parlance of many-body quantum physics, this attractive interaction is viewed as mediated by phonons, that is due to the exchange of virtual quanta of lattice vibrations. One electron emits (creates) a phonon which is absorbed (destroyed) by the other electron. Interaction between material particles by the exchange of some virtual field-quanta is commonplace in physics. Thus, the exchange of virtual photons leads to an interaction between charged particles. We say *virtual* because the exchanged quanta exist only between the times of emission and absorption. It should be clear that these quanta must be bosonic. (Fermions have to be created or destroyed only in pairs.) And phonons are bosons just as photons are. There is a very transparent way of seeing these processes with the Feynman diagram (Fig. 3.10).

It turns out that the attractive interaction mediated by the exchange of phonons is maximum when the two electrons have equal and opposite momenta (velocities). This enables the two electrons to take maximum advantage of the polarization

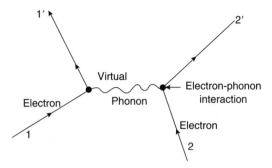

Figure 3.10: Feynman diagram showing emission of a virtual phonon by one electron and its absorption by another electron, giving an effective electron-electron attraction.

created by each other. Also, as the energy exchanged between the electrons is of the order of the energy of the phonon exchanged, which is typically the Debye energy, only electrons lying within the Debye energy of the Fermi surface take part effectively in the process. The Debye energy of lattice vibrations is typically 10^{-4} times the Fermi energy — indeed only a very small fraction of the electrons are affected by this attractive interaction.

3.10   Cooper Pairs and the BCS Theory

Given this attraction between the electrons lying close to the Fermi surface, one is immediately tempted to ask if the electrons are going to bind to form pairs. The pairs so formed will be *bosons* (because of integral total spin, zero or one), and it is not hard to contemplate a Bose condensation that would lead to superfluidity (or rather superconductivity as the pairs will be charged $2e$).

It was Leon N. Cooper of the University of Illinois who first considered such a possibility. Cooper showed that two 'test' electrons having a short-ranged attraction between them and moving in the background of an impenetrable Fermi sea (Pauli exclusion principle) formed by the other 'spectator' electrons, will bind with opposite spins, no matter how weak the attraction is. Pairing with opposite spins means total spin zero — a singlet pairing. There was a hope. Opposite spins allow the electrons to come close enough to take full advantage of attraction. The formidable problem of pairing with all the electrons treated *at par*, gamesters and the spectators alike, was finally solved by John Bardeen, Leon N. Cooper, and J. R. Schrieffer, all at the University of Illinois at that time, in their now famous paper published in *Physical Review* in 1957. This became the celebrated *BCS theory of superconductivity*. The central idea remained that of *pairing* — of the '*Cooper pairs.*' What causes the attraction is a secondary issue. Instead of phonons, other bosonic quanta may be exchanged. One should note, however, that seen from a distance, Cooper pairs will appear and act as bosons. But, it turns out that the size of the Cooper pair is typically 10^{-4} cm (essentially the coherence length), which is much too large compared to the mean electron spacing. Indeed, millions of Cooper pairs overlap. Thus, treating them as compact bosons, like ^{4}He, is an oversimplification. Still the essential physics remains the same. A direct consequence of this pairing is that it costs energy to break the pair. Thus, it turns out that unlike the normal metallic state, we need a minimum energy 2Δ to excite the superconductor. Here, Δ is called the *superconducting energy gap*. It is typically 10^{-4} E_F. That such relatively small energy scales should emerge giving a robust superconductor is an amazing consequence of macroscopic quantum coherence.

The BCS pairing theory with phonon mediated attraction explains many of the puzzles immediately. First, the electron-phonon coupling involved in the pairing mechanism is no different from that causing scattering (resistance) in the normal state. No wonder then that good superconductors are bad normal metals as we had noted at the beginning. Then, there is the isotope effect. Since the pairing mechanism involves motion of the ions, the ionic mass M must be a relevant parameter. Other things remaining the same, the heavier the ion the smaller is the displacement (dynamic polarization) and hence the weaker is the induced pairing. Thus, replacing an ion with its heavier isotope (this being the meaning of the proviso 'other things remaining the same'), T_c must go down. The BCS theory predicts T_c to be proportional to $1/M^{1/2}$. This is indeed observed experimentally. These and many other predictions of the BCS theory have since been confirmed — it is the correct theory of superconductivity.

The connection between superconductivity (or superfluidity) and Bose condensation seems intimate. After all, ^{4}He shows superfluidity at 2.15 K as we remarked earlier while its isotope ^{3}He, which is fermionic (with two protons and one neutron in the nucleus and two outer electrons, and thus with total spin half), shows no sign of it around that temperature. The superfluidity of ^{3}He observed at much lower temperatures in the millidegree Kelvin range is again due to pairing. But this time around, the pairing is with total spin one (triplet) and not zero (singlet) as for BCS superconductors. We should caution, however, that the BCS theory goes far beyond the naive idea of Bose condensation of an ideal Bose gas.

3.11   Some Macroscopic Quantum Effects

As we have remarked several times, particles and in particular electrons have a wave-like nature. Electrons may be reflected and diffracted by a diffraction grating. Indeed, electron diffraction is used to study crystal surfaces, and in the electron microscope. An electron wave can also be made to interfere (with itself) just as the light wave in the Young double-slit experiment. All that we have to remember is that these matter waves are the waves of probability amplitude — they are quantum mechanical waves. A Bose condensate represents a macroscopic wavefunction of all the bosonic Cooper pairs in the condensate. It is, so to speak, a highly coherent superposition of the individual Bosonic amplitudes — something akin to a laser beam (see Chapter 2 on Lasers). The wave nature of this condensate is described by the single complex wavefunction $\psi(r) = |\psi(r)| \exp(i\theta)$ such that $|\psi(r)|^2$ gives the density of the condensate at the point r. It has a global phase which is constant over the whole sample in the absence of any supercurrent. We may treat this complex wavefunction as the superconducting order parameter, much the same way as the magnetization $\mathbf{M}(r)$ is for a ferromagnet. The magnitude of magnetization corresponds to $|\psi(r)|$ while the phase of $\psi(r)$ corresponds to the direction of $\mathbf{M}(r)$. Similarly, all values of the phase from 0 to 2π are energetically equivalent. A fixed global value of θ for a given superconductor is the spontaneously broken symmetry akin to a fixed global direction of spontaneous magnetization (see Chapter 1 on Symmetry). The gradient of θ in space causes a supercurrent to flow.

3.11.1 *Flux Quantization Revisited*

With this picture of the condensate in mind, flux quantization follows in a straight-forward manner. Consider a superconducting ring enclosing a certain amount of flux (Fig. 3.11). Let us reckon the total change of phase as we go round a closed curve C deep in the material of the ring. The total change must, of course, be an integral multiple of 2π. This is because on completing the circuit we must return to the same value of ψ — it must be single-valued, we say. Now, recall that $\exp(i2\pi n) = 1$ for any integer n. Thus, the phase change must be $2\pi n$. Now, it is known from electrodynamics that the change of phase in going around a circuit must be proportional to the magnetic flux enclosed by the circuit. In fact it is $q\phi/\hbar c$, where q is the charge on the basic entity, in our case the Cooper pair. Thus, $q = 2e$. Equating the phase changes computed in these two ways, we get $\phi = n(hc/2e)$. The flux is quantized!

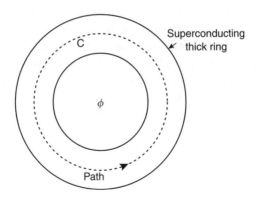

Figure 3.11:   Flux (ϕ) quantization through a superconducting loop.

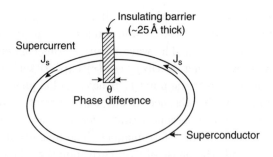

Figure 3.12:   Josephson junction with persistent supercurrent tunneling through it.

3.11.2   *Josephson Tunneling and the Superconducting Interference*

The long-range nature of superconducting coherence shows up dramatically when we interpose a thin dielectric (insulating) barrier between two superconductors. This is called a Josephson junction. We can have, for example, a film of tin oxide (SnO_2) separating the two superconducting tin (Sn) electrodes. No current will flow on closing the circuit if the tin is in the normal state. If, however, the barrier is thin enough (~25Å) and the tin is in the superconducting state, a supercurrent can flow persistently on closing the circuit without any potential drop across the junction (Fig. 3.12). There will be a phase difference θ between the two superconducting contacts on the two sides of the barrier. The supercurrent will vary sinusoidally as a function of θ, *i.e.*, $J_s \propto \sin\theta$. This dramatic effect was predicted by the 20-year-old Brian Josephson at Cambridge in the U.K., for which he shared the Nobel Prize for Physics in 1973. Since the maximum value of $\sin\theta$ is 1, it is clear that the Josephson current must be less than a critical value J_c. Beyond J_c, a voltage appears across the junction and the DC supercurrent drops to zero!

One can use a pair of Josephson junctions in parallel and demonstrate the quantum interference effect (Fig. 3.13). The relative phase of the two superconducting amplitudes along the two arms can be varied by varying the magnetic flux passing through the enclosed area. One obtains the oscillatory pattern reminiscent of the Young double-slit experiment with light. It is indeed possible to count the 'fringes' (the maxima and minima of the current) and thus measure the flux change with unprecedented accuracy. A strange thing to note here is that the effect depends only on the total magnetic flux through the area enclosed by the two parallel superconducting paths. The magnetic field need not touch the superconductors at all. This non-locality again emphasizes the non-classical nature of superconductivity. This is the basic principle of the *superconducting quantum interference device* (acronym, SQUID). Magnetic fields as small as 10^{-9} gauss can be measured. These are the kinds of fields produced by the tiny currents flowing in the human brain, or the rusting fender of your car. Obvious applications are ultra-sensitive sensors.

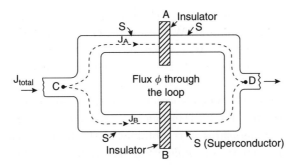

Figure 3.13: Interference of partial quantum amplitudes through two Josephson junctions connected in parallel. The resultant supercurrent oscillates sinusoidally with magnetic flux trapped between alternative paths.

An equally fascinating phenomenon is the AC Josephson effect. If you apply a DC voltage V across the Josephson junction, the phase difference across the junction will begin to increase linearly with time, making the current through the junction oscillate at a frequency 2 eV/h. Thus, a DC voltage of 1 μV (one microvolt) will produce a frequency of 483.6 MHz. You have made an oscillator! This fact can be used to measure the fundamental ratio e/h to unprecedented accuracy. Conversely, we can measure low voltages of the order of 10^{-16} volt! The phase of the superconducting order parameter is perhaps the most important aspect of it.

As an aside, let us mention that superconductivity affords us a means of addressing some very profound fundamental questions of quantum mechanics (see Appendix B). In dealing with microscopic particles like an electron we freely superpose the wave amplitudes. Thus, if there are two possible states labelled 1 and 2 for an electron with wave amplitudes ψ_1 and ψ_2, then the electron can also be found in the superposed state with the wave amplitude $a_1\psi_1 + a_2\psi_2$. Then, $|a_1|^2$ and $|a_2|^2$ give the probabilities of finding the electron in the two states 1 and 2, respectively. The question is if this is true of macroscopic objects with states which are *identifiably (taxonomically) distinct*. This is what Erwin Schrödinger, the founder of quantum mechanics after whom the *Schrödinger equation* is named, expressed rather picturesquely by asking if a cat can be found in a superposed state of being at once dead and alive! A superconductor provides us with a Schrödinger cat of sensible magnitude! This question remains as yet unsettled.

3.12 The Superconductor Comes out of the Cold

Until about 1986, superconductivity belonged in the domain of liquid helium temperatures. The age of *low-temperature superconductors* (LTSC), and with this the pre-occupation with low temperatures, came to an abrupt end with the discovery of the ceramic cuprate superconductor (La-Ba-Cu-O, T_c about 35 K) by J. G. Bednorz and K. A. Müller in 1986. These ceramics were made by mixing oxides of lanthanum, barium and copper, and heating them to high temperatures so that they react in the solid state to form the compound. With this, the era of *high-temperature supercon-ductivity* (HTSC) had begun. After some initial skepticism, confirmations poured in from many laboratories around the world. Superconductors with still higher transition temperatures (T_c about 90 K to 125 K) were soon announced. One could discern a systematic dependence of T_c on the chemical composition. Thus, substituting Y (yttrium) for La (lanthanum) gave a $T_c = 90$ K for Y-Ba-Cu-O. Higher transition temperatures were obtained with Bismuth (Bi), Thalium (Tl) and Mercury (Hg). Reports of *unidentified superconducting objects* (USO's) continue! Room-temperature superconductivity is awaited, though with some studied nonchalance. This is not the time nor the place to tell the story of this break-through, except to re-affirm that the excitement generated by it among physicists, chemists, technologists, metallurgists, material scientists and the common man has no parallel in the recent and the not-so-recent history of science and technology. It also marks the culmination of a prolonged scientific ascent (see Fig. 3.14).

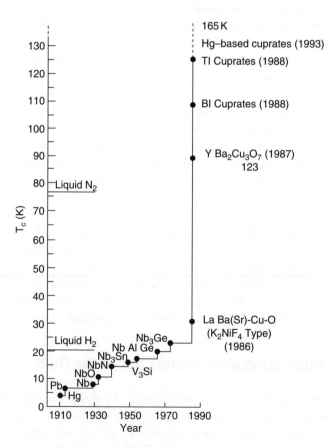

Figure 3.14:   The record of superconducting transition temperatures: The ascent of superconductivity.

What is unusual about these high-T_c superconductors? Are they really different from conventional BCS superconductors? There are, of course, the obvious facts of high T_c (35–165 K) and the high critical field H_{c2} (about 100 tesla). These are strongly Type II superconductors. But there is much more to it than just that. The point is that normally one thinks of a superconductor as being derived from a metal. The new superconductors seem more like to be derived from a parent insulating antiferromagnet. These materials are transition-metal oxides — in fact, earthy ceramics — and are very poor conductors, almost insulators in the normal state. But they are insulators with a difference. It is not that the energy bands of the allowed one-electron states are completely filled — as in the case of band insulators, with electrons immobilized by the Pauli exclusion principle. In fact, the bands are half-filled as in a typical metal. This can be readily verified by doing the electron count right for the parent material, *e.g.*, La_2CuO_4. It is the partial replacement of the trivalent lanthanum (La) by the divalent strontium (Sr), say (a process called doping, borrowed from semiconductor physics), that destroys the parental antiferromagnetism and makes the material metallic, and then, of course, superconducting with the T_c maximum for an optimal doping ($x = 0.15$) for $La_{2-x}Sr_xCuO_4$. In fact, these materials are insulators because of the very large repulsion between the electrons that prevents them from occupying a state doubly — one with spin up and the other with the spin down as in normal metals. In technical terms, this is called electron-correlation, and the materials are said to be strongly correlated. This single fact makes them behave abnormally even in the *normal* state above T_c. A structural feature common to all HTSCs is their layered structures, namely that they comprise weakly coupled layers of CuO_2 (hence layered cuprates). This two-dimensionality seems crucial to their high-temperature superconductivity. These materials hardly show any *isotope effect*. But *pairing* is not in doubt as confirmed by flux quantization experiments. One strongly suspects that pairing may not be due to phonons. It is believed to be electronic in nature, and magnetism seems involved in the superconductivity of these materials in an essential way. It is, however, too early to make any definite claims. There are just too many theories around. What one can definitely say is that the novel ideas generated by these superconductors will have a profoundly enriching effect on our thinking for years to come. In fact, we may have to re-write solid-state physics texts. One thing is clear: These high-T_c cuprate superconductors are complex and rather *chemical*, unlike the low-temperature BCS superconductors, which are simple and rather *physical* (see Fig. 3.15).

On the practical side, however, the possibilities are clearly enormous. High-T_c superconductors can do whatever conventional superconductors do, and obviously do it much cheaper, for the simple reason that liquid helium costs a few dollars a litre while liquid nitrogen costs just a few cents a litre. You save on the cost of cooling. Besides, liquid nitrogen is much more efficient as a coolant.

Several applications come to mind. Some are large scale applications involving high currents, high current densities and high magnetic fields as for power

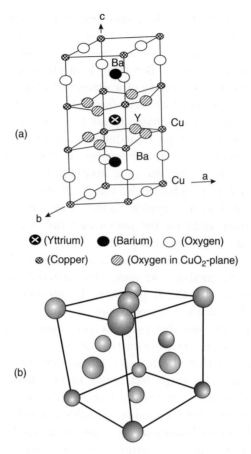

Figure 3.15: (a) Crystal structure of layered cuprate high-T_c superconductor $YBa_2Cu_3O_7$; (b) Crystal structure of a low-temperature BCS superconductor, *e.g.*, aluminium (face-centred cubic) (schematic).

generation, transmission and energy storage. Others are small scale applications as in electronics involving low currents, but the current densities can still be very high. The zero electrical resistance and hence the absence of dissipation (heat loss) helps in two ways. It obviously makes the system more energy efficient and provides cheap magnetic flux. Less obvious, however, is the fact that it allows compact designs as we do not require large surface areas, cooling fins say, to remove heat as none is generated. This, for example, makes it possible to achieve the highest packing density of electronic components by making all the interconnects on a silicon chip superconducting — packing density of logic circuits comparable to that of the human brain is realizable!

Compact and high-field superconducting magnets can be used in giant particle accelerators to confine, store and direct the beams of charged particles. Just think of one such giant machine, the late and lamented six-billion dollar Superconducting Supercollider (SSC) having 10 000 such magnets! The cheap magnetic flux should come in very handy here. The same is true of fusion reactors (TOKAMAK) where the hot thermonuclear plasma has to be confined by a large toroidal magnetic field.

Large superconducting coils carrying persistent currents could be used for non-polluting energy storage. Such Superconducting Magnetic Energy Storage (SMES) systems would have to be built underground for reasons of excessive magnetic stresses on the structure. Thus, the magnetic energy stored in one cubic meter of space with a magnetic field of about 50 tesla could supply power at the rate of 1 kW for a period of about a month! The energy stored is proportional to the square of the magnetic field and, of course, to the volume. Magnetically levitated trains (maglevs) are possible and already being contemplated for operation soon. These ultrafast bullet trains (speeds of about 500 kmph) will be, for one thing, free from the 'click-click' of conventional trains running on the 'permanent ways,' and far safer too. One can also think of large amounts of DC electrical power distribution using superconducting transmission lines, or better, still, underground cables, and cut the *copper-loss* that can easily amount to 5% of the power generated — in 1985 this loss had cost America US\$9 billion! Most of this could have been saved by using DC superconducting transmission lines. In all these cases, however, there is at present the technological problem of drawing wires of these highly non-ductile materials. But, encouragingly enough, 1 km long, 0.3 mm diameter fibers of these HTSC materials are already being made. These may be clad in copper and made into multifilament cables several meters long for actual use. The other limiting factor, almost as important as the critical temperature, is of a more fundamental nature — the critical current density J_c. High J_c is a *sine qua non* for most applications, but requires efficient pinning of the flux lines. Remember that for HTSCs, the lower critical field H_{c1} is very low, about 10–100 gauss and, therefore, the flux lines enter the material rather easily. This is one of the most active areas of research in this field today. We are, however, close to having usable current densities normally used in copper conductors.

There are then the low-power applications. For example, sensors like SQUIDS can be made much cheaper with HTSCs for mapping biomedical magnetic fields of the human brain (10^{-8} gauss) and the heart (10^{-5} gauss). Even gravitational waves, the weakest signal of all, may be detected with the help of these SQUIDS. Studies were under way to use very large Josephson junction arrays for faster, more compact fifth generation supercomputers with very large memory. Here, the junction acts as a binary logic element that can be made to switch between the on-state (corresponding to zero junction voltage) and the off-state (corresponding to finite or normal junction voltage) in less than a picosecond (10^{-12} seconds) by passing a current in excess of the critical current. This is a thousand times faster than conventional switching devices based on transistors. This has, however, been abandoned now as something not quite feasible.

It may be some years before these low-power devices become fully commercially competitive. It may take longer still for large scale applications to become feasible. We may not be able to buy a spool of superconducting wire or ride a superfast maglev train for another decade. But there are no serious doubts about these things. In fact, the possibilities are so diverse and so numerous that in any comprehensive planning for the future along these lines, we would do well to involve a science fiction writer. Just think of the curious toys that become possible with room temperature superconducting materials — magnetically levitated gyros or spinning tops, to name just two.

After all, room temperature superconductivity is a distinct possibility in the decades to come. And in any case, in the foreseeable future when coal and oil have nearly run out, humans may have no choice but to turn to this *perpetual motion* — of the second kind.

3.13 Summary

Superconductivity is the complete disappearance of the electrical resistance of a material at and below a certain critical temperature which is characteristic of that material. The change of state occurring at the critical temperature is a thermo-dynamic phase transition. The superconductor is, however, not merely a perfect conductor, it is also a perfect diamagnet: the magnetic flux is expelled from its bulk as the material is cooled in the presence of an applied magnetic field, through its transition temperature. This is the well-known Meissner effect. This tendency to exclude magnetic flux makes it possible to trap the flux threading through a superconducting ring permanently, or to levitate a magnet above the superconductor. But, if the magnetic field exceeds a certain critical value, the super-conducting state is destroyed and the material reverts to its normal resistive state. The flux expulsion is, however, not complete: the magnetic field does penetrate the superconductor up to a certain depth called the London penetration depth which is typically a few hundred Ångstroms. As the critical temperature is approached from below, the penetration depth tends to infinity (*i.e.*, it equals the size of the system), while the critical field tends to zero. All this is true for the so-called Type I, or soft, superconductors. For Type II, or hard, superconductors on the other hand, the Meissner effect is observed up to a certain lower critical field, which is rather small, while the superconductivity is destroyed above a certain upper critical field, which is much greater. In between, there is the mixed phase where the magnetic flux enters the bulk of the material in the form of flux tubes that have a normal core whose diameter is of the order of the coherence length, varying from a few to several thousand Ångstroms. The latter is the smallest length scale over which the superconducting order may vary appreciably. When a supercurrent flows through the material it exerts a force on these flux tubes making them flow sideways, causing dissipation. It is necessary, therefore, to pin down these flux tubes to achieve high critical currents that can flow without dissipation. At low enough temperature these flux tubes can order as a triangular flux lattice called the Abrikosov flux lattice.

Superconductivity is a purely quantum phenomenon on a macroscopic length scale. This strangely ordered electronic superfluid state is described by a complex order parameter whose phase leads to observable effects, such as the quantization (wholeness) of flux trapped in a superconducting ring, or a flux tube, when measured in certain natural units, or the ability of the supercurrent to flow across an insulating layer, several Ångstroms thick, separating two superconductors. This is the famous Josephson junction effect.

The correct microscopic theory of superconductivity, the celebrated BCS theory named after J. Bardeen, L. N. Cooper and J. R. Schrieffer was proposed in 1957, almost half a century after the phenomenon was discovered by Kamerlingh Onnes in 1911. The basic idea is the formation of loosely bound electron pairs, Cooper pairs, due to an indirect attraction induced by the polarization of the background lattice. Cooper pairs behave roughly as bosons and undergo Bose–Einstein condensation at low enough temperatures. The condensate has the superfluid property.

Superconductivity occurs widely among elements, compounds and alloys. The transition temperatures are, however, abysmally low, typically a few degrees above the absolute zero of temperature. This necessitates the use of liquid helium as coolant, which is both inefficient and expensive. The recent discovery by J. G. Bednorz and K. A. Mueller in 1986 of high temperature superconductivity in certain oxides with transition temperatures now as high as 165 K — much above the boiling point of liquid nitrogen — has changed all this. Liquid nitrogen can now be used as the coolant, which is much more efficient and costs much less.

Most applications depend on the Type II superconductors because of their higher critical temperatures, critical fields and critical currents. Superconducting magnets are already in use. Superconducting cables, energy storage devices, magnetically levitated trains and Josephson-junction based devices are becoming feasible. It seems that the revolution initiated by these novel high temperature superconductors may well be comparable to that brought about by the transistor.

3.14 Further Reading

Books

- C. Kittel, *Introduction to Solid State Physics*, Sixth Edition (Wiley, New York, 1988).
- F. London, *Superfluids*, Vol. 1 (Superconductivity) (Wiley, New York, 1954, reprinted by Dover, New York).

Popular Books

- P. Davies (Ed.), *The New Physics* (Cambridge University Press, 1989). Contains a chapter on superconductivity and superfluidity.
- B. Schecter, *The Path of No Resistance: Story of the Revolution in Super-conductivity* (Simon and Schuster, New York, 1990).

Bose–Einstein Condensate: Where Many Become One and How to Get There

4

4.1 Introduction

Normally, the atoms and molecules of a gas, for example the air at room temperature, are distributed randomly and sparsely in energy and momentum with each particle doing its own thing. This is described by the well-known Maxwellian distribution of velocities with its familiar Gaussian hump. This state of affairs can, however, change dramatically at low enough temperatures and high enough densities for the gas of a certain type of atoms, Bose atoms or bosons, *e.g.*, rubidium atoms (^{87}Rb), provided of course that the gas is still dilute enough so as to bypass temporarily its eventual condensation to the liquid state, or freezing to the solid state. Consider thus a dilute gas of rubidium containing about 10^{12} atoms per cm^3, say, which is some ten orders of magnitude rarer than the typical solid or the liquid that we know of. When cooled to and below an abysmally low *critical* temperature, T_{BEC}, typically only a few billionths of a degree above the absolute zero ($-273.15°$C), the gas enters a new phase in which a finite fraction, close to unity, of the ultra-cold atoms suddenly drops into the lowest energy single-particle state, that now becomes occupied macroscopically. This is most unlike the sparsely occupied higher energy single-particle states that occur at room temperature. Such a many-body state with a macroscopic occupation of the lowest energy single-particle state is called the Bose–Einstein Condensate (BEC). It is an object with coherent wave-like properties that are strange and totally counter-intuitive to our classically imprinted mind. The BEC has aptly been claimed to be a new state of matter. Its strangeness derives directly from the single fact that all the atoms in a BEC move in unison — with every atom doing precisely the same thing. Because of this co-ordinated lockstep motion, the BEC acts as a single giant atom despite the fact that it may actually contain 10^3–10^7 atoms, or more. It is as though the atoms have lost their individual identity — a coherent wholeness of which we may never make a true mental picture. Figure 4.1 is a schematic depiction of this order. In fact, the BEC is a quantum phenomenon in which the wave nature of matter, of the atoms, is manifested on a macroscopic scale — where the microscopic atomic wave-amplitudes add up

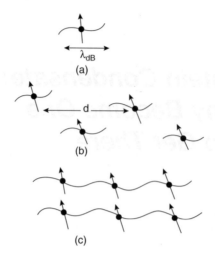

Figure 4.1:   (a) de Broglie wavelength $\lambda_{\mathrm{dB}}(T) = h/\sqrt{(2\pi M k_{\mathrm{B}} T)}$ of an atom of mass M in a gas at temperature T. (b) For $T > T_{\mathrm{BEC}}$, $\lambda_{\mathrm{dB}}(T) < d$, the interatomic spacing. (c) For $T < T_{\mathrm{BEC}}$, $\lambda_{\mathrm{dB}}(T) > d$ and the thermal de Broglie waves overlap in lockstep to give Bose–Einstein condensate (BEC). (Schematic).

coherently to give a giant wave of macroscopic amplitude that has all the wave-like properties such as interference and diffraction, much as the photons of light in a laser beam or the overlapping Cooper pairs of electrons in a superconductor do. Indeed, a BEC atom laser has already been demonstrated.

Such a BEC was experimentally realized first in a *dilute gas* of ^{87}Rb, a bosonic isotope of rubidium, by Eric Cornell of JILA (Joint Institute for Laboratory Astrophysics) and NIST (National Institute of Standards and Technology) Boulder, Colorado, USA, jointly with Carl Wieman of JILA and the University of Colorado, Boulder, USA, in 1995, and independently a few months later by Wolfgang Ketterle of MIT, Cambridge, USA, in sodium (^{23}Na). This opened up a whole new field of atomic, molecular and condensed matter physics with far-reaching potential for applications including: metrology (ultra-precise and stable atomic clocks for global positioning and space navigation); nanotechnology (nanolithography on semiconducting chips using sharply focused atomic lasers); sensitive gravitational wave detection (using atom interferometers); and fundamental physics (detection of subtle and minute high-energy physics effects, *e.g.*, the time-reversal symmetry breaking effect in atoms at low energy). Some, like the atom laser have already been demonstrated. In due recognition of their achievements, the three physicists, Eric Cornell, Carl Wieman and Wolfgang Ketterle were awarded the 2001 Nobel Prize in Physics. Of course, this extraordinary feat of creating the dilute-gas BEC at nanokelvins was made possible by the earlier research of many others on the physics of laser cooling and electromagnetic trapping of atoms that had culminated in the work of Steven Chu (of Stanford University, USA), Claude Cohen-Tannoudji (of Collège de France and École Normale Supérieure, France), and William D. Phillips (of NIST, Gaithersburg, USA) on laser cooling for which they were awarded the 1997 Nobel Prize. More than 20 laboratories around the world have now realized the BEC in various alkali-atomic vapors, and notably also in spin-polarized atomic hydrogen, which was long awaited.

A brief historical note at this point seems in order. The idea and the reality of BEC was, of course, not new and certainly not in doubt. Then why all this fuss about the BEC, one may rightly ask. Let's try to understand this. In fact such a condensed state was predicted for liquid helium (^{4}He) by Einstein back in 1925, as a necessary consequence following from the new statistics proposed in 1924 by the Indian physicist Satyendra Nath Bose — the new rule for counting correctly the distinct arrangements of the indistinguishable photons, the quanta of light. Bose had successfully derived from this the celebrated Planck law of black body radiation. Einstein extended this new statistics to include also the other particles that belonged in the same class as the photon as having an integral spin (intrinsic angular momentum measured in certain natural units), but which, *unlike* the photon, were permanent, *i.e.*, of fixed (conserved) number, such as the atoms of the noble gas helium (^{4}He). After all, de Broglie had already proposed that atoms too are waves — matter waves like photons of light — only that the atoms may have much shorter wavelengths ($\lambda = h/p$) depending on their momentum (p). And, of course, Einstein was right, except that the strong inter-particle interactions of the dense liquid helium (^{4}He), unlike those in the dilute nearly ideal gas of sodium or rubidium, all but deplete the condensate fraction — to even less than 10%. (Ironically, Einstein himself didn't quite believe in the reality of BEC, though he liked the idea of it as something pretty. In fact, after 1925 he never ever returned to the subject of BEC!). But, the new statistics has since been known as Bose–Einstein statistics, and the particles obeying this statistics as bosons. Bose statistics is *inclusive* in that any number of bosons of a kind can, indeed are *encouraged* to, occupy the same single-particle state. It is again a profound result of physics that a particle, be it elementary like the photon or composite like the ^{4}He atom, having an integral spin (total intrinsic angular momentum measured in units of $\hbar$, Planck's constant h divided by 2π) is a boson. (And those with a half-integral total spin, such as the ^{3}He atom or the commonplace electron, are fermions. Fermions obey a different, *exclusive* statistics, Fermi–Dirac statistics in which not more than one particle of a kind can occupy the same state. Sure enough, the bosonic ^{4}He does undergo a Bose–Einstein condensation, but not its fermionic cousin ^{3}He. This connection between the spin and the statistics is one of the deepest theorems in all of physics). So once again, given this knowledge, why so much fuss about the BEC?

Much of the novelty of, and the fascination for, the dilute gaseous alkali atomic BEC lies in the fact that this dilute BEC is near-ideal (having ∼100% condensate fraction, and not just <10% as is the case for dense liquid helium, ^{4}He). Besides, it is tunable — over a wide range of density, composition, strength, range, and even the sign of the inter-particle interaction. Its internal electronic state can be conveniently manipulated with light. This is unlike the case of the dense liquid ^{4}He, which is just as given. Before the entry of the dilute gaseous BEC center-stage, the condensate was the exclusive domain of a few exotic systems — primarily the *aristocratic* ^{4}He and the ^{4}He–^{3}He mixture of course, but also, on the side, the condensate of the short-lived electron-hole excitons created optically in the semiconducting Cu_2O; the pion condensate suspected to lie in the interior of neutron stars; the predicted spin-polarized hydrogen BEC now realized; the Bose-like Cooper pairs in a superconductor; and possibly even vacuum as the particle-antiparticle condensate. With the coming of the dilute alkali gas BEC, this helium-centric exclusivity has finally ended. Moreover, alkali atoms are common and highly reactive, while helium is noble and inert, although admittedly, quite Nobel active!

4.2   Bose Statistics: Counting of the Indistinguishables

Classically identical objects become indistinguishable quantum mechanically. (And quantum mechanics is now known to describe Nature correctly and exactly, from the domain of the very small on the atomic and the sub-atomic scale to the domain of the very large on scales that are sensible and beyond, at which it is well approximated by Newton's laws of motion that underlie classical mechanics). Let us try to grasp this idea of the indistinguishability of identical objects. It is no mere nit-picking. It has observable consequences. Two objects are ordinarily said to be identical if you cannot tell them apart. That is to say that if the two objects were swapped while you were not looking, you simply wouldn't know it — we then speak of a symmetry with respect to exchange, or of the permutation symmetry if you will. A little thought will, however, convince you that the two identical objects in question were nevertheless distinguishable, *even if only by virtue of their occupying different positions in space at the given time.* This is so because it is possible then, in principle, to continuously follow the trajectories of the two objects as they were being exchanged. For then you could always tell which was which. Such a *tagging* and *tracking* is, however, not admissible in quantum mechanics. A trajectory requires knowing the position as well as the momentum (velocity) of the particle at each instant of time, but, the Heisenberg uncertainty principle does not allow such a simultaneous fixation of the position and the momentum (velocity) of a particle — there are no trajectories in quantum mechanics! This puts paid to our hope of continuously tracking the notionally tagged identical objects, even in principle. Thus, even if we could identify them by their positions at one instant, their positions will become totally confused at the very next instant. In fact, all we can claim at any future date is that there is one particle here and one out there, say, without ever knowing which was which. Hence the quantum indistinguishability of classically identical particles! The argument is admittedly heuristic, but has a logical appeal that one may not resist.

This indistinguishability has profound consequences for statistics — for how to count the distinct arrangements of the indistinguishable particles. A simple example will illustrate this point. Let us consider four identical objects, apples $\{A_1, A_2, A_3, A_4\}$ say, and ask in how many ways can we distribute them in two boxes with two apples each. Classically, we can count six distinct partitions, namely,

$$\{A_1, A_2\}/\{A_3, A_4\};$$
$$\{A_1, A_3\}/\{A_2, A_4\};$$
$$\{A_1, A_4\}/\{A_2, A_3\};$$
$$\{A_2, A_3\}/\{A_1, A_4\};$$
$$\{A_2, A_4\}/\{A_1, A_3\};$$
$$\text{and } \{A_3, A_4\}/\{A_1, A_2\}.$$

Here the first set of braces $\{\cdots\}$ denotes box 1 and the second set box 2. But now, let the apples become indistinguishable. Then there will be just one such distinct partitioning — with two apples (which two we do not, and in principle, cannot know) in box 1 and the remaining two in box 2 — a reduction in count by a factor of six in going from the identical to the indistinguishable! This reduction factor would be 252 for 10 apples, and thanks to the tyranny of factorials, it would grow quickly many-fold with more apples. Generalizing, it is clear that for the *indistinguishable* objects we can specify only the *occupation* numbers of the *different boxes*, and *not* the *occupants*. Shuffling of the indistinguishables does not generate new arrangements — $\{A_1, A_2\}/\{A_3, A_4\}$ is the same as $\{A_1, A_3\}/\{A_2, A_4\}$ — and, physically, the labels count for nothing. All we have to do now is to replace the apples with our indistinguishable bosons, and the boxes with the single-particle states in which to put the particles, and we are home!

Associated with this indistinguishability, there is an additional feature of bosons, namely, that any number, $0, 1, 2, \ldots$, of the bosons of a kind can be put into a given single-particle state — they obey the inclusive Bose–Einstein statistics. The identity crisis arising from indistinguishability leads to a reduction in the boson-state count, which then together with the inclusive bosonic statistics leads to a crisis of over-population. The BEC is the resolution of this over-population crisis.

As to what maketh a particle a boson, this is a deep question of which we do not have a complete understanding. The great divide of the *social* bosons and the *asocial* fermions is a fact of life. As pointed out earlier, a particle, whether elementary (*e.g.*, the photon) or composite (*e.g.*, ^{4}He), is a boson if it has an integral spin — meaning its total spin, the electronic and the nuclear spins added vectorially together as per the quantum addition of angular momenta. Thus, the helium atom, ^{4}He, with two protons, two neutrons, and two electrons (all spin-1/2 particles) is a boson with a total spin of zero (like the photon), while the other helium isotope, ^{3}He, with two-protons, one neutron and two electrons is not a boson (it is a spin-1/2 fermion). Similarly, rubidium atoms (^{87}Rb) are bosons, but potassium atoms (^{40}K) are not (they are fermions). Most atoms and other composite objects, even Swiss watches if you like, are bosons. The bosonic character, however, may not be manifest macroscopically unless the de Broglie wavelength exceeds the inter-particle distance (ensuring overlap of the atomic waves as in the BEC) or the range of interaction. Thus, the hugely famous molecule of the last century, the fullerene, or C_{60}, may not reveal its bosonic character any time soon — much less Swiss watches!

We will end this section on a somewhat philosophical note. Two bosonic atoms of ^{87}Rb, say, one in the ground state and the other in the optically excited state, are to be treated as the same object but in two different states. This *non-dualist* view leads to calculable, verifiable and, indeed, verified results. The generally askable, but not so frequently asked, question now is *when two objects are to be viewed merely as two different states of one and the same object, and not really as two different objects altogether*. Enough for us to wonder!

4.3   Bose–Einstein Condensation (BEC): The Over-Population Crisis

Let us see now how BEC really comes about. For a system of N bosons in equilibrium at temperature T, the chance or the probability of the system to be found in a state of energy E is proportional to the Boltzmann factor $\exp(-E/k_B T)$ times the number of distinct configurations (microstates) of the particles in the system corresponding to this given energy. It is this latter combinatorial weight factor that gets drastically reduced for the Bose particles discussed above. Thus, the number of Bose particles of a kind that can be statistically accommodated *freely* diminishes as the temperature is lowered. This diminishing capacity to accommodate particles implies a maximum number for the Bose particles that can be normally held in thermal equilibrium at a given temperature. What should become of the excess then? The excess over this maximum number, as was first argued by Einstein, must drop into the lowest single-particle state, which then becomes *macroscopically* occupied leaving the higher lying states populated sparsely. (This is much the same as a saturated vapor supernatant above a condensed liquid with which it is in equilibrium. At a given temperature, the saturated vapor can hold only so much. Any excess simply condenses out into the liquid. A mere analogy though!). We see, therefore, that the BEC is indeed a resolution of an over-population crisis resulting from the identity crisis coupled with inclusive Bose statistics. A direct consequence of, and the evidence for, this condensation is so-called Bose Narrowing: as the number of the Bose particles is increased beyond a critical number for a given temperature, more and more of the particles drop into the lowest energy BEC state, thereby bringing down the average energy per particle as well as its spread below the classical thermal value $(3/2)k_B T$.

The condition for the onset of BEC follows from some fairly general arguments. Consider an ideal gas of non-interacting Bose particles, to which a dilute Bose gas is a good approximation. Let the number density of the gas be n, giving a mean inter-particle spacing of $n^{-1/3}$. At temperature T, the mean thermal (kinetic) energy per particle $p_{th}^2/2M$ would be $(3/2)\,k_BT$ giving $p_{th} = (3Mk_BT)^{1/2}$. But as we know, a particle of momentum p has associated with it a de Broglie wavelength, $\lambda_{dB} = 2\pi\hbar/p$, giving $\lambda_{dB}(th) = 2\pi\hbar/\sqrt{(2\pi Mk_BT)}$. Now, it is reasonable to argue that for $\lambda_{dB}(th) > n^{-1/3}$ (*i.e.*, the mean separation less than the de Broglie wavelength) the matter waves will overlap appreciably and the system will exhibit a macroscopic coherence. The onset condition for this *degeneracy* from the above consideration is that $n\lambda_{dB}^3(th)$ be close to unity. The exact condition turns out to be $n\lambda_{dB}^3(th) = 2.612$. The quantity $n\lambda_{dB}^3(th) \equiv \rho$ can be recognized as the phase-space density (measured in units of h^{-3}) at a given temperature. In a true sense then the BEC is a confinement in phase-space — a product of simultaneous confinements in the space of positions and in the space of momenta. For ordinary gases at room temperature and pressure, we have $\rho \sim 10^{-6}$, a million times smaller than the critical value. Thus, the condition for the onset of BEC is that there be one Bose particle for every phase-space cell (of volume h^3), this being the minimum phase-space cell volume required by Heisenberg's uncertainty principle. (Note that h^3 is the natural unit for the phase-space volume). The corresponding critical temperature for BEC turns out to be

$$T_{BEC} = \left(\frac{2\pi\hbar^2}{Mk_B}\right)\left(\frac{n}{2.612}\right)^{2/3}.$$

Thus, for a dilute gas of rubidium ^{87}Rb, with $n \simeq 5 \times 10^{12}$ atoms per cm^3, we get $T_{BEC} \simeq 100$ nK (nanokelvins), as is indeed observed in experiments. In fact, using this formula boldly for the dense liquid ^{4}He with ten orders of magnitude higher density, we get $T_{BEC} = 3.2$ K against the observed $T_{BEC} = 2.18$ K, which is quite remarkable considering that the liquid ^{4}He is clearly far from being dilute, much less an ideal (non-interacting) Bose gas. In fact, while the fraction condensed as BEC clearly depends on the inter-particle interaction, the onset temperature for BEC is quite insensitive to it. Unlike other phase transitions, the BEC is really a consequence of quantum (Bose) statistics and not of any dynamical interaction.

Figure 4.2 sketches the Bose–Einstein statistical distribution at different temperatures, and the development of BEC as the temperature is lowered. The classical Maxwell–Boltzmann distribution is also shown for comparison. The general statistical features of an ideal, or a weakly interacting dilute Bose gas in thermal equilibrium is best described by giving the average occupation number of the different single-particle energy states. At temperatures above T_{BEC} these states are occupied sparsely, but at T_{BEC} a spike begins to appear at the lowest energy state that grows in strength with the further lowering of temperature. The spike represents the BEC — the macroscopic occupation of the lowest energy single-particle state.

It is important to realize here that for an extended uniform system (such as liquid helium), the lowest energy single-particle state is the zero-momentum state

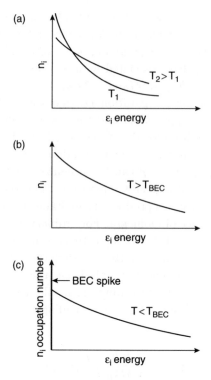

Figure 4.2:   Plot of occupation number n_i of single-particle state i against state energy ϵ_i at different temperatures T: (a) Maxwell–Boltzmann statistics. (b) Bose–Einstein statistics for $T > T_{\text{BEC}}$. (c) Bose-Einstein statistics at $T < T_{\text{BEC}}$. Spike at zero energy shows Bose–Einstein Condensate (BEC). (Schematic).

which is necessarily delocalized over the entire sample volume. Thus, a BEC, if at all there, will occur in this zero-momentum state that is spatially extended. (Indeed, such a zero-momentum BEC in liquid ^{4}He was detected in experiments involving the scattering of neutrons by the BEC.) In the recent realization of BEC in dilute alkali gases, however, the gas is spatially confined in a certain potential trap, and, therefore, the BEC is *localized spatially* in the lowest-energy orbital state in the trap — a much more dramatic object! Another rather subtle point to consider here is regarding the role of the spatial dimensionality. It may appear that the BEC condition for the over-population crisis, namely that the thermal de Broglie wavelength $\lambda_{\text{dB}}(\text{th})$ exceed the mean inter-atomic spacing $n^{-1/3}$, can be satisfied in any dimension (just replace $n^{-1/3}$ for three-dimensions by $n^{-1/2}$ for two-dimensions, for a very thin film, say). A proper counting of states, however, shows that extended BEC is just not possible for fewer than three space dimensions. Indeed, with fewer than three-dimensions there is *plenty of room at the bottom* to accommodate the bosonic population, no matter how low the temperature is. Finally, thermal photons, despite being zero-spin bosons, do not form a BEC no matter how low the temperature is — they simply disappear as the temperature drops, and there is no over-population crisis. Their number is *not* conserved. The situation is quite different for a laser, however, where the macroscopic photon population in a single mode is created and maintained through optical pumping — a non-equilibrium condition! Bose–Einstein Condensation in a gas of bosons of finite but long enough life-time remains an open problem.

4.4 Cooling and Trapping of Atoms: Towards BEC

Cooling means the slowing down of the atoms that are normally in heat motion. This involves removal of their kinetic energy and momentum. It requires forces that depend on the velocity. Trapping means spatial confinement of the atoms. This involves creating a potential minimum or well for them. It requires forces that depend on the position. And all this under conditions of ultra-high vacuum, at about one million millionth of the atmospheric pressure so as to eliminate their scattering out of the trap, or reheating by collisions with the background air molecules that continue to remain at room temperature! For a gas of neutral atoms cooling must precede trapping. This is because the traps are shallow inasmuch as the neutral atoms couple rather weakly to the external electromagnetic fields. And the traps must be contactless to prevent nucleation that would lead to rapid condensation of the vapor to the liquid, or freezing to the solid state. For similar reasons, as we will see below, the gas must be very dilute, *e.g.*, with $n \sim 10^{12}$ per cm^3. This then requires ultra-low temperatures of the order of nanokelvins to obtain the BEC, as discussed in the preceding section. These demands of ultra-low temperatures and of contactless trapping rule out all conventional low-temperature (cryogenic) techniques and material containments. One must use light and magnetic fields only.

Cooling and trapping generally proceeds in two stages. The first involves laser cooling and magnetic trapping, while the second involves magnetic trapping and evaporative cooling. Let us see how. But before we do, let us return briefly to the question of why we must have such a dilution of the gas at all. Why not use instead a denser gas and thus raise the abysmally low temperature required to realize the BEC? (See the expression for T_{BEC} in Section 4.3). The problem really is with the inter-atomic interactions. This involves some interesting physics worthy of a digression. To get an ideal BEC, that is one with $\sim$100% condensate, we must minimize the effect of the inter-particle interactions to almost zero. For this the particles must be kept far apart, away from one another, for most of the time so as not to allow their condensation to the denser liquid droplets; or all is lost. The cold gas must remain gaseous! Now, it is often not well appreciated that such a condensation involves, at the very least, a three-body collision — three atoms must simultaneously approach one another. A two-body elastic collision between two freely moving atoms by itself cannot lead to their coalescence into a bound molecule — the first step towards condensation. This would violate conservation of energy-momentum. One must have a third body around to receive the kick and carry away the excess energy-momentum — one must have an inelastic process, if you like. But, a three-body encounter is relatively improbable and, therefore, rare in a highly dilute gas. It is this statistical rarity of the three-body encounters that keeps the ultra-cold dilute gas gaseous — though metastable against condensation to a liquid drop or freezing to a solid speck — despite the abysmally low temperature. Admittedly, the metastable state has a short lifetime, of the order of seconds, for the dilute alkali-atom gases used for BEC. But this time scale is still long enough on microscopic time scales or for the experimental duty cycles of interest. This is why a dilute gas and the associated ultra-low temperature is a *sine qua non* for realizing a good BEC.

4.4.1   *Laser Cooling and Magnetic Trapping: Down to Microkelvins*

That light carries energy is well known. A short exposure to sunlight and the resultant feeling of warmth amply confirms it. Less well known, however, is the fact that light also carries momentum; and much less that it carries angular momentum too — that it can, therefore, exert pressure and torque on reflection, refraction, absorption, or emission. (In the last case we call it *recoil.*) The pressure due to light is, of course, too slight to be ordinarily sensible. Light pressure was first demonstrated in the laboratory by the Russian physicist P. N. Lebedev back in 1899 by the torque it produced. On a much grander scale, we now know that it is partly due to the pressure of the light from the Sun that the tail of a comet — the tenuous part consisting of the finely divided dust particles and neutral molecules — curves away from the Sun, as first suggested by Kepler. (A much more dominant part of the comet tail, however, comprises ionized gases, and is deflected away from the Sun by the charged-particle solar wind.) On a sunny noon, the pressure due to the sunlight on a reflecting surface is about five micronewtons per m^2. Indeed, giant ultra-light solar sails (light-sails) for spaceships to be propelled by sunlight pressure have been seriously proposed. So, light does carry momentum as well as energy. We can make it work for us. Indeed, a micron-sized optical microrotor, held in an optical tweezer and driven by the flux of light — a veritable windmill — has been demonstrated recently.

Now, light is best described as consisting of photons — packets of energy (E) and momentum (p) related to the wavelength (λ) of the light through $E = h\nu$ and $p = h/\lambda$ with $\nu\lambda = c$, the speed of light. Thus, a photon absorbed by an atom imparts its energy and momentum to it — it can heat and accelerate the absorbing atom. But can it also slow down the atom and thus cool it too? That is the question. And the answer is yes, it can — under certain near-resonance conditions that can be and have been obtained in the laboratory. (Admittedly, it is generally easier to heat than to cool, and we know that, watt for watt, refrigerators cost more than the heaters!) The cooling has indeed been realized through a combination of ingenious

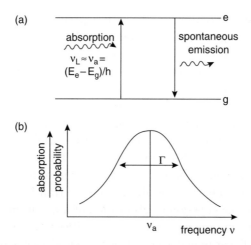

Figure 4.3:   (a) Absorption and spontaneous emission of photons at resonance with atomic transition between ground (g) and excited (e) states. (b) Frequency response of atomic resonance of width Γ centred at ν_a. (Schematic).

ideas that allow *selective* and *adaptive* exchange of energy and momentum between the light and the atoms. The key idea here is that of Doppler Cooling, and the fact that the normally all too weak light-atom interaction gets enhanced many-fold at and about resonance. This light force can indeed accelerate an atom up to $100\,000$ times the acceleration due to Earth's gravity!

4.4.1.1 Doppler Cooling

Doppler effect is the change in the observed wavelength or frequency of light due to the velocity of the observer relative to the source of that light. In the present context, the source is a laser that emits light and the observer is an atom that absorbs (detects or *observes*) the laser light. Stated more precisely, the observed frequency (ν_O) is shifted upwards (downwards) relative to the laser frequency (ν_L) according as the relative velocity (V) of the observer (O) and the laser (L) is towards (away from) each other: $\nu_O = \nu_L(1 \pm \frac{V}{c})$, where c is the speed of light. Understandably, the upward shift is termed the blue-shift and the downward shift the red-shift. (The Doppler effect has the well-known acoustic analogue in the common experience that the sound from a siren or a train whistle has a higher (lower) pitch or shrillness according as the siren or the train is approaching (receding) from us. The Doppler radar used by the highway patrol to check for over-speeding works on the same principle). Indeed, it was the Doppler red-shift of the light from the distant galaxies that gave away the expansion of the universe, as discovered by Edwin Hubble in 1929. It is this very Doppler shift that gives the tunability of light-atom interaction that ultimately allows the laser cooling of atoms, adaptively and selectively.

The atom has an internal structure with a positively charged central nucleus and the negatively charged electrons bound to it by the attractive Coulomb force. The rules of quantum mechanics ordain the system to exist in one of the allowed discrete (quantized) energy states, with the lowest energy (ground) state and many higher lying excited states. These states are labeled by certain quantum numbers that describe the orbital state of the electron — its principal energy level and the angular momentum. Then there is the electron spin whose interaction with the orbital motion gives a spin-orbital fine-structure, while its interaction with the nuclear spin gives a still finer (hyperfine) structure. For a general understanding of laser cooling and trapping, we do not have to enter into this atomic anxiety — the details. It is enough to note that the atom can make transitions selectively between two such states with the emission or absorption of a quantum of light (photon) so as to conserve the energy, among other such quantities that must be conserved, *e.g.*, the angular and the linear momentum. (This is what gives rise to the *line* spectrum characterizing the atom, and used extensively in the laboratory to *fingerprint* them — spectroscopically. Astronomers use it to determine the elemental composition of distant stars from the spectrum of the light received from them.) Now, consider the absorption of an incoming photon of frequency ν by an atom involving its forced excitation from the ground state (g) to an excited state (e). This would require the photon energy to match the excitation energy, *i.e.*, $h\nu = E_e - E_g$, the strict resonance condition (Fig. 4.3a). But not quite so. The excited state actually has a finite lifetime (τ), of the order of 10^{-8} sec, after which it decays spontaneously

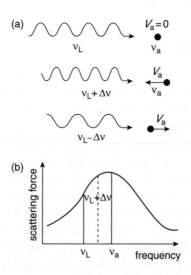

Figure 4.4: Doppler cooling by velocity-dependent detuning: (a) Doppler effect showing blue (red) shift $\Delta\nu = \pm\nu_L V_a/c$ of laser frequency ν_L due to atom moving with speed V_a against or with laser beam. (b) Red-detuned laser light $(\nu_L < \nu_a)$ gets blue-shifted inwards into resonance for atom moving against laser beam. Enhanced scattering rate gives net retarding velocity kick. (Schematic).

back to the ground state re-emitting a photon in the process. This finite lifetime (τ) implies a minimum energy spread (line-width Γ) for the excited state through one of Heisenberg's uncertainty principles ($\Gamma \sim h/\tau$). Thus, the sharp line from the strict resonance condition relaxes to a broader absorption profile for the g–e transition with a peak centered at the atomic resonance ν_a. The resonance condition $h\nu = E_e - E_g$ then need hold only to within this broader sense for a real transition (Fig. 4.3b).

And now for laser Doppler cooling. First, consider an atom placed in the beam of light from a tunable laser with both the atom and the laser at rest in the laboratory. Let the laser be red-detuned with respect to the atomic resonance, *i.e.*, $\nu_L < \nu_a$ with the amount of detuning $\delta = \nu_L - \nu_a \sim \Gamma/h$. We may take the laser beam to be directed along the positive x-axis, west to east, say. The response of the atom will then follow its absorption profile as the laser frequency is varied, with a maximum at the resonance, $\nu_L = \nu_a$, the center frequency.

Now let the atom move with a velocity V_a directed east to west, *i.e.*, along the negative x-axis, and, therefore, moving oppositely to the incident laser beam. The Doppler effect now comes into play. The frequency of the laser light as seen by the moving atom will be blue-shifted *closer into resonance*, enhancing thereby the probability of absorption that would impart a momentum (velocity) kick directed opposite to atom's motion. This causes slowing down, or deceleration of the atom. But what if the atom were moving west to east, *i.e.*, in the direction of the light beam, you may ask. Well, any absorption must then speed up (accelerate) the atom eastwards. But, inasmuch as the light frequency will now be *further red-detuned away from the resonance*, the eastward acceleration will be relatively smaller. All we have to do now is to replace the single west to east laser beam with a pair of counter-propagating red-detuned laser beams aligned east-west. (In practice, a beam from a single laser is split in two using half-silvered mirrors, etc.) A little thought will convince you then that the atom will now be slowed down irrespective of whether it is moving eastwards or westwards, and that the retardation will be proportional to its instantaneous velocity. Next, what if the atom is moving north-south, or up-

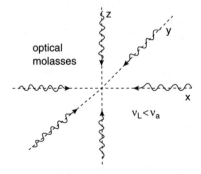

Figure 4.5: Cooling of atoms in optical molasses: Three pairs of counter-propagating mutually perpendicular beams of red-detuned laser light. (Schematic).

down, as indeed it will in a gas. Well, we simply do more of the same: three pairs of counter-propagating red-detuned laser beams aligned along the three mutually perpendicular directions will suffice — after all, any velocity can be resolved into three such orthogonal components (Fig. 4.5). This then is the Doppler cooling scheme. Inasmuch as the retarding force is proportional to the velocity but directed opposite to it, the Doppler cooling system acts as a viscous fluid — very aptly called *optical molasses.*

A nagging doubt, however, remains. The photon absorbed by the atom must eventually be re-emitted *spontaneously*, causing the atom to *recoil.* Will this recoil during the re-emission not undo what was done by the kick received during the absorption? Fortunately not! It turns out that the momentum kicks received from the absorption of the laser photons are highly directed, while the recoil kicks delivered at the spontaneous re-emission are totally un-directed (the spontaneous emission is a random, isotropic process). Therefore, on average, after repeated cycles of absorption and re-emission, the atom suffers a net slowing down. Thus, it is the directed kicks *a priori* followed by the undirected recoils *a posteriori*, that ultimately cause the Doppler cooling.

We conclude this section with a few general remarks that should place the phenomenon of Doppler cooling in a wider context. Thus, *e.g.*, replacing the red-detuned (laser) photons by the blue-detuned photons will turn the Doppler cooling into Doppler heating — the atoms will be speeded up rather than slowed down! Also, the highly directed, coherent laser beams can be replaced by some other radiation environment with a general spectral distribution, and we may still have overall non-trivial Doppler effects. (This will be, admittedly, less effective from the point of laser cooling however.) Indeed, runaway Doppler instabilities due to near-resonant interaction between light and atoms have been considered on the grand astrophysical scales, away from the small laboratory scale setting, where an isotropic background radiation may replace our laboratory laser beams causing a redistribution of the atomic velocities — familiar here as the Compton–Getting effect. Finally, what happens to the (kinetic) energy lost by the atoms as they slow down through Doppler cooling? Well, the mechanical (kinetic) energy of the moving atoms is converted into the electromagnetic energy of the photons. Recall that the red-detuned (low-energy) photons are blue-shifted (to higher energies) in each Doppler cooling cycle of absorption and spontaneous re-emission. There is also an overall increase in total entropy — thus, while the atoms are indeed being cooled down lowering their entropy, the directed photons of the coherent laser light are being converted into an undirected, incoherent radiation through the spontaneous re-emission with a relatively large increase in entropy.

4.5   Doppler Limit and its Break-down

There is, however, a limit to the Doppler cooling. Even the slowest of atoms is forced to continually absorb and re-emit discrete photons in the presence of the laser beams. The lowest temperature is reached when the viscous slowing down (dissipation) is offset by the random recoils (fluctuations) due to the spontaneous re-emissions. Such a connection between fluctuation and dissipation defining a temperature holds quite generally in physics. (The same effect is responsible for the incessant zig-zag motion of a speck of dust, or mote in the air, even in the absence of any convection, as it receives kicks from the colliding molecules of air. There is a name for it: Brownian motion, first observed in 1827 by the Scottish botanist Robert Brown with plant pollen dispersed in water, and explained finally by Einstein in 1905 as due to collisions with the water molecules in thermal motion.) Ideally, the lowest temperature reached by the above Doppler cooling system (called the classical optical molasses) is $T_D = \Gamma/2k_B$, with optimal red-detuning $\nu_L - \nu_a = -\Gamma/2$. This follows from a rather interesting way of looking at things. The atom must be viewed as being in *thermal* equilibrium with the light it nearly resonates with. The width of the resonance ($\sim \Gamma$) may then be viewed as the thermal width $k_B T$, giving a temperature $\sim T_D$. (This is true in general — the energy width of a Maxwellian distribution is indeed $\sim k_B T$. In fact, laser light of spectral line width Δ is rightly viewed as having a temperature $\sim \Delta/k_B$, *i.e.*, the atom is now in thermal equilibrium with the laser light). Experimentally, however, temperatures much lower than the ideal Doppler limit (by as much as six times lower) were realized, which was a real surprise — an anti-Murphy law at work! The explanation for the breakdown of the Doppler limit lay in one of the most remarkable phenomena known in the schemes for laser cooling, namely that of Sisyphus cooling, recognized first by Claude Cohen-Tannoudji and J. Dalibard. The effect is too subtle and minute to be described cursorily. But then, one can not resist its beauty. Let us just get acquainted with the general idea of it.

In Sisyphus cooling, we again begin with the usual scheme for the Doppler cooling involving the ground (g) and the excited (e) atomic states, but now with an additional twist, namely, that the ground state has a finer structure — it comprises two (hyperfine) sub-levels g_1 and g_2. Now, things are so arranged that the sub-level energies E_{g_1} and E_{g_2} vary periodically in space, along the counter-propagating laser beams that now form standing waves. The variations are, however, out of step such that when E_{g_1} is at a maximum, E_{g_2} is at minimum. These two energies act as the two possible potential-energy branches for the moving atom to lie on. Consider now an atom moving along one of these two potential-energy branches, g_1 say, climbing up the potential hill as shown in Fig. 4.6, and thereby losing kinetic energy, which is now stored as potential energy. Now, matters are so arranged that as it approaches the crest of the potential hill, the energy difference $E_e - E_{g1}$ becomes just right for it to absorb a resonant laser photon and get internally excited to the higher lying excited state (e), from which, after a short lifetime τ, it re-emits a photon spontaneously and drops back into the ground state — but this time around into the other branch (sub-level g_2), which is now at its trough. With this switch thus, the atom finds itself once more at the bottom of a potential hill that must be climbed all over again as it continues to move. The energy of the spontaneously emitted photon

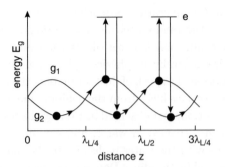

Figure 4.6:   Sisyphus cooling: Atom climbs up the potential hill along g_2 (or g_1) converting kinetic energy into potential energy which is dissipated radiatively by optical absorption g_2 (or g_1) to e at the crest of g_2 (or g_1) followed by atom's transfer to trough of g_1 (or g_2) by re-radiation e to g_1 (or g_2). (Schematic).

$E_e - E_{g_2} > E_e - E_{g_1}$, the energy of the photon absorbed. This energy difference must equal the kinetic energy lost by the atom, and is dumped (lost) radiatively in each cycle of absorption followed by re-radiation. Hence, the Sisyphus cooling, aptly named after the Greek character who was condemned to endlessly roll his stone up the slope only to find that the slope beyond the crest was also an uphill one. Inasmuch as the Sisyphus cooling depends on the finer structure of the ground state, it is very sensitive to any external fields — even Earth's magnetic field may have to be neutralized properly. It may be noted here that the finer sub-level energy structure, g_1 and g_2, of the otherwise degenerate ground state (g), which is of the essence for the Sisyphus cooling, is due to the state-dependent radiative corrections to the energy of the state — also called therefore the light-shift. (It is the change in the energy of a state due to its coupling to the photons, and depends on the intensity and polarization of the light at the place where the atom happens to be. Thus, for a standing light wave with polarization gradient, it varies periodically in space. It is a purely quantum-mechanical effect.) The lowest temperature obtainable with Sisyphus cooling (T_s) is given by $k_B T_s = \hbar\Omega^2/|\delta|$, where Ω (the so-called Rabi frequency) measures the strength of the light-atom dipolar coupling and δ is the detuning, assumed not too small.

The Sisyphus cooling can cool atoms below the Doppler cooling limit, $T_D = \Gamma/2k_B$, because now the atom does not have to equilibrate with the photon in the sense discussed earlier. The kinetic energy is being pumped out and dissipated away radiatively. There is, however, still a lower limit to Sisyphus cooling — the Recoil Limit, $T_R = E_R/k_B$, where E_R is the energy of the one-photon atomic recoil due to the photon momentum (h/λ_L) exchanged with the atom. Thus, for example, for sodium atoms (^{23}Na), $T_R \simeq 2.5$ microkelvin! Laser cooling below the Recoil Limit is also possible by somehow switching off the perturbing light-atom interaction — this is accomplished by pushing the atoms into the so-called Dark State. We do not, however, pursue this line of thought any further here.

It is time now to get some feel for numbers. Consider the case of Doppler cooling for sodium atoms (^{23}Na), where the transition involved is from the ground state (g) to the excited state (e) with the energy difference $E_e - E_g \approx 2$ electron-volt. This is the well-known yellow light from the sodium flame at the wavelength $\lambda \approx 589$ nm. The sodium vapor emerging from a heated source (oven) is at a temperature of about 500 K, with a thermal atomic velocity of about 10^5 cm sec^{-1} (about 4000 km per hour). The momentum loss per collision (photon absorbed) $\sim h/\lambda$, giving a velocity slow down of about 3 cm sec^{-1}. Hence, to slow down from 10^5 cm sec^{-1} to almost zero velocity in steps of 3 cm sec^{-1} will require about 3×10^4 collisions. The lifetime of the excited state $\tau \sim 16$ ns (1 ns $= 10^{-9}$ sec), giving the time required for the Doppler cooling to be $2n\tau \sim 1$ ms (1 ms $= 10^{-3}$ sec). The stopping distance is then about 50 cm. This amounts to an acceleration (actually deceleration) $\sim 10^5 \times$ Earth's gravity! Also, the Doppler limit T_D works out to be ~ 240 μK. As for lasers, typically, the laser power for optical molasses is 10 mW, with a bandwidth of 1 MHz and beamwidth of 2 mm radius. It is interesting to note that Doppler cooling is quite insensitive to laser power. Sisyphus cooling, however, depends strongly on the power of the laser through Ω.

4.6 Trapping of Cold Atoms: Magentic and Magneto-Optic Trap (MOT)

A gas of atoms, no matter how cold, must be trapped spatially for a time long enough to be probed experimentally, or else the atoms will disperse, and most certainly fall freely under Earth's gravity and get lost. Just as laser cooling involved confinement in velocity space centered about the zero of velocity throug the velocity-dependent forces, trapping involves confinement in the positional space centered at the origin through position-dependent forces. Such a force can derive naturally from the gradient of the potential energy of the atom placed in a suitably inhomogeneous electromagnetic field to which the atom couples. Thus, for a neutral atom one can make use of the fact that the atom may have a permanent magnetic dipole moment ($\boldsymbol{\mu}$), which when placed in a magnetic field $\mathbf{B}(\mathbf{x})$ has a potential energy $-\boldsymbol{\mu} \cdot \mathbf{B}(\mathbf{x})$. (This is what makes a floating bar magnet, or the needle of a magnetic compass point due north in Earth's magnetic field.) The same is true of alkali atoms, *e.g.*, the sodium atom (^{23}Na), that has an unpaired electron spin which exhibits an elementary magnetic dipole of moment μ (the Bohr Magneton) associated with its unpaired spin (with $\boldsymbol{\mu}$ antiparallel to the spin). The two possible orientations of the spin, parallel and antiparallel to $\mathbf{B}$ differ in energy by $2\mu B(\mathbf{x})$. Such a splitting, or the shift of an energy level is called the Zeeman effect. It acts on both the spin and the orbital angular momentum, with different strengths though. In an inhomogeneous field $\mathbf{B}(\mathbf{x})$, the potential energy is then position dependent. It has a minimum for $\mathbf{B}(\mathbf{x}) = 0$, with the magnetic moment $\boldsymbol{\mu}$ antiparallel to $\mathbf{B}(\mathbf{x})$. An

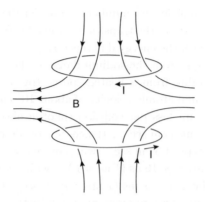

Figure 4.7:   The quadrupolar magnetic field generated by coaxial anti-Helmholtz coils carrying oppositely directed electric currents. The magnetic field is zero at the mid-point. (Schematic).

atom with such a spin orientation seeks out the lowest field and tends to sit at the minimum of $\mathbf{B}(\mathbf{x})$. Such an inhomogeneous magnetic field can be conveniently generated with the help of two co-axial coils carrying electric currents in the opposite sense (the anti-Helmholtz coils). The resulting quadrupolar magnetic field has a zero at the mid-point on the axis as shown in Fig. 4.7. This is the canonical quadrupolar magnetic trap. The magnetic field gradient is typically ~ 10 gauss cm^{-1} over a trap size ~ 1 cm across. The trap potential is ~ 10 mK deep in temperature units. (Room temperature of 300 K corresponds to $\sim 1/40$ eV.) The lifetime of the trapped atom before it gets ejected by collisions is ~ 100 sec at a pressure of 10^{-10} torr (1 atmosphere $= 760$ torr, 1 torr $= 1$ mm of Hg).

This purely magnetic trap can be integrated intelligently into the laser Doppler cooling arrangement discussed earlier. The result is a Magneto-Optic Trap (MOT), which has now become a workhorse for all BEC work. Just as laser cooling is a forced confinement of an atom in momentum (velocity) space towards the zero of momentum, magnetic trapping is a forced confinement of an atom in positional space towards the origin. The Magneto-Optic Trap (MOT) combines the two in a kind of *phase-space* confinement towards the zero of the velocity as well as of the position. In laser cooling we exploit the velocity-dependent Doppler shift of the light from a red-detuned laser into or off the atomic resonance. In a MOT we exploit additionally the position-dependent magnetic Zeeman shift of the atomic levels into or off the resonance with the red-detuned laser light. In all cases it can be so arranged that we have a *selective* and *adaptive* approach towards the zero of velocity and of position. Such a MOT can produce a cold cloud at about 10 μK with 10^{11} atoms per cm^3 corresponding to a phase-space density of about 10^{-5}, which is still far too small for BEC to occur. The cold atoms are huddled in a volume ~ 0.5 mm^3 at the center of the MOT, which is about ~ 1 cm across.

It is to be noted here that unlike the case of a purely magnetic confinement, the confining force in a MOT originates from the scattering of light involving

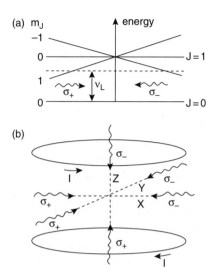

Figure 4.8: (a) Magneto-Optic Trap combining trapping (position-dependent scattering forces due to tuning by inhomogeneous Zeeman shift) and cooling, in one dimension. (b) Three-dimensional MOT. (Schematic).

near-resonant absorption and re-emission — hence called the scattering (or the spontaneous) force. It is just that the condition for the resonance is made position dependent.

In Fig. 4.8*a*, the basic idea of a MOT is illustrated for the case of a one-dimensional geometry, while Fig. 4.8*b* shows the schematic of a three-dimensional MOT. Because the underlying trapping mechanism in the MOT involves the Zeeman effect, it is also referred to as the ZOT (Zeeman Shift Optical Trap). The symbols σ_+, σ_- here denote the circular polarization states of the light beams chosen so as to cause the atomic transitions selectively between the Zeeman-split and -shifted levels.

4.7 Evaporative Cooling: Down to the Nanokelvins

A cold atomic cloud trapped in a MOT and cooled to a few microkelvins is still a factor of 1000 away from (hotter than) the gaseous BEC that demands nanokelvins. This ultimate stage of cooling from microkelvins to nanokelvins — the last mile if you like — involves the simplest of physics, namely that of cooling by forced evaporation! Here a relatively small minority of energetic molecules with relatively large, above-average kinetic energy *per capita*, are allowed to escape over the potential barrier out of the trap, leaving behind the majority in the trap to settle down (re-equilibrate) to a lower average kinetic energy *per capita*, that is to a lower temperature. This is, of course, precisely what happens to a steaming cup of hot tea in an insulating styrofoam left out in the open. It cools mostly by evaporation, the barrier being the work function — the minimum kinetic energy needed to escape from the bulk liquid. The point to note here is that the amount of kinetic energy removed, or the cooling effected, is greatly out of proportion to the mass of the liquid lost by evaporation; a mere 2% loss of molecules can cool the cup of tea by about 20%, *i.e.*, from about 370 K to 300 K (room temperature) in a short interval of time. The effect can be further enhanced by resorting to forced evaporation, *i.e.*, by lowering the barrier to be overcome for the great escape. The evaporative cooling is ultimately traceable to the fact that the atoms in the fluid do not all have the same kinetic energy, or speed. There is a broad thermal distribution — the Maxwellian velocity distribution — that has a tail of very high-speed molecules that, though small in number, can and do escape over the barrier carrying away a disproportionately large amount of the energy. What is crucial, however, for evaporative cooling is the fast process of re-equilibration of the remaining molecules in the trap. This requires some inter-particle interaction which is otherwise inimical to an ideal BEC, as we have argued earlier. This calls for a compromise.

For evaporative cooling it is necessary to switch off all perturbing lasers of the MOT leaving only the confining (trapping) magnetic field on. This, in general, is the loading of a purely magnetic trap with the cold atoms from a MOT, and has been variously achieved. Through some suitable optical pumping it is so arranged that all atoms are put in the same state of the spin, with the magnetic moment (spin) antiparallel (parallel) to the trap magnetic field. As discussed earlier, the atoms with their magnetic moments aligned antiparallel to the magnetic field seek the weak field and are thus attracted towards the center of the trap. The kinetically energetic among them, however, climb up the potential barrier towards the edge of the trap as shown in Fig. 4.9. Now, the trick is to apply a radio-frequency (rf) field of the right intensity, duration and of a frequency resonant with the local Zeeman

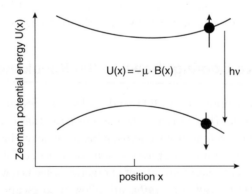

Figure 4.9: Schematic of forced evaporative cooling in a magnetic trap using inhomogeneous Zeeman shift. The upper (lower) potential energy curve is for atomic spin parallel (antiparallel) to local magnetic field and has minimum (maximum) at center. Spin-flip caused by resonant rf field inverts potential upside down forcing energetic atoms nearer edge to fall off out of trap. Remaining slow moving atoms re-equilibrate to lower temperatures. Hence, forced evaporative cooling.

energy-splitting $2\mu B(x)$ so as to flip the atomic spin. This spin-flip inverts the potential barrier as seen by the atom upside down, and the energetic atoms simply fall off the edge of the trap-potential barrier, now becoming downhill. Tuning the radio frequency (rf) progressively downwards, we can effectively move the surface-of-escape-by-spin-flip radially inward, thereby slicing off layers of atoms of lower and lower kinetic energy towards the center of the trap. One very aptly speaks of an rf-scalpel here! The gas of remaining atoms now accumulates near the center of the trap, somewhat depleted but rendered much colder and denser — possibly about 10^7 atoms at 10^{12} cm^{-3}, huddled together in a volume ~ 10 μm across at the center of the magnetic trap ~ 1 cm across, and at an ultra-cold temperature of ~ 20 nanokelvins! Thus, we will have reached the coldest spot in the universe! The entire evaporative cooling cycle lasts the time scale of seconds or minutes.

It is implicit here that the magnetic moment (or the spin) of the moving atom precesses (as a spinning top) fast enough to follow the spatially varying magnetic field so as to remain locally aligned with it — actually antiparallel to it in this case. Inasmuch as the angular speed of spin precision $(2\mu B(x)/\hbar)$ is proportional to the magnitude of the local magnetic field, this *adiabatic* condition must break down at and about the trap center, where the magnetic field has a *pointed zero*, across which the field changes its direction very rapidly. Nothing then energetically prevents the electron spin of the moving atom from flipping its direction relative to the local magnetic field vector, and, therefore, the atom from getting lost through this *hole in the magnetic trap*. This kind of spin flip at the zero of the magnetic field is a subtle effect. It has a name: the Majorana Spin-Flip, said in deep voice. It must be avoided for a magnetic trap to act as a trap. A rather clever solution to this problem turned out to be just this: jiggle the *pointed* zero by the application of a suitable time-varying magnetic field and thus smoothen it out into a time-averaged rounded minimum. This finessing is aptly called a TOP (Time-averaged Orbiting Potential) that does plug this hole in the trap.

Nature uses evaporative cooling on all scales — we have mentioned the humble cup of hot tea as an example on the scale of centimetres, and the not so humble MOT on the scale of millimetres. But, we have evaporative cooling the scale of kilo-parsecs too (1 parsec ~ 3.3 light years). This happens in the globular clusters of stars, where the highly energetic stars get kicked out of the gravitational barrier leaving behind a more compact cluster of less energetic stars. (In a lighter vein, the so-called Brain Drain of the Third World may well be viewed as an evaporative *intellectual* cooling: emigration of the higher-than-average qualified elite leading to the lowering of the average (intellectual) temperature of the population which is left behind). It seems that Nature truly has no architectural excess — it repeats the same design again and again, only the contexts and the scales may vary, and be vastly different!

4.8 BEC Finally: But How Do We Know?

That is the question! There is no thermometer to tell nanokelvins anyway. Once again we have to fall back on light: this time to visualize the condensate, its distribution in velocity and position — its phase-space profile in fact — and to, therefore, indirectly act as a contactless thermometer. For this, all we have to do is to switch off the magnetic confinement too and let the ultra-cold cloud of atoms fall freely under Earth's gravity. Once set free, the atomic cloud expands as it falls. As a result, the velocities translate into distances. In fact, for a parabolic trap the velocity distribution is the same as the spatial distribution, properly scaled. Also, the expanding cloud can be imaged by laser light which is tuned so as to be absorbed resonantly by the atoms, casting thus a shadow on a camera (the so-called Charge Coupled Device, or the CCD camera now in common use). At resonance, the cloud is almost opaque (*i.e.*, optically thick) at the BEC phase-space density ($\rho \approx 1$) and the light can hardly penetrate beyond a depth $\sim$ wavelength of light. But an expanded cloud is rarer and lets some light pass through, and, therfore, casts a shadow on the CCD camera whose shade (dark or grey) then depends on the column density of the cloud traversed by the laser beam. These facts together with the time of flight (TOF) allow us to essentially re-construct the velocity distribution in the cloud at different stages of evaporative cooling (*i.e.*, at different temperatures). Typically, the BEC which is initially about 10 μm across expands to about 200 μm across in about 40 ms of the time of flight.

In Fig. 4.10, we show schematically a cross-section of the velocity (x-component, say) distribution as it evolves with cooling to lower temperatures. At relatively higher temperature, we have the well-known Maxwellian velocity distribution with its single broad Gaussian hump centred at the zero of velocity. This is the thermal cloud. As we cool down just below T_{BEC}, a qualitative change makes appearance.

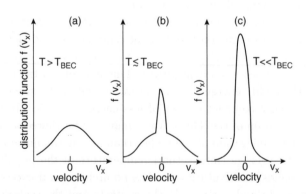

Figure 4.10:   Section of distribution of ultra-cold trapped atoms imaged by expansion method: (a) For $T > T_{\text{BEC}}$ showing Maxwellian thermal distribution. (b) $T \lesssim T_{\text{BEC}}$ showing central spike of BEC emerging out of thermal Maxwellian background hump. (c) Only central spike representing fully formed Bose–Einstein Condensate (BEC). Not shown is elliptical cross-section of spike indicating anisotropy of velocity distribution in BEC in contrast with circular cross-section, or isotropy, of classical thermal background. (Schematic).

The broad Maxwellian hump is pierced by a spike centered at zero velocity —
indicating an accumulation of atoms in the single, zero-velocity state, positioned
at the potential minimum of the magnetic trap, which is now switched off. There
is, of course, the sparse thermal cloud surrounding this central spike, often referred
to figuratively as the Oort cloud by astronomers — that tenuous cloud of matter
that surrounds the solar system, and is believed to be the outlying remnant of the
original molecular cloud that condensed and became our solar system. Finally, far
below T_{BEC} we are left with just the central peak corresponding to the pure BEC
centered at zero velocity.

A remarkable feature of the velocity profile (distribution) is that while for the
background thermal cloud surrounding the central peak it remains isotropic (circular
in cross-section), it is anisotropic (actually elliptic in cross-section) for the central
peak. This characteristic difference gives away the BEC and can be readily understood. The Maxwellian velocity distribution for a gas in thermal equilibrium
(without any bulk motion) is isotropic, independently of the potential it is subjected to — it depends only on the square of the velocity (or momentum). This
explains the isotropic velocity distribution of the classical *thermal* fraction of the
cloud that surrounds the central peak. The situation is qualitatively different for
the BEC in the trap. A BEC occupies the lowest-energy orbital (state) for the
given trap potential. It is a quantum-mechanical ground state. Now, the magnetic
trap potential is really not quite spherically symmetrical; it is actually ellipsoidal
because of the geometry of the coils producing the magnetic field. In fact, the
simple quadrupolar trap potential is elongated — twice as steep along the z-axis
as along the x and y axes. Other highly anisotropic traps with much more sophisticated magnetic field configurations have been used, giving very elongated cigar
shaped condensates. This makes the spatial form of the ground state orbital (wavefunction, if you like) also ellipsoidal. Now, as the trap is switched off, this initial
ellipsoidal spatial form freely evolves into an ellipsoidal velocity distribution (to
which the former is related by Heisenberg's uncertainty principle), centered about
the zero of velocity — that gives away the BEC. It is a quantum-mechanical object,
or goop if you like.

It is of the essence to emphasize here that once the initial BEC has expanded,
as it must, to a phase-space density ρ much below the critical BEC value (of
about unity), it is no longer a condensate in the original strict sense. But, the initial quantum-mechanical velocity distribution remains imprinted on this expanded
cloud, now becoming an almost sensible realization of it! This bears a striking
resemblance to the Big Bang picture of an early universe evolving out of a quantum-
mechanical, point-like initial state into the presently expanded, almost classical realization of it with all the initial-state anisotropies imprinted on it, but now magnified
into the observable anisotropies of the Cosmic Microwave Background Radiation,
that the COBE (Cosmic Microwave Background Explorer) satellite had detected
not so long ago in 1995.

4.9   BEC: What Good is it?

While the making of a BEC is not yet quite a Do-It-Yourself (DIY) proposition, it is certainly very much of a table-top-experimental science that should belong in any university physics department — it is small science, but potentially one with a big impact. Thus, for example, the diode lasers used for the laser cooling and trapping in a MOT are no different from the ones you have in your CD player at home. The big thing about BEC really is that it is a combination of many clever ideas, and that it holds promise for New Physics and novel applications, some of which have already been demonstrated and others are around the corner as noted by the Nobel Committee.

Professionally, for condensed matter physicists, who constitute about two thirds of all physicists, a dilute gaseous BEC is a new state of controlled matter which is coherent over the macroscopic scale, and tunable over a wide range of values of its parameters, including the strength, range and even the sign of interaction between the alkali atoms that compose it; such control is totally inaccessible for its condensed liquid counterpart, namely the exclusively noble and inert liquid helium ^{4}He. The nanokelvin *supergas* can be readily manipulated with light, *e.g.*, it can be made non-uniform. It and its precursor, the microkelvin cold atomic cloud, can be patterned into an optical lattice. Moreover, the gaseous BEC should provide a proper understanding of the relationship between the condensate and its superfluid, or rather its supergas flow properties (of, *e.g.*, zero viscosity), and the infinite thermal conductivity. Indeed, it has opened up an entirely new field of a quantum phase transition where an ultracold supergas of the bosons can change from being a superfluid to being a solid insulator (the so-called Mott insulator) according as the inter-particle repulsion is weak or strong relative to the depth of the potential wells that trap the particles. The latter are patterned by the interference of light so as to form a potential which can be periodic, or quasiperiodic or even random in space. Some of the deeper issues of coherence and decoherence by interactions (*even from within the condensate*) await resolution. So is the case with the question of the bosonic stimulation, namely, that the bosons are preferentially scattered (inelastically) into the state which has the higher pre-existing bosonic population, and its role in the very kinetics of formation of the BEC out of a cloud of cold atoms, which is far from well understood. Estimates of the time-scale of formation vary from micro-seconds to the age of the universe, while the observed time is of the order of seconds. (We should hasten to add, however, that the idea of coherence goes far beyond the usual one of dividing a condensate in two and re-combining it to get the interference fringes as in Young's double-slit experiment. This is so-called first-order coherence. There are higher order coherence effects too, and the gaseous BEC has tested positive to up to coherence of the third order).

Almost all applications of the BEC, practical as well as basic, are traceable to the single essential feature of the BEC, namely, that it has a macroscopic coherence in that it has a finite fraction of all atoms occupying one and the same single-particle state — the macroscopic occupation of the lowest-energy orbital of the trap — and that all the tens of millions of these atoms act in lockstep as one single whole. Let us now consider some of its possible applications. A consequence of this coherence is that a BEC can be used to detect the subtle and the minute such as the suspected violation of the time-reversal invariance in atoms and molecules — too small to be detected otherwise. Here the fact that of the order of tens of millions of the atoms in the BEC act as one giant atom, enhances the tiny effect by a large factor of that very order. Attempts are already afoot to try this with a BEC composed of ytterbium (Yb) atoms. This is a general strategy for detecting small effects, and can ultimately provide a low-energy test, *e.g.*, of some aspects of the Standard Model of high-energy physics.

As for practical applications, the first that comes to mind is the *atom laser*, or more appropriately the *matter-wave laser*, which has already been demonstrated. Recalling the wave nature of matter, with the de Broglie matter-wavelength $\lambda = h/p$, one may view the BEC as a wave of macroscopic amplitude, much as the laser is viewed as a large number of photons occupying the same mode and contributing additively (coherently) to its field amplitude. It has the same interference and diffraction properties — one speaks of *atom optics* here. Except, of course, that unlike photons the atom laser beam cannot pass through, *e.g.*, window glass — it requires ultra-high vacuum, created artifically in the laboratory, or occurring naturally in the outer space. Also, normally, an atom laser may not be as well collimated as light laser, because it is coupled out of a BEC trapped in the localized ground state (unlike in the case of light laser, where the light is coupled out from a higher lying mode of the optical cavity).

The distinctive feature of the condensate, namely that all the atoms in it occupy the same single-particle ground state means that it is free from any inhomogeneous broadening of its atomic spectrum — thus, for example, even as the magnetic field in the magnetic trap varies over the spatial extent of the condensate, the Zeeman splitting in the condensate shall have one single value for all the atoms. Also, somewhat counter-intuitively, the condensate in the trap is free from any random Doppler broadening. All this makes for an exceptional spectral purity. High-resolution spectroscopy with linewidths less than 2 Hz have already been realized using the so-called BEC Atomic Fountain.

All this can be exploited to make BEC based atomic clocks that can beat the best known atomic clocks, and even the celestial clock, namely the rotating neutron star, or the pulsar, by several orders of magnitude in terms of precision and long-term stability, as required for the global positioning system (GPS) and for space navigation. An atom interferometer can replace the laser interferometers such as the LIGO (Laser Interferometer for Gravitational wave Observatory) for detecting the gravitational waves reaching us from the distant gravitational disturbances in our galaxy, the tiny strains of space-time geometry of the order of 10^{-21} — this being the last of the predictions of Einstein's General Theory of Relativity. It can also detect tiny variations or anomalies of a few parts in 10^{10} in Earth's own gravity for use in oil exploration.

A major application of BEC envisaged is in the area of nanotechnology — namely, nanolithography, where an atom laser can *write* nanometric scale features such as electronic components and other devices as integrated circuits on a semi-conducting chip. The latest feat has been to create a BEC of rubidium atoms in a microscopic magnetic trap built into a lithographically patterned microchip, and to move the BEC by as much as a millimetre without loss of coherence. This opens up the new field of *atomtronics*, much as we have electronics or photonics. Even matter-wave holography is being contemplated. Here, the holograms will be real — something out there reified!

A micron-sized Bose–Einstein Condensate in the magnetic trap is in a true sense a giant Single Atomic Particle At Rest In Space (SAPARIS) — *e pluribus unum*.

4.10 Summary

Classically identical particles are quantum mechanically indistinguishable, with different rules for what counts as their distinct re-arrangement. These indistinguishable particles can be either fermions or bosons. Fermions have half-integral spin, and not more than one of a kind shall occupy a given state — they obey the exclusive Fermi–Dirac statistics. Common examples are electrons, protons, neutrons or the helium atoms ^{3}He, etc. Bosons, on the other hand, have integral spin, and bosons of a kind can have, indeed are *encouraged to have*, multiple occupancy of a given state. Common examples are helium (^{4}He), hydrogen (^{1}H), sodium (^{23}Na), etc. This inclusive Bose–Einstein statistics can, for the ideal case of non-interacting Bose particles of a given fixed number, lead to a finite fraction of the total number of particles dropping into the lowest energy single-particle state at a low enough temperature and high enough density, that is for higher than a critical phase-space density of order unity. This macroscopic occupation of the lowest energy state is the Bose–Einstein Condensation (BEC). In a really dense Bose system such as the liquid helium (^{4}He), however, the inter-particle interaction depletes the condensate. This can be avoided by going over to a highly dilute gas of the Bose particles, but that would require going down to ultra-low temperatures, making the system metastable towards a normal condensation to the liquid or the solid state, which would then pre-empt the BEC. Starting in 1995, several laboratories around the world have realized BEC of dilute gases of the bosonic isotopes of alkali metals, *e.g.*, ^{87}Rb, ^{23}Na, etc., cooled down to a few nanokelvins (10^{-9} K). This has involved laser Doppler Cooling in a Magneto-Optic trap (MOT) to microkelvins, followed by Magnetic Trapping and Evaporative Cooling to nanokelvins, all in ultra-high vacuum. The gaseous BEC typically contains about 10^7 atoms, at about 100 nanokelvins and measures about 10 micrometers across, and lasts about a minute or so. The entire process of cooling and trapping is *adaptive*, and most importantly *selective* — thus, *e.g.*, the air molecules surrounding the nanokelvin BEC are essentially at room temperature! The 3 K cosmic microwave background radiation is also very much around. Laser cooling to the BEC, truly marks the ultimate march towards the absolute zero of temperature — starting from room temperature at about 300 K (with atomic velocities at ~ 4000 km/hr) through microkelvins (at ~ 250 m/hr) to nanokelvins (~ 8 m/hr). A triumph of scientific endeavor!

The single most important property of a BEC is that the millions of atoms composing it are all in the same single state — all doing the same thing, in a lockstep, so to speak. This coherence opens up areas of New Physics and of novel applications, basic as well as practical. These include atom lasers for nanolithography (nanotechnology); ultra-precise and stable clocks (metrology and space navigation); atom interferometer for gravitational wave detection; possible detection by coherent amplification of subtle and minute effects, such as violation of time-reversal symmetry of high-energy physics at low energies; high resolution spectroscopy; and many others.

Gaseous BEC is verily viewed as a new state of matter.

4.11 Further Reading

Books

- H. J. Metcalf and P. van der Straten, *Laser Cooling and Trapping* (Springer, New Hork, 1999).

Semi-popular papers

- Eric Cornell, *Very Cold Indeed: The Nanokelvin Physics of Bose–Einstein Condensation*, Journal of Research of the National Institute of Standards and Technology **101**, 419 (1996).
- Carl E. Wieman, *The Richtmyer Memorial Lecture: Bose–Einstein Condensation in an Ultracold Gas*, American Journal of Physics **64**(7), 847 (1996).
- W. D. Phillips, P. L. Gould and P. D. Lett, *Cooling, Stopping, and Trapping Atoms*, Science **239**, 877 (1998).
- Steven Chu, *Laser Trapping of Neutral Particles*, Scientific American, February, 1992, p. 49.
- Steven Chu, *Laser Manipulation of Atoms and Particles*, Science **253**, 861 (1991).

Exploring Nanostructures   **5**

In 1959 Richard Feynman delivered a lecture entitled "There's Plenty of Room at the Bottom,"[1] where the famous physicist speculated about manipulating matter at very small length scales and the enormous effects it portends. A lot has happened since then. A major field in science and technology is being created, and with it we have a new world to explore.

5.1   Towards the Bottom

One of the most wonderful inventions of our times is the laptop computer. This marvel of technology allows us to enter instantly the universe of the internet from almost anywhere in the modern world, letting us exchange gossip with our friends or learn about the latest developments on the human genome project. Its 'brain' is a complete computing engine, called a *microprocessor*, fabricated on a single chip, or thin piece of silicon less than a square inch in size, onto which some fifty million transistors have been etched and which is capable of executing over two billion instructions per second. What a powerhouse when compared with the six-thousand-transistor circuit found in the first home computer in 1974, or a six-transistor radio in 1960! This remarkable progress in harnessing informational power would not have been possible without a parallel advance in fabrication: the smallest wire on the chip, many microns across thirty years ago, now measures less than two tenths of a micron, or five hundred times thinner than a human hair. Other key elements of the computer, for example its data storage media and memory devices, are now made very small too. In fact, we live surrounded more and more by all kinds of finely structured materials designed exactly for our specific needs; ingenious little devices that can do complex tasks have become so common that they are now very much part of our life.

---

[1] Reprinted in *Feynman and Computation: Exploring the Limits of Computers*, edited by Anthony J.G. Hey (Perseus Book, 1998). Also available at the websites: `www.its.caltech.edu/~feynman` and `www.zyvex.com/nanotech/feynman.html`

Since 1974 the number of transistors that can be fabricated on a silicon integrated circuit, following Moore's Law, is doubling every 18 to 24 months, a trend made possible by the incredible shrinking of all electronic components with each chip generation. The squeezing of more transistors and logic gates into integrated circuits shortens the travelling distances of electrons and makes feasible a more efficient circuit architecture (such as pipelining), so more instructions can be executed at a faster rate, consuming less power. Miniaturization saves space and material, making things cheaper and faster.

With this continual size downscaling, quantum effects eventually become dominant, bringing with them both challenges and opportunities which can be exploited for useful purposes. For example, new types of microlasers can be made that have lower threshold currents and a wider choice of radiation wavelengths. Small things just do not look the same, do not act the same as big things. They may change color; bulk gold usually looks yellow, but fine-grained gold appears reddish. They may take on new shapes; graphite normally exists in planar sheets, but when cut up into fragments small enough, it may, under certain conditions, rearrange itself, curl up, and close into hollow spheres or cylinders. As structures become very small, many properties become strongly size-dependent, while others are enhanced by the increased surface areas. By moving to near-atomic scales, scientists can select and determine the properties of the final products right at the atom level, leading to novel or improved materials and devices.

Finally, although strongly driven by technology, this trend towards the very small can produce results that have significant implications in issues of fundamental science: it is a place where one may want to test concepts of quantum theory, or realize working models for complex biological systems.

We refer to that general trend by an umbrella term, *nanoscience*, which covers all research activities centered on objects with defining dimensions on the scale of a nanometer (or 1 nm, which is 10^{-9} m). Nanoscience deals with structures that are actually fabricated with nanometric dimensions (called *nanostructures*), such as carbon fullerenes. It studies bulk materials made from nanosized grains (*nanostructured materials*), such as ultrafine-powdered ceramics. And it also encompasses systems engineered from minuscule electromechanical devices (*nanoelectromechanical systems* or *NEMS*), such as the cantilever used in probe microscopy, or the Digital Micromirror Device invented by Texas Instruments scientists and used in large-screen digital cinemas.

Nanostructures, the basic entities of these studies, range roughly in size from 1 to 100 nm and include both naturally occurring and artificially made objects — metallic or ceramic grains, chemical or biological macromolecules, optical or electronic devices, and protein-based motors (Fig. 5.1). Nanoscience necessarily covers a vast domain that gathers together experts in physics, chemistry, biology, materials science and engineering to work in a collective program whose scope and potential have rarely been seen before.

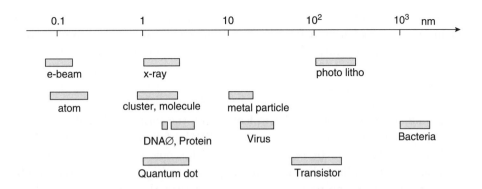

Figure 5.1: Structures in the domain of the small length scales.

5.1.1 Summary

Nanoscience is a multidisciplinary field that deals with structures that have characteristic dimensions on the scale of a nanometer, or with materials and devices constructed with them. These structures may be naturally occurring entities as well as artificially made objects. The objectives of these activities are, on the one hand, to search for improved materials and more efficient ways of fabricating devices and, on the other hand, to explore and eventually exploit new properties that should arise from the small scale.

5.2 The Rise of Nanoscience

Of course, nano-sized particles have been around for a long time, and have become the focus of a systematic though deliberate inquiry at least since Feynman's rallying call. But the recent massive upsurge in interest in the subject that has led to its recognition as a distinct coherent field of endeavor stems from the confluence of three important new developments: novel characterization methods, advanced fabrication techniques and a better understanding of the basic science.

In 1982 Heinrich Rohrer and Gerd Binnig (1986 Nobel Prize laureates) invented a new type of microscope, called the *scanning tunneling microscope* (STM), which is capable of imaging individual atoms. It differs from all previous types of microscope in that it uses no free particles to probe, and so has no need for lenses, and can therefore bypass their inherent limitations. Its probing agent is the tip of an extremely sharp metallic needle, and its operation relies on the *quantum tunneling effect* (see Appendix B). One positions the tip so close to (say, 1 nm above) the surface of a conductive sample that the wave functions of electrons in the tip overlap the wave functions of electrons in the object surface, enabling electrons to flow across the vacuum gap when a small voltage is applied to the probe and sample. The tunneling current thus produced is very sensitive to the tip–sample separation because the electron wave functions decay exponentially with distance in the gap; changing the separation by a fraction of a nanometer causes the current to vary by several orders of magnitude. In an STM operation, you maintain the tunneling current at a set level, and when you scan the probe across the sample, it will follow a constant current path, tracing the contours of the object surface. As the tip is extremely sharp and the current confined to a fine thread connecting the tip end and the sample surface, you can detect features on the surface as small as atoms.

Other types of *scanned probe microscopes* (SPMs) followed close behind. As in the STM case, they work by sensing a local microscopic property of a specimen (*e.g.*, atomic forces, optical absorption or magnetism) with a submicron-scale probe designed to detect the property of interest. You scan the probe back and forth over the surface of the object while measuring the physical quantity, then process and convert its local variations to a three-dimensional real-space representation of the surface. Because of the nanometric- or even atomic-scale resolution that can be achieved and their adaptability to different types of interactions and working environments, including vacuum, air and liquids, these microscopes have become indispensable characterization tools in nanotechnology.[2]

Even more significant is the ease in turning the probe of an SPM into a nanometric-sized mechanical tool by simply changing the strength of the inter-action involved. Increasing the applied potential to a few volts suffices to give the probe enough force to break chemical bonds or initiate a local chemical reaction in the sample. The probe now becomes an exquisitely delicate tool for local modifi-cation and manipulation, capable of removing individual atoms and molecules and repositioning them at selected sites, or forging chemical bonds between individual particles, or even scratching and chiseling surfaces. So, you can alter matter or even build it atom by atom.

Many other techniques of fabrication are found in the nanoscientist's toolbox. They can be loosely divided into two categories. With any method in the first category, one organizes atoms into nanostructures by a careful control of the chem-ical components and environment, and then assembles the building blocks into the final material. It may be referred to as a *chemical* or *bottom-up* approach. In the second category, one relies on some non-chemical force in a significant way to sculpt the functionality of a suitable starting material. It is a *physical* or *top-down* approach. In practice, one often applies a combination of techniques from either or both approaches, and may strive at integrating both steps, the synthesis

---

[2]Besides the STM, other popular SPMs are: the atomic force microscope (AFM), which makes use of the atomic force between tip and sample; the magnetic field microscope (MFM), in which a magnetic tip is used to sense the magnetic domains of the specimen; and the near-field scanning optical microscope (NSOM), in which an optical fiber funnels a light beam onto the specimen, and the locally emitted light yields a spectroscopic image of the sample. Other microscopes with atomic-scale resolutions, but operating on more conventional principles, are the transmission X-ray microscope (STXM) and the transmission and scanning electron microscopes (TEM, SEM).

of nanostructures and their assembly into materials, into a single process.[3] A few of the methods may also be rightly considered *biomimetic* in the sense that they seek to imitate some process in nature, for example in its use of seeds, templates or self-assembly for nucleation and growth. A completely different kind of tool is the *computer*, which now pervades all disciplines. But it is more than just a tool to analyze data or test models; it has become an integral approach to problem solving: it can predict outcomes for experiments too difficult to carry out, test structural designs, or do molecular simulations for materials.

The third factor contributing to the rise of nanoscience is a better understanding of the precise way the macroscopic properties and the functionality of materials and structures are related to the size and arrangement of their components. For example, it is crucial that we know exactly how the size and location of metallic or semiconducting nanocrystals in a material affect its optical and electronic properties if we intend to implant atoms into appropriately sized dots at precisely controlled sites in order to obtain a product with the desired attributes. The new, finely engineered material may lead us to observe novel properties that we would not have otherwise suspected, which in turn may point us to other inventions. It is this kind of cross-fertilization of the fundamental and the applied that makes nanoscience and nanotechnology such an exciting and fast-moving field.

5.2.1 Summary

The recent rise of nanoscience stems from the confluence of three important new scientific and technological developments: novel characterization tools (microscopes capable of atomic-scale resolution), advanced fabrication techniques (deriving from physics, chemistry and biology) and a better understanding of the basic science (aided by mathematics and computing science).

5.3 Confined Systems

In order to gain some understanding of nanostructures, we will find it fruitful to regard them as *confined systems*, that is, systems where the motion of the relevant microscopic degrees of freedom is restricted from exploring the full three-dimensional space. The confining spatial dimensions need not and, in fact, cannot vanish, because of the quantum uncertainty principle (Problem 5.1). An electron or a photon in a semiconducting film with a thickness of the order of the particle's characteristic wavelength is practically confined to a plane, having no freedom of motion in the transverse direction. Particles confined in an ultrathin wire are free

---

[3]Examples of *chemical* methods include chemical reduction, thermal decomposition, electrochemical processes, use of crystalline hosts, molecular-cluster seeding, and self-assembly. Among the *physical* or *physico-chemical* methods, we have lithography, ball-milling, scanning-probe methods and gas-phase condensation. See the Glossary for the definitions of some of these terms.

to evolve in only one dimension, whereas those confined in a nanosized box are constrained in all three directions.

5.3.1 Quantum Effects

Consider a piece of crystalline solid: it consists of a very large number (like 10^{21}) of atoms sitting very close to one another in regular arrays. Each discrete energy level for the electrons in an isolated atom now broadens, by interactions with all other electrons and atoms in the solid, into a band of some 10^{21} closely packed energy levels. The bands become wider with increasing energy and are usually separated from one another by gaps (the remnants of the original atomic energy spacings), although in some materials a pair of adjacent bands may overlap. An electron may have an energy that lies in one of the bands but not in any of the gaps. The occupied band highest in energy is called the *valence band* (and the top filled level is the *Fermi level*), and just above it lies the *conduction band*, which contains allowed but very sparsely occupied levels. The relative locations and the degrees of occupation of these two particular bands determine many of the physical properties of the material. When an electron has an energy near the bottom of the conduction band or near the top of the valence band, it is a good approximation in the absence of applied fields to regard it as a *free* particle propagating with an effective mass, m^*, smaller than its actual mass, *i.e.*, with less inertia than it does in vacuum. As the parameter m^* takes different values with different bands, we have two distinct continuous energy spectra associated with the valence and the conduction electrons, just as experimentally observed.

Now, what happens when space shrinks, say, to an interval d of a few nanometers in the z direction? The electrons move freely as before in the xy plane, but are now confined in the third direction between two energy barriers separated by d, forming what is called a *potential well*. (The conduction and valence electrons evolve in separate potentials.) If we assume the barriers to be infinitely high, the wave functions that represent the particles in the z direction must fit snugly within those confines, vanishing at the interfaces, and hence can only have an integral number of half-wavelengths across the width of the *quantum well*. It follows that the only allowed de Broglie wavelengths are $\lambda = 2d/n$, where $n = 1, 2, 3, \ldots$. So the motion of an electron across the layer is *quantized*, with allowed momenta $p_z = h/\lambda = nh/2d$ and corresponding energies $E_n = p_z^2/2m^* = n^2h^2/8m^*d^2$, where h is Planck's constant. As the energy for the in-plane motion remains essentially the same as in bulk volume, the total energy of an electron in the thin film is given by

$$E(n, p_\parallel) = E_n + \frac{p_\parallel^2}{2m^*},$$

where $p_\parallel$ denotes the in-plane component of the momentum. This energy spectrum consists of a series of overlapping continuous bands (technically, *subbands*, *i.e.*, small

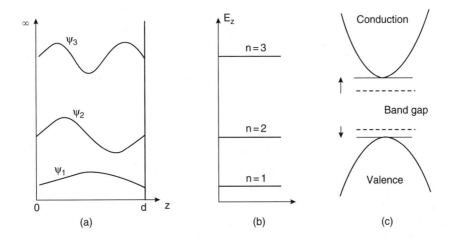

Figure 5.2:   Quantum states in infinite square well. (a) and (b) Wave functions and energies for the first three states. (c) Valence and conduction bands, separated by an energy gap.

bands within broader bands), *each built upon a discrete energy level.* It differs markedly from the completely continuous spectrum $E = p^2/2m^*$ for an electron in a bulk sample (Fig. 5.2). The minimum allowed energy, $E(1,0) = E_1$, does not go to zero (and even increases as d is reduced), in sharp contrast with the unconfined case. The non-vanishing minimum energies of electrons and holes in confined systems raise the conduction-band lower edge a bit, and lower the valence-band upper edge a bit, thus widening the band gap, a result loaded with important physical consequences.

If the number of electrons per unit area is low enough (like 10^{11} per cm^2), they will fill up the lowest subband at zero Kelvin and stay there, provided the thermal energy, kT (the product of the Boltzmann constant k and absolute temperature T), is small enough, *i.e.*, less than the energy spacing of the first two subbands, $E_2 - E_1 = 3h^2/8m^*d^2$. For a range of low energies, the motion in the transverse direction is then described by the same wave function and contributes the same energy; it gets frozen out of the particle dynamics. The electrons have effectively reached the limit of a two-dimensional system and, for this reason, are said to form a *two-dimensional electron system* (2DES). Take, for example, gallium arsenide (GaAs), a common semiconductor. Here, the conduction electron has an effective mass of about 7 percent of the actual mass. For the energy needed for transition to the first excited level to be much greater than the thermal energy at room temperatures,[4] d must be far less than 25 nm. This condition is amply satisfied for a GaAs layer 10 nm thick, since the spacing $E_2 - E_1 = 170$ meV (milli-electron-volt) is too high for thermal excitations (Problems 5.2–5.4).

The effects of confinement on a system become even more transparent when we look at its *density of states* (DOS), which is the number of its possible states

[4]A temperature of $T = 300$ K (or 27 C) is equivalent to thermal energy $kT = 26$ meV.

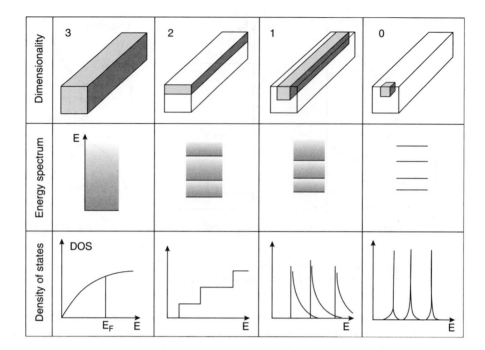

Figure 5.3:   Energy spectra and densities of states of systems with dimensionalities 3, 2, 1 and 0. The symbol E_F stands for the Fermi energy, the energy of the top occupied level.

found at some energy per unit 'volume' per unit energy. It turns out that in three dimensions the density of states starts from zero at zero energy and increases slowly and smoothly with energy; by contrast, in the two-dimensional (2D) case it increases in equal steps, staircase fashion, rising at the successive quantized energies of the confined motion (Fig. 5.3). The DOS is a key factor in calculating transition rates. As it is non-vanishing in 2D even at the bottom level, many dynamical phenomena, such as scattering, optical absorption and gain, remain finite at low energies and low temperatures.

In the preceding discussion, we have assumed ideal confinement, *i.e.*, the walls of the potential to be infinitely high. When they are finite, the results change little, at least qualitatively, for the lowest levels; but whatever changes there are, they may prove significant in certain situations. Then, the particle wave functions do not vanish at the interfaces, but rather continue on and penetrate the barriers, diminishing rapidly in an exponential decay, exactly the effects exploited in scanning tunneling microscopy. The confinement energies decrease with the heights of the potential walls, and the lower the barrier is or the higher the energy level, the deeper the penetration. And, if the barrier walls are thin enough, the wave functions may even go all the way through the walls, providing a mechanism for electrons and holes to escape from the well. Quantum wells and quantum barriers can be combined to build interesting quantum structures and devices.

Quantum wire is the generic name for quasi-one-dimensional (1D) systems with one direction for free motion and two directions of confinement. Its energy spectrum and DOS are qualitatively similar to those for quantum wells, with more subbands in the spectrum and a sawtooth profile for the DOS. The situation changes substantially for *quantum dots*, the quasi-zero-dimensional (0D) structures whose sizes are of the quantum scale all around. Here, the energy spectrum consists entirely of discrete levels, like rungs on a ladder, and the DOS of discrete sharp peaks, like posts in a fence (Fig. 5.3). In general, as the dimensionality is reduced in steps from three to zero, the energies become more and more sharply defined and the densities of states progressively more singular, more condensed at particular energies.

5.3.2 How to Make them

The most common method of making ultrathin films of crystalline solids, usually the first step in the fabrication of many lower-dimensional systems, is a high-vacuum evaporation technique called *molecular beam epitaxy* (MBE). Think of it as spray-painting atoms onto a flat surface. In an ultrahigh-vacuum chamber, place a heated millimeter-thick crystalline (*e.g.*, GaAs, InP) layer; spray it with jets of gases of the substances (Si, Ga, As, In, etc.) you want to deposit. The substrate provides a template for the arriving atoms and mechanical support for the final structure; the high vacuum eliminates unwanted impurities; and the use of regulated jets provides flexibility in the choice of the materials and control of the composition and thickness of the film. In this way, you can produce single or multiple (semiconducting, as well as insulating or conducting) layers of atoms of precise thicknesses and compositions.

To complete the construction of a semiconductor nanostructure in this method, it is necessary to follow growth with *lithography*. This advanced etching technique, similar to that applied in the fabrication of state-of-the-art integrated circuits, is used to transfer a desired pattern from a master copy (mask) into a protective polymer layer (resist) coating the semiconductor surface. After the resist has been irradiated by focused X-rays, electron or ion beams through the mask, its weakened exposed regions are washed off with solvent, and the semiconductor is etched away where it has now been exposed by the removed resist. The patterned semiconductor structure can be further engineered (doped, metallized or etched further) for targeted applications. Insulating tunnel barriers and metallic electrodes, or 'gates,' are often integral parts of the structure and fabricated with the same techniques.

Take, for example, the growth of a layer of aluminum-gallium arsenide ($Al_cGa_{1-c}As$) on a gallium arsenide (GaAs) substrate. This combination of two lattice-matched semiconductors, having nearly identical atom-to-atom spacing, together with a careful crystal growth will guarantee a defect-free, stress-free, and hence high-quality interface. The alloy $Al_cGa_{1-c}As$ consists of periodic arrays of arsenic atoms together with a fraction c of aluminum and $1 - c$ of gallium. The aluminum fraction, controlled during growth, determines the energy band gap of the

alloy as well as the shape of the confining potential. Keeping it constant throughout the deposition leads to an abrupt interface and hence a square energy barrier,
whereas varying it spatially during the fabrication process produces a compositionally graded material and hence a z-dependent potential well.

A quantum-well structure is built by sandwiching a thin (say 10 nm) layer of
GaAs between two thicker layers of $Al_cGa_{1-c}As$ (Fig. 5.4). These two lattice-
matched semiconductors differ slightly in their energy band gaps and so also in
the energies of their free electrons. An aluminum concentration around 30% gives
AlGaAs a band gap larger than that found in GaAs by 360 meV, split roughly 2:1
between the conduction and valence bands, so that electrons see a 240 meV energy
barrier and holes see 120 meV. With potentials that deep, quantum size effects are
easily observable even at room temperatures.

There are no free electrons that move about in pure semiconductors at low
temperatures, all of them being consumed by the bonds that hold the solid together.
However, by adding during the crystal growth a small number of silicon atoms in
the AlGaAs compound at a distance of about 0.1 μm from the interface (a process
known as *modulation doping*), the outer-shell electrons of the Si atoms escape into
the GaAs layer, which has lower-energy electron states, and move freely about in
this highly pure crystal, unimpeded by their then-ionized parent impurities, which
remain behind in the AlGaAs layer on the other side of the barrier. The dipolar

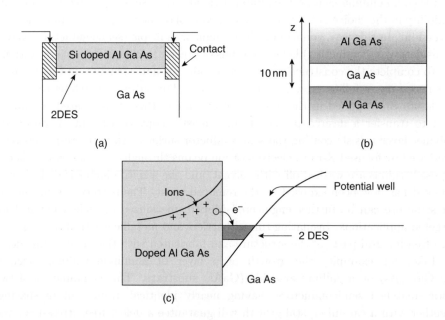

Figure 5.4: (a) Modulation-doped heterojunction. (b) Quantum well in a three-layer double-
heterostructure. (c) Formation of a 2DES in the potential well at the Si:AlGaAs/GaAs interface.

charge distribution formed from the free electrons and the stationary ions creates an electric field that pulls those electrons against the AlGaAs barrier of the interface. The charge carriers effectively lose their freedom of motion in the z direction, being quantum-mechanically bound within the confines of the GaAs film, but remain highly mobile in the xy plane. The mean free path (a measure of this mobility) of such an electron in today's modulation-doped GaAs/AlGaAs heterostructures reaches 0.2 mm at low temperatures, meaning that it flies past one million atoms of the semiconductor before a collision takes place — a *ballistic flight*.

5.3.3 Summary

In order to gain a physical understanding of many of the nanostructures, it is fruitful to view them as confined systems, that is, systems that are reduced to nanometric dimensions in one or several directions. The motion of a particle in these directions is then quantized, leading to effects observable on the macroscopic scale.

5.4 . Quantum Devices

These are the devices that exploit the quantum properties of confined systems. They are built on the basic concepts of the quantum well, quantum wire and quantum dot.

5.4.1 Quantum Wells

Quantum wells have found widespread use in light-emitting diode (LED) and laser diode applications (*e.g.*, displays, compact-disc players, communication systems). While sharing the same double-heterostructure design and operation principles with conventional semiconductor devices,[5] they have several advantages over the latter, all due to the nanometric size of their active region, where electrons and holes recombine to emit light. First, as the light from such devices is emitted with a wavelength determined by the energy of the effective (*i.e.*, broadened by confinement) band gap and as this quantity varies with the chemical composition of the materials and the thickness of the active layer, quantum-well devices can be designed to reach new wavelengths. Secondly, the close proximity of the electrons and holes in the active region implies a vastly improved recombination rate and hence a higher optical gain. And the concentration of states at sharp energy levels gives rise to a narrower gain spectrum, with considerably fewer unused excitations. This means smaller internal losses and far less excess heat to be extracted from the laser die. Add to these factors the electron's high mobility in a modulation-doped semiconductor, and you have devices with much better electrical-to-optical power efficiency functioning at a far lower current threshold.

---

[5] Refer to Chapter 2 for elements of photonics and lasers.

Much of the initial research and development on quantum-well devices concentrated on the GaAs/AlGaAs and GaAs/AlInGaP combinations which operate efficiently in the red–yellow optical region. However, the versatility of MBE and other growth techniques allows us to explore other materials and engage in *band-structure engineering*, where we tailor artificial electronic states and energy bands to our design by programming the growth sequence of the MBE machine. An interesting example is the long search for suitable *blue-light emitting materials*, which has led eventually to the successful development of efficient devices based on gallium nitrides. The new light-emitting diode is a structure consisting of a 3 nm thick layer of indium-gallium nitride ($In_{0.2}Ga_{0.8}N$) sandwiched between a p-type (hole-rich) layer of aluminum-gallium nitride (AlGaN) and an n-type (electron-rich) layer of gallium nitride (GaN), all grown on a sapphire substrate. The InGaN film forms a quantum well into which are fed electrons and holes by the surrounding materials. The charge carriers go into their respective bottom ($n = 1$) subbands, and recombine to emit blue and green light. By combining red-, blue- and green-light emitting diodes of comparable power and brightness, it is possible to make full-color displays and efficient white-light sources that work longer, use less energy and, best of all, can be coupled to computers for novel applications.

While the InGaN LED consists of a single quantum well, the corresponding laser is a *superlattice*, or system of 2–10 coupled quantum wells formed by alternating thin layers of InGaN and AlGaN or GaN. As with most other absorption devices based on quantum wells, it is necessary to couple several wells together in order to boost the absorption rate to a sizeable level. This type of laser can operate continuously at a single wavelength in the 390–500 nm range at room temperatures. These short wavelengths should give decisive advantages to many applications. For instance, as CDs and DVDs now on the market rely on red semiconductor lasers to read the stored information, their data packing density is always limited by the size of the focused laser spot, but it can be increased when shorter wavelengths become more widely available. The data storage capacity of single-layer DVDs, now at 4.7 Gbytes, could be improved at least threefold by moving to blue light, putting high-definition DVDs within the general consumer's reach.

The presence of subbands in quantum wells gives us the possibility of having *intersubband* optical transitions when a conduction electron, excited into an upper subband, relaxes to a lower subband and emits a photon with an energy roughly equal to the energy difference between the two subbands. We expect the photon emitted to be in the mid-to-far-infrared spectrum, which includes wavelengths to which the atmosphere is transparent. A device based on intersubband transitions would differ in a fundamental way from conventional LEDs, because it would rely on only one type of carrier (electrons) and on transitions between energy levels arising from quantum confinement (and not transitions across a band gap).

Such a novel device has been developed by Frederico Capasso and his collaborators at AT&T Bell Laboratories in the US. This beautiful piece of technology,

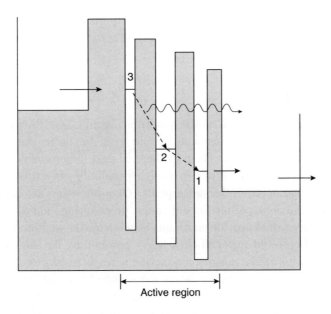

Figure 5.5:   An active region of a quantum-cascade laser.

which they called a *quantum-cascade laser*, is a closely coordinated complex of 500 perfectly calibrated layers of GaInAs (gallium-indium arsenide) and AlInAs (aluminum-indium arsenide) grown by MBE (Fig. 5.5). The central region consists of 25 identical stages, acting like steps in an energy staircase, and each stage is in turn made up of an optically active region and an injection–relaxation region. Each active region consists of three quantum wells, which are nanometer-thick layers of GaInAs flanked by insulating layers of AlInAs. The wells differ slightly in thickness and hence in energy for corresponding levels. When an appropriate electric field is applied, electrons are injected by tunneling into the $n = 3$ excited state of the first well, where they stay for a relatively long time (like 4 ps), before tunneling through an AlInAs layer into the second well, dropping into the $n = 2$ level and emitting an infrared photon (of about 300 meV). The electrons then drop very quickly (after, say, 0.5 ps) to the $n = 1$ subband of the third well (releasing a quantum of heat of about 30 meV), from which they escape equally quickly into the injection–relaxation region, before being re-injected into the $n = 3$ state of the next well to start another stage all over again, and so on down the cascade. Thus, for each electron injected, 25 photons will be produced over the whole cycle. Population inversion between the two upper states responsible for the lasing action is ensured by the long relaxation time of the uppermost state and the extremely short tunneling escape times out of the two lower states. Each injection–relaxation region is a compositionally graded AlInAs–GaInAs superlattice designed so that the energy levels between the steps are arranged in a descending order; its role is to convey the electrons from one active region to the next. All this is achieved by a superb band-structure engineering that allows the designer to control the emitted light by varying the thicknesses and spacings of the quantum wells.

5.4.2 Quantum Wires

If two-dimensional devices, such as the quantum-cascade laser, are promising, then similar structures in fewer dimensions should even be better for the enhanced speed and quantum effects that their further-reduced dimensionality must imply. Electrons scatter even less and attain even higher mobility when traveling through a one-dimensional channel than through a plane. *Quantum wires* act in effect as wave guides for charge carriers, permitting only a few propagating modes. Endowed with such a quality, they would provide enormous benefits for the photonics and electronics industries. They could make efficient lasers powered by far less current, offer a more economical alternative to superconducting wires as resistance-free conductors, or serve as tiny sensors designed to detect the slightest trace of chemicals.

The ballistic propagation of particles in a quantum wire shows up most clearly in *the quantization of the wire conductance* (or inverse of resistance). Electrons that enter empty states at one end of the channel, to which we have applied a small voltage V, pass right through it without loss or gain, provided that the wire length is much shorter than the electron's mean free path. As the current flowing along the channel is independent of which subbands the electrons occupy, it must be proportional to the number of occupied subbands and the applied potential. The proportionality factor turns out to be a universal constant (the conductance quantum), $G_0 = 2e^2/h$, where e is the unit of electric charge, and 2 the number of spin states for each electron. So, the conductance increases in steps of G_0 as the wire

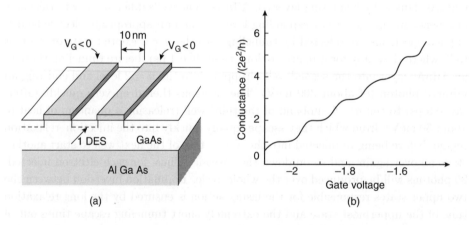

Figure 5.6: One-dimensional quantum system. (a) Quantum wire consisting of electrons confined by an electrostatic potential. (b) Conductance of quantum wire as function of voltage, showing plateaus at multiples of $2e^2/h$.

is made wider, thereby accommodating successive higher-order wave guide modes (Fig. 5.6). Recent experiments indicate, however, a more complex situation that involves the electron spin in some decisive way, leading to an additional conductance structure at $0.7G_0$ (Problem 5.5).

5.4.3 Quantum Dots

Quantum dots, the ultimate in confinement, are solid-state structures that contain mobile electrons in a 'box,' or a very small conducting region surrounded on all sides either by insulating materials or by electric fields which produce the potential needed for confining the electrons. If the confinement is robust enough and the size of the box small enough, quantum effects can be observed macroscopically. Like natural atoms, quantum dots contain a countable number of electrons (ranging from a few to a few hundred thousands) and display a spectrum of sharp energy levels, leading to the frequent reference to them as *'artificial atoms.'* The growing interest in them is of a fundamental nature as well as strongly driven by potential applications where their small size and discrete energy spectrum would be beneficial.

However, the two types of electronic systems also differ on several significant points. First, while in real atoms confinement of electrons is caused by the electrostatic force of the nucleus, in artificial atoms it is accomplished by material boundaries or electrodes whose shape and strength control the shape, size and symmetry of the confining potential. Quantum dots have been created and studied, with or without symmetry and in various geometries, including rods, pancakes and spheres. Secondly, the relevant length and energy scales are not the same, with the typical size and Fermi energy in naturally occurring metal clusters being around 0.5 nm and 5 eV, whereas the corresponding values in semiconductor nanostructures are 50 nm and 10 meV, respectively. It means that both the thermal motion and the electron–electron interaction play a relatively more important role in the much larger artificial atom than in the smaller real atom (Problems 5.6 and 5.7).

One possible way to construct a quantum box is to surround a small region of a two-dimensional electron system with lateral walls. To implement this idea, a crystal wafer is grown layer by layer, starting from the bottom (Fig. 5.7). One begins with a heavily doped crystal of GaAs, which serves as a metal gate, over which one grows by MBE a layer of AlGaAs, thin enough to let electrons leak through it. Next, one grows a layer of pure GaAs, which is where the electrons will accumulate and form a 2DES when a positive voltage V_G is applied to the gate. Finally, a pair of tiny longish specially shaped metal electrodes, which look like two gaping jaws baring their teeth, are deposited on top of the crystal wafer. In addition, two contacts (called *source* and *drain*) are placed at the ends of the electrodes to allow electrons to enter or leave the system. The metal plates are negatively biased and so repel the electrons in the 2DES lying directly underneath, thereby erecting potential barriers that separate a shallow pool of charges within

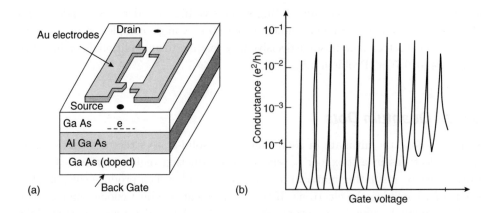

Figure 5.7: Zero-dimensional confined system. (a) Schematic of a quantum dot. (b) Conductance of a quantum dot a as function of the gate voltage, showing sharp peaks at equal intervals (adapted from Marc Kastner, 1993).

the perimeter formed by the electrodes and the split gates from all the electrons left outside. This quasi-zero-dimensional pool of electron gas is our quantum dot, also called *Coulomb island*. For a sufficiently large negative bias voltage, each constriction (or gap in the potential barrier under a pair of opposing metal teeth) is completely closed off, and electrons must tunnel through it. But when the voltage is made less negative, the constriction begins to open up, letting a current pass freely through.

To probe this little box, you may imagine changing its charge content ever so slightly. What work does it take, you'd ask, to add the smallest possible speck of charge? Assuming there is no charge in the box and no applied voltage to begin with, you need an energy $Q^2/2C$ to add a charge Q to a dot with total capacitance C (which is proportional to the island's diameter and of the order of 10^{-16} farad). Since you cannot add less than an electron, a transfer of charge onto the island costs at least $e^2/2C$ in energy: charge quantization requires an electron to have an energy $e^2/2C$ above the Fermi level in the contact in order to hop onto the island, and a hole to have an energy lower than the Fermi level by the same amount in order to cross the tunnel. So you see, in the absence of an applied voltage, no current can flow through the system as long as the thermal energy stays below the minimum charging energy e^2/C, a phenomenon known as the *Coulomb blockade*.

However, when we apply a voltage V_G at the bottom plate, the attractive interaction between the negative charge Q and the positively charged gate must be added to the repulsive interaction among the bits of charge on the dot to give the total electrostatic (charging) energy:

$$E = QV_G + \frac{Q^2}{2C}.$$

This equation tells us that, as a function of Q, the energy E has a minimum at induced charge $Q_0 = -CV_G$ or, equivalently, at gate voltage $V_G = -Q_0/C$. When $Q_0 = -Ne$, or $V_G = Ne/C$, an integral number N of electrons minimizes E, but the Coulomb interaction still demands an amount of energy $e^2/2C$ to either add or remove one electron. Given the control we have over the applied voltage, we may wonder if, near this optimum number N, we can find a value of V_G for which it would cost us less (and hopefully nothing) to add or remove a charge quantum. As V_G varies continuously, it may be written in the form $V_G = (N + \delta)e/C$, where δ is a number between 0 and 1. We will write the energy as $E(N)$ to emphasize its dependence on the charge number $N = -Q/e$, and readily show that the states with N and $N+1$ electrons differ in energy by $E(N+1) - E(N) = (1 - 2\delta)e^2/2C$, a quantity independent of N. We see that as δ is varied from 0 to 1, this energy difference goes from being positive to being negative in passing by zero at $\delta = 0.5$. In other words, if the dot finds it energetically more favorable to have N electrons at a voltage below $V_G = (N + 0.5)e/C$, then for higher voltages it would prefer to have one additional electron. However, *when the voltage is set precisely at that value, the two configurations have equal energies*, and the charge fluctuates between $-Ne$ and $-(N + 1)e$ even at zero temperature. This fluctuation occurs when an electron tunnels onto the dot from the left lead, and later another tunnels off the dot to the right lead (Problem 5.8).

When a very small, fixed voltage is applied between the source and drain, just large enough to coax the electrons to move along, we can measure the tunneling current through the quantum dot as the gate voltage V_G is varied. Sharp current peaks appear periodically at gate voltage $V_G = (N + 0.5)e/C$ and every time the voltage increases by steps of e/C, which is the amount necessary to add one more electron to the confined pool of electrons. So, increasing the gate voltage of an artificial atom is equivalent to moving through the periodic table of real atoms. While a range of voltages produces no current at all, a sharp spectacular current spike appears when the voltage reaches the next critical value, and we know then that another electron has just hopped into the pool. Whereas thousands of charge carriers must flow (and produce a large amount of heat) each time a traditional semiconductor transistor flips, it suffices a single electron to turn a quantum dot on and off again. That is why some people consider, justifiably, a quantum dot a *single-electron transistor* (SET).

This unique transport property of the single-electron transistor is the basis of operation for several devices, such as electrometers and electron turnstiles, which require extreme charge sensitivity. Even more promising in terms of new physics and new applications are *artificial molecules*, consisting of two or more closely spaced quantum dots, and *artificial solids*, made of large arrays of linked quantum dots. The exciting feature here is that the researchers can control the strengths of interactions between neighboring artificial atoms by varying at will the voltages on electrodes, and thereby learn to engineer desired properties into artificial matter.

5.4.4 Summary

Starting from the basic concepts of the quantum well, quantum wire and quantum dot, researchers have made use of the control they have over the fabrication process — the parameters of confinement, the choice of the primary materials, and the composition and architecture of the structures — to design novel or improved devices. The quantum-cascade laser and the single-electron transistor are good examples of the photonic or electronic devices that exploit judiciously the unique electronic and transport properties of systems having lesser dimensions.

5.5 The Genius of Carbon

Nature endows carbon with a wonderful versatility: by exploiting its bonding to the full, it can build structures in all dimensionalities. Diamond is a three-dimensional lattice, graphite is a stack of two-dimensional layers, and other carbon structures are found so small that they may be considered one- or zero-dimensional.

5.5.1 Carbon Fullerenes

The extraordinary all-carbon molecule C_{60} has become one of the best known nanostructures. An interesting story behind its discovery, a pretty name and an alluring shape also help.

The experiments that led to its discovery in 1985 were aimed at simulating in laboratory the conditions under which carbon atoms cluster in the atmosphere of a carbon-rich red giant star and, specifically, at exploring the possibility that long carbon chain molecules could form when carbon nucleates in the presence of hydrogen and nitrogen. In these experiments, an intense pulsed laser beam, focused on a rotating graphite disk, vaporized the material, and the resulting carbon vapor was entrained in a powerful stream of helium, which, being a chemically inert gas, cooled the vapor so that it could condense into small clusters. As the carrier gas expanded through a nozzle into a vacuum, it produced a jet of cold clusters whose masses could be measured by a mass spectrometer. Reactive gases such as hydrogen or nitrogen could also be added to the carrier gas, and the products of the reactions of these gases with the carbon clusters could be similarly analyzed. The experiments showed that species such as HC_7N and HC_9N could indeed be produced, but, unexpectedly, a host of even-numbered clusters with 38–120 carbon atoms were also generated in a roughly Gaussian mass distribution strongly dominated by C_{60} (Fig. 5.8). The leaders of the experiments, Robert Curl, Harold Kroto and Richard Smalley, realized that the unusually high stability of C_{60} could be explained by a molecular structure having the perfect symmetry of a soccer ball, and called the molecule *buckminsterfullerene*, or *buckyball* for short, because its shape was reminiscent of the geodesic domes popularized by architect R. Buckminster Fuller.

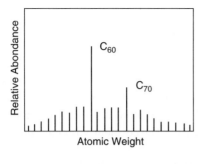
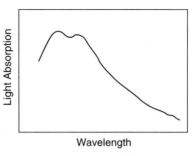

Figure 5.8: Crucial graphs in the discovery of C_{60}. In 1985 the cluster-beam generator showed the presence of many even-numbered carbon clusters, especially C_{60}. In 1990 Krätschmer and Huffman showed that this humped ultraviolet absorption spectrum revealed the presence of C_{60}.

Although they did not have direct evidence to support their conjecture, subsequent experiments in spectroscopy and crystallography have proven them right, which earned them the 1996 Nobel Prize in Chemistry.

The proposed structure for C_{60} looks like a hollow truncated icosahedron obtained from an icosahedron by snipping off each of its twelve vertices. As a result, each vertex of the Platonic polyhedron is replaced by a pentagon, and each triangular face converted into a hexagon (Fig. 5.9). With the sixty atoms evenly distributed among its vertices, the truncated icosahedral structure has exceptional strength and stability (Problem 5.9).

The carbon atom has four valence electrons occupying an outer shell, which is only half full. By combining with other elements which have similar, incomplete shell structures, and by sharing with them one or more pairs of valence electrons (thus creating single, double or triple bonds), it can form a stable molecule. It

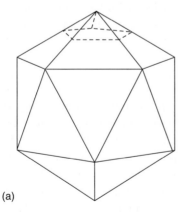
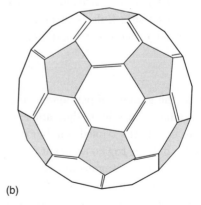

(a)   (b)

Figure 5.9: Structure of C_{60}. When the vertices of a regular icosahedron are snipped off, one obtains a truncated icosahedron, which gives the framework for C_{60}. The carbon atoms are located at the vertices, and the double bonds at the two-hexagon fusions.

can also bond with other carbon atoms in different ways to create structures with entirely different properties. When the four valence electrons of a carbon atom are shared equally with four other carbon atoms sitting at the vertices of a tetrahedron whose center is occupied by the given atom, we have, assuming an infinite lattice, isotropically strong diamond. When only three are shared between neighbors, with the fourth moving freely among atoms, we have graphite, a layered material possessing highly anisotropic properties. Strength, elastic modulus, electric conductivity, and thermal conductivity are much higher within the covalently bonded planes than across them. In bulk solid phase, graphite is thermodynamically stable up to very high temperatures under normal pressure. But in a finite chunk, atoms at the surfaces, having no carbon neighbors, must tie up their unsatisfied (dangling) bonds with hydrogen in the air. Similarly, in a synthesis process where there are fewer than a few hundred carbon atoms available at any given time, the growing structures will seek energetically favorable, finite configurations — linear chains, rings or closed shells — that keep dangling bonds to a minimum.

Once monocyclic rings are formed, they can grow by adding further atoms or small chains, and fold into polycyclic networks, which resemble pieces of hexagonal graphite lattice or fragments of chicken wire. The general physical tendency to reach the lowest-energy level induces the planar sheets to eliminate the unsatisfied valences of the edge atoms by curling up. To curl up, or to produce a convex curved structure, the sheets rearrange their bonding so that five-membered rings are formed, enabling the dangling bonds of some of the edge atoms to tie up into good carbon–carbon bonds. As the process continues, there is a high probability that the graphitic sheets curl until the opposing edges meet to create closed cage-like structures of nanometric dimensions.

A closed cage-like molecule that contains only hexagonal and pentagonal faces is called a *fullerene*. This definition leaves out heptagons and octagons, which are responsible for the concave parts and treated as defects. Like any simple polyhedron, a fullerene satisfies Euler's theorem, which relates the numbers of vertices (carbon atoms) v, edges (covalent bonds) e, and faces f:

$$v - e + f = 2 \,.$$

Let p be the number of pentagons, and $h = f - p$ the number of hexagons. As each vertex radiates into three edges and each edge joins two vertices, the doubled number of edges equals the tripled number of vertices, $2e = 3v$, which is also $5p + 6h$ since each edge belongs to two adjacent faces. A simple calculation yields $p = 12$, meaning that exactly 12 pentagons are required to provide the curvature needed to close a hexagonal lattice into a defect-free fullerene. The addition of each hexagonal face increases the total number of vertices and hence the number of atoms by two, so the total number of carbon atoms in a fullerene must be even, according to $v = 2(h+10)$, which is consistent with the observed mass spectra. The smallest fullerene, C_{20}, corresponds to $h = 0$, but no closed cage can be formed with $h = 1$, and C_{22} has never been observed. Otherwise, any number of hexagons is allowed, and one could imagine an elongated fullerene with exactly 12 pentagons and millions of hexagons. However, not all such constructions are stable. For example, two adjacent pentagons would be energetically unfavorable compared with isolated pentagons, because they would give higher local curvature and hence more strain to the structure. That is why unsaturated organic molecules with contiguous pentagonal rings are rarely found; a chemically stable molecule cannot have adjacent pentagonal rings. The molecules C_{60} and C_{70} are the two lowest-mass carbon fullerenes that satisfy both Euler's 12-pentagonal closure principle and the isolated-pentagon rule and, for this reason, appear prominently in the observed mass distributions of carbon clusters (Problems 5.10 and 5.11).

C_{60} is now produced in quantities large enough to make solids (fullerite) of weakly bound molecules. Pure C_{60} solid is an insulator. When it is doped with alkali atoms, these impurities contribute electrons to the lowest empty band of the fullerite, and the compound thus formed becomes insulating or conducting (*i.e.*, metallic), depending on the degree of electronic occupation. At low temperatures, not only some of these compounds (such as A_3C_{60}, where A = K, Rb, Cs), but also alloys C_{60}-alkaline-earth (*e.g.*, Ba_6C_{60}, Sr_6C_{60}) and C_{60}-rare-earth (*e.g.*, $Yb_{2.75}C_{60}$) are superconducting, with a transition temperature T_c that is surpassed only by the cuprates (Chapter 3). Thus, T_c is 33 K for $RbCs_2C_{60}$ and 40 K for Cs_3C_{60} under pressure. The trend seems to suggest that even higher T_c could be obtained by incorporating larger cations. Other workers, taking a different approach, showed that fullerite doped with holes, rather than electrons, became superconducting at 52 K.

Within days of the discovery of C_{60}, Smalley and his collaborators found that metal atoms could be trapped inside fullerenes by impregnating graphite with the desired metal before exposing it to laser vaporization. A carbon fullerene C_n encaging a metal atom M is called an *endohedral metallofullerene* and written $M@C_n$ or, alternatively, fullerene-*incar*-metal and written iMC_n. All such complexes are of the ship-in-a-bottle type, and many have been synthesized, including radioactive species, which raises the prospect of their applications in materials, biological and medical science. In 1992 Daniel Ugarte announced that under high-energy electron irradiation, nested fullerenes could be generated. These '*graphitic onions*' are of considerable size, with diameters up to 47 nm, and contain thousands, even millions, of carbon atoms in spherical shells separated by spacings of 0.335 nm, just as layers in graphite. As nested structures, they are a special form of multiwalled nanotubes.

5.5.2 *Carbon Nanotubes*

A carbon nanotube is a single molecule of many carbon atoms arranged in a hexagonal network that curls into a long, slender tubule capped at each end, and is comparable in all respects to a very elongated fullerene. It exists in singles, known as *single-wall nanotubes*, or in nested multiples, called *multiwalled nanotubes*. The latter — the first of the two to be discovered, by Sumio Iijima in 1991, in samples created by arc discharges between carbon electrodes immersed in a noble gas — consist of several concentric cylindrical shells of graphitic sheets co-axially arranged around a central hollow with a constant separation between the layers. They have diameters ranging from 2 to 25 nm and lengths up to many micrometers. Single-wall nanotubes are cylinders made of single sheets of graphene (one-atom thick layers of graphite), with diameters distributed in a narrow range (1–2 nm). When produced in the vapor phase, single-wall nanotubes self-assemble into larger bundles (ropes) that consist of tens of nanotubes.

Suppose you want to build an ideal single-wall carbon nanotube. First, take a rectangular sheet of graphene (chicken wire), then roll it up into a cylinder such that the dangling bonds (open wire fragments) at opposing edges match perfectly to form a seamless tube, and, finally, close the open ends of the tube with the hemispheric domes obtained from a fullerene by cutting it evenly in half. Depending on the width of your rectangle, you may roll in one of several directions relative to some fixed row of bonds, so that on the curved surface of the tube the hexagonal arrays of

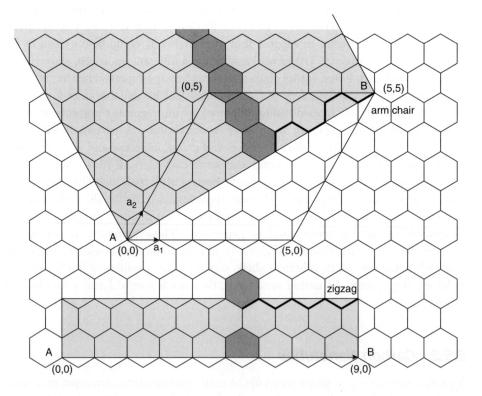

Figure 5.10: Construction of an armchair nanotube $(5,5)$ and a zigzag nanotube $(9,0)$. Colored hexagons indicate the direction of the tube axis. The rolling is along the vector AB.

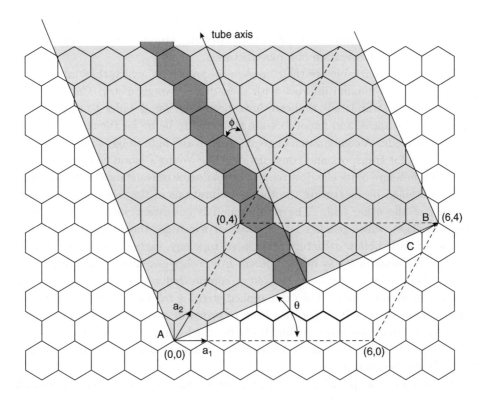

Figure 5.11: Construction of a chiral nanotube $(6, 4)$. Also indicated is the winding angle ϕ defined with respect to the tube axis.

carbon atoms may wind around in a helical fashion with a constant winding angle with respect to the tube axis. To give a more precise description, let us define a vector joining two sites, A and B, on the lattice by a pair of integers n and m, which record the numbers of steps starting from point A and going along two reference axes. This vector defines the wrapping into a tube of circumference AB if after folding along its direction, points $A = (0, 0)$ and $B = (n, m)$ coincide, so that (n, m) contain all information on the structure (diameter and winding angle) of the tube, and hence may serve to characterize it. Thus, the indices $(n, 0)$ designate *zigzag nanotubes*, and (n, n) *armchair nanotubes* (Fig. 5.10), whereas all other possibilities are termed *chiral* or *helical* (Fig. 5.11). Because of their simple and well-defined structure, single-wall nanotubes serve as models for theoretical calculations and key experiments (Problem 5.12).

What makes the carbon nanotubes fascinating is their wide-ranging superior properties, which result from a unique combination of dimension, topology and structure. For example, their light weight (1 g/cm^3) results from a hollow cage-like architecture, their relatively large surface area $(10 \text{ m}^2/\text{g})$ is due to nanometric dimensions, and they owe their unusual capillary behavior to the smooth, straight, one-dimensional channel in their center. All these properties make nanotubes useful as catalysts, components of novel materials, atomic-sized storage systems and templates for fabrication of nanostructures.

Even more remarkable are their *mechanical properties*, which they inherit from their graphitic frame and improve with their distinctive geometry. Carbon fibers, which are long strands of graphene, have been used for decades to strengthen a variety of substances, and nanotubes are predicted to be by far the strongest fibers that can be made. A combination of great strength and light weight make them ideal as materials for transport and construction. The Young's modulus of a nanotube, which is a measure of its elastic strength, or its degree of resistance to deformation, can be obtained by measuring the thermal vibrations of the free end of the tube clamped at the other end. The result found is consistent with the value for a graphene sheet, which reaches 1 terapascal, or five times the value for steel. But, because of their hollow structure and closed topology, carbon nanotubes respond to extreme strains in a way quite different from other graphitic structures. Unlike carbon fibers, they can be bent, twisted or compressed without breaking, and will regain their original shape when the applied stress is released. The reversibility of deformations, such as buckling, has been recorded in electron microscope images, and indicates that nanotubes are highly elastic. Such superior mechanical properties make them promising in reinforcement applications and for use as tips or tools in proximity probe microscopy.

Apart from their special structural attributes, nanotubes possess equally intriguing *electronic properties*. In graphite, there is no band gap between the empty and full states, but there are also very few free electrons (one per 10^4 atoms, compared to one per atom in copper) capable of carrying charges along the graphene sheets. Graphite therefore is not quite a conductor: it is a *semimetal*. In nanotubes, we have the same electronic structure but an entirely different situation.

The differences stem from the fact that the free electrons in a nanotube are confined to the one-dimensional geometry of a thin cylinder. The electrons can propagate freely in only one direction, along the tube, rather than in the two directions that were available in the graphene plane, being constrained in the transverse direction to move around the tube. The periodic boundary condition imposed on the wave function by this confinement means that only a whole number of wavelengths can fit around the tube, and so the electron wave vector around the circumference is quantized. Since this quantization depends on the circumference and winding angle, it follows that the electronic states and energies must depend on the indices n and m of the nanotube.

Calculations predict, and experiments concur, that armchair tubes have valence and conduction bands crossing at the Fermi level and are, therefore, *metallic*. Among the other (chiral and zigzag) tubes, one third (those for which $n - m = 3l$, where l is an integer) are *metallic*, and two thirds (for which $n - m \neq 3l$) *semiconducting* (Problem 5.13).

As with all other confined systems, the tube's energy levels do not spread out into a wide continuous band, but instead group into subbands with band onsets at the discrete energies of the confined degrees of freedom. For single-wall nanotubes, these subbands are well spaced out, which suppresses thermal excitations even at room temperatures. In the metallic type, only two subbands cross the Fermi energy, so that a current through such a tube is carried by only these two subbands. Since each subband can support a quantum conductance of $G_0 = 2e^2/h$, we expect a perfect metallic nanotube to have a conductance of $2G_0$.

Experiments based on scanning-tunneling microscopic and spectroscopic studies have borne out all these predictions about the electronic structure and the strongly one-dimensional character of conduction of carbon nanotubes. These structures behave as true quantum wires. But, oddly enough, they also possess, at low temperatures (like 1 K), features that belong more naturally to quantum dots. This is so because, although nanotubes are much longer than they are wide (by a factor of 10^4), they still have a finite length, so that the boundary conditions at both ends impose a limit on the number of allowed wave vectors in the longitudinal direction — which means a completely quantized motion. So a nanotube is effectively a zero-dimensional quantum system, with all the interesting consequences for electron transport and conduction — single-electron tunneling, resonant tunneling through molecular orbitals — that such a complete confinement implies.

This wealth of electronic properties is the source of a potential diversified technology. Because they are so thin yet so strong, carbon nanotubes, when stood on end and electrified, emit electrons from their tips at prodigious rates and at lower voltages than any other kind of known electrodes. They have all the desirable attributes of electron guns for electron microscopes, and field emitters for vacuum-tube lamps or flat-panel display. Cees Dekker and others at Delft University of Technology in the Netherlands have further shown that they could build working electronic devices out of carbon nanotubes. For example, they have made a current flow through a semiconducting nanotube lying over two electrodes and switched it on and off by applying voltages to a nearby gate electrode, just as in a transistor. From the intrinsic properties of nanotubes, they expected this switching device to be more sensitive, run faster, and use much less power than a silicon-based transistor.

There is also the exciting possibility of engineering nanotube complexes by cutting, joining, and bending individual tubes. For instance, a junction of two tubes, one metallic and one semiconducting, should behave as a diode, permitting electricity to flow in only one direction, while other combinations of nanotubes with different band gaps could operate like light-emitting diodes and perhaps even nanolasers. And because carbon nanotubes conduct heat as well as diamond or sapphire, and have chemical bonds much stronger than those found in any metal, they can carry awesome amounts of electricity without overheating and vaporizing the wire. All these studies raise the hope that soon one could build tiny circuits entirely with these functionalized molecules.

Scientists have also made brave attempts to coax nanotubes into holding data. In one approach, they set up arrays of perpendicular tubes and inserted spring-like molecules at the junctions where the wires crossed so as to create the on- and off-states necessary for data storage. In another, they used an electric field to bend metallic tubes towards or away from a transverse semiconducting tube, thereby creating the on- and off-states. In short, nanotubes are amazingly versatile molecules that can *conduct, switch electric current* and *store information*.

To build single-molecule devices that can perform functions identical or anal-ogous to those of the conductors, transistors, diodes, memory devices and other key components of today's microcircuits — that is the dream of the proponents of *molecular electronics*. But before that dream can turn into reality, either with nanotubes or some other more exotic molecules, we must gain a deeper under-standing of their quantum behavior, an important aspect of which resides in that quintessential quantum property called *spin*.

5.5.3 Summary

Carbon fullerenes and nanotubes are among the best-known naturally occurring nanostructures. They are appealing to theorists because of their high degree of symmetry, which allows detailed calculations. For the experimentalists, their distinctive hollow structure, unique electronic and mechanical properties offer a rich potential for studying quantum phenomena and developing applications in diverse areas.

5.6 Spintronics

The spin of the electron has its origins rooted deep in quantum mechanics and relativity. Yet, because it imparts an orientation to the charge carrier's magnetic moment, it gives rise to a physically observable phenomenon, magnetism, which is apparent to every schoolchild and which is essential to the functioning of many common appliances. An electron does not spin like a top, but it can nevertheless be described by an intrinsic angular momentum (or spin), which is a permanent characteristic of the particle, as fundamental as charge. The quantization of angular momentum measured in a specified direction implies that the magnetic moment in that direction is also quantized. It confers on the electron two possible spin states, called spin-up and spin-down states. The movement of spin, like the electric current, can carry information among devices, but the existence of two controllable spin states suggests even more: a new kind of binary logic of ones and zeros. As spin — or its alter ego, the magnetic moment — can be readily manipulated by external magnetic fields, we may look forward to the development of a new generation of materials and structures based on the flow of spin in addition to the flow of charge that can perform much more than is possible with today's electron-charge-based microelectronics devices. Some call this young field '*magnetoelectronics*,' others '*spintronics*.'

5.6.1 Spin Flow

As a general rule, the most energetically stable state in an atom is the one in which most or all of the electron spins cancel out in pairs, with the two spins in each pair pointing in opposite directions. The moments that have not got so canceled out could all line up and produce magnetism under the influence of some internal force. The most obvious candidate — the purely magnetic force between the unpaired electrons on neighboring atoms — turns out to be too weak to generate the kind of spin alignment observed in magnetic metals. Quantum theory predicts that there should be a much stronger interaction between neighboring atoms mediated by electrons. This '*exchange*' interaction, strong but short-ranged, depends critically on the relative alignment of the two unpaired magnetic moments. For iron, cobalt, nickel, and the other materials that we qualify as *ferromagnetic*, parallel alignment of neighboring spins is favored, and the configuration of lowest energy in a 'domain' (small region) will be the one with all the spins pointing in the same direction. Other materials, known as *antiferromagnets*, see all neighboring spins in pairwise antiparallel alignment and, therefore, have no overall magnetic moment.

Let us note that magnetism is a cooperative behavior among atoms which derives from a purely quantum-mechanical process effective only over distances of a few atomic spacings. That is why magnetic materials structures are much harder to design and control than semiconductors; they must be manipulated at the length scale of a nanometer or less to have any impact on their behavior, whereas semiconductors can exhibit novel properties already at carrier lengths of tens of nanometers.

Spin is more than magnetism: it can flow along with charge. What are the spin carriers? Where and how do they show up? How do they move about in materials?

The relevant electronic states in ferromagnets, just as in conductors or semiconductors, are those lying close to the Fermi energy, at the top of filled levels, and their densities of states explain to a large extent the transition to ferromagnetism. With spin present, each energy band splits in two, corresponding to the two possible spin orientations. In a *nonmagnetic metal*, such as copper, the spin bands remain lined up at the same energy level, and are equally occupied by the free electrons. So copper has no net magnetic moment, and the conduction electrons at the Fermi level are unpolarized, that is, equally distributed between the two spin orientations (Fig. 5.12). However, in a *ferromagnetic metal*, the spin-up and spin-down states are shifted in energy with respect to one another by the exchange interaction. This shift leads to an *unequal filling* of the spin bands, which is the source of the net magnetic moment for the metal, and also causes the spin-up and spin-down charge carriers to be unequal in number, character and mobility.

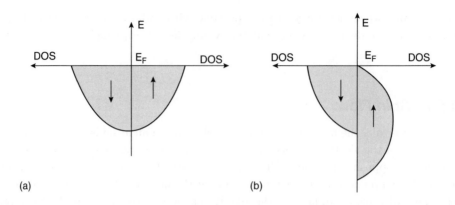

Figure 5.12: Density of states (DOS) that are available to electrons near the Fermi level (E_F) in (a) a normal metal and (b) a ferromagnetic metal whose spin-up states are completely filled.

Let us take cobalt as an example. In the atom, the d-level (so called because its orbital angular momentum equals $2\hbar$) is incompletely filled, and the electrons that normally occupy it are weakly bound. In the metal, the tendency of the system to march towards the lowest-energy level demands that the spin-up (majority) states, shifted down by the exchange splitting, get all filled up first, so that the d-electron states at the Fermi level contain only spin-down (minority) electrons. There arises then an imbalance in spin populations at the Fermi level. The magnetic moment of cobalt is simply proportional to the difference between the occupations of the two spin bands. At the same time, the highly polarized d-electrons, together with a few other electrons with different orbital angular momenta at the Fermi level, produce a partially spin-polarized flow of charge in which the majority of electrons are in a spin-aligned state.

Thus, a ferromagnetic metal may be used as a source of spin-polarized carriers injected into a normal metal, a semiconductor or a superconductor, or made to tunnel through a nonmagnetic insulating barrier. Although Fe, Co, Ni and their alloys are only partially polarized (with about 70% of the carriers in a spin state) and adequate for useful devices, there are continuing efforts to find 100% spin-polarized materials, which would strictly have only one occupied spin band at the Fermi level and, when used as filter, would permit true on–off operation, with an essentially infinite ratio of impedance between the two polarization states.

In the absence of any applied fields, magnetic materials have a complex large-scale texture that results from several competing influences in submicron-sized elements. While the exchange interaction attempts to make the whole of an element magnetized in the same way, the magnetostatic force tries to split it into small domains of independent magnetizations, so as to form closed magnetic loops and keep stray fields to a minimum. However, the situation becomes simpler at the nanometric scale. For instance, in a small needle-shaped *grain*, the magnetization prefers to align along the long axis and occupy the whole element as a single domain; so it can remain highly stable against the influences of external fields. The existence of two stable magnetization states make such structures suitable for data storage applications. *Thin films* with a uniform composition and a proper thickness (like 1 or 2 nm) also have a simple field pattern, with magnetization lying in the plane of the layers rather than perpendicular to it. When made of *soft* magnetic materials, such as permalloy ($Ni_{0.8}Fe_{0.2}$) (whose magnetization is very sensitive to external fields, in contrast to the magnetically *hard* materials), those films have proved to be useful for a wide range of applications in sensors, memory elements and data storage. For this reason, thin film structures will play a central role in our discussion.

Now, let us see how a thin layer of magnetized ferromagnet affects the flow of electrons (Fig. 5.13). The particles that can pass through the film are those having spins oriented in the same direction as the spins of the states available at the Fermi level; all others are reflected back at the surface. If the incident current is *unpolarized*, electrons passing through the magnetized material acquire this same spin bias, and the film acts like a spin polarizer. On the other hand, if the current is completely *polarized*, it will pass through if the spins of the carriers are aligned with those of the atoms in the layer; otherwise, its passage will be seriously hindered. The film operates then like a spin analyzer. So, for a 100%-polarized current, a magnetized ferromagnet can function either as a conductor

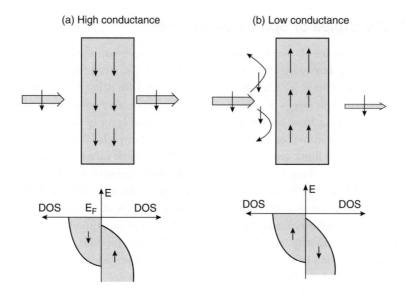

Figure 5.13: Spin-polarized transport through a magnetized ferromagnetic layer. (a) When the current is 100% polarized and its polarization has the same orientation as the magnetization of the layer, the transmission is maximum. (b) When its polarization is opposite to the material magnetization, the conduction is severely reduced. Sketches of the densities of states at the Fermi level are also shown.

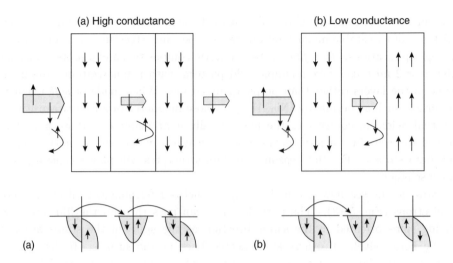

Figure 5.14: Spin-valve action. Spin-polarized transport through a sandwich consisting of layers of ferromagnetic metal, normal metal and ferromagnetic metal (a) when the magnetic moments in the ferromagnetic layers are aligned, and (b) when they are anti-aligned.

or an insulator, depending on whether its direction of magnetization is parallel or antiparallel to the spin polarization of the current.

5.6.2 *Principles of Spintronic Devices*

These basic principles of spin-dependent transport can be applied to a simple structure made of two ferromagnetic (*e.g.*, permalloy or Co) thin films separated by a nonmagnetic metallic (*e.g.*, Cu) spacer layer, whose role it is to stop any magnetic coupling between the layers but not to hinder the spin movement itself (Fig. 5.14). Electrons originating from one spin state at the Fermi level in the first film will be accepted by empty states of the same spin at the Fermi level in the second film. If the two magnetic films are magnetized *parallel* to each other, the minority (*e.g.*, down-spin) electrons from the first film will go into minority unfilled states in the second, and the majority (up-spin) electrons from the first film will seek empty majority states in the second. But if the magnetization of the second film is now reversed so that the two films are magnetized in *opposite* directions, the identity of majority and minority will be reversed in the second film, and the minority (down-spin) electrons from the first film will look for empty majority (down-spin) states in the second, while the majority electrons will try to find empty minority states. Keeping in mind that the rate of any transition at some energy varies with the density of states at that energy, we can see from Fig. 5.14 that the conductance of the system increases (the resistance is reduced) when the magnetizations are in an aligned state, and decreases (the resistance is higher) in the anti-aligned state. We may aptly compare this effect to that seen when light passes through an optical polarizer–analyzer. However, in the optical case, crossing the polarizer axes at 90° prevents light transmission, whereas here minimum conduction is obtained when the moments of the two ferromagnets are rotated 180° away from parallel. (The difference in the rotation angles comes from the different spins of the electron and photon.)

We refer to the change in the electrical resistivity of a material due to the introduction of a magnetic field as *magnetoresistance* (MR), and measure it by the percentage change of the initial resistivity (when the field is absent). Most metals have very small MR. For instance, copper has MR $\sim 1\%$ at a high magnetic field, $H = 10$ T (tesla), or 10^5 gauss (for comparison, the earth's magnetic field is 0.7 gauss). But magnetic-layered structures, such as those described earlier, exhibit large MR $\sim 25\%$ at $H = 50$ gauss and room temperatures, and even larger at higher fields. For this reason, we refer to the effect observed here as *giant magnetoresistance* (GMR). As we have seen, it arises from a difference in the carrier scattering rates when the magnetizations in the adjacent layers change their relative orientations.

The geometry we have considered, in which the current flows perpendicular to the plane of the layers, though conceptually simple, is much less practical (owing to the extremely low resistance across the nanolayers) than the more common arrangement in which the current flows in the layer plane. In this case, we still can observe the suppression of the spin-dependent scattering by the interfaces when the films go from an antialigned state to the aligned state, but the effect is harder to visualize.

This simple metallic trilayer structure is called a *spin valve*. It is constructed so that the moment of a hard magnetic film is held fixed (pinned) by an antiferromagnetic (*e.g.*, FeMn) overlay and serves as a reference, whereas the moment of the other layer can be easily flipped back and forth by an external magnetic field, and so will act as a control valve. We can design the system so that initially, when there is no field, the magnetization in the free layer lies antiparallel to the pinned layer, but when the field is applied, it turns to parallel orientation. This will lead to a decrease in resistance.

Magnetoresistance manifests itself also in another process characteristic of two-dimensional magnetic nanostructures — spin-polarized tunneling. It occurs in trilayer systems, technically *magnetic tunneling junctions* (MTJs), that consist of two (one hard, one soft) ferromagnetic layers separated by an insulating (*e.g.*, aluminum dioxide) film, typically 1 nm thick, which acts as a tunneling barrier. When a voltage is applied, electrons can quantum-mechanically tunnel through the spin-dependent energy barrier, with a probability proportional to the density of states available at the Fermi level in the acceptor layer, so that, when the moments of the two ferromagnetic layers are aligned parallel, there is a lower impedance than when they are antiparallel. The tunneling current, which flows through the junction perpendicular to the plane of the layers, is generally much lower than the current in the all-metal spin valve. However, the high resistance typical of tunneling devices proved to be unattractive in terms of response time and noise, the more so because it increased as the junction dimensions decreased. For this reason, magnetoresistance in MTJs was a small effect at room temperatures for a long time, until better growth techniques were developed for atomically sharp metal–insulator interfaces. Now the observed effect comes close to the theoretical limit, and the resistance change can be as large as 47% in small fields. As they can operate at low current and exhibit high magnetoresitance in weak fields, MTJs would be unbeatable devices if they could be manufactured in large number at an affordable cost.

5.6.3 Magnetic Recording

Both spin valves and magnetic tunnel junctions have been considered for applications, especially in the data storage industry where they are starting to play a meaningful role. The first application was for the read/write heads in magnetic recorders, which are components of every computer. At the heart of a state-of-the-art recording head is a magnetoresistive sensor element, which can pick up weak magnetic signals from magnetized domains (Fig. 5.15). To *write* information on the medium (a magnetically hard material coating on a tape or disk), a tiny electromagnet, also built into the recording device, magnetizes small segments of very narrow tracks on the medium, each magnetized segment containing a unit of data, or bit. Although there is no magnetic field emanating from the interior of a magnetic domain itself, flux lines may extend out of the medium at the 'wall' separating two adjacent domains. Whenever two adjacent domains are magnetized in opposite directions, there are 'transitions' at the wall, meaning that there are field lines extending out of or returning back into the medium, generated by uncompensated magnetic poles. When two bits are written successively in the same direction, there are no net transversal fields and hence no transitions at the domain wall. The presence of a transition indicates a binary 1, just as its absence, a binary 0.

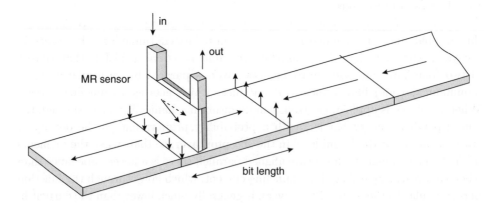

Figure 5.15: Magnetoresistive sensor element reading bits written on a track of a magnetized recording medium.

The magnetoresistive sensor element is fabricated so that the moment in the soft magnetic layer lies parallel to the plane of a disk in the absence of any applied fields, but the moment in the fixed magnetic layer is always perpendicular to and directed towards the disk. When the head passes over the disk, *reading* the data as it goes, the moment at the transitions causes the moment in the soft layer to rotate towards a more anti-aligned or a more aligned direction with respect to the fixed magnetic moment, producing an increase or a decrease in resistance in the sensor element and signaling that a '1' has been read.

Tens of millions of spin-valve read heads have already been made for the most advanced disk drives, and many more will soon come off the production line to replace the more traditional pick-up devices. Not far behind, MTJ read heads, which are capable of delivering greater signals at lower sensed fields, have been moving through the prototype state and will reach the manufacturing stage soon.

5.6.4 Data Storage and Processing by Spin

The second application that is expected to have a large economical impact is found in *nonvolatile memory*. ('Nonvolatile' means permanent, or retaining stored data even when power is switched off.) Magnetic disks provide the most widespread form of nonvolatile information-storage media, because of their low cost and long storage lifetime. The coating of a magnetic disk consists of grains of magnetic nanoparticles, which generally segregate randomly in size, coercitivity and shape. A recording density of 1 Gbit per square inch requires about a thousand grains per bit in today's commercial disks, with typical grain sizes in the 10–20 nm range. There are continuing efforts to increase the storage density — by reducing the grain size and the number of grains per bit, and by engineering new materials — with the ultimate goal of quantized or single-particle-per-bit recording. We are still far from that goal, but tremendous advances have been made in recent years, which have allowed the density to double and the cost per bit to halve every three years (Problems 5.15 and 5.16).

Although the position of hard disks in data storage is not threatened by any means, their usefulness remains limited by the difficulty in a fast retrieval of the stored information. We are rather more interested in examining possible improvements in another form of computer memory, an electronic storage known as *dynamic random access memory* (DRAM), which keeps temporarily most of the useful data, ready for instant access by the computer processing unit during a working session. Similar to a microprocessor, a memory chip is an integrated circuit made of millions of electronic components, organized in pairs. Each pair, consisting of a transistor and a capacitor, is a *memory cell* capable of containing a single bit of data. The *capacitor* holds the bit, a '1' when the capacitor is more than half filled with electrons and a '0' when it is not, whereas the *transistor* acts as a switch that lets the control circuitry on the memory chip interrogate the capacitor or change its state.

Useful as it is, DRAM has a serious drawback: charge leaks continually from the charged capacitors. It means that they have to be refilled before they discharge, an operation repeated thousands of times per second for every register, at a great cost in time and power, and generating a large amount of heat. It also means that every time you turn your computer on, its basic input/output system takes several seconds to perform a series of functions to check every memory address and transfer the basic operating codes from the hard disk to live memory.

Do you want to avoid that tedium of boot-up and inefficiency of refresh? Then MRAM (*magnetic random access memory*) is the answer to your wish because magnetism rather than electricity would be used to store data; and it would be more economical to operate because it requires just a small amount of electricity to switch the polarity of each memory cell. The idea of making a *magnetic memory cell* is rather simple: a layered structure, whether a spin valve or an MTJ, will be used to store the information, replacing the capacitor in current designs. In one scheme, the free layer is switched relative to the pinned layer to store a '0' or a '1', and measuring the resistance of the element indicates its magnetic state and hence the bit it holds. In another scheme, high-current pulses are used to write bits in the magnetically hard layer. To read the information, lower-current pulses are used to flip the magnetization in the soft layer, first one way and then the other. Comparing the change in measured resistance in each direction reveals the orientation of the magnetization in the hard layer and hence the stored data.

Each memory circuit consists of an array of multilayered elements fabricated with standard lithographic processes. In the chip based on spin valves, these elements are arranged in series and electrically connected by lithographic wires to form a 'sense' line, which stores the information. Additional current lines above and below the sense lines cross at right angles in an xy grid pattern, intersecting at each of the spin valves. These currents produce the magnetic fields needed to manipulate the magnetization of the data-storing elements (Fig. 5.16). A similar

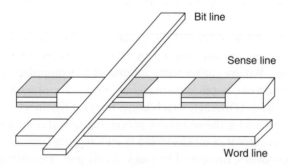

Figure 5.16: Schematic representation of a data storage scheme, in which the memory cells are constructed of GMR elements connected in series along the sense line. The currents passing through the bit and word lines generate the magnetic fields that manipulate the GMR elements for writing and reading.

array architecture may be constructed with magnetic tunnel junctions. But, in this case, as the sense current has to flow perpendicular to the layers and so can proceed through many elements and not just the one at the intersection, a diode is placed at every intersection so that the current is forced through the desired path and only in one direction.

While the development of MRAM chips is racing ahead, spurred on by a huge global market, there have been several other directions of research in materials and spin transport that suggest possible applications. The search for useful materials exhibiting enhanced polarization continues apace, and the range of materials studied has significantly increased, including novel ferromagnetic semiconductors, high-temperature superconductors, and carbon nanotubes. It was found that a family of perovskite materials (Chapter 3) exhibit very large magnetoresistance, up to 100 000%, but at very high magnetic fields (6 T), an effect termed '*colossal*' (CMR). Even more interesting is the *extraordinary magnetoresistance* (EMR) exhibited by *non-magnetic* narrow-gap semiconductors with an embedded non-magnetic metallic inhomogeneity. Room temperature EMR in excess of 100% at 0.05 T and 3 million percent at 5 T has been obtained. Researchers have already demonstrated that mesoscopic read-heads made of nonmagnetic silicon-doped indium antimonide, operating on the EMR principle, were sensitive enough to read data at 116 Gbits per square inch, auguring well for the future of terabit-per-square-inch recording media.

Several schemes for spin transistors, in which the flow of spin-aligned electrons is controlled by a magnetic field, have been proposed. The general intent is to search for ways to construct smaller, more rugged multifunctional devices, that not only could function as switches or valves and amplify signals, but would also possess intrinsic memory, and could be seamlessly integrated with traditional electronic technology.

Beyond perfecting existing technology, there is an even more ambitious vision for spintronics of fully exploiting the quantum nature of spin. Central to this picture is the tantalizing possibility of building spin-based quantum computers. The basic idea of quantum computing (to be explained in more detail in Chapter 6) is to exploit the laws of quantum mechanics to process information. Whereas the basic unit of information in a conventional computer is a binary digit, either a '0' or a '1', a quantum computer processes information by quantum bits, or *qubits*, which are representations of arbitrary linear combinations of both values, thus vastly expanding the power of computing. To be useful as carriers of information, the states of many qubits, which represent pieces of data, must be controlled precisely and must remain coherent, or undisturbed by interactions with their environment, for a long time. This makes electron spin an ideal candidate for the qubits. Off hand, spins should have long coherence times because they are unaffected by the long-range electrostatic interactions between charges, which are the most pervasive kind of force in solid-state surroundings. Experiments have verified that it is indeed the case. David Awschalom and co-workers have demonstrated that electron spins could survive in a coherent state for more than 100 nanoseconds, sufficient for the standard operations on a memory chip, and coherently precessing electronic spins could travel without a substantial increase in decoherence (losing information) over distances exceeding 100 μm, comparable to the size of a typical electronic device.

5.6.5 Summary

Traditional electronics is based on the flow of the electron charge. A new field, called spintronics, intends to use both the flow of charge and the flow of the electron spin to develop new materials and devices that can perform more than is possible with today's products. In particular, two magnetic-layered structures — spin valves and magnetic tunnel junctions — exhibit a large magnetoresistive effect, which has been exploited to make sensors, recording heads and memory cells.

5.7 Nanos at Large

In our discussion, we have focused on only three classes of nanostructures, at the exclusion of many other cases of equally great interest. Even in the optoelectronics field, which we have considered in some detail, we left out the important invention of the *photonic band-gap crystals*. These structures are based on periodic variations of the dielectric constant, and can produce many of the same phenomena for photons as an ordinary crystal does for electrons. This is a good example of how scientists can exercise control over the optical properties of materials and, in so doing, engineer materials that reflect light of any polarization incident at any angle, or allow its propagation only in certain directions at certain frequencies, or localize light in specified areas.

Structural materials too can be improved by a control over their make-up at the nanometric level. Where conventionally produced materials tend to be gross and irregular in structure and composition, nanostructured materials can be created in regular and flawless shapes, or with high strength and low weight, or with a controlled brittle behavior. These materials, being more finely grained, have a greater surface-to-volume ratio than conventional materials and, therefore, find many applications in paint and coatings, and in catalysis. They can even be designed so that they contain pores that admit particles of a particular size, thus opening the way to 'smart' membranes that can selectively block out certain molecules.

Polymers are long-chain molecules in which a molecular unit repeats itself along the length of the chain. We are in daily contact with them in the form of adhesives, plastics and fibers (*e.g.*, nylon and rayon). Silk, wool, and the molecule of deoxyribonucleic acid (DNA) are examples of naturally occurring polymers. The interest of the physicist in these materials is aroused particularly by the discovery that polyacetylene could be made conductive by suitable doping, thus opening the possibility of controlling conductivity in materials normally regarded as good insulators (they are then called conjugated polymers). *Block copolymers* result from two reactant oligomer species that form polymeric chains having segments (or blocks) that are attracted to one another. They can give rise to nanoscale phases (which may, for example, be present as spheres, rods or sheets) and provide the framework for manufacturing a wealth of materials — including catalysts, ceramics and insulators — with unique properties. Proteins are an example of block copolymers with two phases, in the form of helical coils and sheets. Polymers have also been used in medical applications, say, to produce artificial skin, dental fillings, and high-density polyethylene for knee prostheses.

These polymers are examples of materials belonging to an interesting and vast class called biomolecular materials, or *biomaterials*. Biological molecules, that nature has been perfecting for millions or even billions of years, have important lessons to teach and inspire us, especially for applications to nanotechnology. Biological sources have presented us with proof that proteins fold into precisely defined three-dimensional shapes, and nucleic acids assemble according to well-understood rules; and that antibodies are extremely specific in recognizing and binding their ligands, and molecular motors can perform specialized tasks in the cell.

A key feature of biomaterials is their ability to undergo *self-assembly*, a process in which aggregates of molecules and components arrange themselves into ordered, functioning entities without external intervention. It is inextricably linked to the idea of *molecular recognition*, according to which subunits, entrusted by nature with sets of instructions, recognize each other and bind to each other, selectively. From this observation, the chemist has constructed a model of biomembrane, affectionately called SAM for *self-assembled monolayer* (Fig. 5.17). It is a one- to two-nanometer thick film of organic molecules that form a two-dimensional crystal

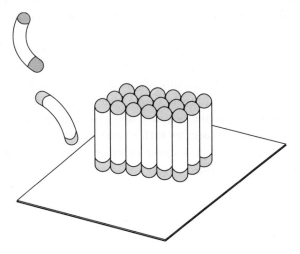

Figure 5.17: Schematic illustration of the molecular structure of a self-assembled monolayer on a surface substrate.

on an adsorbing, typically metallic, surface. The molecules in a SAM are longer than they are wide, and have an atom or atomic grouping at one end that 'sticks' spontaneously (by chemical affinity) to the substrate. When attached, they protrude from the substrate, like a vast forest of identical trees planted in a perfect array. At the other end, scientists can attach selected molecular fragments to give the SAM a well-defined chemical surface property. For example, the molecular layer can be made to attract or repel water, which in turn can affect its adhesion, corrosion and lubrication. SAMs find applications in biological sciences, for example, in the study of interactions of cells with surfaces, neural synaptic integration in planar neural arrays. A long-term goal of research in organic thin films is to find ways of making electronic devices in which the components are individual molecules that self-assemble on substrates from solution or by deposition from interfaces.

Biosensors, devices that couple a biological or biologically-derived sensing element with a physico-chemical transducer, have been known for many decades (for example, in applications to analytical problems in health care, environmental monitoring, defense and security). They are all based on the observation that biological species (from dogs and snakes to enzymes and microbes) may sense the presence of certain molecular species with extreme sensitivity and selectivity. When miniaturized to nanoscale size, they could be implanted in the patient's body, and regulate a controlled release of drugs in phase with the body's changing demands. Recent advances in nanotechnology are broadening their already considerable range of utility. For example, nanotubes filled with enzymes or coated with DNA could be used as electrodes for biosensors. They are so small that an array of them containing different enzymes could be integrated with a single microelectrode enabling many simultaneous analyses.

In the development of sensors and, more generally, in bioelectronics and other biologically related fields, the *biomimetic* approach plays an increasingly important role. It involves directly mimicking biological systems or processes to produce improved materials, or applying techniques observed in nature in a different context or using different materials. For instance, neural networks have arisen from attempts to reproduce the architecture of the human brain, but are implemented using standard electronic and optoelectronic components. Bioengineers, following this path, have succeeded in synthesizing new chain molecules, which then may self-assemble into desirable structures with new or improved properties. They have obtained in this way a variety of designer polymers, such as natural proteins (like silk or collagen) and their modified forms, and synthetic proteins that have no close natural analogues.

As we all know, Nature is not only a genius in physics and chemistry, but also a superb engineer: witness the many sophisticated molecular motors she has built. It has been known for some time that much of the molecular transport in biological systems proceeds not by diffusion but by transport. These biomotors are proteins that use the energy of ATP (adenosine triphosphate) hydrolysis to shuttle along individual fibers: myosin, responsible for muscle contraction, moves on an actin filament; and kinesin, responsible for cellular transport, moves on a microtubule. Recent single-molecule experiments have shown that motor proteins like these act in a discrete, stepwise fashion with very high efficiency, much like a thermal ratchet. A number of researchers have proposed schemes by which such molecular motors could be harnessed to deliver molecules, one at a time, and assemble nanoscale devices in a sort of Lilliputian assembly-line factory. In fact, some micro-organisms have already been employed in making semiconducting quantum dots: when introduced to a potentially lethal concentration of cadmium, the organisms respond by synthesizing crystalline spheres of cadmium sulphide coated with a peptide molecule.

A final but not least important example of the marriage of electronics and molecular biology centers on the 'molecule of life' itself. During the last half-century, researchers have concentrated in studying its biological properties. But many of the same methods they have painstakingly developed — to identify and extract fragments of DNA, to recombine them with other sections to create a new genetic material, to modify the molecule's ends for anchorage to appropriate surfaces — can be applied, together with the tools of nanotechnology, to investigate its remarkable physical properties as well. They have studied, for instance, its electrical conductivity ('Is it a conductor or an insulator?') and the mechanisms of electron transfer within the molecule ('Is it a single-step tunneling process or a multi-step hopping?').

As is generally known, the DNA carries with it an incredibly complex quatranary code. But can it compute? In a pioneering experiment, Leonard Adleman and colleagues showed that a computer constructed of specially encoded strands of DNA could solve a very difficult computational problem (of the 'traveling salesman' type) with 20 variables and find the only correct answer from over a million possible solutions. This is a very exciting result, which could turn out to be a watershed in DNA computation. The DNA, with its unique assembly and recognition properties together with its exceptional stability and adaptability, is bound to become one of the key components in the future molecular electronics.

This chapter is just a snapshot of a young field in vigorous growth. Most of the results discussed in these pages are less than a decade old; yet, at the pace advances are being made, some results will rapidly date, but many will last, with enduring value. The next few decades will see explosive waves of scientific and technological development that will transform our lives to a far greater extent than we have seen in the past, especially in the areas where nanoscience overlaps information science and molecular biology, the two other most promising major areas of scientific activities of our times.

5.7.1 Summary

We have discussed in detail three specific areas in nanoscience: optoelectronics, carbon fullerenes and nanotubes, and spintronics. But there are other areas just as active, such as structural and biological materials, bioelectronics and molecular electronics, from which we expect important developments near term.

5.8 Further Reading

Nanoscience and Nanotechnology

- G. Binning and H. Rohrer, *In Touch with Atoms*, Rev. Mod. Phys. **71** S324–320 (1999).
- R. Service, *Atom-scale Research Gets Real*, Science **290** 1523–1531 (2000).
- M. Rourkes, *Plenty of Room Indeed*, Scientific American, September, 2001, pp. 48–57.
- http://www.research.ibm.com/nanoscience/ (IBM nanoscience site).

Quantum Devices

- J. Faist *et al.*, *Quantum Cascade Laser*, Science **264** 553–556 (1994).
- F. Capasso *et al.*, *Quantum Cascade Lasers*, Physics Today, May, 2002, pp. 34–40.
- Marc Kastner, *Artificial Atoms*, Physics Today, January, 1993, pp. 24–31.
- Mark Reed, *Quantum Dots*, Scientific American, January, 1993, pp. 118–123.
- R.C. Ashoori, *Electrons in Artificial Atoms*, Nature, **1** February, 1996, pp. 413–419.

Fullerenes and Nanotubes

- Robert Curl, *Dawn of Fullerenes*, Rev. Mod. Phys. **69** 691–701 (1997); Harold Kroto, *Symmetry, Space, Stars and C_{60}*, *ibid.* 703–722; Richard Smalley, *Discovering Fullerenes*, *ibid.* 723–730.
- C. Dekker, *Carbon Nanotubes as Molecular Quantum Wires*, Physics Today, May, 1999, pp. 22–28.
- R. Saito, G. Dresselhaus and M.S. Dresselhaus, *Physical Properties of Nanotubes* (Imperial College Press, London, 1998).
- P. Harris, *Carbon Nanotubes and Related Structures* (Cambridge U. Press, Cambridge, 1999).
- http://buckminster.physics.sunysb.edu/ (SUNY Stony Brook site).

Spintronics

- G.A. Prinz, *Spin-polarized Transport*, Physics Today, April, 1995, pp. 58–63; *Magnetoelectronics*, Science **282** 1660–1663 (1998).
- D. Awschalom, M. Flatte and N. Samarth, *Spintronics*, Scientific American, June, 2002, pp. 67–73.
- http://www.almaden.ibm.com/st/projects/magneto (IBM magnetoelectronics site).

Biomolecules

- G.M. Whitesides, *Self-Assembling Materials*, Scientific American, September, 1995, pp. 146–149.
- D. Goodsell, *Biomolecules and Nanotechnology*, American Scientist, May, 2000.
- M. Reed and J. Tour, *Computing with Molecules*, Scientific American, June, 2000, pp. 86–93.

5.9 Problems

5.1 Using the position–momentum uncertainty relation (Appendix B), explain why it is not possible to have confinement in a thin film of vanishingly small thickness.

5.2 Consider an electron trapped in an infinitely deep square potential of width d, for which the energy spectrum is given by $E_n = (nh)^2/(8m^*d^2)$, as stated in the text. We assume m^* is 7% of the electron mass (given by $m_e = 0.5 \times 10^6$ eV/c^2, where c is the speed of light). Calculate the excitation energies $\Delta E_{21} = E_2 - E_1$ and $\Delta E_{32} = E_3 - E_2$ for $d = 10$ nm. If these are converted into photons, what are their wavelengths? Repeat the calculations for $d = 5$ nm; which region of the light spectrum are we in?

5.3 The alloy cadmium–selenium can be fabricated as powder of crystallites, each a few nm across. It is observed that the powder with the larger-sized grains appears red, whereas it appears yellow with the smaller-sized grains. Explain why it is so.

5.4 In the model of confinement in one dimension by an infinite potential of width d, the excitation to the first excited state requires the energy ΔE_{21}, as defined in Prob. 5.2. In order to eliminate this dimension from the particle dynamics, we must require $\Delta E_{21} > kT$. Find the corresponding condition on d, and calculate its limiting values for $m^* = m_e$ and $m^* = 0.07 m_e$, where m_e is the mass of the electron. Assume $kT = 0.026$ eV.

5.5 Consider electrons flowing through a 1D channel to which we have applied a small voltage V. Using the uncertainty relation $\Delta E \Delta t \sim h$ and the Pauli principle, derive an expression for the conductance quantum G_0.

5.6 The Coulomb energy for a single electron in a sphere of radius R in a surrounding medium with dielectric constant ε is given by $E_c = e^2/\varepsilon R$ (cgs units). Using the data $e = 4.8 \times 10^{-10}$ esu, 1 eV $= 1.6 \times 10^{-12}$ erg and $\varepsilon = 12$ (for silicon), find the limiting value of R from the condition $E_c > kT$ for room temperature.

5.7 Why is the electron–electron interaction relatively more important in artificial atoms than in natural atoms?

5.8 The charging energy of a quantum dot of capacitance C holding charge Q is given by $E(Q, V_G) = QV_G + Q^2/2C$, where V_G is the voltage responsible for the charging. We assume that $V_G = (N + \delta)/C$, where N is an integer and $-0.5 \leq \delta \leq 0.5$, and take $e = 1$ to simplify.

(a) Calculate for given δ (hence V_G) the charging energy $E(N, \delta)$ for a quantum dot with charge $Q = -N$, and the energies it would cost us to add or remove one electron, defined by $\Delta_+(N, \delta) = E(N + 1) - E(N)$, and $\Delta_-(N, \delta) = E(N - 1) - E(N)$. Calculate $\Delta_+ + \Delta_-$. Remark on how these quantities depend on N, V_G and C.

(b) Show that when V_G takes the values $1/2C$, $3/2C$, $(2N + 1)/2C$, it costs absolutely no energy to add an electron to a dot containing 0, 1, N electrons, or to remove an electron from a dot holding 1, 2, $N + 1$ electrons.

5.9 How many kinds of vertices and how many of each kind are there in C_{60}? How many hexagons are there? Given that the average edge is 0.142 nm long (which is the carbon–carbon bond length), calculate the diameter of C_{60}.

5.10 The C_{70} molecule may be considered a rugbyball-shaped cavity, constructed by inserting a ring of ten atoms between the split halves of a C_{60} structure. How many pentagonal and hexagonal faces are there in C_{70}? Given that the average edge is approximately 0.142 nm long, estimate the width and length (from end to end) of the cavity.

5.11 For a polyhedron containing only pentagonal and hexagonal faces, the number of pentagons is given by $p = 12$. How is this relation modified when the structure includes, in addition, heptagons and octagons?

5.12 On a hexagon, one may always define two vectors, called a_1 and a_2, separated by an angle of 60°, starting from the same vertex and going to the opposite vertices. Given a the edge length, what is the lengths of a_1 and a_2? On a hexagonal lattice, let a vector be given by $C = na_1 + ma_2$. Show that the length of C and the angle between C and a_1 are given by

$$|C| = \sqrt{3}a(n^2 + nm + m^2)^{1/2}$$

and

$$\theta = \arctan[\sqrt{3}m/(2n + m)].$$

The winding angle defined in the text is given by $\phi = 30° - \theta$.

5.13 Assuming the C_{60} radius is 0.34 nm, identify the indices of the nanotubes that can be capped by the split halves of that molecule. Which of them are metallic?

5.14 Explain why in copper (and other nonmagnetic metals), the up and down spin bands line up at the same energy level, and so there is no population imbalance.

5.15 Consider the problem of storing data on a magnetic medium. If it is given that the data-storage density is x Gbits per square inch, what is the area (in square nanometers) occupied by a bit? Assuming that a bit contains N grains, each a square of size L, what is the storage density x? For a density of 1 Gbit/in^2 and $L = 15$ nm, how many grains are required to store one bit? Recall that 1 inch equals 2.54 cm.

5.16 A standard compact disk has 12.8 square inch of usable surface. As in the previous exercise, each bit contains N square grains of sides L. Assuming that $L = 10$ nm, how much data can a CD store if $N = 3000$, and if $N = 1$? Let us now assume, following the late Richard Feynman, that all the books ever published amount to 1 Pbit $= 10^{15}$ bits, how many CDs do we need to store all that, in either case?

A grain of size $L = 10$ nm contains about 1500 atoms. This is the limit in today's technology, below which the magnetic noise makes signals unreadable. Suppose however that somehow this limit can be lowered to, say, $L = 3.5$ nm. How many CDs do we need to store 1 Pbit, assuming $N = 1$?

Quantum Computation and Information

6

6.1 Introduction

With continuing miniaturization, as discussed in Chapter 5, computers will soon reach a point where the quantum effect begins to affect their basic operations. Instead of fighting to circumvent this problem, a more constructive approach is to use it to our advantage to design a new breed of computers and algorithms, capable of doing things that a conventional classical computer cannot do. This is theoretically feasible because quantum mechanics allows states to be coherently superimposed, which translates into a coherent superposition of numbers in the computer context. This capacity of operating on many numbers all at once allows a highly parallel processing, which in turn can be used to design computer algorithms that run much faster. Unfortunately, quantum computers needed to carry out such manipulations are difficult to build, so much so that as of December, 2001, the largest quantum computer is a 7-bit computer built by IBM. Worse still, as will be discussed in Sec. 6.9.3, this particular technology is really not a pure quantum technology, and much larger computers cannot be built with this method. A sizable quantum computer that can beat conventional computers in real operations is still quite some time away. Nevertheless, when they finally arrive, they will be very powerful. To illustrate how powerful we shall discuss two famous quantum algorithms: the algorithm for sorting an unstructured database by L. K. Grover, and the algorithm for factorizing a large integer into its prime components by P. W. Shor. Both are fast, but the second is even faster than the first. It takes a classical computer $O(N)$ steps[1] to locate a specific item in a database with N entries, whereas Grover's algorithm allows a quantum computer to do so in only $O(\sqrt{N})$ steps, a vast saving of time for large N. A more dramatic saving occurs in Shor's algorithm for factoring a large integer of the form $N = pq$ into its prime factors p and q. If n is the number of bits in N, i.e., if $2^{n-1} < N \leq 2^n$, the fastest classical algorithm takes $O(\exp[(64/9)^{1/3} n^{1/3} (\ln n)^{2/3}])$ steps to get it done. Shor's algorithm

---

[1]$O(N)$ stands for 'of the order N,' or equivalently, cN for some constant c which does not grow with N.

can accomplish the task in fewer than $O(n^3)$ steps. For example, a 129-digit number known as RSA129 has 64-digit and 65-digit prime factors. It requires something like 1.5×10^{17} (150 million billion) instructions by an ordinary computer to get it factorized. This was carried out in 1994 using 1000 work stations over a 8-month period! For a quantum computer with a 100 MHz clock and Shor's routine, the factorization of RSA129 could be done in a few seconds.

There are practical consequences when a quantum computer can be built large enough to run Shor's algorithm. The RSA encryption scheme commonly used in the Internet to encode transactions, whose security depends on the difficulty for a classical computer to factorize a large number N, will no longer be reliable.

While quantum mechanics can be used this way of break a cipher, it can also be used to increase security in another way. These topics of encryption will be discussed towards the end of this chapter.

Quantum mechanics can also be used to send information on the quantum state of a system, without physically sending the system itself. This is known as *quantum teleportation*. It might remind you of what happens in the science fiction movies 'Star Trek,' but at this point we are merely talking about 'beaming up' a single quantum mechanical number, not the quantum state of a whole human being. This topic will also be discussed briefly at the end of the chapter.

In order to understand how quantum algorithms really work, there is no way of skipping a certain amount of detailed discussion. These discussions are relatively dry and harder to understand, and to emphasize this fact they are printed in a smaller font. For a first reading and for readers not interested in the details, those parts can be skipped.

Quantum computation and quantum information is a fast growing field, so clearly this chapter can provide only a very elementary introduction to the subject. Readers wishing to know more can consult the books and articles, as well as the web-sites, quoted at the end of this chapter.

6.1.1 Summary

Superposition of states allowed in quantum mechanics can be used to design a new breed of computers, quantum computers. Algorithms designed using this coherent property can run much faster than the corresponding classical algorithms. However, the hardware needed to run them is difficult to build.

6.2 Classical Computers

In order to understand the outstanding features of a quantum computer, let us first sketch how a usual (classical) computer works.

A computer relies on a central processing unit (CPU) to carry out its basic operations. It is the brain of the computer, although it really cannot think by itself. It must be told what to do, step by step, in a set of instructions known as a 'program.' These operations are paced by a clock so that the next step will not start before the current step is finished. The speed of the clock, and the efficiency of the CPU, are some of the factors that determine the power of the computer. A computer also requires devices to store its program, and the input and output of the computation. These are: the internal random access memory (RAM), and the permanent storage devices like a hard drive or a CD ROM. In the rest of this section, we will concentrate on some of the most basic operations of the CPU.

A computer works on integers. As we learn in elementary school, a floating point number like 3.2095764 can be added or multiplied as if it were an integer 32095764. All that we need to do is to put back the decimal point at the right place at the end of the calculation. For that reason it is sufficient to consider operations of integers when we deal with numbers. Alphabetical characters can also be coded in terms of integers. For example, in the ASCII code, the lowercase letter 'a' is represented by the integer 97, and the uppercase letter 'A' is represented by 65. Thus, only integers need to be considered by a computer.

The integers we are familiar with are the *decimal numbers*, with each *digit* running from 0 to 9. In a computer, for ease of storage and operation, integers are expressed as *binary numbers*, each *bit* of which being either '0' or '1'. The conversion between a decimal number and a binary number goes as follows. The decimal number 1 is also 1 in binary, decimal 2 is 10 in binary, 4 is 100 in binary, 8 is 1000 in binary, etc. The binary number 1 followed by m zeros is the decimal number 2^m. More generally, if $(x_n \cdots x_3 x_2 x_1)$ is an n-bit binary number corresponding to the decimal number y, with each x_i being either 0 or 1, then

$$y = x_1 + 2\, x_2 + 2^2 x_3 + 2^3\, x_4 + \cdots + 2^{n-1}\, x_n \,. \tag{6.1}$$

For example, the 6-bit binary number 101101 is equal to the decimal number 45.

Hardware-wise, 0 and 1 are represented in a computer by the two states of a *transistor*. A transistor is an electronic device with three leads, an incoming lead, an outgoing lead, and a third one connected to the ground. If the incoming lead carries a distinctly positive current or a distinctly positive voltage, the transistor conducts, and a current flows from the outgoing lead via the transistor and its third lead to the ground. If the incoming lead carries a negative or no current or voltage, then the transistor stops conducting. In short, the transistor is a switch, it is either 'on,' to allow current to flow through, or 'off,' to stop current from flowing through. The number 0 is represented by the 'off' state of the transistor, and the number 1 is represented by its 'on' state. The output current or voltage of one transistor can be fed to the incoming lead of another transistor to switch it on and off, and so on down the line. The basic operations in a CPU are implemented by hooking up the transistors in an appropriate way. Different operations correspond to different ways of connecting the transistors.

The basic operations in a CPU can be built up from relatively simple units known as *gates*. For example, there is a gate to do the addition of two bits without 'carry,' and there is another gate to do the multiplication of two bits. One advantage of using binary numbers is the simplicity of its arithmetic rules. If a, b are two bits, each being either 0 or 1, then there are only 4 rules of multiplication $a \times b$ that the computer has to be told: $0 \times 0 = 0$, $0 \times 1 = 0$, $1 \times 0 = 0$, $1 \times 1 = 1$. Compare this with the decimal multiplication table that we were made to memorize in elementary school! There are also only four rules of addition, $a + b$: $0 + 0 = 0$, $0 + 1 = 1$, $1 + 0 = 1$, $1 + 1 = 10$. The last rule appears a bit complicated, because the result 10 is a 2-bit number rather than a 1-bit number like the other seven rules. This can be rectified by using $\oplus$, addition without carry. In that case, $0 \oplus 0 = 0$, $0 \oplus 1 = 1$, $1 \oplus 0 = 1$, and $1 \oplus 1 = 0$. The carry bit of $a + b$ is simply $a \times b$.

In computer or logical jargon, the multiplication operation is known as AND. It is often denoted as $\cdot$ rather than $\times$. The $\oplus$ operation is called XOR, standing for exclusive OR. By definition, a AND b is 1 if and only if both a and b are 1; a OR b is 1 if either a or b, or both, is 1; a XOR b is 1 if either a or b is 1, but excluding the case when a and b are both 1. With these definitions, it is clear that AND is the same as $\times$, and XOR is the same as $\oplus$.[2] Another useful logical operation is called NOT, which interchanges 0 and 1. Therefore, NOT applied to 0 becomes 1 and NOT applied to 1 becomes 0. The compound operation of AND followed by a NOT is called NAND.

All classical computer calculations can essentially be obtained by combining the above simple logical operations, together with the rather trivial operations FANOUT and SWAP. FANOUT of a bit a is simply to make a copy of it, changing a to (a, a). SWAP is to exchange a pair of numbers, replacing (a, b) by (b, a). Moreover, these logical operations are not all independent. In addition to SWAP and FANOUT, only two more are needed to produce the rest. For more details, consult a book on digital computers.

It is convenient at this point to establish some conventions for later use. We will call the leftmost (the largest) bit of an n-bit binary number the first bit, and the rightmost (the smallest) bit the nth bit.

Registers are places where binary numbers are stored. Every n-bit binary number can be stored in a single n-bit register, or it can be separated into two (or more) portions, with the first k bits stored in a k-bit register and the remaining $(n - k)$bits in an $(n - k)$bit register. In the latter case, the register containing the largest (leftmost) bits is called register I, the one containing the next largest bits is called register II, etc. If necessary, parentheses are used to group the bits in the same register. For example, $(10)(110)$ means the first two bits of 10110 is stored in register I, and the last three bits of 10110 is stored in register II.

A gate is denoted graphically by a rectangular box, with its name written inside. Data are fed into the gate from the left and extracted on the right. A gate may work on only some of the bits of an input number, in which case those bits will be

---

[2] The OR operation can be related to arithmetic operations by the formula a OR $b = (a \oplus b) + (a \times b)$. In electronic literature, a OR b is often denoted as $a + b$. We shall avoid using that notation lest it get confused with the addition sign used above.

written as subscripts to the symbol for the gate. A Latin subscript means a bit, a Roman subscript means all the bits in that particular register.

For example,

$$10110 \rightarrow \boxed{\text{AND}_{12}} \rightarrow \mathbf{0}110$$

$$(\mathbf{10})(110) \rightarrow \boxed{\text{AND}_I} \rightarrow (\mathbf{0})(110)$$

$$10\mathbf{110} \rightarrow \boxed{\text{XOR}_{34}} \rightarrow 1000$$

$$(\mathbf{10})(110) \rightarrow \boxed{\text{XOR}_I} \rightarrow (\mathbf{1})(110)$$

The bits being operated on are printed in bold face.

6.2.1 Summary

Classical computers work on binary numbers through elementary gates. Some of the usual gates are NOT, AND, OR, XOR, NAND, FANOUT, and SWAP.

6.3 Quantum Computers

6.3.1 Introduction

What makes a quantum computer different is the presence of 'schizophrenic states,'[3] and gates to manipulate them. To understand that, let us first review the basic facts of quantum mechanics needed to understand this section. (See also the discussion in Appendix B.)

In quantum mechanics, 'particles'[4] are regarded as waves. Different states (energy, spin, position, etc.) of a particle are described by different complex-valued wave functions $\psi(\mathbf{x}, t) = |\psi(\mathbf{x}, t)|e^{i\theta(\mathbf{x}, t)}$. It is conventional to adopt Dirac's notation and denote the wave function as a whole by $|\psi\rangle$. The intensity of the wave, or equivalently, the probability of finding the particle at position $\mathbf{x}$ at time t, is proportional to $|\psi(\mathbf{x}, t)|^2$.

The sum of two waves is another wave. Such a *coherent* superposition of two waves may produce very strange effects. If ψ_1 is the wave function of a particle in a certain state, and ψ_2 is the wave function of the same particle in another state, then the wave function $\psi = \psi_1 + \psi_2$ represents the particle in a '*schizophrenic*' state, with finite probabilities of finding it in either of the two 'personalities.' Since $|\psi_1 + \psi_2|^2 \neq |\psi_1|^2 + |\psi_2|^2$, the intensity of the combined wave is not equal to the combined intensities of the individual waves. This is a quantum effect that has no classical counterparts. In other words, because quantum mechanical particles are

---

[3]'Schizophrenic state' is not an official name in quantum mechanics. We use this term to describe a linear combination of (single or multi-particle) basis states.

[4]We use the generic term 'particle' to mean any isolated quantum mechanical system. Thus, a particle can be a photon, an electron, a nucleus, or an atom.

waves, interference effects may be present. The difference of the intensity of the sum, and the sum of the intensities, is given by the cross term, which depends on the phase difference $\Delta\theta = \theta_1 - \theta_2$ of the two waves.

Unless the particle is completely isolated, its interaction with the environment produces a phase shift. For a complex environment this additional phase shift could become random, destroying the original phase difference. In that case the cross term averages to zero, and the probability of the sum is then equal to the sum of the probabilities, reverting it back to the classical scenario. This is known as *decoherence*. The operation of a quantum computer relies on coherence being maintained, which means that the effect due to environmental interactions must be small during the course of quantum mechanical gate operations. This turns out to be one of the most difficult problems to solve in practice. Nevertheless, we shall assume in what follows that this can be achieved and that decoherence effects can be ignored. Just like the transistor in a classical computer whose on-off positions represent the basic bits 1 and 0, in a quantum system we must find two convenient states $|1\rangle$ and $|0\rangle$ to act as the quantum mechanical bits, otherwise known as *qubits*. For example, the two spin states of a spin-$\frac{1}{2}$ electron could be used provided that their energy levels are not the same. More generally, any two states of a single particle could be used in principle, but in practice, they must be relatively long lived to allow quantum mechanical operations to be carried out before they decay. Moreover, practical techniques must be available to put the particle in any one of these two *basis states*, as well as any normalized schizophrenic state $|\psi\rangle = \alpha|0\rangle + \beta|1\rangle$, for any complex numbers α and β satisfying $|\alpha|^2 + |\beta|^2 = 1$. The picture below depicts the basis states and one such schizophrenic state.

A transistor in a classical computer can either be on, or off, but nothing in between. A particle in a quantum computer can be in $|1\rangle$ or $|0\rangle$, but it can also be in a schizophrenic state. This difference between a qubit and a classical bit is the fundamental reason why a quantum computer can do things that a classical computer cannot do. To work on a linear combination is like working on the basis states all at once, so a quantum computer is like a highly parallel classical computer. The problem is, a particle in a schizophrenic state is known only probabilistically, so it is not generally possible to extract a definitive answer at the end of a computation, which we need. To overcome this trade-off, a fast operation against a muddy outcome, we must find ways to exploit the schizophrenic advantage, yet ending up in an almost pure basis-state to avoid probabilistic uncertainty. Ingenuity is required to make this happen, and even so it works only in certain problems. The famous algorithms by Deutsch, by Shor, and by Glover discussed below are some examples of these ingenious algorithms.

6.3.2 Multiple Qubits

In a classical computer, a binary number of n bits is implemented by n transistors, one for each bit. In a quantum computer, it is represented by n non-interacting particles, each in the state $|0\rangle$ or $|1\rangle$, or a linear combination of them. The binary number 11001, for example, is represented by a 5-particle state, $|1\rangle|1\rangle|0\rangle|0\rangle|1\rangle$, also written simply as $|11001\rangle$. We shall often label an n-bit binary number by its decimal equivalent. In that notation, the state $|11001\rangle$ is $|25\rangle$, and the state $|1000000\rangle$ is $|64\rangle$. Every state of an n-particle system can be stored in an n-(qu)bit register. The picture below shows a 10-qubit register with the state $|0000100101\rangle = |37\rangle$ stored in it.

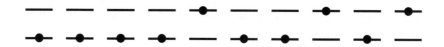

Like single-particle states, we are allowed to make linear combinations of the multi-particle basis states. Some of these linear combinations can be factorized into a product of single-particle states, e.g., $|00\rangle + |01\rangle + |10\rangle + |11\rangle = (|0\rangle + |1\rangle)(|0\rangle + |1\rangle)$. There are others which cannot, e.g., $|00\rangle + |11\rangle$, or $|01\rangle - |10\rangle$. These latter states are known as *entangled states*. They contain a non-trivial and counter-intuitive correlation between the particles. It is states of this kind that lead to the EPR (Einstein–Podolsky–Rosen) 'paradox'[5] of quantum mechanics.

In a quantum computer, two-particle entangled states are produced by 2-bit 'controlled gates,' which take an uncorrelated 2-qubit input and convert it into a correlated entangled output. The necessary correlation makes these 2-bit controlled gates difficult to build, but their presence is indispensible if we want to obtain non-trivial outputs from a quantum computer.

6.3.3 Summary

Each qubit in a quantum computer is represented by two states, $|0\rangle$ and $|1\rangle$, of a quantum particle. These states can be superimposed and manipulated by quantum

---

[5]Suppose a spin-0 particle at rest decays into two spin-$\frac{1}{2}$ particles, flying away in opposite directions. By spin angular momentum conservation, the spin wave function of the final state is $\frac{1}{\sqrt{2}}(|01\rangle - |10\rangle)$, where $|0\rangle$, $|1\rangle$ are respectively the spin-down and spin-up states of the particle. This is an example of an entangled state. If a measurement of particle 1 shows that it is in the spin-up state, then we automatically know that particle 2 is in the spin-down state, and vice versa. This fact has been confirmed by experiments, but it violates usual intuition, and hence it is thought to be a 'paradox.' Since the two particles travel in opposite directions, and could be very far apart when a measurement is made, this correlation seems to show that the result of the measurement of particle 1 is instantaneously transmitted to observer 2, with a speed faster than the speed of light. That this is not the case has been demonstrated by showing that we cannot use the effect to transmit a message from observer 1 to observer 2 at a speed faster than the speed of light.

gates. One may consider the basis states in the superposition to be manipulated simultaneouly, so in that sense, a quantum computer works in a highly parallel manner. If the final result remains a superposition of basis states, then the answer is only probabilistic, and the advantage gained by parallel processing is lost. However, for certain problems, the final state can be put in a single basis state (to be continued).

6.4 Quantum Gates

A gate in a classical computer is a basic component which turns a number into another one, according to some function $y = F(x)$. The AND gate, for example, takes a 2-bit number (ab) and converts it into the 1-bit number $a \times b$. The XOR gate takes (ab) and converts it into $a \oplus b$.

A gate in a quantum computer converts a state $|x\rangle$ into another state $F(|x\rangle) \equiv |F(x)\rangle$. This is accomplished by switching on suitable quantum mechanical interactions inside the gate. Interactions in quantum mechanics must be *linear* and *unitary*, so quantum gates must be linear and unitary as well. If $|\psi\rangle = \alpha|x_1\rangle + \beta|x_2\rangle$, linearity demands $|F(\psi)\rangle = \alpha|F(x_1)\rangle + \beta|F(x_2)\rangle$. In practice, this means that once we know how a gate transforms the basis states, we know how it transforms all their linearly combined states.

Since it is linear, the function F may be thought of as a matrix, mapping the 2^n n-particle orthonormal states $|x\rangle$ to the 2^n states $|F(x)\rangle$. Unitarity requires this matrix to be unitary, meaning that the 2^n states $|F(x)\rangle$ must also be orthonormal.[6] A unitary matrix has an inverse, hence two different states $|x\rangle$ must map to two distinct states $|F(x)\rangle$. In particular, if the gate has an n-bit input, it must also have an n-bit output. For that reason, none of the classical gates AND, XOR, OR, and NAND can be a quantum gate, because these gates map a 2-bit number into a 1-bit number.

Although the classical gates AND, XOR, etc. are not unitary, we can design in each case a unitary gate to encompass their functions. What we need is to include additional bits to pad it into a unitary gate. More generally, from a function $y = f(x)$ mapping an n-bit register into an m-bit register, an $(m + n)$-bit unitary gate $\mathcal{U}_f$ can be constructed. Unlike the function F, which has to be linear and unitary, the function f here only has to be linear. The unitary gate $\mathcal{U}_f$ maps $|x\rangle|y\rangle$ onto $|x\rangle|y \oplus f(x)\rangle$, with the first register carrying n bits and the second m bits:

$$|x\rangle|y\rangle \rightarrow \boxed{\mathcal{U}_f} \rightarrow |x\rangle|y \oplus f(x)\rangle. \tag{6.2}$$

---

[6]The inner product $\langle\psi_1|\psi_2\rangle$ of two wave functions is defined to be $\int d^3x \psi_1^*(x)\psi_2(x)$. If this inner product is zero the two states are said to be orthogonal. If $\langle\psi_1|\psi_1\rangle = 1$, the state is said to be normalized. A set of n wave functions $\psi_i(x)$ is said to be an orthonormal set if $\langle\psi_i|\psi_j\rangle$ is 1 when $i = j$, and 0 when $i \neq j$. The set of basis vectors are always chosen to be orthonormal.

The symbol $\oplus$ here means addition modulo[7] 2^m. It can be shown that this operation is unitary but we shall skip the proof. In particular, $\mathcal{U}_f$ maps $|x\rangle|0\rangle$ onto $|x\rangle|f(x)\rangle$ so we can read out $f(x)$ from the second output register.

For example, the $\mathcal{U}_{\text{AND}}$ gate is a 3-bit gate, converting the 3-bit state $|x\rangle|y\rangle = |ab\rangle|y\rangle$ into the 3-bit state $|ab\rangle|y \oplus a \times b\rangle$. In particular, it converts $|ab\rangle|0\rangle$ into $|ab\rangle|a \times b\rangle$, so the result of the AND operation can be read out in the third output bit.

We turn now to the question of how these quantum gates can be constructed.

The most commonly used gates are 1-bit gates. Happily, they are also the easiest to construct. A 1-bit gate U changes $|0\rangle$ into some specific normalized state, and $|1\rangle$ into an orthogonal one. If $|0\rangle$ and $|1\rangle$ are two levels separated by an energy $\hbar\nu$ ($\hbar$ is Planck's constant), then an electromagnetic pulse with frequency ν and an appropriate phase, applied to the particle for a suitable length of time, can produce any unitary transformation on the two basis states $|0\rangle$ and $|1\rangle$, and hence any U gate. If the two basis states are the two polarizations of a photon, then we can use standard optical equipment to rotate and combine these polarizations to produce a desired 1-bit gate. It is also possible to have $|0\rangle$ and $|1\rangle$ represented by light beams from two orthogonal directions, as shown in the diagram below. Then two prisms sandwiching a partially silvered mirror can be used to mix the two beams (to get outputs a and b), and a dielectric material inserted in the path of a beam (a) can be used to shift the phase of the beam (to get c). The amount of mixing is determined by the amount of silvering of the mirror, and the amount of phase delay is determined by the thickness of the dielectric material. In this way one can again construct any 1-bit unitary gate U.

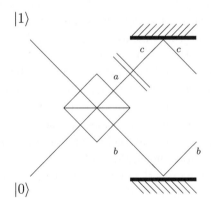

---

[7]a mod c is an integer between 0 and $c-1$. It is obtained by subtracting from a a suitable integral multiple of c. For example, 3 mod 15 = 3, but 31 mod 15 = 1 because $31 = 1 + 2 \times 15$. We can apply this rule to addition, so $57 + 13$ mod 15 is 70 mod 15, which is 10. An equivalent way of thinking about 'mod c' is to imagine counting numbers in the following way. We count $0, 1, 2, \ldots$, up to $c - 1$, then we circle back and count from 0 up again. Hence, c is 0, $c + 1$ is 1, etc. The binary addition without carry, discussed in Sec. 6.2, for which we use the symbol $\oplus$, is the same as addition mod 2.

The most frequently used 1-bit gates are the NOT (N) gate and the HADAMAD (H) gates. The N gate interchanges $|0\rangle$ and $|1\rangle$. It promotes the particle at the ground state $|0\rangle$ to the excited state $|1\rangle$.

The H gate changes the basis states into an equal mixture of the two. Specifically,

$$|0\rangle \rightarrow \boxed{\text{H}} \rightarrow \frac{1}{\sqrt{2}}\left(|0\rangle + |1\rangle\right)$$

$$|1\rangle \rightarrow \boxed{\text{H}} \rightarrow \frac{1}{\sqrt{2}}\left(|0\rangle - |1\rangle\right).$$

When H is applied to a 1-bit ground state $|0\rangle$, it changes into the state $|h\rangle = (|0\rangle + |1\rangle)/\sqrt{2}$. When H is applied to every bit of an n-bit ground state, it changes every bit into this state h. The result is a sum of 2^n n-bit states, where every bit can be either a 0 or a 1. Converting that into decimal notation, it becomes

$$|0\rangle \rightarrow \boxed{\mathcal{H}} \rightarrow \left(\frac{1}{\sqrt{2}}\right)^n \sum_{x=0}^{2^n-1} |x\rangle \equiv |\Upsilon\rangle, \tag{6.3}$$

where $\mathcal{H} \equiv \prod_{i=1}^{n} \text{H}_i$, and H_i acts on the ith bit of $|0\rangle$. We shall refer to the gate $\mathcal{H}$ as the n-bit HADAMAD gate. The ordinary H gate is simply the 1-bit HADAMAD gate.

The uniform state $|\Upsilon\rangle$ is a very useful state in quantum computation. It allows maximal parallel processing because a quantum gate acting on it acts simultaneously on all the 2^n states $|x\rangle$.

A quantum computer requires also 2-bit CONTROLLED-U (CU) gates to function. One of the two incoming bits is a control bit (c), and the other is the data bit (d). The outgoing control bit is always the same as the incoming control bit. The outgoing data bit d' is also the same as the incoming data bit if $c = 0$, but d is to be transformed by the 1-bit U gate to get d' if $c = 1$.

These gates establish a correlation between the two bits. Such correlations are clearly needed for computations, for without them one cannot even add or multiply two numbers. Unfortunately these gates are difficult to manufacture in practice precisely because of the required correlations. In fact, getting these gates working is one of the big obstacles in the construction of a practical quantum computer. We will discuss the problem more thoroughly in Sec. 6.9.

The most common CU gate is the CONTROLLED-NOT (CN) gate where U = N. In this case, $d' = c \oplus d$, so it can be used as an adder without carry.

The correlation inherent in a control gate can also be used to produce entangled states from unentangled ones. For example, starting from the 2-bit ground state $\sqrt{2}|00\rangle$, apply H to the control bit to change it to $|0\rangle + |1\rangle$, and apply N to the data bit to change it to $|1\rangle$. Now pass that through a CN gate. The outgoing state $|cd'\rangle$ is then the entangled state $|01\rangle + |10\rangle$.

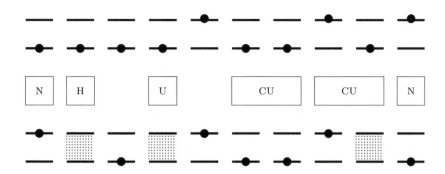

Some of these 1-bit and 2-bit gates are summarized in the following graph. The incoming bit(s) are on top and the outgoing bit(s) are at the bottom, with the gate in between. Schizophrenic states are indicated by dots between the two basis states.

For better or for worse, the linearity of a quantum gate makes it impossible to copy states faithfully. To see why, let us copy the one-particle states $|0\rangle$ and $|1\rangle$ onto a second particle, thus forming the states $|00\rangle$ and $|11\rangle$, respectively. This can actually be accomplished using the CN gate if its control bit c is the bit to be copied:

$$|c\,0\rangle \rightarrow \boxed{\text{CN}} \rightarrow |c, c \oplus 0\rangle = |c\,c\rangle\,.$$

If the control qubit is now $|\psi\rangle = \alpha|0\rangle + \beta|1\rangle$, then the input qubits to the CN gate is $|\psi\rangle|0\rangle = \alpha|00\rangle + \beta|10\rangle$. By linearity, the output qubits of the CN gate are $\alpha|00\rangle + \beta|11\rangle$, and not the desired $|\psi\rangle|\psi\rangle = (\alpha|0\rangle + \beta|1\rangle)(\alpha|0\rangle + \beta|1\rangle) = \alpha^2|00\rangle + \alpha\beta(|01\rangle + |10\rangle) + \beta^2|11\rangle$. In other words, even if we can design a device to copy the basis states, by linearity the same device will not be able to copy their linear combinations faithfully. This is sometimes referred to as the 'no-cloning theorem.' The proof can be carried out in a completely general context.

This inability to copy is both good and bad. It is good because data cannot be copied and stolen without our knowing it. It is bad because classical computer algorithms often call for making copies of some data for later use, so such algorithms must be modified.

Let us finally mention a famous 3-bit unitary gate, which can actually be constructed from 1-bit and 2-bit controlled gates. It is the CONTROLLED-CONTROLLED-NOT (CCN) gate, otherwise known as the TOFFOLI gate. The incoming state $|cc'd\rangle$ of this gate contains two control bits c and c', which remain unchanged after passing through the gate. The data bit d is also unchanged unless both c and c' are 1, in which case the outgoing data bit is NOT operating on the incoming data bit d. This gate can be used as a multiplier of the two control bits, because it converts $|cc'0\rangle$ into $|c, c', c \times c'\rangle$.

6.4.1   *Summary*

The magic of a quantum computer lies in its ability to manipulate superpositions of qubits. The gates used to manipulate them are linear and unitary, quite different from the gates used in a classical computer. One-bit gates are relatively easy to construct, with the NOT (N) and HADAMAD (H) gates being some of the most useful ones. Two-bit controlled gates such as the CONTROLLED-NOT (CN) and more generally the CONTROLLED-U (CU) gates are also needed, but they are much more difficult to manufacture because of the correlation requirement.

6.5   Deutsch's Algorithm

Quantum computers can process a mixture of 'numbers' in the form of a schizophrenic state. With this special feature efficient quantum algorithms can be designed. If S is the (large) number of steps needed to do a job in a classical algorithm, then we will consider the corresponding quantum algorithm efficient if it takes only $O(S^a)$ steps, with some $a < 1$ (power saving) or, better still, $O(\ln S)$ steps (exponential saving).

The first successful quantum algorithm that provides a substantial (actually exponential) saving is Deutsch's algorithm. It is a toy program in the sense that it is not that useful in practice. Nevertheless, it is important because it is simple, and it illustrates how quantum savings can be accomplished by the use of schizophrenic states, plus a judicious choice of a final state to yield a definitive rather than a probabilistic answer.

Let us now state the problem to be solved. It concerns a class of functions $y = f(x)$, mapping an n-bit number x onto a 1-bit number y. Every function in this class is either *constant*, mapping all 2^n values of x onto the same number, or *balanced*, mapping half the values of x onto 0 and the other half of the values of x into 1. A function of this class is given to us in a black box, also known as an *oracle*. When x is fed into the black box, $y = f(x)$ will appear as the output.

We are told that constant functions and balanced functions occur with equal probability, but we do not know which it is in a specific black box. We are now asked to design an efficient algorithm, incorporating this black box, to decide precisely whether the function given is constant, or balanced.

In a classical computer, the only way to decide is to feed into the black box, one x after another, and compare the outputs. If the function turns out to be constant, then it takes exactly $2^{n-1} + 1$ (half the number of x plus one) steps to recognize it as such. This is so because even if the first 2^{n-1} values of f turn out to be the same, it can still be balanced if the next 2^{n-1} output has opposite values. On the other hand, if the function is balanced, then we will be able to tell as soon as a different value of f turns up, because then f obviously cannot be constant. Since a balanced function has as many 0 values as 1 values, the chance that an opposite value of f turns up within the first few steps is very high. For that reason, even if n is very, very large, it can be shown that the average number of steps, a, it takes to tell that it is balanced is finite. Since we are told that balanced functions and constant functions occur with equal probability, the average number of steps for a classical algorithm to decide is $S = (a + 2^{n-1} + 1)/2 \simeq 2^{n-2}$ steps, when n is large.

Deutsch's quantum algorithm can accomplish this task in $O(n) = O(\ln S)$ steps, thus gaining an exponential advantage compared to the classical algorithm. Here is how it works.

First of all, we must realize that the black box itself is not a unitary gate, because it has an n-bit input but only a 1-bit output. However, the gate $\mathcal{U}_f$ in Eq. (6.2), which maps $|x\rangle|y\rangle$ to $|x\rangle|y \oplus f(x)\rangle$, is unitary. Note that $\mathcal{U}_f$ maps $|x\rangle(|0\rangle - |1\rangle)$ onto itself if $f(x) = 0$, and it reverses the sign of this state if $f(x) = 1$. In other words, it produces a prefactor $(-1)^{f(x)}$ when operated on this state, no matter how many bits x carries. What Deutsch did was to make use of this feature to determine what kind of function f is.

In the notations explained at the end of Sec. 6.2, Deutsch's algorithm can be written schematically as follows:

$$|0\rangle|0\rangle \rightarrow \boxed{\text{NOT}_{II}} \rightarrow |0\rangle|1\rangle$$

$$\rightarrow \boxed{\mathcal{H}_I\text{H}_{II}} \rightarrow \left(\frac{1}{\sqrt{2}}\right)^n \left(\sum_{x=0}^{2^n-1} |x\rangle\right) \frac{1}{\sqrt{2}} (|0\rangle - |1\rangle)$$

$$\rightarrow \boxed{\mathcal{U}_f} \rightarrow \left(\frac{1}{\sqrt{2}}\right)^n \left(\sum_{x=0}^{2^n-1} (-1)^{f(x)}|x\rangle\right) \frac{1}{\sqrt{2}} (|0\rangle - |1\rangle) \tag{6.4}$$

$$\rightarrow \boxed{\mathcal{H}_I} \rightarrow |a\rangle \frac{1}{\sqrt{2}} (|0\rangle - |1\rangle).$$

It starts from the $(n+1)$-bit ground state $|0\rangle$, with its first n bits stored in Register I and the remaining bit stored in Register II. A NOT gate followed by a HADAMAD gate on II converts the last qubit into the state $(|0\rangle - |1\rangle)/\sqrt{2}$, while the n-bit HADAMAD gate converts the content of the first register into the uniform state. Now the $(n+1)$-bit state is passed through the gate $\mathcal{U}_f$ constructed from the black box. The result is an extra sign factor $(-1)^{f(x)}$ applied to each state $|x\rangle$ in the first register. The resulting state in the first register is finally passed through an n-bit HADAMAD gate to convert it into a state $|a\rangle$. If f is constant, then since $\mathcal{H} \cdot \mathcal{H}|0\rangle = |0\rangle$, $|a\rangle$ is $\pm|0\rangle$. The sign is plus when all $f(x) = 0$, and minus when all $f(x) = 1$. On the other hand, if f is balanced, then the state $\sum_x (-1)^{f(x)}|x\rangle$ is orthogonal to the uniform state. Since the unitary gate h_I preserves inner products, the resulting state $|a\rangle$ is orthogonal to the n-bit ground state $|0\rangle$, meaning that at least one of the n-bits in $|a\rangle$ is 1. If we count as one step the query of a bit to determine whether it is 0 or 1, then it takes n steps to decide whether a 1 is present in at least one bit, and therefore whether the function is balanced. In counting the number of steps perhaps we should include the number of operations needed to get $|a\rangle$, and that is of order n as well. Hence, the final number of steps is $O(n)$, which provides an exponential saving compared to the classical algorithm.

If we were to ask what precisely f is, rather than just whether it is constant or balanced, then quantum algorithm or not, it requires $O(2^n)$ steps.

6.5.1 *Summary*

Deutsch's algorithm is designed to find out whether a given function $y = f(x)$ is constant or balanced. It makes use of the property that the unitary gate $\mathcal{U}_f$ associated with these functions maps $|x\rangle(|0\rangle - |1\rangle)$ onto itself when $f(x) = 0$, and minus the state when $f(x) = 1$. A uniform state is passed through this gate to allow the Deutsch algorithm to gain an exponential saving over the corresponding classical algorithm.

6.6 Finding the Period of a Function

6.6.1 *Introduction*

What allows the Deutsch algorithm to succeed is the special nature of the function $f(x)$, being either constant or balanced, as well as the limited nature of our enquiry, whether f is constant or balanced, and not precisely what the function f is. In this section we will discuss another algorithm with similar characteristics. This time the function f is periodic, and we only want to know what its period r is. The quantum algorithm for this task again offers an exponential saving over the classical algorithm. Moreover, it will be needed in Shor's factorization routine to be discussed in the next section.

Consider an m-bit periodic function $y = f(x)$ of an n-bit variable x, with a period r, so that $f(x + r) = f(x)$ for every x. We shall assume that $f(x)$ takes on different values at different x within a period, and that r divides $S = 2^n$. Again f is handed to us in a black box, or an oracle. Our aim is find out the period r of this function.

The most straightforward method is again to pass each x value through the black box and record its outcome y, from which we determine the period r. This takes $r + 1$ steps. The period r could be as small as 1 and as large as 2^n, so on average it would take roughly 2^{n-1} steps to find the period this way.

Since f is periodic, we may also use the Fourier transform to find the period. However, this turns out not to be a good way with a classical computer, because it takes $O(S^2) = 2^{2n}$ steps just to compute all the values of the Fourier transform,[8] many more than what is required in the naive method discussed in the preceeding

---

[8]There are S values in the Fourier space and each can be computed by summing the S terms in the Fourier transform.

paragraph. The number of steps can be reduced to $O(S \ln S)$ by using a fast Fourier transform, but that still does not beat the naive approach.

The situation is different with a quantum computer. The Fourier transform $\mathcal{F}$ of a state $|x\rangle$ is a state given by

$$|x\rangle \rightarrow \boxed{\mathcal{F}} \rightarrow \frac{1}{\sqrt{S}} \sum_{k=0}^{S-1} e^{2\pi i k x / S} |k\rangle . \tag{6.5}$$

If $|x\rangle = |0\rangle$, the output is the uniform state $|\Upsilon\rangle$ of Eq. (6.3), which can be generated from $|0\rangle$ using the n-bit HADAMAD gate $\mathcal{H}$ in n steps. If $|x\rangle \neq |0\rangle$, additional phase factors are present, which requires 2-bit controlled gates to obtain. The number of steps required is then $O(n^2)$, which still gains an exponential saving of steps compared to the naive method. This is the reason why it pays to find the period r using quantum Fourier transform if a quantum computer is present.

In the next subsection we shall discuss in more detail how to implement a quantum Fourier transform and why it can be done in $O(n^2)$ steps. In the subsection after we shall discuss how the quantum Fourier transform can be used to find the periodicity of the function $f(x)$. These two subsections are more technical, so readers who are not interested in such details may skip now to the next section.

6.6.2 *Implementing a Quantum Fourier Transform*

Both the left-hand side and the right-hand side of Eq. (6.5) can be factorized. Their ℓth qubits are related by

$$|x_\ell\rangle \rightarrow \boxed{\mathcal{F}} \rightarrow \frac{1}{\sqrt{2}} \left(|0_\ell\rangle + e^{2\pi i x / 2^{n-\ell+1}} |1_\ell\rangle \right),$$

which can be shown by expressing $|k\rangle$ in (6.5) in binary notation. If we expand x in the exponent into binary form like Eq. (6.1), then every bit of x will enter into the exponent on the right-hand side. In other words, $\mathcal{F}|x_\ell\rangle$ depends not just on the ℓth bit of x, but it also depends on every other bit of it. Such a correlation clearly requires control gates to implement. The resulting gate to transform $|x_\ell\rangle$ turns out to be $H_\ell \Phi_\ell$, where Φ_ℓ is made up of a product of $(\ell - 1)$ 2-bit control gates CU for some appropriate U's. Counting each control gate and each H gate as one step, a qubit can therefore be transformed with ℓ steps. The total number of steps for the quantum Fourier transform is therefore $\sum_{\ell=1}^{n} \ell = n(n+1)/2$. This is $O(n^2)$ as promised.

6.6.3 *Period of the Function from the Fourier Transform*

To find the period of the function $f(x)$, construct the $(n+m)$-qubit state

$$|F\rangle = \frac{1}{\sqrt{S}} \sum_{x=0}^{S-1} |x\rangle |f(x)\rangle .$$

This can be obtained by applying the n-bit gate $\mathcal{H}$ to the ground state $|0\rangle$ of the first (n-bit) register, and then the $\mathcal{U}_f$ gate of Eq. (6.2) to both registers. These two operations are summarized as

$$|0\rangle |0\rangle \rightarrow \boxed{\mathcal{H}_I} \rightarrow \frac{1}{\sqrt{S}} \left(\sum_{x=0}^{S-1} |x\rangle \right) |0\rangle \rightarrow \boxed{\mathcal{U}_f} \rightarrow |F\rangle .$$

It can be accomplished in $n+1$ steps: n steps for $\mathcal{H}$ and one step for $\mathcal{U}_f$. This is negligible compared to the $O(n^2)$ steps needed later to carry out the quantum Fourier transform.

Organize the components of $|F\rangle$ according to the value $|y_0\rangle = |f(x_0)\rangle$ of the second register. Since the period of $f(x)$ is r, every $f(x)$ with $x = x_0 + jr$ (j is an integer) gives the same value y_0. In this way we can re-write the state as

$$|F\rangle = \frac{1}{\sqrt{r}} \sum_{x_0=0}^{r-1} |g(x_0)\rangle |f(x_0)\rangle ,$$

where $|g(x_0)\rangle = \sum_{j=0}^{K-1} |x_0 + jr\rangle/\sqrt{K}$. This definition of $|g(x_0)\rangle$ can be extended to all x_0 between 0 and $S-1$, provided we interpret the sum $x_0 + jr$ as the sum mod S. With this extension, $|g(x_0)\rangle$ is periodic in x_0, with a period r. In this way, the periodicity of f is transferred to the periodicity of g, which is easier to handle because it is independent of other details of f. Since g has a period r, its Fourier transform, Eq. (6.5), has a spectrum which vanishes except when k is an integral multiple of $K = S/r$. By performing measurements in the first register of the state $\mathcal{F}_I |F\rangle$, we can find out K and hence $r = S/K$.

6.6.4 *Summary*

An efficient algorithm with exponential saving can be designed to measure the period r of a periodic function $f(x)$, by using a quantum Fourier transform. The number of steps needed is $O(n^2)$, if x ranges from 1 to $S = 2^n$. In contrast, this requires $O(n2^n)$ number of steps to accomplish by using a classical Fourier transform, and $O(2^n)$ steps to do it naively.

6.7 Shor's Factorization Algorithm

Suppose we know that an n-bit number N is made up of the product of two prime factors p and q. Given N, how difficult is it to find p and q? Naively, it takes an average of $O(\sqrt{N})$ steps by trial and error to get them. Starting from 2, we test whether it divides N. If it does, we have found one of the prime numbers, and the other can be obtained by dividing N by it. If it does not, we proceed to the next prime number 3, and try again. Since one of p and q must be smaller than $\sqrt{N}$, we can certainly obtain the answer after at most $\sqrt{N}$ tries.

If N is large, say with 128 bits so that it is of the order of $2^n \simeq 3 \times 10^{38}$, then it will take a long time to succeed with this naive method. Fortunately, there are classical algorithms which can do this much faster. The fastest known takes $O(\exp[(64/9)^{1/3} n^{1/3} (\ln n)^{2/3}])$ steps. For $n = 128$, this is of the order the of $e^{27.6} \simeq 9.6 \times 10^{11}$, still a very large number of steps. This difficulty in factorization has been exploited to design a coding system, the RSA public key encryption scheme, which can be broken only when one succeeds in finding p and q from a given N. Since it takes such a long time to do so, by the time one succeeds, N and hence the coding scheme will have been changed, so one's effort is wasted and this encryption method is safe. We will discuss exactly how the public key encryption works in a later section.

As soon as the first quantum computer is turned on, this encryption system will become obsolete. A quantum algorithm invented by Peter Shor in 1994 allows the factorization to be done in less than $O(n^3)$ steps. What took months to do will now be accomplished in seconds.

We will outline in this section how Shor's algorithm works. Essentially, it turns the factorization problem into a problem of finding the periodicity of a function, so the technique discussed in the last section can be used to gain an exponential saving.

To start with, pick a number $a < N$ at random. If a divides N, we are done, for we have found either p or q. If a does not, then according to Euler's theorem in number theory, a number r can be found so that

$$a^r \equiv 1 \bmod N. \tag{6.6}$$

The number r is not unique, but we will take the smallest of them.

For example, if $N = 15$, and we picked $a = 4$, then $r = 2$ because $4^2 = 1 + 15$. Note that we also have $a^4 \equiv 1 \bmod 15$ because $4^4 = 256 = 1 + 17 \times 15$, but we will take r to be the smallest, so it is 2 and not 4.

We will discuss in a moment how r can be found using the algorithm from the last section. If r turns out to be odd, then we must start over again, pick another a, and find another r. If r is even, then the condition (6.6) can be written in the form:

$$(a^{r/2} - 1)(a^{r/2} + 1) = kN = kpq,$$

for some integer k. Call the two factors on the left b and c. Unless one of them is a multiple of N, otherwise the largest common divisor of b and N, or c and N, will yield p or q. By the way, the largest common divisor of c and N is by definition the largest number that divides both numbers. It can be found by elementary algebraic means (see Problem 6.9).

Taking again the example when $N = 15$, the choice of $a = 4$ produces $r = 2$, so $b = a^{r/2} - 1 = 3$ and $c = a^{r/2} + 1 = 5$. The largest common divisor of 3 and 15 is 3, and the largest common divisor of 5 and 15 is 5. So in this way we have found the prime factors $p = 3$ and $q = 5$. Of course $N = 15$ is hardly a large number, and we do not need this large machinery to find its prime factors. Nevertheless, it illustrates what the procedure is for any N.

It can be proven that the probability of r being even and b or c not being a multiple of N is at least $\frac{1}{2}$. If we get unlucky the first time, getting either an odd r of b, c being multiples of N, chances are that we will be lucky when we try again.

We will now discuss how to find r for a given a.

Define a function $f(x) \equiv a^x \bmod N$, where x ranges from 0 to $2^n - 1 \equiv S - 1 \equiv S_1$, for some $S > N$ to be specified later. Using Eq. (6.6), we see that $f(x) = f(x + r)$ so the function f is periodic with a period r, if $S/r \equiv K$ is an integer. In that case, we may use the period-finding algorithm of the last section to obtain r. However, as illustrated in the diagram above, generically K will not turn out to be an integer, so the period-finding algorithm of the last section must be refined. We will not describe the details here, though the idea is quite simple. We must pick S to be large enough so that S/r is a large number, in which case K is so large that the remainder from an integer is not too important. Detailed analysis shows that an S of the order of N^2 is sufficiently large for that purpose.

6.7.1 Summary

The problem of finding the prime factors p and q from $N = pq$ can be reformulated as a problem of finding the period of the function $f(x) = a^x \bmod N$, for any number a which has no common factor with N. The method explained in the last section can then be used to find the period of $f(x)$, and from there the prime factors p and q.

6.8 Grover's Search Algorithm

Given an unstructured database with N records, we want to pick out one with some specific properties. What is the most efficient way of doing that?

For simplicity we will assume only one record in the database has exactly these properties. This is not a necessary condition for the algorithm, but we will not discuss the additional details needed to remove this restriction. There are also other variations of the algorithm but here we will stick only to the basic version.

For example, suppose there is a database containing all the essential data of every person living in a country. Suppose we know that one and only one person in the country is 120 years of age, but we do not know anything else about the person — not the name nor the address. We would like to pull that record from the database to find out possible clues to her/his longevity. What is a good and fast way to find the record?

Naively we simply have to pull and examine every record until we find the right one. That takes on the average $N/2$ trials, so the number of steps required is $O(N)$.

Note that if the database were structured, then there are much faster ways to find the desired record. For example, if all the records are already arranged by age, then no algorithm is needed because we simply pull the last few records. The sorting problem is the hardest when the database is unstructured.

In this section we shall describe a quantum algorithm invented by Lov Grover, which accomplishes the task in $O(\sqrt{N})$ steps. If the country has 400 million people, this can be accomplished in of the order of 20,000 steps, rather than 200,000,000 steps.

By inserting empty records if necessary, we may assume the length of the database to be $N = 2^n$ for some n. In that case we can label the records by an n-bit integer x. Suppose the desired record is located at x_0, then the purpose of the algorithm is to find this number x_0.

To determine whether a record is the one we want, we must examine it for the desired properties. Such an examination can be carried out by a subroutine designed specifically to look for the properties we want, and we shall assume that such a subroutine is already available, and is given to us in the form of a black box, or an oracle. When we feed the xth record into the black box, it will examine it and spit out either a 1, when it is the desired record, or a 0, when it is not. In other words, the black box can be replaced by a 'yes-no function' $f(x)$, with $f(x_0) = 1$ and $f(x) = 0$ for all other x.

The idea of the algorithm is as follows. Start with a uniform state,

$$|\Upsilon\rangle = \frac{1}{\sqrt{N}} \sum_{x=0}^{N-1} |x\rangle \,,$$

and use the oracle repeatedly to filter out the desired state $|x_0\rangle$. As indicated before, this can be done in $O(\sqrt{N})$ steps. We will now proceed to describe how this is accomplished. Readers not interested in such details can skip to the next section.

We start from the uniform state $|\Upsilon\rangle$ obtained by applying the n-bit HADAMAD gate $\mathcal{H}$ to the ground state $|0\rangle$. A GROVER gate (G) is then used repeatedly to filter out from $|\Upsilon\rangle$ the desired state $|x_0\rangle$.

The GROVER gate is defined by

$$G = -\mathcal{H}I_0\mathcal{H}I_{x_0} ,$$

where $\mathcal{H}$ is the n-bit HADAMAD gate mentioned above, and I_{x_0} is a 1-bit gate that reverses the sign of the state $|x_0\rangle$ but leaves all other states $|x\rangle$ unchanged. In other words,

$$I_{x_0}|x\rangle = (-1)^{f(x)}|x\rangle .$$

Similarly, I_0 is a gate that reverses the sign of the state $|0\rangle$, but leaves all other states unchanged.

I_{x_0} can be constructed from the $\mathcal{U}_f$ gate of Eq. (6.2). In addition to the n-bit register which holds $|x\rangle$, we need an auxiliary 1-bit register which holds $|0\rangle$ at the beginning. Then

$$|x\rangle|0\rangle \rightarrow \boxed{H_{II}} \rightarrow |x\rangle\frac{1}{\sqrt{2}}\left(|0\rangle - |1\rangle\right)$$

$$\rightarrow \boxed{\mathcal{U}_f} \rightarrow |x\rangle\frac{1}{\sqrt{2}}\left(|0 \oplus f(x)\rangle - |1 \oplus f(x)\rangle\right)$$

$$= \quad (-1)^{f(x)}|x\rangle\frac{1}{\sqrt{2}}\left(|0\rangle - |1\rangle\right) \rightarrow \boxed{H_{II}} \rightarrow (-1)^{f(x)}|x\rangle|0\rangle ,$$

so that the output of $I_{x_0}|x\rangle$ appears at the end in the first register. In arriving at this result, the fact that the property function $f(x)$ is either 1 or 0 has been used.

Carrying out the algebraic computation, which we skip, it can be shown that passing the uniform state $|\Upsilon\rangle$ through the GROVER gate k times will result in the state:

$$|\Upsilon_k\rangle = \frac{1}{\sqrt{N-1}}\left(\cos[(2k+1)\theta]\sum_{x \neq x_0}|x\rangle + \sin[(2k+1)\theta]|x_0\rangle\right) ,$$

where θ is determined by $\sin\theta = 1/\sqrt{N}$. For large N, the angle θ is small. We see from this formula that the weight of the uniform state is decreased through each pass, and the weight of $|x_0\rangle$ is correspondingly enhanced. After k passes through the GROVER gate, where $(2k+1)\theta$ is chosen to be as close to $\pi/2$ as possible, namely,

$$k \simeq \frac{1}{2}\left(\frac{\pi\sqrt{N}}{2} - 1\right) ,$$

then $|\Upsilon_k\rangle \simeq |x_0\rangle$. Even if the right-hand side of this equation is not an integer, there is still a very high probability of finding $|\Upsilon_k\rangle$ in the state $|x_0\rangle$ when N is large. In that case we simply have to run the search algorithm several times to determine the exact value of x_0.

6.8.1 Summary

An unstructured database of N records can be sorted in $O(N)$ steps classically, but only $O(\sqrt{N})$ steps quantum mechanically. The Grover algorithm for doing that starts with the uniform state of all records to enable maximal parallel processing, then it uses the sorting criterion to design a filter to weed out the unwanted records. When this filtering procedure is carried out a suitable number of times, what is left is the desired record.

6.9 Hardware and Error Correction

All these nice quantum algorithms are useless unless we can build quantum gates to process them. That unfortunately is a difficult task for control gates, so much so that the largest quantum computer available, as of December 2001, is a 7-qubit system built by the IBM group, using the NMR scheme (to be discussed later). Earlier, in 2000, IBM also announced a 5-qubit device, each qubit being supplied by a fluorine atom in a single molecule.

To build a quantum computer, we must choose a medium to store the qubits, find ways to prepare an initial state, and means to read out the final state. We must also construct the quantum gates needed to manipulate the data, while maintaining coherence throughout their operations. Some of these problems have been solved in small prototypes, but not yet for large computers.

The amount of time it takes to carry out a gate operation is also an important parameter. The speed of the computer depends on it, and whether the state can stay coherent during that time may also depend on it. The ability to maintain coherence over a sufficiently long time depends on the careful choice of the system and the basis states. Detailed discussion of this type tends to be technical so we will skip it.

For most systems, initial states and 1-bit gates are relatively simple to construct (see Sec. 6.4).

The final state of the particle can be read bit-by-bit using fluorescence, as depicted in the picture below. To do so an auxiliary state $|aux\rangle$ which couples strongly to $|1\rangle$ is required. Suppose $|aux\rangle$ is an energy $\hbar\nu_f$ above $|1\rangle$. Shining an electromagnetic pulse of frequency ν_f on the particle in state $|1\rangle$ causes it to jump to $|aux\rangle$, and then it will relax into the ground state by emitting a fluorescent photon. Thus a florescent photon is present for $|1\rangle$ but not for $|0\rangle$. If the particle is in a schizophrenic state, then the probability of detecting a photon is proportional to its probability to be in state $|1\rangle$. Since the act of measurement generally changes the state of the particle, to get the probability we must repeat the calculation a sufficient number of times. This will not change the estimate of the number of steps for an efficient algorithm if the number of repetitions is insignificant compared to that number of steps.

Needless to say, we may also take the occupation of $|0\rangle$ rather than $|1\rangle$ to fluoresce.

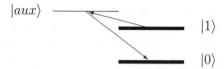

2-bit controlled gates are much harder to construct, because of the necessary correlation between the control qubit and the data qubit. Its fabrication depends on the system as will be discussed later. In fact, the ability to produce 2-bit control gates is an important criterion for choosing the system.

We come to the question of errors. Errors are unavoidable, even in a classical computer. The problem is worse in a quantum computer because additional errors can creep in through decoherence and the quantum gate operations. To keep the calculation reliable, there must be ways to recognize the errors and the means to correct them. If errors occur too frequently, then there is nothing one can do except to build a better computer. If errors occurs infrequently, then there are well-known mechanisms in classical computers to recognize and correct for them. These go under names such as *parity* and *majority voting*; duplication and repetition are the keys. In a quantum computer, there are more ways for errors to occur, so a built-in error correction algorithm is even more crucial. We cannot do it quite like the classical way, because of the no-cloning theorem, and because we must correct for a continuous change between $|0\rangle$ and $|1\rangle$, not just flips between 0 and 1. Nevertheless, the ideas are very similar, though the details are far more complicated, and beyond our present scope.

The number of steps in a quantum algorithm given previously assumes an idealized situation where no error correction is necessary. Since quantum error correction codes are longer than classical error correction codes, taking them into account will reduce the advantage we gained for a quantum algorithm over a classical algorithm. How much that sets us back depends on the quality of the quantum computer, which is inversely related to the length of the error correction code.

In the remainder of this section, we shall describe some systems used to experiment with the construction of qubits and quantum gates. We will concentrate on 2-qubit correlation, in some way the most difficult objective to achieve, but absolutely needed for the construction of 2-bit control gates.

6.9.1 *Trapped Ions*

This system consists of a string of ions placed in a linear Paul trap. A linear Paul trap is a device using static and radio-frequency electric fields to confine charged particles along a line. Each ion, depicted below by a round dot, represents a qubit with its two selected levels. The entire chain of ions is laser cooled and vibrates together, either in its vibrational ground state, or the first excited state, thus creating another qubit (called *phonons*) which is depicted by a pair of thick lines. The phonon is clearly coupled to the qubit of each ion because an ion is part of the chain.

Suppose the ground state $|0\rangle$ and the excited state $|1\rangle$ of each ion is separated by an energy difference $\hbar\nu$, and the ground state $|0\rangle$ and the first excited state $|1\rangle$ of the phonon is separated by $\hbar\nu_p \ll \hbar\nu$. The energy levels of the 2-bit state $|\textbf{phonon},$ ion$\rangle$ can be taken to be 0, $\hbar\nu_p$, $\hbar\nu$, $\hbar(\nu + \nu_p)$, respectively for the states $|00\rangle$, $|10\rangle$, $|01\rangle$, $|11\rangle$. Notice the difference between this level structure and that of two ions. The energy difference between $|1\rangle$ and $|0\rangle$ depends on what state the phonon is in. In other words, there is a correlation between the two qubits. This correlation can be used to construct control gates:

$$
\begin{array}{ll}
\rule{2cm}{0.4pt} & |aux\rangle \\
\hbar(\nu + \nu_p)\rule{1.5cm}{0.4pt} & |11\rangle \\
\hbar\nu \quad \rule{1.5cm}{0.4pt} & |01\rangle \\
\\
\\
\hbar\nu_p \quad \rule{1.5cm}{0.4pt} & |10\rangle \\
0 \quad \rule{1.5cm}{0.4pt} & |00\rangle
\end{array}
$$

Let us illustrate this point by constructing a CU gate in which U leaves $|0\rangle$ unchanged but adds a minus sign to $|1\rangle$. In this construction an auxiliary level $|aux\rangle$ of the ion must be present, say with an energy $\hbar\nu_{aux}$ above $|11\rangle$.

A pulse of frequency ν_{aux} shone on the ion for a suitable amount of time can cause $|11\rangle$ to jump to $|aux\rangle$ and then back to itself, gaining a minus sign. None of the other states are affected because that particular frequency cannot excite any others. In this way it accomplishes what the desired CU gate sets out to do.

This scheme works for a $|\textbf{phonon},$ ion$\rangle$ state but not a $|$ion, ion$\rangle$ state, because of the lack of direct coupling between the two ions. To see this more clearly, assume again that there is an auxiliary state $|aux\rangle$ in an ion. Since the two ions are identical, this state will be present in the other ion as well. Now a pulse shone on one ion can cause its $|1\rangle$ state to jump to $|aux\rangle$ and back, but that would be the case whether the other ion is in state $|0\rangle$ or $|1\rangle$. This lack of correlation makes it impossible for either ion to act as a control bit.

6.9.2 Photons

In this system, 0 is represented by a photon coming in one direction, and 1 is represented by a photon coming from an orthogonal direction. The 1-bit gate of this system is shown in the first graph in Sec. 6.4.

For detection of photons, one can use charge coupled devices (CCD) and photo-multipliers. However, single photon detectors are not yet well developed.

Production is more tricky, because we need to have exactly one photon in the channel. A laser pulse produces many photons, but if we make it sufficiently weak, then the probability of having two or more photons can be made arbitrarily small. However, in that case, there is also a higher probability of having no photon, than having one. This would severely affect the speed of the computer, but even without worrying about that, we must also have a non-destructive way to find out when a photon is present to do the calculation. This can be achieved through non-linear optics. Material exists that can convert one photon into two less energetic ones. One of these two photons is detected destructively in the usual way, but the other is untouched so it can go on to do its job.

What eventually fails this simple system is the great difficulty in constructing 2-bit gates. For that, we need two photons to interact to produce correlations. Photons hardly interact directly, but in principle, this can be achieved through some non-linear material by using the Kerr effect.

The Kerr effect refers to a dependence of the index of reflection of a material on the intensity of the beam. In this system, the 2-bit states $|01\rangle$ and $|10\rangle$ are represented by two photons traveling in opposite directions, and the 2-bit states $|00\rangle$ and $|11\rangle$ are represented by photons traveling in the same direction. This causes the intensity to go up and a shift in the index of refraction of the material to occur, thus causing a relative phase shift between the states $|00\rangle$ and $|11\rangle$ on the one hand, and $|01\rangle$ and $|10\rangle$ on the other hand. This correlation can be exploited to construct 2-bit controlled gates.

The problem is, all known Kerr materials have a weak dependence on the intensity. Moreover, these materials cause a lot of absorption. To overcome the first problem, the Kerr material the photons go through has to be thick, but then we run into the second problem, which nobody knows how to solve at the moment.

6.9.3 Nuclear Magnetic Resonance (NMR)

In the presence of a static magnetic field, the two degenerate levels of a spin-$\frac{1}{2}$ *nucleus* split up. The size of this *Zeeman effect* depends on the strength of the magnetic field, as well as the magnitude of the nuclear magnetic moment. Since the magnetic moment of a particle is inversely proportional to its mass, nuclear magnetic moments are small because nuclei are heavy. The resonance frequency between the Zeeman doublets of a hydrogen nucleus is typically around 500 MHz in a 11.7 tesla magnetic field. This radio-frequency pulse has a wavelength much larger than the size of an atom, so it is very difficult to target the pulse at a single nucleus unless the atoms are very far apart. The other difficulty is the weakness of the signal generated by a single nucleus, because of the smallness of its magnetic moment. The only recourse then is for the pulse to address many molecules and their nuclei all at once, usually in a solution to allow free movements of the molecules.

There is another difficulty, arising from the small energy difference between the Zeeman doublet, because it is far smaller than the thermal energy at room temperature. As a result, the system is no longer in its ground state, as thermal excitation causes both $|0\rangle$ and $|1\rangle$ to be occupied, though the occupation number of the ground state $|0\rangle$ is slightly larger than that of the $|1\rangle$ state. We shall refer to this effect as *thermal pollution*. How can we separate a $|0\rangle$ from a $|1\rangle$ then? And, since we are not targeting individual nuclei, how can we get several qubits out of the system?

The second problem is solved by using, in the solution, molecules with several atoms. In the ideal situation, (the nucleus of) each atom of the molecule represents a different bit. The reason why this works is the same reason why NMR is useful in chemistry. The magnetic field felt by the nucleus is shielded by the electronic cloud in the molecule. Thus, depending on its local electronic environment, the NMR frequency may differ from nucleus to nucleus. In chemistry this is used to probe the environment of the nucleus, namely, the molecular structure. For our purpose, we use this property to set up several distinct qubits, if their corresponding nuclei have sufficiently different NMR frequencies. By using (or detecting) radio pulses of the right frequency, we can selectively address (or detect) a particular qubit. For example, if the Zeeman frequency of the ith nucleus is ν_i, then a $|1\rangle$ state in the ith bit will yield an NMR line of frequency ν_i when it decays, whereas such a line will be absent if this bit is $|0\rangle$. By counting which of these lines is present, we will be able to tell which bits are 1 and which bits are 0.

As mentioned before, a 7-qubit NMR quantum computer has been successfully constructed by IBM. Unfortunately, we cannot expect a sizable quantum computer to be built this way because it is very hard to scale up NMR computers indefinitely. To do so, we need to find a molecule large enough whose distinct NMR signals for different nuclei can be controlled. Even so, thermal pollution, which increases with the number of qubits, will eventually drown out the signal and render the computer unworkable.

Still, pretending there is no thermal pollution, let us see how gates can be constructed in this system. 1-bit gates are constructed in the usual way by suitable radio pulses. 2-bit controlled gates require interactions between two nuclei, which fortunately is naturally present in the form $\hbar s_1 s_2 \nu_N$, where $s_1 = \pm\frac{1}{2}$ and $s_2 = \pm\frac{1}{2}$ indicate the $|1\rangle/|0\rangle$ state of each nucleus. This interaction is small — $\nu_N \ll \nu_1$, ν_2 — nevertheless it causes the energy of the first nucleus to depend on the state of the second nucleus, and vice versa, as is the case between the ion bit and the phonon bit in the ion-trap system. We can therefore exploit this correlation in a similar way to construct control gates.

Finally, let us consider how thermal pollution is dealt with. Suppose n_1 and n_0 are respectively the number of nuclei in the states $|1\rangle$ and $|0\rangle$, with $n_0 > n_1$. We can consider this as having $(n_0 - n_1)$ nuclei in the quantum state $|0\rangle$, in the midst of an incoherent background of n_1 nuclei found equally in the two states. It can be shown that this incoherent background is not affected by unitary gate operations, so in the idealized situation we can ignore it and concentrate only on what the $|0\rangle$ state is doing as it goes through the gates. At the end, the incoherent background has to be subtracted out to get the final answer. Since the incoherent background is much larger than the signal, this gets to be very difficult when too many qubits are involved.

6.9.4 Other Systems

None of the systems considered above seem to be ideal for scaling up the size of the quantum computer. Solid state devices like quantum dots (see Chapter 5) and superconductors may have a better chance in that regard. There is now an active program in these areas, but the success is still fragmentary. Only more research will be able to tell what the best system is going to be.

6.9.5 Summary

Practical issues in the construction of a quantum computer are briefly discussed. Maintenance of coherence and the ability to construct 2-bit control gates are two of the most important issues governing the selection of a practical medium. Among them, systems using ion traps, photons, and nuclear magnetic resonance are separately discussed.

6.10 Cryptography

We might think that cryptography is for spies — not so in this electronic age. If you do not want people to snoop on your communications, if you do not want information on your credit card to be stolen when shopping on the Internet, and if you do not want your transaction with the bank to be known to others, you have to encrypt your message before you communicate. In modern browsers this is done for you automatically, using the RSA public key system. This and a private key system will be discussed in this section, with special attention to how quantum technology may affect their outcome.

To begin with, we shall always assume the original message (known as the plaintext) to be expressed in integers. For example, we may use a two-digit decimal number to represent every key on the computer keyboard. The numbers 0 to 9 could be represented by 01 to 09, the 26 lower case letters 'a' to 'z' could be represented by 11 to 36, and the 26 upper case letters 'A' to 'Z' could be represented by 41 to 66, leaving 99 to represent space, and all other numbers up to 98 to represent the punctuation marks and the rest. The plaintext message 'Send me 1000 dollars' will appear in numerical form as 59 15 24 14 99 23 15 99 01 00 00 00 99 14 25 22 22 11 28 29.

This *ad hoc* system of convention is not in use; it is presented here just for simple illustration. There are more standard conversion schemes, like the ASCII code, which are usually used. We shall always assume that the sender and the receiver have agreed on which conversion scheme is being used, so that when the plaintext is sent directly, they can understand each other. The method of encryption discussed below is independent of the conversion scheme.

For security reasons, instead of sending the message over an open channel, the sender A (often known as Alice) converts the plaintext numbers into coded numbers, which only the intended receiver B (often known as Bob) can decipher. An eavesdropper E (often known as Eve) may be able to intercept the string of coded numbers, but she will have absolutely no idea what they mean.

6.10.1 Private Key System

In this system, both A and B possess a highly guarded string of numbers, known as a private key. A codes her message by combining the numbers in the plaintext with the numbers of the private key in some reversible manner. When B receives the coded message, he simply reverses the process, combining the coded message with the private key to get back the plaintext.

Let us illustrate this with the *ad hoc* conversion scheme above, taking as the private key k the numerical value of π with the decimal removed:

$$k = 31415926535897932384626433832795028841971693993 75\ldots$$

In practice we would never use such an obvious private key that every scientist can guess at, but for illustrative purpose this would do.

Since each character of the *ad hoc* system is given by a number between 0 and 99, we shall separate the string of numbers in the plaintext p and the private key k into pairs. We can create the coded number c by adding the corresponding pairs in p and k, modulo 100. For example, the coded message c for the plaintext 'Send me 1000 dollars' appears as follows:

```
      S  e  n  d     m  e       1  0
p 59 15 24 14 99 23 15 99 01 00
k 31 41 59 26 53 58 97 93 23 84
c 90 56 83 40 52 81 12 92 24 84

      0  0     d  o  l  l  a  r  s
p 00 00 99 14 25 22 22 11 28 29
k 62 64 33 83 27 95 02 88 41 97
c 62 64 32 97 52 17 24 99 69 26
```

When B receives the coded message c, all he has to do is subtract k from it (modulo 100) to recover the plaintext p.

The private key system is absolutely secure if the key is as long as the message, kept confidential, and used only once. In practice that is almost impossible to achieve. To compromise, we might have a key of finite number of digits, $2L$ say, and we can use it to code blocks of plaintext L characters long. We can also use the same key in several messages, but we must make sure that it is not used too often, for if many messages coded the same way have been intercepted, then there may be ways to figure out the key because some letters are statistically used more frequently than others. For that reason the private key k has to be changed often, but then the difficulty is how to communicate the new private key between A and B in a secure way. It is here that quantum properties can help.

According to the no-cloning theorem discussed in Sec. 6.4, quantum states cannot be copied faithfully without altering them. It is essentially this property that allows a private key to be transmitted securely between A and B. To illustrate how this is done, let us discuss the system of C.H. Bennett and G. Brassard using polarized photons. In practice, it is not easy to communicate quantum information over a large distance, because of decoherence and absorption. However, using optical fibers, this scheme has now been carried out successfully over many kilometers.

In this scheme, A sends B her private key over an open channel. The key is sent in binary using linearly polarized photons to represent 0 and 1. The trick is to transmit them randomly between two polarization bases. In one basis, to be denoted by $\oplus$, photons polarized along the x-direction represent a 1 and photons polarized along the y-direction represent a 0. In the other basis, to be denoted by $\otimes$, photons polarized along the 45° line represent a 1, and photons polarized along the 135° line represent a 0. The polarized states of one basis is an equal superposition of the states in another basis, so according to the probability rules of quantum mechanics, a 0 in the $\oplus$ basis has equal probability of being read off as a 0 or a 1 in the $\otimes$ basis.

To begin with, A sends the private key she invented to B using randomly one of these two basis for each bit. The random basis provides the coding and the security, as we shall see. Not knowing what basis A uses to send out a bit, B will detect the coded message using another random set of bases of his choice. For a given bit being sent out, say 0, B will detect it as a 0 if he happens to use the same basis she uses, but B will detect it with equal probability either as a 0 or a 1 if he happens to use the other basis.

Now A and B talk over an open channel to tell each other the bases they used for each bit, and then they keep only those bits where they happen to use the same basis. This shortened message is known as the *sifted key*. In doing so of course they have to agree on a way to authenticate themselves so that no imposter can get into the act at this point.

Suppose the signal sent from A is intercepted by E before it gets to B. Depending on the basis E uses, she may or may not get the right bit that A sends out, and she will not know. After intercepting a bit, E either resends it in the state she detects it, or not. If she does not resend it, then B will have a lost bit and that bit will be discarded in the sifted key. If she resends it to B, and that happens to be the wrong bit because E is using the wrong basis, then the sifted key determined by B may not be identical to the sifted key possessed by A. To guard against this, A and B have to get on the open channel again to compare a portion of the sifted key. If every bit of this test portion coincides, chances are that there are no eavesdroppers. In that case they discard this tested portion (which has become open knowledge) and use the remainder of the sifted key as their final private key. If some of the bits of the test portion do not coincide, then there could either be a transmission error, or an indication of the presence of an eavesdropper. If the number of disputed bits is small, they can still use the sifted key as the private key after discarding the tested portion. In that case if E is present, she will only get a small portion of the private key which is likely to be useless to her. The details of this can be worked out mathematically to determine how small a portion is tolerable.

Let us illustrate this with an example, in which no disputed bits occur in the tested portion. The original message from A is given in the first two lines, and the received message in B's basis is given in the next two lines. The sifted key is indicated in the fifth line, and the tested bits are indicated by a $\times$ in the sixth line, leaving the final private key in the last line:

A's basis	$\oplus$ $\oplus$ $\otimes$ $\oplus$ $\otimes$ $\otimes$ $\otimes$ $\otimes$ $\oplus$ $\otimes$ $\oplus$ $\oplus$ $\oplus$
A's message	1 1 0 1 0 0 0 0 1 1 0 1 0
B's basis	$\oplus$ $\otimes$ $\otimes$ $\oplus$ $\oplus$ $\otimes$ $\otimes$ $\otimes$ $\oplus$ $\oplus$ $\otimes$ $\oplus$ $\oplus$
received message	1 1 0 1 1 0 0 0 1 1 1 1 0
sifted key	1 0 1 0 0 0 1 1 0
test portion	$\times$ $\times$ $\times$ $\times$ $\times$
private key	1 0 1 0

To summarize, a raw key in binary is transmitted from A to B via the two linearly polarized states of a photon. To ensure security, each bit is sent out randomly using one of the two polarization bases, $\oplus$ or $\otimes$. B detects the bits in a set of random bases of his choice, so the received bit coincides with the transmitted bit when

A and B happen to use the same basis for that bit. Subsequent to the message, A and B compare their bases openly and retain only those bits for which their bases are the same. This constitutes the sifted key. To test whether an eavesdropper is present, they take the same portion from each sifted key to compare them over an open channel. If they are almost identical, a private key is obtained by discarding this test portion from the sifted key. Otherwise, the method fails.

6.10.2 The RSA Public Key System

In the private key system, the sender and the receiver both possess the same private key. In the public key system, B supplies everybody with a public key which can be used to code their messages to B. To decode these messages, B keeps a secret key to himself. This system is convenient because the secret private key does not have to be transmitted. Moreover, everybody can send B a coded message that only B can decipher. The drawback is that this system is not *absolutely* secure, though it can be designed so that it is *almost* impossible to break, not with the present computational power anyway. We shall illustrate this system with a very famous scheme invented by R. Rivest, A. Shamir, and L. Adleman, known as RSA encryption. It works as follows.

B picks two large prime numbers, p and q, and computes from them two public keys N and k, which he broadcasts, and a secret key s, which he retains. The public key N is simply $N = pq$, and the second public key k is any number which has no common factor with $p-1$ nor $q-1$. The secret key is computed using the following formula:

$$s = k^{\phi(\phi(N))-1} \bmod \phi(N),$$

where $\phi(N)$ is the Euler function defined below.

Given a positive integer n, $\phi(n)$ is defined to be the number of integers less than n which has no common factor with n. In this regard the number 1 is always considered to have no common factor with n, so $\phi(n) \geq 1$. If n is a prime number, then no factor less than n has a common factor with it, so $\phi(n) = n - 1$. If $N = pq$ where both p, q are prime numbers, then $\phi(N) = (p-1)(q-1)$ for the following reason. The only number less than N that has a common factor with N is either an integral multiple of p, and there are $q-1$ of them whose product with p is less than N, or, an integral multiple of q, and there are $p-1$ of them. So the number of integers smaller than N with no common factor with it is $(pq-1)-(q-1)-(p-1) = (p-1)(q-1) = \phi(N)$.

As an example for the computation of these three keys, let us suppose $p = 11$ and $q = 13$. Then $N = 143$. We may pick k to be any number without a common factor with $p-1 = 10$ and $q-1 = 12$. Let us pick $k = 7$ in this example. From the formula above, $\phi(N) = (p-1)(q-1) = 120$, and one can find out from tables that $\phi(\phi(N)) = \phi(120) = 32$. Thus, $s = k^{\phi(\phi(N))-1} = 7^{32-1} \bmod 120$, which is 103. This is the secret key.

We will now describe how to use the public keys N and k to code a message, and how to use the secret key s to decode a message.

Coding

We shall assume the plaintext message to be given by a string of numbers t. Using the public keys k and N, each number t of the string is changed into a coded number c by the formula:

$$c = t^k \bmod N.$$

For example, if the message 'Send me 1000 dollars' is given by the string of numbers 59 15 24 14 99 23 15 99 01 00 00 00 99 14 25 22 22 11 28 29 discussed before, then the coded string of numbers using $k = 7$ and $N = 143$ is

p	59	15	24	14	99	23	15	99	01	00	00	00
c	71	115	106	53	44	23	115	44	1	0	0	0

p	99	14	25	22	22	11	28	29
c	44	53	64	22	22	132	68	94

Decoding

Using the secret key s, the coded number c can be decoded into the original number t using the formula:

$$t = c^s \bmod N.$$

It is straightforward using this formula to check from the table above that this does recover the plaintext t from the coded text c. For example, if $c = 44$, using $s = 103$ and $N = 143$, we get for $c = 44$ a value of p equal to $44^{103} \bmod 143 = 99$.

We shall now show why this decoding works. The reader who is not interested may skip this.

The decoding formula works because of Euler's theorem in number theory, which states that $x^{\phi(n)} = 1 \bmod n$, whenever x and n have no common factor. We may apply this theorem to $n = \phi(N)$ and $x = k$, because k is chosen to have no common factor with $\phi(N) = (p-1)(q-1)$. Then we get $k^{\phi(\phi(N))} = 1 \bmod \phi(N)$. Since s is calculated using the formula $s = k^{\phi(\phi(N))-1} \bmod \phi(N)$, we conclude from Euler's theorem that $ks = k^{\phi(\phi(N))} \bmod \phi(N) = 1$. Since the coded number c is equal to $p^k \bmod N$, hence $c^s \bmod N = p^{ks} \bmod N$. But we know that $ks = 1 \bmod \phi(N)$, or equivalently $ks = a\phi(N)$ for some integer a, it follows that $c^s \bmod N = p \cdot [p^{a\phi(N)}$

mod N]. Using Euler's theorem again on $x = p^a$ and $n = N$, the number in the square parenthesis is 1, and hence $(c^s \bmod N) = p$, as claimed.

Quantum Breaking

RSA cryptography relies on the fact that it is hard to factorize the public key N into its prime factors p and q, because once we know p and q we can compute the secret key to break the code. As mentioned previously, once a quantum computer is turned on and Shor's algorithm for factorization is used, we will get an exponential advantage in factorization so it is no longer difficult to factorize N, unless it is so huge that it becomes impractical to use the public and private keys generated by it to code and decode.

6.10.3 Summary

We considered two encryption systems. The difficulty with the private key system is the secure transmission of the private key. If coherence can be maintained over a long distance, this problem can be solved by using the Bennet–Brassard scheme.

The security of the RSA public key system relies on the difficulty of factorizing a large number into its two prime components. With a quantum computer and Shor's algorithm, this task is rendered much easier. As a result the security of the RSA public key system will become woefully inadequate.

6.11   Quantum Teleportation

In this last section, we consider briefly a scheme to transmit an image of a 1-qubit quantum state $|\psi\rangle = \alpha|0\rangle + \beta|1\rangle$ from A to B, without physically carrying the state across. This is known as quantum teleportation. We assume that although A is in possession of the state $|\psi\rangle$, and she does not know what the coefficients α and β are, for otherwise she can simply send over those two coefficients.

The scheme invented by Bennett, Brassard, Crepeau, Jozsa, Peres, and Wootters (BBCJPW) makes use of the correlation of an auxiliary entangled state to send $|\psi\rangle$ across. From now on we use subscripts to distinguish the different particles in the scheme. In addition to particle 1 which is in the state $|\psi\rangle \equiv |\psi_1\rangle$, we need the service of two more particles, 2 and 3. These two are prepared in an entangled state, *e.g.*, $|\phi_{23}^I\rangle = (|0_2\rangle|1_3\rangle - |1_2\rangle|0_3\rangle)/\sqrt{2}$. Particle 2 is given to A and particle 3 is given to B, but with the coherence of the state $|\phi_{23}^I\rangle$ maintained even when A and B are separated by a long distance. In practice this is the most difficult part to achieve, but by using low loss optical fibers, this can be done over a distance of tens of kilometers.

Now A is in possession of particles 1 and 2, whose wave functions are linear combinations of the four basis states $|0_1\rangle|0_2\rangle$, $|0_1\rangle|1_2\rangle$, $|1_1\rangle|0_2\rangle$, $1_1\rangle|1_2\rangle$. For the present scheme we should express the wave function as a linear combination of four orthonormal entangled states,

$$|\phi_{12}^I\rangle = \frac{1}{\sqrt{2}} \left(|0_1\rangle|1_2\rangle - |1_1\rangle|0_2\rangle\right),$$

$$|\phi_{12}^{II}\rangle = \frac{1}{\sqrt{2}} \left(|0_1\rangle|1_2\rangle + |1_1\rangle|0_2\rangle\right),$$

$$|\phi_{12}^{III}\rangle = \frac{1}{\sqrt{2}} \left(|0_1\rangle|1_2\rangle - |1_1\rangle|0_2\rangle\right),$$

$$|\phi_{12}^{IV}\rangle = \frac{1}{\sqrt{2}} \left(|0_1\rangle|1_2\rangle + |1_1\rangle|0_2\rangle\right).$$

Now the three-particle system is in the quantum state $|\psi_1\rangle|\phi_{23}^I\rangle$, and it is just a matter of simple algebra to rewrite this as

$$|\psi_1\rangle|\phi_{23}^I\rangle = \frac{1}{2} \left[|\phi_{12}^I\rangle(-\alpha|0_3\rangle - \beta|1_3\rangle) + |\phi_{12}^{II}\rangle(-\alpha|0_3\rangle + \beta|1_3\rangle) \right.$$

$$\left. + |\phi_{12}^{III}\rangle(+\alpha|0_3\rangle + \beta|1_3\rangle) + |\phi_{12}^{IV}\rangle(+\alpha|0_3\rangle - \beta|1_3\rangle)\right].$$

Now suppose A can carry out a measurement to tell in which of the entangled states $|\phi_{12}^I\rangle$, $|\phi_{12}^{II}\rangle$, $|\phi_{12}^{III}\rangle$, $|\phi_{12}^{IV}\rangle$ particles 1 and 2 are in. Once that is measured she picks up a telephone or send an email to tell B about it. Once he knows about this he can figure out what the state $|\psi_1\rangle$ is, thus succeeding in having the desired information transmitted from A to B.

For example, if her two-particle state is $|\phi_{12}^{II}\rangle$, then B immediately knows that his state is $-\alpha|0_3\rangle + \beta|1_3\rangle$, so all that he has to do is to pass his state through a simple 1-bit gate to reverse the sign of $|0_3\rangle$ to get himself an image of the state that A possesses, namely, $+\alpha|0_3\rangle + \beta|1_3\rangle$. If A reports getting another state, say $|\phi_{12}^{IV}\rangle$, then he will have to pass his state through another 1-bit gate, which reverses the sign of $|1_3\rangle$ instead.

6.11.1 Summary

The BBCJPW scheme for transmitting the quantum state of particle 1 requires the help of an extra pair of particles in an entangled state. Particle 2 of the entangled state is sent to A and particle 3 of the entangled state is sent to B. A then goes ahead to determine which of the four possible entangled states particles 1 and 2 are in. Once this is communicated to B, he can use this knowledge to produce an image of the original quantum state of particle 1.

6.12   Further Reading

Books

- R. P. Feynman, *Feynman Lectures on Computation*, edited by A. J. G. Hey and R. W. Allen (Addison-Wesley, 1996).
- D. Deustch, *The Fabric of Reality* (Viking Penguin Publishers, London, 1997).
- H.-K. Lo, S. Popescu and T. Spiller, *Introduction to Quantum Computation and Information* (World Scientific, 1998).
- A. J. G. Hey, *Feynman and Computation*: *Exploring the Limits of Computers* (Perseus Book, 1999).
- D. Bouwmeester, A. Ekert and A. Zeilinger, *The Physics of Quantum Information* (Springer, 2000).
- M. A. Nielsen and I. L. Chuang, *Quantum Computation and Quantum Information* (Cambridge University Press, 2000).
- S. Braunstein and H.-K. Lo (eds.), *Scalable Quantum Computers* (Wiley-VCH, 2000).

Review articles

- C. Bennett, *Quantum Information and Computation*, Physics Today, October, 1995, p. 24.
- A. Ekert and R. Jozsa, *Quantum Computation and Shor's Factoring Algorithm*, Review of Modern Physics **68** (1996) 733.
- J. Preskill, *Battling Decoherence: The Fault-tolerant Quantum Computer*, Physics Today, June, 1999, p. 24.
- D. Gottesman and H.-K. Lo, *From Quantum Cheating to Quantum Security*, Physics Today, November, 2000, p. 22.
- A. Ekert, P. Hayden, H. Inamori and D. K. L. Oi, *What is Quantum Computation?*, International Journal of Modern Physics **A16** (2001) 3335.

Web-sites

- http://lanl.arxiv.org/archive/quant-ph
- http://qso.lanl.gov/qc/
- http://www.ccmr.cornell.edu/ mermin/qcomp/CS483.html
- http://www.iqi.caltech.edu/index.html
- http://www.qubit.org
- http://www.rintonpress.com/journals/qic/index.html

6.13 Problems

6.1 Two theorems and one game:

1. Convert the decimal numbers 128 and 76 into binary numbers.
2. Suppose we have k binary numbers, $s_1, s_2, \ldots, s_k$, the largest of which has n bits. Add up these k numbers, *bit by bit, without carry*, and express the sum of each bit as a decimal number. In this way we get k decimal numbers $d_1, d_2, \ldots, d_n$.

    (a) Prove that if all d_i $(i = 1, 2, \ldots, n)$ are even, then it is impossible to decrease one and only one number s_j so that every number of the new bit-sum is still all even.
    (b) Prove that if at least one d_i $(i = 1, 2, \ldots, n)$ is odd, then it is possible to decrease one s_j so that every number of the new bit-sum is even.

There is an interesting game, NIM, whose winning strategy is based on these two theorems. It is a two-person game, played with k piles of coins, with s_i coins in the ith pile. The two players take turns to remove any number of coins from any one pile of his/her choice. The person who succeeds in removing the last coin wins.

Suppose we express s_i in binary numbers. The winning strategy is to remove coins so that every bit-sum after the removal is even. According to (a), his/her opponent must leave at least one bit-sum odd after his/her removal. Since the winning configuration is 0 in every pile, whose bit-sum is even, his/her component can never win.

For example, if there are three piles, with 3, 4, 5 coins each. Then $s_1 = 11$, $s_2 = 100$, $s_3 = 101$, so that $d_1 = 2$, $d_2 = 1$, $d_3 = 2$. The person who starts the game will be guaranteed to win by reducing the first pile of 3 ($s_1 = 11$) coins to 1 coin (01). This changes the bit-sum to $d_1 = 2$, $d_2 = 0$, $d_3 = 2$, so according to (a), his/her opponent can never win if he/she keeps on following the same strategy to the very end.

6.2 Show how to obtain the NAND gate from the AND and XOR gates.

6.3 Suppose the input register of a CN gate contains the number $|cd\rangle$, where c is the control bit and d is the data bit. Show that the output of the gate is $|c, d \oplus c\rangle$.

6.4 Compute 1000 mod 21.

6.5 These are exercises in relating the different gates:

1. Construct the NOT gate from the CN gate.
2. Construct the AND gate from the CCN gate.
3. Construct the XOR gate from the CN gate.
4. Construct the OR gate from the AND, XOR, and FANOUT gates.

6.6   Let θ be a real number, R_X be a gate that changes $|0\rangle$ to $\cos\theta|0\rangle - i\sin\theta|1\rangle$, and $|1\rangle$ to $\cos\theta|1\rangle - i\sin\theta|0\rangle$. Let R_Z be a gate that changes $|0\rangle$ to $e^{-i\theta}|0\rangle$ and $|1\rangle$ to $e^{i\theta}|1\rangle$. Show that $R_Z = \text{H}R_X\text{H}$, where H is the HADAMAD gate.

6.7   Let U be a 1-bit gate which leaves $|0\rangle$ unaltered, but changes $|1\rangle$ to $-|1\rangle$. Verify that the CONTROLLED-NOT gate CN can be obtained from the HADAMAD gate H and the CU gate as follows: $(\text{CN}) = \text{H}_2(\text{CU})\text{H}_2$, where H_2 represents the H gate operating on the second (data) bit.

6.8   This is an exercise on Deutsch's algorithm. Suppose x varies from 0 to 15, and suppose $f(x) = 0$ when $0 \le x \le 7$, $f(x) = 1$ when $8 \le x \le 15$. Find out what the state $|a\rangle$ is in Eq. (6.4).

6.9   The greatest common divisor of two numbers R_0 and R_1 can be obtained in the following way. Let R_0 be the larger of the two numbers. First, divide R_0 by R_1 to get the remainder R_2. Then, divide R_1 by R_2 to get the remainder R_3, and so on down the line. At the nth step, R_{n-1} is divided by R_n to get the remainder R_{n+1}. Continue this way until the remainder $R_{m+1} = 0$. Then R_m is the greatest common divisor of R_0 and R_1.

Use this method to find the greatest common divisor between

1. $R_0 = 124$ and $R_1 = 21$.
2. $R_0 = 126$ and $R_1 = 21$.
3. $R_0 = 21$ and $R_1 = 7$.
4. $R_0 = 21$ and $R_1 = 9$.

6.10   This is an exercise to factorize N by finding an even number r such that $a^r = 1 \bmod N$, where a is any number which has no common factor with N (see Sec. 6.7). Unless $b = a^{r/2} - 1$ or $c = a^{r/2} + 1$ divides N, otherwise the prime factors p and q of N can be obtained by finding the greatest common divisor between b and N, and between c and N.

Suppose $N = 21$.

1. Find out what r, b, c are when $a = 5$, then find out the greatest common divisor between b and N, and between c and N.
2. Find out what r, b, c are when $a = 2$, then find out the greatest common divisor between b and N, and between c and N.

Chaos: Chance Out of Necessity

7

7.1 Introduction: Chaos Limits Prediction

Physics is ultimately the study of change — of becoming. Changes are determined by the laws of physics that be. For example, we have Newton's three laws of motion. The laws themselves are, of course, believed to be changeless. The necessary connection of events implied by the deterministic nature of the physical laws leaves us no chance of freedom except for that of the choice of initial conditions, that is, the initial positions and velocities of all the elementary subunits that make up our system. Once this set of initial data is entered, the future course of events, or the process, is uniquely determined in detail and, therefore, predictable in principle for all times, indeed just as much as the known past is. Thus, for instance, the trajectories of all the 10^{19} molecules that belong in each cubic centimetre of the air in your room, suffering some 10^{27} collisions each passing second, are in principle no less predictable than those an oscillating pendulum or an orbiting planet — only much more complex.

You may recall that in order to specify the initial conditions for a single particle, taken to be a point-like object, we need to enter a set of three numbers, its Cartesian coordinates x, y, z, say, to fix its position. We say that the particle has three dynamical degrees of freedom. Another set of three numbers is required in order to fix the corresponding components of its velocity (momentum). For N particles these add up to $6N$ independent data entries. The state of the system can then be conveniently represented as a point in a $6N$-dimensional abstract space called *phase space*. The motion of the whole system then corresponds to the trajectory of this single representative phase point in this phase space. (The nature of phase space, or more generally speaking *state space*, of course, depends on the context. Thus, in the case of a chemical reaction we may be concerned with the concentrations x, y, z, say, of three reacting chemical species whose rate of change (kinetics) depends only on these concentrations. In that case the state space will be a three-dimensional one.) Deterministic dynamics or kinetics implies that there is a unique trajectory through any given phase or state point, and it is calculable in principle. In our example of the air in the room, $N = 10^{19}$! Behold the tyranny of large numbers! The complexity here is due to our having to store large amounts of input information and to solve as many equations of motion — the computational complexity of information processing. We will do well to remember here that a molecular dynamicist of today with free access to the fastest supercomputer available, capable of performing a billion floating point operations per second, can barely simulate the digitized motion of some 10^4 particles, and then only approximately. But these practical limitations are besides the point. In principle the motion is calculable exactly and hence a predictable claim.

It is true that we speak of chance and probability in physics, in statistical physics to wit, where we have the Maxwell–Boltzmann distribution of velocities of molecules of a gas, the Gaussian distribution of errors in a measurement, or the random walk of a Brownian particle (a speck of pollen or a colloidal particle floating in water, for example) and various averages of sorts. But these merely reflect an incompleteness of our knowledge of the details. The apparent randomness of the Brownian motion is due to its myriads of collisions, about 10^{21} per second, with the water molecules that remain hidden from our sphere of reckoning. In point of fact, even if we could calculate everything, we wouldn't know what to do with this fine-grained information. After all, our sensors respond only to some coarse-grained averages such as pressure or density that require a highly reduced information set. It is from our incompleteness of detailed information as also, and not a little, from our lack of interest in such fine details, that there *emerges* the convenient intermediate concept of chance, and of probability. But strictly speaking, everything can in principle be accounted right. There is, strictly speaking, truly no game of chance: the roll of the dice, the toss of the coin or the fall of the roulette ball, can all be predicted exactly but for the complexity of computation and our ignorance of the initial conditions. This absolute determinism was expressed most forcefully by the 19th century French mathematician Pierre Simon de Laplace. Even the whole everchanging universe can be reduced to a mere unfolding of some initial conditions, unknown as they may be to us, under the constant aspect of the deterministic laws of physics — *sub specie aeternitatis.*

But, in a truly operational sense, this turns out not to be the case. Laplacian determinism, with its perverse reductionism, is now known to be seriously in error for two very different reasons. The first, that we mention only for the sake of completeness, has to do with the fact the correct framework theory for the physical happenings is not classical (Newtonian) mechanics but quantum mechanics (See Appendix B). There is an Uncertainty Principle here that limits the accuracy with which we may determine the position and the velocity (momentum) of a particle simultaneously. Try to determine one with greater precision and the other gets fuzzier. This reciprocal latitude of fixation of the position and the velocity (momentum to be precise) allows only probabilistic forecast for the future, *even in principle.* Quantum uncertainty, however, dominates only in the domain of the very small, *i.e.*, on the microscopic scale of atoms and molecules. On the larger scales of the 'world of middle dimensions' of common experience and interest, deterministic classical mechanics is valid for all practical purposes. We will from now on ignore quantum uncertainty. We should be cautioned though that the possibility of a fantastic amplification of these quantum uncertainties to a macroscopic scale cannot be ruled out.

Macroscopic uncertainty, or rather the unpredictability, that we are going to talk about now, emerges in an entirely different and rather subtle manner out of the very deterministic nature of the classical laws. When this happens, we say that we have *chaos*, or rather *deterministic chaos* to distinguish it from the thermal disorder, or the *molecular chaos* of stochastic Brownian motion.

7.1.1 The Butterfly Effect

But how can a system be deterministic and yet have chaos in it. Isn't there a contradiction in terms here? Well, the answer is *no*. The clue to a proper understanding of deterministic chaos lies in the idea of *Sensitive Dependence on Initial Conditions.* Let us understand this first. As we have repeatedly said before, the deterministic laws simply demand that a given set of initial conditions lead to a unique and, in principle, calculable state of the system at any future instant of time. It is implicitly understood here, however, that the initial conditions are to be given to infinite precision, *i.e.*, to an infinite number of decimal places if you like. But this is an ideal that is frankly unattainable. Errors are ubiquitous. What if the initial conditions are known only approximately? Well, it is again implicitly assumed that the approximately known initial conditions should enable us to make approximate predictions for all times — approximate *in the same proportion*. That is to say that while the errors do propagate as the system evolves, they do not grow inordinately with the passage of time. Thus, as we progressively refine our initial data to higher degrees of accuracy, we should get more and more refined final-state predictions too. We then say that our deterministic system has a predictable behavior. Indeed, operationally this is the only sense in which prediction acquires a well-defined meaning. In terms of our state-space picture, this means that if we started off our system from two neighboring state points, the trajectories shall stay close by for *all* future times. Such a system is said to be well behaved, or *regular*. Now, the point is that deterministic laws do *not* guarantee this regularity. What then if the initial errors actually grow with time — that too exponentially? In our phase space picture then, any two trajectories that started off at some neighboring points initially, will begin to diverge so much that the line joining them will get stretched exponentially (as $e^{\lambda t}$, say) with the passage of time. Here, λ measures the rapidity of divergence (or convergence) according to whether it is positive (or negative). It is called the Lyapunov exponent. The initial instant of time can, of course, be taken to be any time along the trajectory. The condition $\lambda > 0$ is precisely what we mean by the sensitive dependence on initial conditions. It makes the flow in phase space complex, almost random, since the approximately known initial conditions do not give the distant future states with comparable approximation. The system will lack error tolerance, making long-time prediction impossible. This is often referred to picturesquely as the Butterfly Effect: the flap of a butterfly's wings in Brazil may set off a tornado in Texas. Some sensitivity! When this happens, we say that the dynamical system has developed chaos even though the governing law remains strictly deterministic. We might aptly say that chaos obeys the letter of the law, but not the spirit of it.

There is yet another way of expressing the sensitive dependence on initial conditions, without comparing neighboring trajectories. After all, a given dynamical evolution is a one-shot affair and it should be possible to express this characteristic sensitivity in terms of that single one-shot trajectory *per se*. It is just this: one cannot write down the solution of the dynamical equations in a closed, smooth (analytic) form valid for all times, since this would mean that the state variable at any time must be a smooth function of the initial conditions, which negates the sensitive dependence. This means that the evolution equation (*i.e.*, the algorithm for change) must be solved step by step all the way to the final time and the calculated values of the state variable catalogued. There is no short cut. There is a kind of computational complexity (or irreducibility) in this, despite the simplicity of the algorithm that generated the change.

Now, what causes this sensitive dependence on initial conditions? Does chaos require a fine tuning of several parameters, or does it persist over a whole range of the parameter values? Is chaos robust? How common is chaos? Is every chaotic system chaotic in its own way or is there universality — within a class maybe? How do we characterise chaos? Can a simple system with a small number of degrees of freedom have chaos, or do we need a complex system with a large, almost infinite number of degrees of freedom? Is the claim of distinction between a chaotic and a statistically random system mere nitpicking, or one of physical consequence. These are some of the questions that we will address in the following sections in a somewhat intuitive fashion. Collectively, these problems are studied under the forbidding heading of 'Dynamical Systems.' Deep results have been obtained in this field over the last three decades or so. Some questions still remain unanswered or only partially answered. But a physical picture of chaos has emerged, which is already fairly complete. It is likely to improve greatly with our acquaintance with some examples of chaos.

7.1.2 Chaos is Common

One may have the impression that the sensitive dependence on initial conditions necessary for chaos must require a fine tuning of control parameters, which can happen only accidentally in nature. This would make chaos an oddity that can rarely occur in our sensible world and hence an object only of mathematical curiosity. This notion is, however, erroneous. Chaos is a robust phenomenon that persists over a wide range of values of the control parameters. The sensitive dependence on initial conditions does not wash easily. Chaos is, therefore, not rare. Indeed, the contrapositive is true. Nature is full of it. Almost any real dynamical system, driven hard enough, turns chaotic. Looking for a chaotic system is like looking for a non-elephant animal! Most animals are non-elephant.

The most celebrated example of chaos is, of course, fluid dynamical turbulence — the last unsolved problem of classical physics. Great physicists of the 20th century have bemoaned, 'Why turbulence'? We see turbulence in jets and wakes, in water flowing through pipes and past obstacles. As the flow rate (the control parameter) increases beyond a critical value, and this is the meaning of the phrase 'driven hard enough,' the smooth laminar flow becomes unstable. It develops waviness and quickly turns into a complex, almost random pattern of flow, live with swirling eddies of all sizes as we move downstream. The flow pattern is aperiodic — it never repeats itself. This is fully developed turbulence. We can demonstrate this easily by injecting a marker dye in the flow tube, that makes the flow pattern visible to the eye (See Fig. 7.1a).

At the onset of turbulence the otherwise fine and straight thread of marker dye undulates wildly and quickly disrupts into a complex ramified pattern, down to

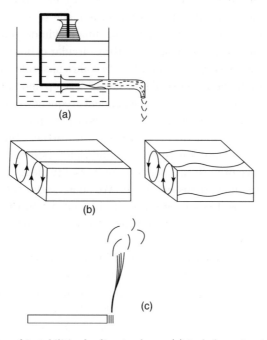

Figure 7.1: Examples of instabilities leading to chaos: (a) turbulence in pipe flow marked by a dye; (b) convective instability showing cylindrical rolls and their wobbling; and (c) rising smoke column developing whorls.

length scales too fine for the eye to discern, making the water appear uniformly colored as we move downstream — as if through thorough mixing. The threshold condition for the onset of this instability can be conveniently expressed in terms of a dimensionless control parameter called the *Reynolds number* (R), after the great fluid dynamicist Osborne Reynolds. The Reynolds number $R = UL\rho/\eta$, where U is the typical flow velocity (or rather variation of velocity), L the tube diameter (or length scale over which the velocity variation takes place), ρ the fluid density and η the fluid viscosity (treacliness). Turbulence sets in for R greater than a critical value R_c, which is about 2000 for water, but can be as high as 10^5 if the flow is increased very gently and the pipe is very smooth. (This is something like the superheating of water beyond its boiling point.) It is clear that the critical Reynolds number R_c is not a universal number. It depends on the geometry of the problem, whether the tube cross-section is circular or elliptical, for instance. But importantly, the sequence of flow patterns, or the route to turbulence turns out to be universal — within a class. (This is very reminiscent of the universality class of a second order phase transition. Thus, different ferromagnets have different Curie (critical) temperatures. But the behavior of magnetization near the critical point is universal within a wide class. These are deep problems.)

Another example of chaos that has played a decisive role in the experimental and theoretical study of turbulence-as-chaos is the *Couette–Taylor instability* problem. Here, the fluid is contained in the annular space between two long co-axial circular cylinders, one of which, usually the inner one, is made to spin. At low spin rates the fluid is dragged viscously around by the inner cylinder and there is a time independent laminar flow in the form of rotating circular cylindrical sheets. Beyond a critical spin rate, however, this laminar flow becomes unstable (the Couette–Taylor instability) and the circular cylindrical flow sheets develop undulations in the form of a stack of doughnuts or tori. A single independent frequency of oscillation appears. At a still higher spin rate another independent frequency appears making the system doubly periodic. As the spin rate is increased still further, the pattern suddenly jumps to a chaotic one. Again the threshold condition is non-universal but the route to chaos, the sequence of patterns, defines a universality class. This seems different from that of the pipe flow discussed above. We should note that the Couette flow is a closed one: the flow feeds back on itself. In contrast to this, the flow in a jet or wake is an open flow where the turbulence develops as we move downstream. It does not close on itself.

Yet another example of chaos is provided by the convective flow of a fluid mass which is heated from below. In fact this was the example that E. N. Lorenz studied so intensively as a model of his 'toy weather' that he had simulated on his computer, a Royal McBee, at MIT way back in 1963. His paper entitled '*Deterministic Non-periodic Flow*' (published in the *Journal of Atmospheric Sciences, Vol. 20, pp. 130–41, in 1963*), laid the foundation of the modern science of *chaos*. It was here that he had observed the sensitive dependence on initial conditions that makes long-range weather forecast impossible. The great French mathematician Henri Poincaré had hinted at such a sensitive dependence in 1903! (It is a sobering thought that Lorenz did all his simulations on a Royal McBee, a vacuum-tube based computer capable of doing just about 60 multiplications a second — a snail's speed compared to the modern Cybers and Crays capable of hundreds of millions of floating point operations a second.)

Returning now to our convection cell, nothing much happens if the temperature difference is small enough. Heat is simply conducted to the cooler top surface through molecular diffusion. At a higher critical temperature difference, this steady state becomes unstable (the Rayleigh–Bénard instability) and cylindrical rolls of convection develop in our cell, assumed to be rectangular for convenience (see Fig. 7.1*b*). This happens when the forces of buoyancy overcome the damping effects of viscous dissipation and thermal diffusion. Heated fluid at the bottom expands, and thus made lighter, ascends along the center to the cooler top layers, delivers heat there and then moves out and down the sides, completing the roll. On further heating beyond a threshold, this pattern becomes unstable and the rolling cylinders begin to wobble along the length. More complex flow patterns appear successively, cascading to a point of accumulation at which the flow becomes turbulent. The route to chaos (turbulence) is the so-called *period-doubling* bifurcation that we will discuss later. Incidentally, the control parameter in this case is the dimensionless *Rayleigh number* $R_a = g\rho\alpha h^3 \Delta T/\eta\kappa$, where g = acceleration due to gravity, κ = thermal conductivity, α = coefficient of thermal expansion, h = height of cell, and ΔT the temperature difference. The first instability sets in when R_a exceeds a critical number that depends on the geometry, *e.g.*, the aspect ratio of the cell. Convectional instability is common in Nature. On a hot summer's day one cannot fail to observe the pattern of clouds imprinted on the sky by the rising convectional currents.

There is an amusing example of turbulence that you can readily observe if you are a smoker, or have a smoker friend willing to oblige. The column of smoke rises from the lighted tip straight up to a height and then suddenly disrupts into irregular whorls (see Fig. 7.1*c*). Here again the hot smoke rises faster and faster through the cooler and so denser air until it exceeds the critical Reynolds number and turns turbulent. It is really no different from an open jet.

The examples of chaos cited above are all from fluid dynamics — turbulence. But chaos occurs in the most unexpected places. It is revealed in the ECG traces of patients with arrhythmiac hearts and the EEG traces of patients with epileptic seizures. It is suspected in the apparently random recurrence of certain epidemics like measles. It lurks in the macroeconomic fluctuations of commodity and stock prices. It is seen in chemical reactions and in the populations of species competing for limited resources in a given region. The irregular pattern of reversals of Earth's magnetic field is suspected to be due to the chaotic geodynamo. Chaos is, of course, well known in nonlinear electrical oscillators — the classic Van der Pol oscillator

and lasers, where chaos masquerades as noise. Chaos is implicated in the orbits of
stars around the galactic center. The list is endless. But most of these examples
suggest that chaos requires an interplay of a large number of degrees of freedom. Is
this really so?

7.1.3 Can Small be Chaotic?

Small here means smallness of the number of degrees of freedom that the dynamical
system may have. A number as small as three, say. Can such a small system exhibit
the apparent randomness that we associate with chaos? Chaos, as we have seen in
the case of turbulence, means on aperiodic flow with a long-range unpredictability
about it. The flow pattern must not repeat itself at the very least. Can a system
with a small, indeed *finite*, number of degrees of freedom be so 'infinitely inventive'
of the flow patterns as to go on surprising us forever without exhausting its stock
of possibilities and without having to repeat itself? That is the question. And the
answer that has emerged slowly but surely over the last three decades or so is that it
can, provided there is nonlinearity in it. This is a necessary condition. Dissipation
will help — it is there in all real systems anyway. This view disagrees sharply with
the ideas about turbulence that have prevailed since the time of the great Soviet
physicist L. D. Landau. Landau's paradigm for turbulence visualized a sequence
of an *infinite* number of competing oscillations with incommensurate frequencies,
emerging one at a time as the control parameter crossed the successive thresholds of
instabilities. So it is 'infinite' versus 'finite.' To appreciate this fully let us examine
some of the examples of chaos mentioned above more carefully.

The examples of chaos we have described so far are mostly about turbulence
in fluids. It is clear that the fluid flow is described completely by Newton's laws
of motion. Of course, they have to specialize, to the case of the fluid continuum
where they take on the form of the macroscopic equations of fluid dynamics —
the Navier–Stokes equations. This equation simply expresses the law that mass
times acceleration of an element of fluid equals the net force acting on it due to the
difference of pressures fore and aft, and the viscous drag on the sides. (Viscosity of
course, implies dissipation of energy that will cause the motion to die down, unless
kept alive by an external driving agency, the rotating inner cylinder in the Couette
flow for instance.) How can such a deterministic system exhibit the apparent randomness of turbulence. The idea that goes back to Landau is this. The fluid has
infinitely many degrees of freedom. How? Well, at each point of space the fluid can
move up and down, right and left and forward and backward. That is three degrees
of freedom. There are infinitely many points in any finite volume and that adds up
to an infinite number of degrees of freedom.

It is often convenient to combine these local point-wise degrees of freedom
distributed in space into extended oscillations or waves, called modes, of different
wavelengths and talk in terms of these infinite number of modes instead. It is just

as a wave on the surface of water is made up of movements of the elements of water all over the surface. Conversely, the movement of the fluid at any given point can be re-constituted from the superposition of these waves. The two descriptions are completely equivalent and are related by what mathematicians call the (spatial) Fourier Transform. Likewise, a flow at a point fluctuating in time can be regarded as a superposition of periodic oscillations of different frequencies through a temporal Fourier transform — giving a frequency spectrum. The strength of the different Fourier components gives the power spectrum of the flow.

This then is the infinite number of oscillations or modes of the fluid that Landau invoked to describe turbulence. As the frequencies are incommensurate, the flow is only quasiperiodic, never quite repeating itself. Two frequencies are said to be incommensurate if their ratio is irrational, *i.e.*, it cannot be expressed as a ratio of two whole numbers. A fully developed turbulence has infinitely many such frequencies making the flow aperiodic for all practical purposes. The power spectrum is then a continuous one with no dominant sharp frequencies in it. This is what chaos is. This aperiodicity can be roughly appreciated with the help of a simple example. Imagine a system with a large number, 26 say, of simple pendulums of periods $0.5, 0.6, 0.7, \ldots, 2.8, 2.9, 3.0$ seconds and let them go simultaneously from the right of their equilibrium positions. Some thought and arithmetic will show that it will take 7385 years before the system repeats its initial state! And that when the frequencies (or the periods) *are still commensurate*, and finite in number. Incommensurate frequencies will push this recurrence time to eternity. (This is much like the problem of finding the recurrence time for a planetary alignment in the solar system — the *syzygy* problem of finding the lowest common multiple, of course).

This scenario for turbulence, and for chaos in general, seems eminently appealing but has really never been put to the test. No-one has seen this route to turbulence via the successive emergence of these infinite number of mode frequencies. Perhaps it applies to open flows — jets and wakes. On the contrary, there are known examples of turbulence that refute it. The Couette flow as discussed above is a case in point. So is the case with the convective flow which was modelled by Edward Lorenz by a finite number, in fact three, of interacting macroscopic degrees of freedom. It showed chaos in no uncertain terms. And the route to this convective chaos was studied experimentally in an ingenious experiment on a liquid helium cell by Albert Libchaber, and was found to be of the period-doubling kind, quite different from that of Landau, as we shall see later. In any case there are a number of computer models with small numbers of degrees of freedom that show chaos. The question is what makes them chaotic. The answer lies in nonlinearity. The Navier–Stokes equations are nonlinear.

Nonlinearity is a mathematical way of saying that the different dynamical degrees of freedom act on each other and on themselves so that a given degree of freedom evolves not in a fixed environment but in an environment that itself changes with time. It is as if the rules of the game keep changing depending on the present state of the system. A simple example will illustrate what we mean. Take $y = ax$, a linear relation. Changes in y are proportional to those in x, the constant of proportionality being 'a.' In this game the effect is proportional to the cause, and the controlling parameter 'a' is fixed. Now consider $y = ax^2$. Here the effect is proportional to the square of the cause. We may rewrite this in the old form as $y = (ax)x$ and say that the parameter controlling the game, namely (ax) itself depends on x. This is nonlinearity, or nonlinear feedback if you like. In point of fact, interactions that are linear in the state variables are no interactions at all. One can re-define new state variables as linear combinations of the old ones such that these new state variables do not interact at all. This is called normal mode analysis. It is the nonlinearity that makes for the complex behavior — in particular it can generate the sensitive dependence on the initial conditions. It can amplify small changes. One does not need an infinite number of degrees of freedom.

But you may turn around and say that the fluid dynamical system does have an infinite number of degrees of freedom anyway. Well, this is where dissipation (friction) comes in handy. It turns out that all but a few of these get damped out rather quickly by this friction, and the system settles down to the few remaining, macroscopic degrees of freedom that really define its state space. As we will see later, we need a minimum of three to have chaos in a continuous flow.

A final worry now. A dynamical system with a small number of degrees of freedom will have a low dimensional state space. Moreover, the degrees of freedom are expected to have a finite range — we don't expect the velocities to be infinite for example. Thus the phase point will be confined to a finite region of the low dimensional state space. This raises a geometrical question of packing. How does the phase trajectory, confined to a finite region of the low dimensional state space, wind around forever without intersecting itself or closing on itself? There is an ingenious geometrical construction that illustrates how this is accomplished. We know that two trajectories from neighboring points diverge, stretching the line joining them exponentially. But this cannot go on indefinitely because of the finiteness of the range. The trajectories must fold back, and may approach each other only to diverge again (Fig. 7.2).

This will go on repeating. This process of stretching and folding is analogous to a baker's transformation. He rolls the dough to stretch it out and then folds it, and then repeats the process again and again. We can drop a blob of ink on the dough to simulate a dust of phase points. The process of stretching and folding will generate a highly interleaved structure. In fact after a mere two dozens or so iterations we will have a 2^{24} layers and thus a fine structure down to atomic scales. The inky spot would have spread out throughout the dough coloring it apparently uniformly, suggesting thorough mixing. Yet, actually it is finely structured. The neighboring points on the inky spot would have diverged out and become totally uncorrelated after a few rounds of the baker's transformation. Indeed, stretching with folding is a highly nonlinear process that generates the above sensitivity. The

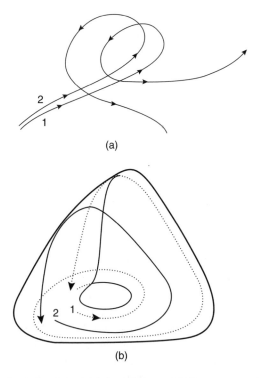

Figure 7.2: Sensitive dependence on initial conditions: (a) divergence and, (b) divergence-cum-folding-back of neighboring trajectories.

baker's transformation seems to be a general algorithm for chaotic evolution.

Our discussion so far has centered on some deep philosophical questions about and a general qualitative understanding of chaos. We will now turn to a simple but real, in fact household experiment with chaos that should be refreshing.

We are going to experiment with a leaking faucet! We will be following here the idea of Robert Shaw of the University of California, Santa Cruz.

7.2 Lesson of the Leaking Faucet

Leaky faucets are known universally. They are perhaps best known as a very effective form of torture as many an insomniac waits attentively for the next drop to fall. It is, however, less well known that there is a universality to their pattern of dripping as the flow rate is turned on gradually and sufficiently. There is some pretty physics involving this. It also happens to be rather easy to experiment with. All you need is a leaky faucet, preferably one without a wire mesh, and a timing device to monitor the time intervals between successive drops as the flow rate (our control parameter) is gradually increased. Unfortunately, the time intervals can get rather short, of the order of a fraction of a second, down to milliseconds and even

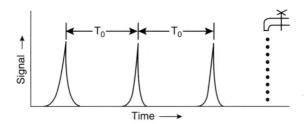

Figure 7.3:   Leaking faucet dripping with single period T_0 at low flow rate.

microseconds, just when interesting things begin to happen. We will, therefore, have to employ detectors other than our unaided eyes and ears. We could, for instance, use a microphone to pick up the sound signals of the falling drops, or better still, arrange to have the falling drop interrupt a beam of laser (or ordinary) light and detected by a photodiode. The data can then be acquired by and stored on a PC, which has now become part of the science kit in most high schools. At low enough rate of flow, the faucet leaks with a monotonous periodicity — drip, drip, drip, ... The successive drops fall at equal intervals, T_0 say (Fig. 7.3).

This dripping pattern persists up to a fairly high threshold flow rate. The time interval T_0, of course, gets shorter and shorter. At and beyond this threshold, however, this pattern becomes unstable and a new pattern emerges — pitter-patter-pitter-patter $\cdots$ (Fig. 7.4). The intervals between successive drops become unequal now. We have a short interval T_1 alternating with a long interval T_2, generating thereby a sequence $T_1, T_2, T_1, T_2, \dots$. We say that the period has doubled. The original single period T_0 has bifurcated into a pair of unequal periods T_1, T_2. The repeat 'motif' now consists of a pair of two successive periods, one long and the other short. Note that the period-doubling refers to this two-ness, and not to the absolute value of the time interval, which, if anything, only gets shorter as the flow is turned up progressively. This new pattern in turn persists up to a still higher threshold, and then becomes unstable. There is a period doubling again: Each of the intervals T_1 and T_2 bifurcates into two unequal intervals, leading to the pattern $T_3, T_4, T_5, T_6, T_3, T_4, T_5, T_6, \dots$. Let us summarize our findings so far. We

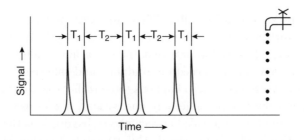

Figure 7.4:   Leaking faucet dripping with doubled period (T_1, T_2) at a higher flow rate.

started with period 1 ($= 2^0$), which bifurcated to period 2 ($= 2^1$) and which in turn bifurcated to period 4 ($= 2^2$). The trend continues. At the nth bifurcation, we get the period 2^n. It turns out that the successive bifurcations come on faster and faster in terms of the control parameter (the flow rate) and they pile up at a point of accumulation as $n \to \infty$ (infinity). This is the critical value of our control parameter. At this point, the period becomes 2^∞, which is infinite. The dripping pattern never repeats itself. It has become aperiodic. It is as if the drops fall to the beats of an *infinitely inventive drummer*. This is *chaos*. We have just discovered a period-doubling route to chaos!

The chaos is robust and it has a universality about it that we will discuss later. Thus, you can change the faucet; you can replace water with beer if you like; you can add some surfactants to reduce its surface tension or whatever. The period-doubling sequence will still be there, and many numbers characterizing the approach to chaos will remain unchanged. This is reminiscent of the universality familiar from the critical behavior of second-order phase transitions.

What happens if we push our control parameter beyond this critical value. Well, the aperiodic chaos persists but there are now finer structures such as the appearance of narrow windows of periodicities, that we cannot pause to discuss here.

It is amazing that a simple-looking system such as a leaking faucet should reveal such complexity, or conversely, that chaos should hide such simplicity. We will now turn to a more quantitative description of chaos.

7.3 A Model for Chaos

Now that we have experimented with chaos, let us make a simple model of it that we can play around with, and hopefully solve it in the process. Solving here requires no more than elementary operations of addition, multiplication and raising to power (exponentiation), done repeatedly (that is iteratively), all of which can be entered easily on a hand-held calculator. In fact this is precisely what was done by Mitchell J. Feigenbaum of the Los Alamos National Laboratory in the early seventies using his HP-65 hand-held calculator. And that made the study of chaos such a refreshing exercise with numbers, and, of course, led to many a breakthrough.

7.3.1 The Logistic Map

The model we are going to study is the so-called *logistic map* or equation: $X_{n+1} = bX_n(1 - X_n)$. It is an algorithm, that is, a rule for the growth of a quantity 'X' controlled by a growth factor 'b.' Given the value X_n of this quantity at the nth instant, or *go-round*, all you have to do is to substitute this value in the right-hand side expression, evaluate it and voila! You have the value of X at the $(n + 1)$th go-round, namely X_{n+1}. You can iterate this process and thus generate the value of X at any future instant, starting from an initial value X_0, called the seed. This is all the mathematics that there is to it. Simple and yet it holds in it the whole complexity of chaos. But let us first examine how such a *logistic equation* may arise in a real physical situation.

Consider a savings bank account with a compound interest facility. So, you deposit an amount X_0 initially and a year later it becomes $X_1 = bX_0$. The growth factor 'b' is related to the rate of compound interest. Thus, for an interest rate of 10%, the control parameter $b = 1.1$. Two years after the initial deposit we will have $X_2 = bX_1 = b^2 X_0$ and so on. In general we have $X_{n+1} = bX_n$. This linear rule, or algorithm, gives an unlimited growth, in fact an exponential growth. Of course, if b were less than unity, our algorithm would lead to total extinction too and your deposit would dwindle ultimately to nought. (This could happen if X represented the *real* value of your money and the rate of inflation exceeded the rate of interest.) The quantity X could equally well denote the value of your stocks in a stock market. A much more revealing example for our purpose, however, will be that X_n represents the population of a community at the nth round of annual head count, or census. It may be the population of fish, or gypsy moths, or the Japanese beetles or even that of cancer cells. This is a matter of great interest to population biologists, and was studied extensively by Robert May at Princeton University, using the logistic equation. But why this logistic equation? Let us see.

The rate of growth of a population is obviously controlled by the natural birth and death rates. The growth factor b depends on their difference. Thus, if the birth rate exceeds the death rate, then b exceeds unity and we have the classical Malthusian scenario of unlimited exponential growth. If the inequality is reversed, then b will be less than unity and we face total extinction. So, very aptly, Robert May called b the 'boom-and-bustiness' parameter. Real life is, however, different. Communities live in a finite ecosystem competing for the common resource (food) which is limited. They often have the predator-prey relationship. (They may also be self-limiting because of moral constraints or ritualistic cannibalism.) Now, if the population of a community grows too large, it faces death by starvation and the growth rate declines automatically. Thus there is a logistic limit to growth. (In such a case it is convenient to express the population X as a fraction of its maximum possible value. Then X will lie between zero and unity.) This slowing down of the growth rate is a kind of negative feedback. We can easily simulate it by replacing our growth factor b by $b(1-X)$. Clearly then the effective growth rate declines with increase in the population X. The result is that our Malthusian growth scenario, represented by the linear rule $X_{n+1} = bX_n$ gets modified to the logistically limited growth scenario $X_{n+1} = bX_n(1 - X_n)$, which is nonlinear. Most real systems are nonlinear. This makes all the difference.

7.3.2 Iteration of Map

Having thus convinced ourselves of the reasonableness of the *logistic map* as a model for self-limiting growth, we have set the stage for action — on our programmable calculator. Select some value of the control parameter 'b.' Enter then a seed value X_0 as input on the right-hand side of the logistic equation, evaluate the expression $bX_0(1 - X_0)$ and just output it as X_1. To get X_2, we have to use the just evaluated X_1, as the new input and out comes X_2. The procedure can be iterated n times to get X_n at the nth round, and so on for the whole sequence. Now repeat this numerical exercise with another value of the seed X_0, and look for any change in the pattern of the sequence of numbers that come out. (Note here that our X's by definition lie between zero and one. It is readily verified that this constrains the control parameter b to lie between zero and four.) Is the pattern periodic? Does the period change with b? Or otherwise, is there a pattern at all? These are the questions to be answered. We are zeroing on chaos.

All this can be viewed live on the screen of your PC with the help of a few lines of statements in BASIC. But, for a clearer understanding of what is really going on, it is best to resort to the following graphical construction. Just plot X_{n+1} (vertically) against X_n (horizontally) using them as the Cartesian coordinates. This is our phase space if you like: a finite phase space, a unit square. On this plot, our logistic function $bX_n(1 - X_n)$ against X_n is nothing but a parabola standing on the unit horizontal base (Fig. 7.5).

Now start with the input seed X_0 marked on the horizontal axis. To get X_1, just move vertically up (or down as the case may be) meeting the parabola at the point P_0 on it. It is clear that P_0 has the coordinates (X_0, X_1). In order to get X_2 now, we have to use X_1 as the new input on the horizontal axis and repeat the above procedure. It is much more convenient, however, to draw a 45° straight line, diagonally across the phase space on which $X_n = X_{n+1}$. With this, to get X_2 from X_1, all we have to do is to move horizontally from the point P_0 across up to the diagonal and then move vertically meeting the parabola at a point P_1, say. Clearly, P_1 will have the coordinates (X_1, X_2). The procedure can now be iterated to generate the whole sequence, $X_0, X_1, \ldots, X_n, X_{n+1}, \ldots$, *ad infinitum* (Fig. 7.5).

7.3.3 The Period Doubling Bifurcation

With the graphical construction in hand, we can at once make several observations by mere inspection. Thus, for sufficiently small values of the control parameter b, the parabola lies entirely below the diagonal line. It is readily seen that starting with any seed value X_0 whatsoever, we quickly cascade down and finally converge to the origin (Fig. 7.5a). This makes the origin an *attractor*. Let us denote this point by P_0^* and the corresponding X value by X_0^* ($= 0$, of course). It is clear that on further iteration $X_0^* \to X_0^*$. Hence it is a *fixed point*, in fact a stable fixed point. That is, starting with any X_0, we end up there.

This situation persists as we increase our control parameter b until the diagonal is just tangential to the parabola at the origin. This threshold value, b_1 say, of the control parameter can be obtained in a straightforward manner by equating the

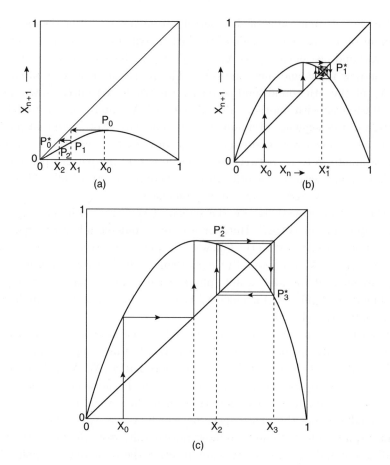

Figure 7.5: Graphical construction of logistic map: (a) stable fixed point P_0^* at origin; (b) stable fixed point P_1^* away from origin; (c) period-two attractor P_2^*, P_3^*.

two slopes at the origin. We get $b_0 = 1$. For larger values of b, the parabola gets steeper and intersects the diagonal once again. This alters the situation completely since now, starting with any seed X_0, other than zero, we converge not to P_0^* but to P_1^* (Fig. 7.5b). The corresponding stable fixed point value is readily found to be $X_1^* = 1 - 1/b$. The earlier fixed point X_0^* has become unstable now — it is a repeller rather than an attractor. It is readily appreciated that the question of stability under iteration is related to the slope of the parabola at the point of intersection with the diagonal. A slope steeper than $+1$ or -1 means 'unstable,' otherwise it means stable. This slope-stability relation is a standard result in the so-called linear stability analysis. But graphically, it is obvious by construction.

As we hike up the control parameter further, things become really interesting when the control parameter exceeds a second threshold $b_1 (= 3)$. The fixed point X_1 becomes unstable! Starting with a seed value X_0, the sequence X_0,

$X_1, \ldots, X_n, X_{n+1}, \ldots$ fails to converge to any fixed point at all. Instead it settles down to a *2-point periodic* cycle $X_2^* \to X_3^* \to X_2^* \to X_3^* \to \cdots$. We call X_2^* and X_3^* elements of the cycle (Fig. 7.5c). We say that the earlier fixed point X_1^* has *bifurcated* to the alternate pair (X_2^*, X_3^*) that forms an attractor of period 2, or a 2-cycle. This is the famous *period doubling*. The alternating sequence $X_2^* \to X_3^* \to X_2^* \to X_3^* \to \cdots$ is reminiscent of the pair of alternating long and short intervals of the dripping faucet discussed in the last section. The general trend should be clear now. At the next threshold value b_2, say, of the control parameter b, the 2-cycle in turn becomes unstable and we have instead a *4-point periodic* cycle $X_3^* \to X_4^* \to X_5^* \to X_6^* \to X_3^* \to X_4^* \to X_5^* \to X_6^* \to \cdots$. So we now have a 2^2-cycle. It is simply that each element of the 2-cycle has bifurcated to two elements (see bifurcation plot in Fig. 7.5c). And so on to the 2^3-cycle, and in general to the 2^k-cycle at the kth threshold b_k. What you get looks like a *cobweb* plot for the logistic map. It turns out that the successive thresholds of instability come on faster and faster and converge to a point of accumulation b_∞ as $k \to \infty$, where we have a 2^∞-cycle: the period becomes infinite! The pattern never repeats itself. It has become aperiodic. This is the onset of *chaos*, and the entire sequence is known as the Feigenbaum scenario of the period-doubling bifurcation route to chaos. The critical value of the control parameter for the onset of chaos turns out to be $b_\infty = 3.569\ldots$, the *Feigenbaum constant*.

7.3.4 *Universality*

The approach to criticality is subtle and interesting. The ratio of the successive intervals between the threshold values of the control parameter, the so-called bifurcation ratio $(b_k - b_{k-1})/(b_{k+1} - b_k)$ tends to a limiting number $\delta = 4.669\cdots$ as k tends to infinity. The entire sequence of events leading up to chaos seems to simulate the dripping patterns of the leaking faucet surprisingly well. Could this be a mere coincidence? Well, it could have been suspected to be so but for the fact that the behavior observed for the logistic equation actually turns out to be universal within a whole class. As was shown by Feigenbaum, we can replace the parabolic (quadratic) map by any other single-hump map, but one with a smooth maximum, without changing the period doubling bifurcation sequence or the associated limiting ratio δ and other critical scaling exponents characterizing the onset of chaos. On the other hand, replacing the parabola with a 'tent' having a sharp triangular apex at the maximum point is a different matter. This would be in a different class. But $X_{n+1} = r \sin(\pi X_n)$ belongs to the same class as the logistic map. This is something familiar from the modern theory of second-order phase transitions due to Kenneth G. Wilson. It is this universality of the critical behavior that makes an algorithm as simple-looking as the logistic map capture the essential physics of diverse systems close to a crisis. After all, it is the crisis that brings out the intrinsic character common to a whole class of *individuals*.

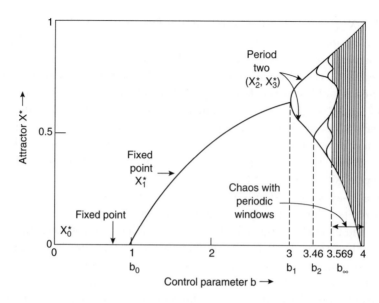

Figure 7.6: Bifurcation plot of attractors X^* against control parameter 'b' for period-doubling route to chaos — after Feigenbaum (schematic).

7.3.5 Fully Developed Chaos

What happens if we push our control parameter b beyond b_∞? The behavior continues to be generally chaotic except for the windows of periodic oscillations for narrowly tuned values of the parameter. There is the phenomenon of *intermittency* where apparently periodic behavior appears in the midst of chaos. But these are structures too fine for us to discern here. It is, however, very revealing to consider the extreme case where b takes on its maximum possible value, *i.e.*, $b = 4$. For then, we can simplify matters tremendously by a simple change of variable. Let $X_n = \sin^2(2\pi\Theta_n)$, where Θ is an angle in radians. Remember that the trigonometric function $\sin(2\pi\Theta)$ is periodic in its argument with a period 1 ($\pi = 3.1415\cdots$ is the ratio of circumference to diameter of a circle). Our logistic equation written in terms of Θ's now reads $\Theta_{n+1} = 2\Theta_n \pmod 1$. This simply means that to obtain Θ_{n+1} from Θ_n, multiply Θ_n by two and drop the integral part of it so that the remainder lies in the interval 0 and 1. This is the meaning of the notation 'mod 1.' The remainder is then Θ_{n+1}. Thus, for example, if $\Theta_n = 0.823$, then $\Theta_{n+1} = 2\Theta_n \pmod 1 = 0.646$, and $\Theta_{n+2} = 2\Theta_{n+1} \pmod 1 = 0.292$, and so on. Thus, starting from a seed value Θ_0, say, we can generate the entire sequence $\Theta_0, \Theta_1, \ldots, \Theta_n, \Theta_{n+1}, \ldots$. You see now that the scale factor 2 in the relation $\Theta_{n+1} = 2\Theta_n \pmod 1$ will 'stretch' the interval between any two neighboring seed values on iteration. The condition (mod 1), however, *folds* back the values as soon as they try to get out of the unit interval. This is the nonlinear feedback. The above stretching and folding operations are precisely the ones we had discussed as baker's transformation in the previous section. This is what gives the *sensitive dependence on initial conditions*. It can be shown that under this stretching and folding, the Θ_n values hop over the entire unit interval at random, eventually covering it densely. The entire unit interval is the *attractor* — the *strange attractor*, as we will see later. This is fully developed *chaos*! Similar things happen in the range $b_\infty < b < 4$ also. However, here the Θ_n values hop randomly on a subset of the unit interval whose *dimension* is less than unity, as we will see later: it is a *fractal* (see Section 7.5). Indeed, at the onset of chaos ($b_\infty = 3.569$), the fractal dimension of the strange attractor can be shown to be 0.537. For $b = 4$, however, the dimension is actually an integer ($= 1$), and X_n covers the entire unit interval densely.

The *logistic map* is a kind of 'Feigenbaum Laboratory' in which we can do numerical experiments with chaos and learn a great deal. One wonders, however, how such a discrete map can approximate reality when all real processes happen in continuous time. We address this question next.

7.3.6 *Poincaré Sections: From Continuous Flows to Discrete Maps*

A discrete map of the kind we have discussed above can be constructed for a real dynamical system quite naturally by taking the so-called *Poincaré section* of the continuous flow, or the trajectories, in the phase space of the system. Figure 7.7 shows such a Poincaré section for a dynamical system having a three-dimensional phase space. The *surface of section* has been taken to be a plane perpendicular to the z-axis. But other choices are possible. Here, instead of watching the trajectory continuously in time, we simply record the sequence of the points $P_0, P_1, \ldots$ of intersection of the trajectory with the chosen plane as the trajectory crosses it in a given direction — of the negative z-axis, say. In effect we have replaced the continuous phase flow in time described by the *differential* equation, by a discrete mapping relating the (x, y)-coordinates of the successive points P_n, P_{n+1} of intersection — an algebraic, *difference* equation. We may have, for example $X_{n+1} = 1 - cX_n^2 + Y_n$

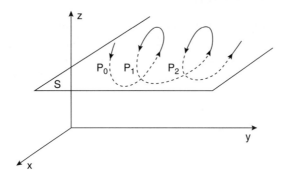

Figure 7.7: Poincaré section $(P_0, P_1, \ldots)$ of a phase space trajectory by the plane S.

and $Y_{n+1} = bX_n$, where '*c*' and '*b*' are adjustable parameters, *e.g.*, we may have $c = 1.4$ and $b = 0.3$. This two-dimensional discrete map, called the *Hénon map* after the French astronomer Michel Hénon, describes the Poincaré section of a *chaotic attractor* in a three-dimensional phase space, of which more later. The Poincaré section of a three-dimensional flow will be, in general, a two-dimensional surface. But if the dissipation is large and the areas contract rapidly, then the section will consist of points distributed along a curve. In other words, we now have a one-dimensional map $X_t \rightarrow X_{t+1}$. This is often called the *first-return map*.

Taking Poincaré sections is a very revealing technique for studying, or rather visualizing the geometrical forms, such as the orbits and their limits in the phase space (the *phase portraits*). It certainly reduces the dimension by one. And yet it retains the all important qualitative, global features of the dynamics, *e.g.*, the divergence or convergence of neighboring trajectories, expansion or contraction of phase-space volume elements, the periodicity or aperiodicity of the orbits and many other signatures of dynamics. Also, the algebraic difference equations are easier to handle than the original differential equations. One can, for example easily implement the Hénon map on a programmable calculator. The Poincaré section is particularly useful for the study of attractors in higher dimensions. We will look at these attractors next.

7.4   Strange Attractors and Routes to Chaos

Attractors are geometric forms in the phase space to which the phase trajectories of the dynamical system converge, or are attracted and on which they eventually settle down, quite independently of the initial conditions. The idea of an attractor is quite simple but very powerful for a global qualitative understanding of the motion, both regular and irregular, without having to solve the equations of motion — which is seldom possible anyway. It is best illustrated through examples.

7.4.1   *Stable Fixed Point*

The simplest attractor is a *fixed point*, or rather a *stable fixed point*. Consider the phase portrait of a damped linear oscillator, a pendulum with friction for example. The phase space is two-dimensional comprising the velocity (momentum) and the position coordinates. Because of damping the phase point spirals in onto the origin and rests there (Fig. 7.8*a*).

The origin is thus a fixed point. It is obviously stable — displaced from it, the system returns to it eventually. Also, all trajectories, no matter where they begin, are attracted towards it — hence the name attractor. It is also readily seen that any element of the phase space will contract in its extension as it is attracted towards the fixed point. As there is contraction along the two independent directions in the two-dimensional phase space, both the Lyapunov exponents are negative. This 'contraction' is the meaning of the term 'dissipative flow' in phase space language. The most interesting flows in Nature are dissipative and have to be maintained by external driving forces, such as stirring, heating, pumping, kicking, etc. We call them open systems.

7.4.2 Limit Cycle

Next comes the *limit cycle*. It is a closed loop in the phase space to which the trajectories converge eventually (Fig. 7.8*b*). The limit cycle corresponds to a stable oscillation. Here, one Lyapunov exponent is zero (along the loop) and the rest are negative. Its Poincaré section is just a point. Again, the flow is dissipative. For a two-dimensional phase space, the limit cycle is the only attractor possible, aside from the fixed point. This is a direct consequence of the condition that the phase trajectories cannot have self-intersection. To have anything more complicated, the trajectory must escape in a third dimension. (Recall that from a given point a unique trajectory must pass.) The limit cycle is at the heart of the most simple periodic processes in Nature — the beating of the heart, the 'circadian rhythms' of period 23 to 25 hours in humans and animals, the cyclic fluctuation of populations of competing species in an ecosystem, oscillating chemical reactions like the Beluzov–Zhabotinsky reaction, marked by colour changes every minute or so, etc. The limit cycle thus is a natural clock.

7.4.3 The Biperiodic Torus

The next most complicated attractor has the geometric form of a doughnut or anchor ring — the *torus* (Fig. 7.8*c*). The trajectories converge on the surface of the torus, winding in small circuits around the axis of the torus (at a frequency f_2) while orbiting in large circles along the axis (at a frequency f_1). Here, two of the Lyapunov exponents are zero and the rest are negative. This corresponds physically to a compound or biperiodic oscillation resulting from the superposition of two independent motions. If the frequencies f_1 and f_2 are commensurate, that is if the ratio f_1/f_2 can be expressed as the ratio of two integers, then the trajectory on the torus closes on itself — frequency locking. The motion is actually periodic then. A Poincaré section of the trajectory will be a finite set of points traversed by the successive go-rounds. If, on the other hand, the frequencies f_1 and f_2 are incommensurate, the trajectory will cover the torus densely and the Poincaré section will be a continuous closed curve. The motion will be quasiperiodic. Such a biperiodic torus attractor is known to show up in the Couette flow. But the simplest example will be two coupled oscillators of different frequencies. Incidentally, if the frequencies are close enough and the coupling is nonlinear, the oscillators may get *locked into* a common frequency mode. This is called 'frequency entrainment,' lock-in, noted first by the great 17th century Dutch physicist Christiaan Huygens. He

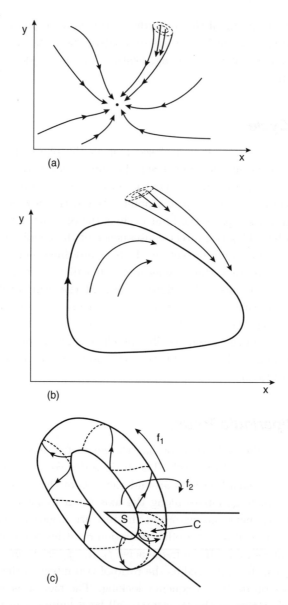

Figure 7.8:   (a) Stable fixed point; (b) limit cycle; (c) biperiodic torus.

was surprised that two proximate church lamps hanging from the ceiling should oscillate at exactly equal frequencies.

7.4.4 *The Strange Attractor*

The zero-dimensional *fixed point*, the one-dimensional *limit cycle*, and the two-dimensional *torus* are all examples of low-dimensional attractors that characterize dissipative flows which are regular, that is, stable and predictable to any degree of accuracy. These flows are essentially periodic. There is, however, an entirely different type of low-dimensional attractor that characterizes flows which are irregular, that is, unstable and unpredictable in the long term. Such an attractor is a phase-space non-filling set of points or orbits to which all trajectories from the outside converge, but on which neighboring trajectories diverge. Thus, at least one Lyapunov exponent has to be positive. It is '*strange*' in respect of the geometry of its form as well as in terms of its manner of traversal of this finite region of phase space. It is called the *strange attractor* and is the engine that drives *chaos*. It was discovered jointly by the Belgian mathematician David Ruelle and the Dutch mathematician Floris Takens around 1971.

The *strange attractor* is the answer to the question we had posed earlier: how can a trajectory remain confined forever to a finite region of phase space, without ever intersecting itself as it must not, and without closing on itself, as again it must not for that would mean a periodic motion? To make matters worse, the system happens to be dissipative — eventually the phase-space volume must contract to zero. You see the conflict! Just think of a one-meter long strand of the contortionist DNA packed within a cell measuring one millionth of a meter across, and you will get some idea of the packing problem, which is actually infinitely worse. The conflict of demands for zero volume and eternal self-avoidance is resolved by making the attractor into a *fractal* — a geometric form with fractional dimension lower than the phase space in which it is embedded. Fractals are an ingenious way of having surface without volume if you like. We will now illustrate all this with two celebrated examples of the strange attractor: the *Hénon attractor* of the discrete two-dimensional Hénon map for its simplicity, and the *Lorenz attractor* of the three-dimensional dimensional convective flow for its complexity. The latter is also historically the first known example of a strange attractor.

7.4.5 The Hénon Attractor

As already mentioned, the Hénon map, discovered by Hénon in 1976, is given by $X_{n+1} = 1 - cX_n^2 + Y_n$ and $Y_{n+1} = bX_n$. Thus, starting with any point in the (X, Y)-plane, one can generate the entire attractor by iterating the map a large number (usually 10^5–10^6) of times. It can be shown that the map is dissipative for a magnitude of b less than unity, *i.e.*, for $|b| < 1$. (It is conservative (phase volume preserving) for $|b| = 1$.) For $b = 0$, the dissipation is so large that the map contracts to a one-dimensional quadratic map of the kind we have discussed earlier. For a convenient value of $b = 0.3$ (moderate dissipation), we can now study the map as a function of the control parameter c. The map follows a period doubling route to chaos. For $c = 1.4$ we have a strange attractor, shown schematically in Fig. 7.9.

Two crucial features are to be noted here. One of these has to do with the geometry and the other with the motion. First, the attractor has a self-similar

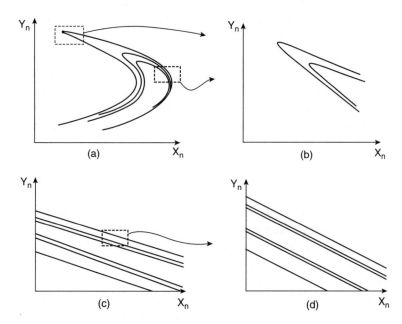

Figure 7.9: The Hénon attractor and its fractal structure: (a), (b) and (c) reveal self-similarity under magnification.

microstructure: on magnification, an element of it reveals details similar to the whole. This self-similar geometry leads to a fractional dimensionality for the attractor. It is a fractal. The second crucial feature is the manner in which the fractal is traversed by the phase point. The pattern of points in Fig. 7.9 tends to guide the eye along certain lines. This is, however, merely the 'closure tendency' of the human eye. Actually, the successive points on these lines are *not* traversed successively by the phase point. In fact, the phase point hops almost randomly over the entire attractor. This is, of course, best seen on your computer screen as the points appear iteratively, one by one, at totally unexpected places, but eventually a pattern of points is generated that looks like that in Fig. 7.9.

7.4.6 The Lorenz Attractor

We have already discussed convection in a mass of fluid heated from below and the chaos that results when the temperature difference exceeds a critical value. Of course, such a system has infinitely many degrees of freedom, but all except three get damped out and the chaotic regime is just a three-dimensional phase space (X, Y, Z). Roughly speaking, X measures the rate of convective overturning, Y the horizontal temperature variation and Z the vertical temperature variation. Just to give a flavor of things, we write down the equations obeyed by these variables:

$$\dot{X} = \sigma(Y - X)$$

$$\dot{Y} = \gamma X - Y - XZ$$

$$\dot{Z} = XY - bZ .$$

These are the equations that Lorenz studied in 1963 as his model for weather. The overhead dot denotes the rate of change with time. Here, σ (sigma), γ (gamma) and b are parameters defining the conditions of the experimeter. Thus, b is a geometry factor (depending on a certain aspect ratio). Sigma (called the *Prandtl number*) measures the ratio of viscosity to thermal conductivity-times-density, and γ is the control parameter — the Rayleigh number normalized in a certain way. In our experiment, as you increase the temperature difference from zero, you increase γ from zero. For γ between zero and one, we have the steady diffusion of heat. For γ greater than 1 but less than about 24.74, steady convective rolls develop. Eventually, for γ greater than about 24.74 we have motion on a strange attractor giving chaos (the Lorenz attractor).

A strange attractor in three dimensions is hard to visualize. One could either resort to making Poincaré sections, or just look at the projection on a plane (see Fig. 7.10).

One can readily see that the trajectory winds around an unstable fixed point C, spirals outward and after reaching out to a critical distance from C, crosses over to wind around the other unstable fixed point C'. The number of circuits the trajectory makes around any one of the unstable fixed points before it switches to the other, generates a sequence which is as random as the tossing of a fair coin.

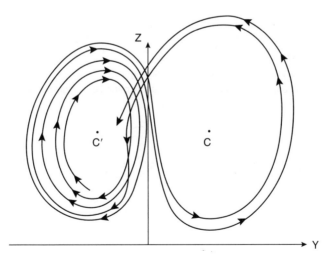

Figure 7.10: The Lorenz (strange) attractor projected on YZ-plane. C and C' are two unstable fixed points. The strange attractor depicts chaos (schematic).

It turns out that the trajectories are actually confined to a sheet of small but finite thickness, within which is embedded the entire complex geometry of the attractor; it is a fractal.

The Lorenz system turns out to describe other systems as well, for example the irregular spiking of laser outputs. The irregular reversal of Earths' magnetic field, in the geodynamo dynamo model, may be linked to the random jumping of the trajectory between C and C'.

7.4.7   Routes to Chaos

We have seen that chaos is described by strange attractors. In most of the physically interesting cases these attractors are the low-dimensional ones. We will now summarize two commonly observed routes to chaos described by such low-dimensional strange attractors.

The first route is that of an infinite sequence of period-doubling bifurcations as the control parameter is varied. We begin with a stable fixed point (stationary point) that becomes unstable and bifurcates to a limit cycle (aperiodic orbit), which undergoes a period-doubling sequence leading to a pile-up, that is, an accumulation point (periodic orbit) — the onset of chaos. This is the Feigenbaum scenario that we have already come across in the case of the leaky faucet. It is also the route followed by convective turbulence (Lorenz attractor). This route has already been shown schematically as the bifurcation plot for the logistic map. In Fig. 7.11*a* we show this route in terms of the projection of the phase trajectory in a plane.

There is another route to chaos that is qualitatively different. Here, we start with a stable fixed point (stationary point) that becomes unstable and bifurcates to

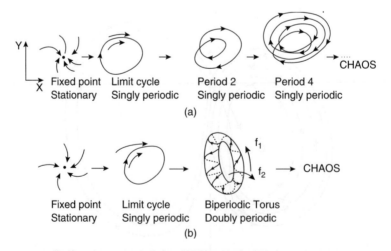

Figure 7.11:   Routes to chaos involving (a) an infinite number of period-doubling bifurcations; (b) a finite number of topological transitions.

a limit cycle (a periodic orbit). The limit cycle in turn becomes unstable and bifurcates to a biperiodic torus (doubly periodic orbit) with incommensurate frequencies, *i.e.*, quasiperiodic. Finally, from the biperiodic torus we jump to chaos — without any intervening sequence of appearances of new frequencies. This is the sequence leading to turbulence observed in the Couette flow discussed earlier. Here, the fluid is confined to the annular space between two coaxial cylinders, one of which is made to spin. In Fig. 7.11*b* we have shown schematically this route to chaos.

It is important to appreciate a qualitative difference between these two routes to chaos. In the first route (period-doubling route), it is simply the period of the limit cycle that changes at bifurcation. No new states are being generated. In the second route, the bifurcation from the limit cycle to the torus is a topological change involving new states.

There are other routes to chaos possible. But there is no experimental evidence for the Landau scenario involving the appearance of incommensurate frequencies, one at a time, leading eventually to a confusion that we would call chaos.

7.5 Fractals and Strange Attractors

Fractals are geometrical objects with shapes that are irregular on all length scales no matter how small: they are self-similar, made of smaller copies of themselves. To understand them, let us re-examine some familiar geometrical notions.

We know from experience that our familiar space is three-dimensional. That is to say that we need to specify three numbers to locate a point in it. These could, for example, be the latitude, the longitude and the altitude of the point giving respectively, its north-south, east-west, and the up-down positions. We could, of course, employ a Cartesian coordinate system and locate the point by the triple (x, y, z). Any other system of coordinates will equally suffice. The essential point is that we need *three* numbers — more will be redundant and fewer won't do. Similarly, a surface is two-dimensional, a line one-dimensional and a point has the dimension of zero.

These less-dimensional spaces, like the two-dimensional surface, the one-dimensional line, or the zero-dimensional point, may be viewed conveniently as embedded in our higher three-dimensional space. All these ideas are well known from the time of Euclid and underlie our geometrical reckoning. So intimate is this idea of dimension that we tend to forget that it has an empirical basis. Thus, it seems obvious that the dimension has to be a whole number. Indeed, the way we have defined the dimension above it will be so. (We call it the *topological dimension*, based on the elementary idea of neighborhood.) There is, however, another aspect of dimensionality that measures the spatial content or the capacity. We can ask, for instance, how much space does a given geometrical object occupy. Thus, we speak of the volume of a three-dimensional object. It scales as the cube of its linear size —

if the object is stretched out in all directions by a factor of 2, say, the volume will become $2^3 = 8$ times. Hence, it is three-dimensional. Similarly, we speak of the area of a surface that scales as the square of its linear size, and the length of a line that scales as the first power of the linear size. As for the point, its content, whatever it may be, stays constant — it scales as the zeroth power of the linear size, we may say. In all these cases, the '*capacity dimension*' is again an integer and agrees with the topological dimension.

But this need not always be the case. To see this, let us consider the measurement operation in some detail. Consider a segment of a straight line first. To measure its length we can use a divider as any geometer or surveyor would. The divider can be opened out so that the two pointed ends of it are some convenient distance apart. This becomes our scale of length. Let us denote it by ϵ. Now all that we have to do is to walk along the line with the divider marking off equal divisions of size ϵ. Let $N(\epsilon)$ be the number of steps (divisions) required to cover the entire length of the line. (There will be some rounding off error at the end point which can be neglected if it is sufficiently small so that $N(\epsilon) \gg 1$.) Then the length of the segment is $L = \epsilon N(\epsilon)$. Let us now repeat the entire operation but with ϵ replaced by $\epsilon/2$, that is, with the distance between the divider end points halved. The number of steps will now be denoted by $N(\epsilon/2)$. Now we must have $L = \epsilon N(\epsilon) = (\epsilon/2)N(\epsilon/2)$. This is because we have covered exactly the same set of points along the straight line skipping nothing and adding nothing. Thus, halving the measuring step length simply doubles the number of steps required. This can be restated generally as the proportionality or scaling relation $N(\epsilon) \sim \epsilon^{-1}$, that is, the number of steps $N(\epsilon)$ required is inversely proportional to the size ϵ of the linear step length for small enough ϵ. This seems obvious enough. The '1' occuring in the exponent gives the *capacity dimension* (D) of the straight line. In a more sophisticated language, $D = -(d \ln N(\epsilon)/d \ln \epsilon)$ as $\epsilon \to 0$. In our example this indeed gives $D = 1$. Here, ln stands for the natural logarithm. Thus, finding the capacity dimension is reduced to mere counting. It is indeed sometimes referred to as the box counting dimension.

So far so good. But what if we have a curved line? For, in this case as we walk along the line with our divider, we will be measuring the chord length of the polygon rather than the true arc length of the curved line, as can readily be seen from Fig. 7.12a.

Are we not cutting corners here? Well, no. The point is this. If the curve is a *smooth* one, it will be straight locally. Thus, as we make our step length ϵ smaller and smaller, the chords will tend towards the arcs, and the polygon will approximate the curve better and better, until it eventually hugs the curve as $\epsilon \to 0$. (This is the significance of the proviso $\epsilon \to 0$ in our logarithmic formula.) In the end, therefore, there is no distinction between the case of the straight line and that of a curved but *smooth* line: in both the cases, we will get $D = 1$. The smooth curve is rectifiable! Thus, all smooth curves have the same capacity dimension $D = 1$.

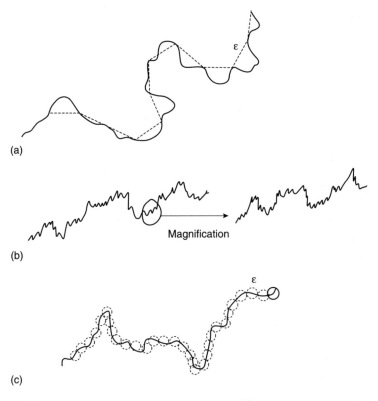

Figure 7.12: (a) Covering a line with scale length ϵ; (b) self-similarity of a fractal curve; (c) covering a line with ϵ discs or balls.

Curves that are smooth, that is, straight locally (*viz.*, have a well-defined tangent) are said to be regular. Now, what if the curve is irregular — that is to say, if it has a zig-zag structure down to the smallest length scale? You take a small segment of it that looks smooth enough at the low resolution of the naked eye and then view it again under a magnifying lens of higher resolution. It may reveal an irregularity that looks just the same as that of the original curve. We say that the irregularity is self-similar (Fig. 7.12*b*). It has no characteristic length scale below which it is smooth. It is like the infinite 'matryoshka' with a doll within a doll within a doll, *ad infinitum*, all exactly alike but for the size. How do you find the dimension and the length of such an irregular curve? Intuition is no guide now. But we can get back to our divider and walk the curve as before. The polygon traversed by us will now never quite hug the irregular curve. But we still have our ϵ and $N(\epsilon)$, and we can hopefully find the scaling $N(\epsilon) \sim \epsilon^{-D}$ as $\epsilon \to 0$. The exponent D is then the capacity dimension of the curve by our definition. However, D is now not necessarily an integer. It will, in general, be a fraction! The zig-zag curve is then called a *fractal* and D its *fractal dimension*. For example, we may survey the irregular coastline of England and find that it is a fractal, with a fractal dimension of about 1.3. What about the length of this fractal coast line? Well, we still have $L \simeq \epsilon N(\epsilon)$. Thus, as $\epsilon \to 0$, the length of the coastline tends to infinity! The length depends on the scale of resolution. It was the study of such irregular shapes and forms that had led Benoit Mandelbrot of IBM to the discovery of fractals and fractal geometry in 1960.

A fractal line is something intermediate between a one-dimensional line and a two-dimensional surface. The fractal line with its self-similar irregularity structure down to the finest length scale has an amusing but revealing aspect to it. If we were to run the needle of a gramophone on such a fractal track (groove), it will produce a symphony that is independent of the speed of the needle! The scale invariance in space gets translated into a scale invariance in time via the constant speed of traversal of the groove by the needle.

The same ideas can now be extended to an irregular (fractal) surface. In this case, of course, we will employ elementary squares or plaquettes measuring ϵ on the side, to cover the surface. If $N(\epsilon)$ is the number of plaquettes required to cover it, then we expect $N(\epsilon) \sim \epsilon^{-D}$, where D is the fractal dimension. In all these examples, our geometrical objects, the lines (topological dimension 1) and the surfaces (topological dimension 2), are assumed to be embedded in the background Euclidean space of topological dimension 3, which is our familiar space. But we can always imagine our objects to be embedded in a background Euclidean space of a higher dimension denoted by d. We will call this the *embedding dimension*. It is important to realize that the fractal dimension of the geometrical object is its *intrinsic* property. It does not depend on the space in which it may happen to be embedded. Of course, the embedding space must have a Euclidean dimension greater than the topological dimension of the object embedded in it. Thus, for example, a zig-zag curve embedded in our three-dimensional Euclidean space may be 'covered' with three-dimensional ϵ-balls or ϵ-boxes instead of one-dimensional ϵ-segments (Fig. 7.12c). The intrinsic fractal dimension will work out to be the same, and it shall be greater than unity, even up to infinity. Such an *intrinsic fractal dimension* is relevant for reckoning the *weight* of an irregular (zig-zag) thread, a linear polymer such as a highly random coiled DNA strand, say. We could also define an *extrinsic* fractal dimension for the irregular line — it now measures the *basin* wetted by the line. But, this extrinsic fractal dimension can at most be equal to the embedding dimension — here the line fills the space!

And now we are ready for an important generalization. We will ask for the fractal dimension of a *set of points* embedded in a Euclidean space of dimension d. We need this generalization for our study of chaos. Because here the background Euclidean space is the state space that can have any integral dimension. In this state space the set of points, or more like a dust of points is embedded forming the attractor whose fractal dimension (D) we are after. The dimension d of the state space is the embedding dimension. Our method remains the same. We cover the set of points with d-dimensional balls of radius ϵ and then fit the scaling relation

$N(\epsilon) \sim \epsilon^{-D}$ or use the equivalent logarithmic formula discussed earlier. All this can be done rather quickly on a PC. It is, however, important to appreciate that the scaling relation, *e.g.*, $N(\epsilon) \sim \epsilon^{-1}$, ceases to have any physical validity in the limit when ϵ becomes comparable with the microscopic size of the atoms or the size of the macroscopic object. The scaling holds only in the interesting region of *intermediate asymptotics*. The whole point is that this intermediate scaling region covers a sufficiently wide range $\sim 10^2$–10^4. Hence the operational validity of the notion of the fractal. Of course, mathematicians have constructed a few truly fractal curves. But nature abounds in physically fractal objects — *e.g.*, lightening discharges, cloud contours, coastlines, mountain ranges, fractures, crumpled paper balls, diffusion-limited aggregates (spongy masses), biological sponges, foams, human and animal lungs, human cortex foldings, protein surface folds. The clustering of galaxies suggests a large scale mass structure of the universe with a fractal dimensionality $D \simeq 2$, that may account for the missing mass that astronomers and cosmologists are very much concerned with. As to why Nature abounds in fractals, the answer may lie in what is known as *self-organized criticality* (SOC). Forms in Nature often emerge from recursive growth building up to the point of instability and breakdown, giving a landscape which is rough on all length scales — a critical state, a self-similar fractal!

Before we go on to the fractal attractors, let us get acquainted with some of the fascinating fractals that have become classics. We will actually construct them.

7.5.1   The Koch Snowflake

Take an equilateral triangle. Now divide each side in three equal segments and erect an equilateral triangle on the middle third. We get a 12 sided Star of David. The construction is to be iterated *ad infinitum*. Smaller and smaller triangles keep sprouting on the sides and finally we have a zig-zag contour that is irregular on all length scales in a self-similar fashion, resembling a snowflake (Fig. 7.13).

To find its fractal dimension, all we have to note is that at each iteration the scale of length gets divided by 3 and the number of steps required to cover the perimeter increases by a factor of 4. This gives the fractal dimension $D = -\ln 4/\ln(1/3) = 1.26$. It is interesting to note that while the actual length of the perimeter scales as $(4/3)^n$ to infinity as expected, the area enclosed by the figure remains finite, close to that of the original triangle.

7.5.2   Cantor Dust

Take a segment of a straight line of unit length, say. Trisect it in three equal segments and omit the middle third, retaining the two outer segments (along with their end points). Now iterate this construction by trisecting these two segments and again omitting the middle thirds, and so on *ad infinitum*. Eventually, you would

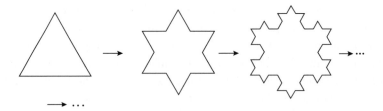

Figure 7.13:   Koch snowflake. Construction algorithm for this fractal of dimension 1.26.

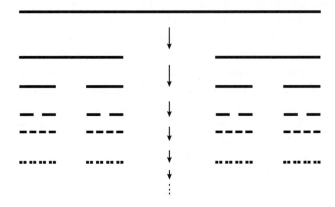

Figure 7.14:   Cantor dust. Construction algorithm for this fractal of dimension 0.63.

be left with a sparse dust, called the *Cantor set*, of uncountably infinite points (Fig. 7.14). What is its capacity dimension? Well, in each iteration, the length of the segments gets divided by 3 while the number of such segments is multiplied by 2. Hence the fractal dimension $D = -\ln 2/\ln(1/3) = 0.630$. Of course the total length of the set at the nth iteration is $(2/3)^n$, which tends to zero as $n \to \infty$.

Aside from these and many other fascinating examples of fractal sets, which are just mathematical constructs, we do have fractal objects occurring in Nature. We have already mentioned several of them occurring naturally. Indeed, a fractal geometry should be Nature's strategy for enhancing the surface to volume ratio.

The fractal dimension D is a purely geometrical measure of the capacity of a set. Thus in 'box counting,' all the boxes, or rather the neighborhoods, are weighted equally. In the context of chaos, however, a fractal is a set of points in the state space on which the 'strange attractor' lives. The different neighborhoods (boxes) of the fractal are visited by the phase point with different frequency in the course of its 'strange' aperiodic motion; the fractal set is physically inhomogeneous. The fractal dimension D as defined above fails to capture this physical fact. It is possible to define yet another fractal dimension that weights different neighborhoods according as how frequently they are visited. In fact a whole spectrum of fractal dimensions are defined in what is called *multifractal analysis* of the fractal set. This is a highly technical matter and we will not pursue it further here. Suffice to say that the multifractal analysis combines geometry with physics and thus reveals more information.

7.6 Reconstruction of the Strange Attractor from a Measured Signal: The Inverse Problem

A dynamical system may have a number of degrees of freedom. This whole number is the dimension 'd' of its phase space. The dynamics will eventually settle down on an attractor. If the dynamics is chaotic, it will be a strange attractor (a fractal) of fractional dimension D, imbedded in the phase space of (imbedding) dimension $d > D$. Both d and D are important specifications of the nature of the system and its strange attractor. Now, in general, these two dimensions are not given to us. All we have is an experimentally observed, seemingly random signal representing the measurement of a single component of the d-dimensional state vector in continuous time. It could, for instance, be the temperature in a convection cell, or a component of fluid velocity of a turbulent flow at a chosen point. This can be measured, for example, by the Laser Doppler Interferometry, in which the frequency of a laser light scattered by the moving liquid at a point is Doppler shifted. The shift is proportional to the instantaneous velocity of the fluid at that point. The question is, can we reconstruct these important attributes such as D and d from such a single-component measurement in time? It was proved by Floris Takens that we can indeed retrieve some such vital information about the system from this stream of single-component data.

The philosophy underlying this claim is quite simple. In a dynamical system with interacting degrees of freedom, where everything affects everything all the time, it is conceivable, indeed highly plausible, that a measurement on any one of them should bear the marks of all others, however implicitly. The real question now is how to recover this information. This is a rather technical matter. The technique is, however, very straightforward to apply and is worth knowing about.

Let us take $x(t)$ to be the measured signal. The first step is to discretize time. For this, pick a convenient time interval τ and form the sequence $x(t), x(t + \tau)$, $x(t + 2\tau), \ldots$. This is known as a *time series*. Next, we are going to construct vectors out of this series. First, construct the two-dimensional vectors $(x(t), x(t + \tau))$, $(x(t + \tau), x(t + 2\tau)), (x(t + 2\tau), x(t + 3\tau)), \ldots$. Each vector has two Cartesian components, for example $x(t)$ and $x(t + \tau)$, as any two-dimensional vector should. Plot these vectors in a two-dimensional phase space, so as to generate a dust of points. Now, count how many neighbors a point has, on average, lying within a radius of it. Let the number be $C(\epsilon)$. For small ϵ, we will find that $C(\epsilon)$ scales as $C(\epsilon) \sim \epsilon^\nu$. Then, ν is said to be the *correlation dimension* of our set. We will use a subscript and call it ν_2 because we constructed it from a two-dimensional phase space. Now repeat this procedure, but this time with the three-dimensional vectors constructed

out of our time-series, *i.e.*, $(x(t), x(t+\tau), x(t+2\tau)), (x(t+\tau), x(t+2\tau), x(t+3\tau)), \ldots$.
Each vector has three Cartesian components. We will, therefore, get the correlation-dimension ν_3 this time. And so on. Thus, we generate the numbers $\nu_1, \nu_2, \nu_3, \ldots$.
We find that ν_n will saturate at a limiting value, a fraction generally, as n becomes sufficiently large. That limiting value is then the Correlation Dimension of the Attractor of our system. It is usually denoted by D_2 and is less than or equal to the fractal dimension D of the attractor. Also, the value of the integer n, beyond which ν_n levels off, is the minimum imbedding dimension of our attractor. That is, it is the minimum number of degrees of freedom our dynamical system must have to contain this attractor.

This technique, due to P. Grassberger and I. Procaccia, seems to work well with low dimensional attractors. It has been used extensively to characterize chaos in all sorts of systems including the human cortex. Thus, the EEG time series for various brain-wave rhythms, *e.g.*, with eyes closed, with eyes open, with and without epileptic seizure have been analyzed with startling results. The epileptic brain is less chaotic (*i.e.*, it has lower fractal dimension) than the normally active brain!

The time-series analysis and reconstruction of the attractor provides other important information as well. It can distinguish deterministic chaos from truly random noise. Thus, for example, the correlation dimension D_2 is infinite for random noise, but finite and greater than two for chaos. (For a limit cycle, it is one and for a torus (quasiperiodic) it is two). Thus, we know, for example, that the noisy intensity output of a laser can be due to a low dimensional strange attractor in its dynamics.

7.7 Concluding Remarks: Harvesting Chaos

That a simple looking dynamical system having a small number of degrees of freedom and obeying deterministic laws can show chaos is a surprise of classical physics. It has made chaos a paradigm of complexity — one that comes not from the statistical law of large numbers, but from the highly fine-structured self-similar organization in time and space. There is no randomness put in by hand. Chaos is due to the *sensitive dependence* on *initial conditions* that comes from the stretching and folding of orbits. Mathematically, this is represented by the all important nonlinear feedback. Thus, chaos can amplify noise, but it can also masquerade as 'noise' all by itself. It is robust and occurs widely. It has classes of universality.

The theory of chaos rests on a highly developed mathematical infra-structure that goes by the forbidding name of *nonlinear dynamical systems*. But high-speed computing has brought it to within the capability of a high school student. The contemporaneous ideas of fractal geometry have provided the right language and concepts to describe chaos.

There are still many open questions. For instance, we have discussed chaos for open dissipative systems only. There are closed dynamical systems that are conservative (non-dissipative) — the problem of chaos in a non-integrable Hamiltonian system. A celebrated example is that of a three-body problem in astronomy, the problem of three planets gravitationally attracting each other. This is a non-integrable system and chaos lurks there. This is, in fact, very relevant to the long-term stability of the solar system — on the time scale of hundreds of millions of years. Then there is the problem of two nonlinearly coupled oscillators for example. These can be treated quantum mechanically. Does chaos survive quantization? What is the signature of quantum chaos? There are other rather technical issues that remain still unresolved. But it can be fairly claimed that for most dissipative systems, the theoretical framework of chaos is complete. It is now time to harvest chaos.

Undoubtedly, the most outstanding application of our knowledge of chaos is to understanding fluid dynamical turbulence — the last unsolved problem of classical physics. For several types of closed flow, routes to turbulence are now understood. The problem of turbulence in open flows such as wakes and jets is, however, still unsettled. Do we have here, after all, the Landau scenario of confusion of incommensurate frequencies, appearing one at a time?

Next, comes the problem of short and intermediate range weather forecasting that started it all from the time of the early works of E. N. Lorenz. Chaos, of course, rules out long-range forecast. But can we extend our present 3–4 day forecasts to a 10–12 day forecast? One must estimate the time over which memory of the initial data is lost due to folding back of the phase-space trajectories. This ultimately limits the range of prediction.

Reconstruction of the strange attractor, if any, at the heart of an apparently random behavior, is a powerful diagnostic tool that is finding application in diverse fields of study including neurology, psychology, epidemiology, ecology, sociology and macro-economics.

There is yet another way in which we can make chaos really work for us. Thus, chaotic encryption can be used for secure communication — this is based on the idea that two similar chaotic oscillators can be synchronized through a fairly general kind of coupling. Also, if we want to explore a large portion of phase space efficiently, as for example, for mixing things in a chemical reactor, it may be best to use a chaotic trajectory. Nature seems to have made use of this chaotic access to, or exploration of, these possibilities in the vital matters of evolution, and in the workings of our immune system. Chaos in our neural network is also an important subject for study.

There are other highly speculative aspects of chaos. Chaos is physically intermediate between the 'thermal bath' and macroscopically ordered motion. What happens when a chaotic system is coupled to a 'thermal bath'? Can we make 'intelligent heat engines'? What is the thermodynamic cost of driving a system chaotic? Or an even more speculative question such as whether chaos is ultimately responsible for the well-known quantum uncertainties. We do not know.

It has been claimed that the scientific revolution, or rather the shift of paradigm caused by 'chaos' is comparable to that caused by quantum mechanics. To us this comparison does not seem very apt. In any case, judging by their respective impacts, the claim is overstated. But what can definitely be claimed is that chaos has definitely been a surprise of classical physics. It has the subtlety of number theory. You can certainly have fun with chaos if you have a personal computer, or even a hand-held calculator.

7.8 Summary

A dynamical system is said to be chaotic if its long time behavior cannot be computed and, therefore, cannot be predicted exactly even though the laws governing its motion are deterministic. Hence the name deterministic chaos. A common example is that of turbulence which develops in a fluid flowing through a pipe as the flow velocity exceeds a certain threshold value. The remarkable thing is that even a simple system, such as two coupled pendulums, having a small number of degrees of freedom can show chaos under appropriate conditions. Chaotic behavior results from the system's sensitive dependence on the initial conditions. The latter amplifies uncontrollably even the smallest uncertainty, or error unavoidably present in the initial data as noise. This sensitivity derives essentially from the nonlinear feedback, and often dissipation inherent in the dynamical equations. The dynamical evolution may be described by continuous differential equations or by discrete algebraic algorithms, and can be depicted graphically as the trajectory of a point in a phase space of appropriate dimensions. The representative point may converge to a stable fixed point, or a limit cycle, or a torus, and so on. These limiting sets of points in the phase space are called *attractors* and signify respectively a steady state, a singly periodic motion, or a doubly periodic motion, and so on. Chaos corresponds to a *strange attractor*, which is a region of phase space having a fractional dimension (a fractal) and its points are visited aperiodically.

There are several routes to chaos. We have, for example, the period doubling route of Feigenbaum in which the dynamical system doubles its period at successive threshold values (bifurcation points) of the control parameter, until at a critical point, the period becomes infinite and the motion turns chaotic. These different routes define different universality classes of common behavior.

Chaos has become a paradigm for complex behavior. Many apparently random systems are actually chaotic. Powerful techniques have been developed to reconstruct the strange attractor from the monitoring of a single fluctuating variable. This helps us make limited predictions and diagnostics for diverse problems, such as the weather forecast and turbulence, epileptic seizures, populations of competing species, macroeconomic fluctuations, and the recurrence of epidemics.

7.9 Further Reading

Books

- G. L. Baker and J. P. Gollub, *Chaotic Dynamics*: *An Introduction* (Cambridge University Press, New York, 1990).
- P. Bergé, Y. Pomeau and C. Vidal, *Order Within Chaos* (Wiley, New York, 1986).
- R. C. Hilborn, *Chaos and Nonlinear Dynamics*: *An Introduction for Scientists and Engineers* (Oxford University Press, Oxford, 1994).

Semi-popular articles

- J. P. Crutchfield, J. D. Farmer, N. H. Packard and R. S. Shaw, *Chaos*, Scientific American **255** (December 1986), p. 38.
- Joseph Ford, *How Random is a Coin Toss?*, Physics Today **36** (April 1983), p. 40.

Bright Stars and Black Holes   **8**

On a clear night, if you look up at the sky you will likely see a broad band of faint distant stars stretching from horizon to horizon which the ancient Greeks likened to a river of milk and accordingly called a *galaxy*. This band is an edge-on view of our galaxy, which we simply refer to as the Galaxy or Milky Way. If we could observe our galaxy far away from outside, it would appear as a gigantic thin disk with a central bulge. The disk is about 10^{21} m across, $(0.5–1) \times 10^{19}$ m thick, packed mostly with stars traveling in nearly circular orbits about the galactic center with very small up-and-down motions. The central bulge itself contains stars moving with small circular speeds in random directions, which accounts for their roughly spherical distribution. Above and below the disk, a spherical swarm of *globular clusters* with large random velocities forms a halo concentrated toward the center of the Galaxy. If we could look at our galaxy face-on, as we can with external systems such as the galaxy NGC4622, we would see in the plane of the galactic disk a beautiful spiral pattern composed of several spiral arms outlined by very brilliant stars. These bright stars, strung along the arms like 'beads on a string,' contrast sharply with the very dark inside edges of the arms and the faint, more diffuse patches throughout the disk. These irregular formations suggest the presence of matter — gas and dust — in the space between the stars.

But it is the stars that play the most important role in a normal galaxy: they provide most of its mass and hence are responsible for the gravitational forces that bind the galaxies into stable associations. Not all stars are alike. They come in a wide variety of forms: massive and not so massive, large and small, bright and faint. Why is it so? This is a question central to all astronomy, which we will try to answer in this chapter. We will find that stars constantly change in composition and appearance: starting with different initial masses, they evolve at different rates from one form to another, and metamorphose into different final objects. Stars are born, live and die. Underpinning this study is the assumption that the laws of physics, discovered and verified here on earth, are universal: they can be applied to all the physical universe. That such an assumption is well founded has led to the creation of one of the most exact of the astrophysical sciences.

8.1   The Basic Properties of Stars

In the night sky, the 'dog star' (Sirius A) appears to be the brightest star, not because it radiates the most energy, but because it is among the closest to us. What we perceive in our observations of the sky is the *apparent brightness*, which depends on the star's distance from us and does not express its true energy output. This true energy output, a property intrinsic to the star, is measured by the *absolute brightness*, or *luminosity* — the total amount of energy radiated each second in all forms at all wavelengths.[1] As a star radiates in all directions, the fraction of its power reaching a detector with receiving surface area A a distance d away is $A/4\pi d^2$. So the star's *luminosity* L is given in terms of its *apparent brightness* $\mathcal{L}$ and distance d by $L = 4\pi d^2 \mathcal{L}$. In SI metric units, L is given in watts, or W (1 W is 1 joule (J) of energy radiated each second), d in meters (m), and $\mathcal{L}$ in W/m^2. A star's apparent brightness can be measured by photometry.

Distances to nearby stars are measured by triangulation, whereas distances to nearby galaxies can be determined by the brightness and periodicity of their Cepheid variable stars.[2] For these vast distances, astronomers use two basic units: the *light-year*, abbreviated as ly, which is the distance light travels in one year, *i.e.*, 9.46×10^{15} m; and the *parallax-second*, or *parsec*, abbreviated to pc (1 pc = 3.26 ly). The nearest star beyond the sun, Proxima Centauri, is about 4.3 ly or 1.3 pc away, whereas the brightest star, Sirius A, is 8.8 ly or 2.7 pc away. The sun could be said to be eight light-minutes away. We see that once the distance d and apparent brightness $\mathcal{L}$ of a star are determined, we also have its intrinsic luminosity L.

Some stars, like Sirius A, are bluish; others, such as Antares in α Scorpii and Arcturus in α Bootes, tend to be reddish or orangish. Although sunlight appears to us as yellow, we have learned elsewhere (*e.g.*, in Chapter 2) that it also contains, to a lesser extent, all other colors, or wavelengths, spread over a continuous *spectrum*. Let us now see how this applies to stars and what starlight can tell us.

In studying its general features, we may regard a star during most of its lifetime as a ball of gas in thermal equilibrium — a large collection of electrons and atoms in random motion, jostling each other, at a uniform temperature throughout. The random motions of the free particles in matter give rise to a form of energy we call *heat*, which is measured by the *temperature* of the object: the faster the thermal motions, the higher the temperature. As the electrons (and atoms and molecules) are constantly and randomly disturbed by thermal motions, getting pushed and pulled by other electrons and atoms, they emit a photon at each deflection with an energy precisely matching the kinetic energy lost by the interacting electron. Since the electrons move at random speeds and the radiative processes occur randomly, photons may be produced practically at any wavelengths. What we have then is a *continuum spectrum of thermal radiation.*

---

[1] It is more precisely called *bolometric luminosity*, to distinguish it from *visible luminosity*, which is the energy radiated each second in the *visible* part of the spectrum.

[2] More details on the definitions and measurements of cosmic distances can be found in Chapter 10.

As a thermal object, our ball of gas obeys known statistical physical laws.[3] In particular, the thermal radiation it emits exhibits an energy distribution over wavelengths with a profile uniquely defined by its temperature T (see Fig. C.3 of Appendix C). This distribution curve has a single maximum located at wavelength $\lambda_{\max}$ such that $\lambda_{\max}T = 0.0029$ m K, a relation known as *Wien's displacement law*, which we have stated in SI metric units. Hotter sources radiate more blue light than cooler sources.

Another important consequence of the photon distribution is the *Stefan–Boltzmann law*, which says that the *energy flux f* (or the energy of radiation emitted at all wavelengths per unit area and per unit time) depends only on the temperature T of the radiating object and is given by $f = \sigma T^4$, where σ is the Stefan–Boltzmann constant.[4] As the temperature increases, the energy flux rises even more steeply. If we regard a star as a sphere of radius R, its *luminosity*, or radiant power, is given by the energy flux f generated over its whole surface area, $4\pi R^2$ — that is, $L = 4\pi R^2 f$, or $L \propto R^2 T^4$. The hotter an object is, the brighter it shines.

In applying these laws to stars, we must remember that T is the star's *effective surface temperature*, which is the effective average temperature of the various light-emitting layers of gas. So, once we've determined a star's color, we use Wien's law to obtain its surface temperature T. With T and L known, we can calculate the stellar radius R from the Stefan–Boltzmann law written in the form $L \propto R^2 T^4$.

In the above discussion, we considered radiation by free electrons.[5] The nature of radiation changes completely when the radiating electrons are confined in atoms, because then they (and hence the atoms themselves) may exist only in a limited number of *discrete states* defined by the rules of quantum mechanics. The energies of these states, plotted in increasing order like rungs in a ladder, give us a graphical representation of the *atomic energy spectrum*, as unique to an atomic species as the fingerprint to an individual. When an atom interacts with an electromagnetic field, it may pass from one allowed state to another by absorbing or emitting a photon, provided the photon energy exactly matches the energy difference of the two atomic states. As the atomic energy spectrum is discrete, only discrete values are allowed for the photon energy in emission or absorption. If a few of these values are somehow known, then we can identify the atomic element involved.

When a beam of light passes through a cloud of atoms, many photons of certain colors may be absorbed by the atoms and removed from the beam. The absence of these ('line') photons in the outgoing beam reveals itself in *absorption lines*, which appear as dark vertical lines on a photographic spectrum or valleys on a graph (Fig. 8.1). Similarly, when excited atoms de-excite by radiative transitions, photons of specific colors are emitted and added to the outgoing light, giving rise to *emission lines*, which appear as bright vertical lines or sharp peaks.

---

[3]See Appendix C for more details.
[4]In SI units, $\sigma = 5.67 \times 10^{-8}$ W/(m^{2}K^4).
[5]Of course, they are not really free since interaction fields exist all around. What we mean is that they are not bound in atoms.

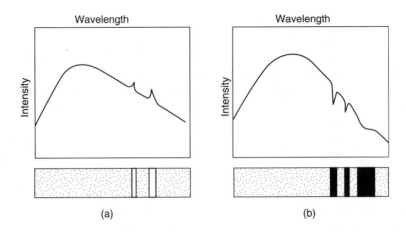

Figure 8.1: (a) Emission lines and (b) absorption lines on a background radiation continuum.

For stars, absorption spectra come about through the following processes. As thermal photons of all wavelengths rise up from the stellar depths to the surface, those with the 'right' energies (for which atomic transitions are allowed) will be absorbed by atoms in the cooler upper layers. The excited atoms then de-excite either by non-radiative collisions or by radiative transitions. In the first case, the surplus energy is converted into kinetic energy for the particles involved in the collisions, and no radiation is produced. In the second case, when radiation accompanies the de-excitation, the photon emitted is equally likely to travel in any direction. Because the hot stellar surface is surrounded by a dark absorbing space, there are more photons originally traveling toward us being scattered out of our line of sight than into it, and so there is a net loss of line photons traveling in our direction. In either case, photons are removed from the underlying radiation field at the allowed transition energies, leading to the appearance of *absorption lines* in the stellar spectrum.

The pattern of the spectral lines tells us *what elements* are present in the star, and their relative strengths (*i.e.*, widths and degrees of darkness) tell us *what fraction of atoms* of an element is in each excited state. This is a crucial result. Why? Because with it we gain direct information about the composition of the star and, from the composition, determine the temperature, density and pressure conditions in the far-away gas.

In a classification system invented by astronomers Edward Pickering and Annie Cannon, stars are divided into *spectral classes*. These may run in any order, but the sequence finally adopted by astronomers[6] is O, B, A, F, G, K and M, with the O stars showing ionized helium lines in their spectra, the B stars showing neutral helium lines, etc. But since the observed spectral properties of a star reflect its effective temperature, spectral classes form a temperature sequence. The adopted

---

[6] Just remember — Oh, Be A Fine Girl, Kiss Me!

sequence corresponds to an order of decreasing temperatures. The O stars are so hot, at 40 000 K, that even the most tightly bound atoms, such as helium, become ionized, losing some of their electrons. In class G stars, at around 5 500 K, lines of ionized metals appear along with lines of neutral hydrogen. By M class, the temperature has dropped so low, at 3 000 K, that atoms can stick together to form molecules.

Finally, what about the star's mass? Let us first consider a double-star system, consisting of two visible stars in orbit around each other. Classical mechanics gives the period of revolution (or the time it takes to complete an orbit) in terms of the distance between the two stars and the sum of their masses. So by measuring the period of revolution and the distance between the two stars, we can calculate the sum of their masses. Again, from classical mechanics, we know that each star orbits around an imaginary point called the center of mass. By measuring the distance of each star from the center of mass, we can determine the ratio of the masses. Once the sum and the ratio of the masses are known, we can get each individual mass. For example, the bright star Sirius A and the faint star Sirius B form a binary system, close enough to us to permit a detailed analysis of its motion. With the data available, the preceding steps show that the sum of the masses[7] is 3 $M_\odot$, and A is twice as massive as B. It follows immediately that Sirius A has a mass of 2 $M_\odot$, and Sirius B, 1 $M_\odot$.

The study of binary systems shows that ordinary stars with nearly identical spectra usually have nearly identical masses (and vice versa). This fact allows the masses of many stars, not necessarily in binary systems, to be estimated from their spectral properties.

8.1.1 Summary

A star has characteristic luminosity, temperature, mass and spectral lines. Once its luminosity and temperature are known, its size can be inferred from the Stefan–Boltzmann law, and its dominant color determined by Wien's displacement law. There are no simple relations for the mass, which must be determined by observations or estimated from the star's spectroscopic properties, which also tell us about the internal composition of the star.

8.2 The Hertzsprung–Russell Diagram

Observed stars come in a bewildering variety: masses range from 0.1 $M_\odot$ to 60 $M_\odot$ or more; luminosities from one millionth to a million times the sun's luminosity

---

[7]It is sometimes convenient to express astronomical quantities in *solar units*. The solar mass is 1 $M_\odot = 1.99 \times 10^{30}$ kg; the solar radius, 1 $R_\odot = 6.96 \times 10^8$ m; and the bolometric luminosity of the sun, 1 $L_\odot = 3.84 \times 10^{26}$ W. The astronomical unit (AU) is the mean radius of the earth's orbit about the sun. 1 AU $= 1.50 \times 10^{11}$ m.

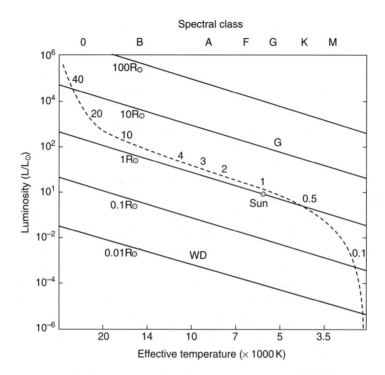

Figure 8.2: Hertzsprung–Russell diagram. The dashed line represents the main sequence; along this line, numbers give stellar masses in solar units. G indicates the general region of red giants, and WD that of white dwarfs.

($L_\odot$); surface temperatures from one third to ten times the solar temperature; and sizes from less than one hundredth to more than 100 times the solar size. And there are blue stars and red stars. But when we represent every star by a point on a two-dimensional logarithmic plot, luminosity versus surface temperature, we find that the stars do not fall haphazardly on the diagram, but occupy quite specific regions, which implies definite relationships between the various forms of stars. This type of plot is called the *Hertzsprung–Russell* (H–R) *diagram* (Fig. 8.2) in honor of its creators.[8]

If such a two-dimensional plot is found to be useful, it is because we need only two parameters to differentiate various types of stars: of the basic stellar properties, only two are independent. Let us try to understand this point.

First, the color and temperature of a star are tied together by Wien's law. And so in an H–R plot with the temperature scale increasing from right to left, the right part of the diagram contains redder, cooler stars, and the left part, bluer, hotter stars.

Next, any point on an H–R diagram corresponds to a specific luminosity, temperature and radius, according to $L \sim R^2 T^4$; and if the scales of the coor-

[8]Danish astronomer Ejnar Hertzsprung and American astronomer Henry Norris Russell.

dinates are logarithmic, as is the case for the standard H–R plot, stars of different radii fall on different straight lines. If we move upward from the sun's (or any star's) location on the diagram, we encounter brighter, bigger objects. If we move downward, we find fainter, smaller stars. To the right of the sun's location, stars with a similar luminosity are cooler and bigger; but those to its left are hotter and smaller. (But different masses may coexist in the same region, except along the main sequence band, as we'll see below.)

Once the luminosities and temperatures for individual stars are determined, we can construct the H–R diagram for the observed *luminous stars*. The outstanding feature of the diagram so produced is that most of the stars fall in a narrow band, called the *main sequence*, which runs diagonally from the top left, where the most massive stars lie, to the bottom right, where the lowest-mass stars lie. Our sun, a type G star, occupies about the middle of the main sequence and is a very ordinary star.

Extending from about the sun's position toward the top right is a region called the *giant branch*, inhabited by large, cool stars with very low mass density. Aldebaran in α Tauri — a spectral type K star, forty times bigger and a hundred times brighter than the sun — is a *red giant*; Betelgeuse in α Orionis — a spectral class M star, seven hundred times the size of the sun and radiating ten thousand times more energy — belongs to the category of *supergiants*. Extending across the diagram from the giant branch toward the left at constant luminosity is the so-called *horizontal branch*, where one finds very unstable configurations. Some pulse with a constant period; others flare up sporadically. The Cepheids belong to this region, although they are not usually plotted in observational H–R diagrams, precisely because of the variability in their apparent brightness.

In the lower left-hand corner of the H–R diagram, below the main sequence, we find a group of bluish-white, compact and very faint stars known as *white dwarfs*. Sirius B, the faint companion to the brightest star in the night sky, is such a star: five hundred times less luminous than the sun, yet only 2.7 pc away. Procyon B, also found in a binary system, radiates about ten times less than Sirius B and belongs to spectral type F. White dwarfs have typically the mass of the sun, but the size of the earth.

White dwarfs are one of the three forms of 'dead stars.' The other two are even more massive and compact. They emit no light at all, and so cannot be seen directly by optical means and would fall outside the standard H–R plot: they are *neutron stars* and *black holes*.

We see that the H–R diagram is more than just a plot of luminosity against temperature. It is a systematic representation of the correlations between the surface properties of the luminous stars. It separates stars into distinct groupings: the main-sequence stars, the giants, the supergiants and the white dwarfs. What causes these groupings? Are they somehow related to one another? Or more aptly: What happens in the *interior* of the stars to make them *look* the way they do?

It has been known for a long time that stars evolve from one form to another and that stars we see in different forms are stars at different ages. The English astrophysicist Arthur Eddington already understood in 1926 that stars exist because the two opposing forces at work in the systems — *gravity* which pulls inward on the stellar gas while *gas pressure* and *radiation pressure* which push outward — are just balanced. At about the same time, H. N. Russell and H. Vogt established an important result: the *mass* and *chemical composition* of a star determine uniquely how it looks at equilibrium. Simply put, give me a mass and a composition, and I shall tell you what the star looks like. It means that the main-sequence stars of similar masses must have similar pressures, radii, temperatures and other physical attributes. It also implies that, if some red giants and white dwarfs have a similar mass, it is just because they have evolved and acquired different compositions, and hence reached different equilibrium structures. And so, different forms of stars mean different internal compositions corresponding to different stages of evolution. How does the stellar composition change? The answer is, by *nuclear reactions*: these processes provide not only the mechanism for the star's internal *changes*, but also the energy necessary for maintaining its *stability*. In the final analysis, the key to a good understanding of the stellar evolution lies in nuclear physics, particle physics and, eventually, general relativity.

8.2.1 Summary

One of the most useful tools in astronomy is the Hertzsprung–Russell diagram, a two-dimensional plot of luminosity versus temperature of luminous stars. It gives a systematic representation of the correlations between the stars basic properties and hence of the stars at different stages of development. The path a star takes in the H–R diagram is the history of its evolution from its birth in an interstellar cloud to its end as a dead star.

8.3 Bright, Shining Stars

The space between stars is filled with rarefied and cold matter whose chemical composition reflects that of the universe as a whole. The most striking feature of our galaxy is its spiral pattern. The spirals are probably formed by matter-density waves that the Galaxy builds up as it rotates rapidly and differentially about its own axis. Half of the interstellar matter is distributed very thinly throughout the Galaxy in the equatorial plane, with the rest found in the spiral arms as cloud formations of various sizes and densities. Of greatest interest for their roles in the theory of stellar evolution are the Small Molecular Clouds and the Giant Molecular Clouds, which can grow big and dense enough to contract through their own weight.

A clump of cloud becomes self-gravitating when it gains enough gravitational energy to overcome its dispersive internal turbulent motions. It can get gravitational energy either by adding mass from the surrounding medium or by being suddenly compressed by some powerful external force. Within a short time, the collapsing globule of gas will build up a dense central nucleus, which tends to fall away from the outer parts but which may continue to pick up infalling material from the envelopes to form a progressively opaque core having a typical stellar mass. The evolution of this object, or *protostar*, will take it through several successive phases (free fall, convection and radiation), characterized by increasingly higher internal temperatures, before it reaches its position on the main-sequence line of the Hertzsprung–Russell diagram.

8.3.1 Nuclear Sources of Stellar Energy

In the pre-main-sequence period, only two kinds of forces are involved — the *gravitational* and the *electromagnetic force* — but now, as the star reaches the main sequence, the other two fundamental forces[9] of Nature enter the scene: the *strong* (*nuclear*) and the *weak force*, so called because the former is the strongest of all known forces, and the latter, much weaker than the electromagnetic force. In contrast to the first two, they have a very short range, 10^{-15} m or less — they will not act until the two interacting particles come that close together — and, for this reason, we cannot directly see their effects in our everyday life. Yet, they are essential to the structure and the dynamics of matter at a deeper level and on the grander cosmic scale, and are, therefore, of primary importance to humans. These are precisely the two forces thought to be capable of producing the amount of energy necessary to sustain the long-lasting luminosity observed in most stars.

Except for the nucleus of hydrogen, which consists of a single proton, the nuclei of all other atoms, including isotopes of hydrogen, contain both protons and neutrons, collectively called *nucleons*.[10] Nucleons interact with one another through a mainly attractive nuclear force and a repulsive electric force. Since the former is a hundred times stronger than the latter, it contributes the largest part to the potential energy in a nucleus.

The mass of a bound nucleon system always totals less than the sum of the constituent masses. This *mass deficit* represents[11] the *binding energy B* of the

---

[9]Explanations on the notions of forces and particles we use here can be found in Chapter 9.

[10]The nucleus of hydrogen (H) consists of a single proton; the nuclei of its isotopes, deuterium (^{2}H) and tritium (^{3}H), contain, besides the requisite proton, one and two neutrons, respectively. Helium (^{4}He) has two protons and two neutrons, and its isotope ^{3}He two protons and a neutron. Generally, a nucleus is designated by the symbol ^A_ZX, where X stands for the name of the element, A the mass, or nucleon, number and Z the charge number, which is often omitted. Thus, ^A_ZX is the same as ^{A}X.

[11]Recall that a mass m is equivalent to energy mc^2, according to the famous Einstein equation, $E = mc^2$.

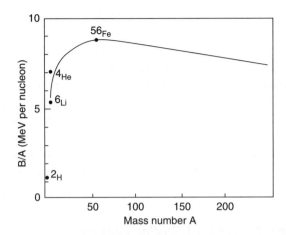

Figure 8.3: The nuclear binding energy per nucleon. Note that ^{4}He is more strongly bound than the neighboring-mass nuclei and that ^{56}Fe has the largest average binding energy among all of the nuclei.

nucleus, *i.e.*, the energy that must be provided to split the nucleus apart so that its separated nucleons are at rest and out of reach of one another's forces. Defined in this way, the binding energy is a *positive quantity*.

In Fig. 8.3 we present the average binding energy in nuclei (total binding energy B of each nucleus divided by its nucleon number A) as a function of A. The curve exhibits two outstanding features. First, the binding energy in most nuclei averages about 8 MeV per nucleon. Second, after reaching a maximum around iron, B/A slightly but steadily drops off. Let us try to understand these features. While the dominant nuclear interaction makes the major contribution to the nuclear binding, other smaller effects must be present that tend to work in the opposite directions. Which are they? In the very light nuclei, where practically all nucleons are on the surface, it is the surface tension that weakens the binding. At $A = 56$ and beyond, as the number of protons rises, their mutual Coulomb repulsion becomes an increasingly important factor that opposes the preference for tighter structures.

That the binding energy per nucleon deviates systematically from its average suggests that energy can be extracted by 'climbing' the energy curve, and this in two ways. The first proceeds by *nuclear fusion*: when two light nuclei combine into a single nucleus less massive than iron, the final product has a greater average binding energy than the reacting nuclei, and energy is released. For example, in controlled-fusion reactors and fusion-based weapons, a deuteron and a triton (nuclei of deuterium and tritium) are forced to fuse into an α-particle (nucleus of helium) plus a neutron, producing 3.5 MeV per nucleon. In the second method, known as *nuclear fission*, a nucleus much heavier than iron splits up into smaller fragments and, again, the energy unlocked is converted into kinetic energy for the final products. In controlled-fission reactors and fission-based weapons, an isotope of uranium (^{235}U) is often used as fuel; it breaks up into a variety of lighter nuclei, yielding about 0.9 MeV per nucleon. Needless-to-say, the total energy of the system is conserved throughout the reactions: there is no gain nor loss, only a transfer of one form of energy into another.

Hydrogen is by far the most abundant material in the universe: more than 90% of the atoms in the universe are hydrogen, and all but less than 1% of the remainder is helium. As the protostar is steadily contracting gravitationally, the density and the temperature near its center rise steeply (7.5×10^{31} protons/m^3 and 15×10^6 K), and the conditions are set for hydrogen to 'burn.' Initially, the basic process is the fusion of two protons into deuteron. Since the two reacting particles carry positive charges, they tend to repel one another electrically. For them to come close enough together to allow the attractive part of the nuclear force to take over and create a bound nuclear state, they must have a sufficiently high relative speed, which can arise naturally in large random thermal motions in a very hot gas (or it can be produced artificially in terrestrial accelerators). The proton–proton fusion reaction and all other common fusion reactions, which involve charged light nuclei, require a high temperature to proceed; for this reason, they are called *thermonuclear reactions*.

At the high temperatures prevailing in the core of the star at this stage, the particles have just about a thousand electron-volts in kinetic energy,[12] which is insufficient for two protons to overcome their mutual electric repulsion (in excess of 1 MeV). However, they can still traverse the Coulomb barrier by 'tunneling' under it, a mechanism allowed by quantum mechanics (Fig. 8.4; see also Appendix B). The higher energy they have, the more vigorously they can tunnel and the more likely they are to complete the fusion process. And since there always exist some sufficiently energetic particles in a thermal gas, the reaction is bound to occur, more or less frequently. The rate at which it actually occurs and hence the rate of energy production itself depend very critically on the temperature: fusion turns on and off quickly with the slightest changes in temperature. This is why nuclear fusion can ignite and maintain itself only in the deep interior of stars.

8.3.2 On the Main Sequence

With the onset of nuclear fusion, the protostar becomes a *star* — a temporarily stable object making a long pause on the main sequence as it traces its evolutionary path through the luminosity–temperature space. The star owes its exceptional stability to a constant interplay between the nuclear and gravitational forces (Fig. 8.5). As the rate of nuclear reactions is strongly influenced by the temperature, the production of *nuclear energy* might for an extremely short time exceed that of *gravitational energy* in the star's central regions, and so the heated stellar

---

[12]Thermal energy is proportional to temperature. Particles in a thermal distribution at temperature 12 000 K have an average kinetic energy of 1 eV per particle.

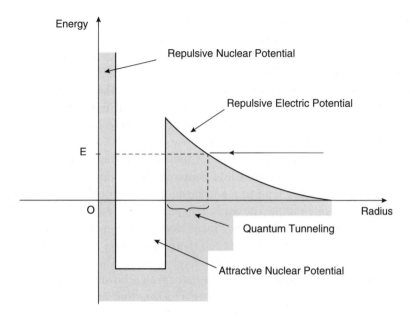

Figure 8.4:   The potential for two protons (or any two nuclei) is a combination of a repulsive elec-tric potential and a mainly attractive nuclear potential. To get close together, the two interacting particles must traverse a classically forbidden region (the Coulomb barrier) by quantum tunneling.

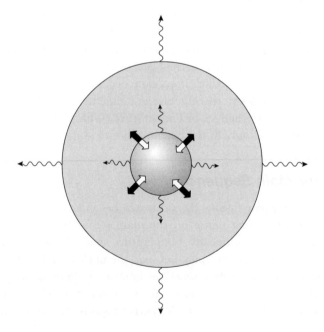

Figure 8.5:   Normal stars owe their exceptional stability to the interplay between the inward pull of gravitation and the outward push of nuclear pressure.

material might expand locally to some infinitesimally small degree. Since the whole system is in quasi-equilibrium, the expansion will cause pressure and temperature to drop ever so slightly, which in turn immediately dampens the nuclear fusion rate. The gravitational pull will quickly take over as the dominant force, and the resulting *compression* instantly raises the temperature, restoring the system to its original equilibrium state which has been temporarily broken by local heating. This harmonious interplay between contraction and expansion, together with an abundant supply of hydrogen as fuel, contributes to maintain the star in an overall *stable structural equilibrium*, at constant radius and temperature, that can last for a considerable length of time. Most luminous stars we see in the sky belong to this period of their evolution, by far the longest in their active life.

The American physicist Hans Bethe (winner of the Nobel Prize for Physics in 1967) found that the most important mechanism for generating nuclear energy in stars, at this stage, is nuclear fusion, which produces, in net, helium from four hydrogen nuclei in two main reaction sequences — the *proton–proton chain* and the *carbon–nitrogen cycle*.

8.3.2.1 The Proton–Proton Chain

The first step in this sequence, which gives the chain its name, is the combination of two protons into the only stable two-nucleon system, the deuteron, accompanied by a positron (e^+) and a neutrino (ν_e):

$$p + p \rightarrow {}^2_1H + e^+ + \nu_e \,. \tag{8.1}$$

Even at high temperatures, this reaction occurs very slowly because it is mediated in part by the weak interaction, which must occur to turn a proton into a neutron, in a process inverse to the more familiar β decay. Because of the Coulomb barrier, only protons in the high-energy tail of the thermal distribution can have sufficiently high relative speeds to penetrate the barrier and come close enough together to interact and form a bound state. For these reasons the collision rate remains low, about 5×10^{-18} per second per proton. It is the slowest and least probable step (the bottleneck) in this chain of reactions. Nevertheless, there are so many protons around (like 10^{56}) that the reaction can in fact take place at a sufficient rate to get the chain rolling.

Once a deuteron is formed, it is most likely to combine with another proton to produce the helium isotope ^{3}He. Once created, there are several ways for ^{3}He to go, the most likely being to follow the '*main branch*' and combine with another ^{3}He to produce ordinary helium:

$$^3_2He + {}^3_2He \rightarrow {}^4_2He + 2p + \gamma \,. \tag{8.2}$$

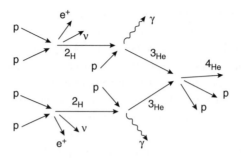

Figure 8.6:   Sequence of reactions in the main branch of the proton–proton chain of fusion.

The complete sequence, known as the *proton–proton chain*, is shown in Fig. 8.6. In net, we have four protons converted to helium:

$$4p \rightarrow {}^4_2\text{He} + 2e^+ + 2\nu_e \,. \tag{8.3}$$

Part of the energy generated is carried away as kinetic energy by the reaction products, leading to a net heating of the surrounding medium. The positrons will eventually annihilate with existing electrons, producing more gamma rays. All the gamma rays produced will interact with the medium to give additional heating; they will take some hundred thousand years to reach the surface. The neutrinos, in contrast, interact so weakly with the surrounding matter that they emerge directly from the core without contributing to the heating of the *photosphere* (the outer region of the star where the energy released in nuclear reactions is converted into light). Except for the small energy carried off by neutrinos, the mass deficit of the reaction product (26.7 MeV) becomes an effective heat input for the star. In other words, of 1 kg of hydrogen, 7 g are converted into light, an amount equivalent to 180 GWh.

There are other ways for ${}^3\text{He}$ to react: it could combine with protons, deuterons or α-particles. But reactions with protons are not possible because they would produce unstable systems which immediately break up on formation, while fusion with deuterons is highly improbable because deuterons are so scarce, being converted into ${}^3\text{He}$ as soon as formed. The third alternative is the only one possible, when a ${}^3\text{He}$ nucleus encounters a ${}^4\text{He}$ nucleus to form the nucleus of ${}^7\text{Be}$, which is followed next either by the sequence,

$$ {}^7_4\text{Be} + p \rightarrow {}^8_5\text{B} + \gamma \,, \quad {}^8_5\text{B} \rightarrow {}^8_4\text{Be} + e^+ + \nu_e \,, \quad {}^8_4\text{Be} \rightarrow 2{}^4_2\text{He} \,; \tag{8.4}$$

or by the sequence,

$$ {}^7_4\text{Be} + e^- \rightarrow {}^7_3\text{Li} + \nu_e \,, \quad {}^7_3\text{Li} + p \rightarrow 2{}^4_2\text{He} \,. \tag{8.5}$$

The net reaction is the same for all *three possible paths*, but calculations identify the main branch (8.2) as the most important in solar-mass stars. We can test this

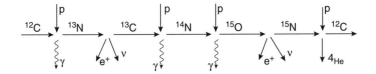

Figure 8.7:   Sequence of reactions in the carbon–nitrogen fusion cycle.

prediction and discover the relative importance of the three alternatives by detecting the incoming neutrinos and analyzing their *distinctive energy profiles*. We have in the first case (8.2) a continuous distribution of neutrinos with a maximum energy of 0.42 MeV, and similarly in the second case (8.4) with a maximum of 14 MeV. In the third case (8.5), the two-body ^{7}Be electron capture produces monoenergetic neutrinos mainly at 0.86 MeV.

8.3.2.2   The Carbon–Nitrogen Cycle

Bethe also found that higher-charge species can act as *catalysts* for alternative fusion schemes, involving heavier elements, that might compete with the proton–proton chain. The most important of these paths follows the *carbon–nitrogen (CN) cycle*, which consists of a series of proton captures and beta decays, as shown schematically in Fig. 8.7. The net process is again the production of a helium nucleus from four hydrogen nuclei, exactly as in the p–p chain, with an associated release of energy. Carbon is needed to start the chain, but reappears at the end of the cycle to be used again in the first reaction of the subsequent cycle. Once initiated, the cycle can proceed more rapidly than the p–p chain because it does not have a bottleneck analogous to the deuteron formation. However, as the Coulomb barrier is six times higher for proton and carbon than for two protons, the CN cycle will not be competitive with the p–p chain until the central temperature rises well above 15 million K. Therefore, while the p–p chain is the main route for the synthesis of helium in stars less massive than 2 $M_\odot$, the CN cycle becomes the dominant source for energy production in more massive, much hotter stars, because the energy production in effect then, with a rate that depends on temperature as $\sim T^{16}$, will completely dominate the weaker source from the p–p chain, whose rate goes only as $\sim T^4$.

8.3.2.3   Properties of Stars on the Main Sequence

The position of a star on the main sequence (MS) varies little with its age on the MS, or with the amount of hydrogen burned up, as long as the fuel in the core (containing 10% of the star's mass) has not been completely exhausted.

   The most important factor in determining a star's position on the MS is its mass. The empirical fact that normal stars have masses in the range 0.1–60 $M_\odot$ can be understood as follows. A protostar with a mass less than 0.1 $M_\odot$ is called a *brown dwarf*, it does not have enough self-gravity to compress its center to the high temperatures needed to complete the fusion process; after deuterium is burned out, the star will cool to a fully degenerate configuration without achieving the stellar status. On the other hand, an object with a mass much greater than 60 $M_\odot$ will see its centers raised to temperatures so high that nuclear burning can proceed at furious rates, enabling radiation pressure ($P \sim T^4$) to dominate over matter pressure ($P \sim T$). Eventually, the star ejects enough matter to fall below the critical mass. However, this does not mean that *supermassive* stars might not form and live, albeit for a short time. Stars with masses in the thousands or even millions of solar units may have an exceedingly short life, but, with their prodigious energy release, they would not pass without disturbing the universe.

Basically, a star shines because heat can leak out from its interior. The rate of this leakage determines the star luminosity L, which turns out (on general physical grounds) to vary with its mass as M^4. This dependence means that an MS star of 10 $M_\odot$ produces 10 000 times more power than the sun. And yet, with only 10 times more fuel to burn, it has an MS life-span 500 times shorter than the sun's — 20 million years to the sun's 9 billion years.

For a star to remain in equilibrium, it must adjust its configuration to have an exact balance between its kinetic and potential energy. This implies that its radius and internal temperature must be related, $T(\text{core}) \sim M/R$. Since nuclear fusion is ignited at a specific threshold temperature, regardless of other properties, all stars reaching the MS should have roughly the same core temperature and hence the same mass-to-radius ratio. In other words, $R \propto M$ on the MS. Putting this result together with the two relations for luminosity ($L \sim M^4$ and $L \sim R^2T^4$), we see that the surface temperature of stable stars increases (weakly) with mass ($T = \text{constant} \times \sqrt{M}$). Massive MS stars not only must have higher luminosities but also hotter surfaces than low-mass MS stars. This is the reason for the distribution of normal stars along the diagonal band in the H–R diagram.

Not all stars on the MS have the same internal structure, though. For one thing, small stars are denser than big stars, which can be understood if we just remember that M/R is roughly constant on the MS, and so the mass density must vary as $\rho \sim M^{-2}$.

Stars also differ in their internal make-up. Hot, massive stars have a convective inner zone surrounded by a radiative region, where energy is transported by radiation. For example, for a star ten times the mass of the sun and four to five times as hot, the convective core may occupy one fourth of the star's volume and contains the same fraction of its mass (Fig. 8.8a). The temperature there may reach 25×10^6 K and the carbon–nitrogen cycle provides most of the star's energy. As this process generates energy at a prodigious rate, radiative transfer alone becomes inadequate to remove the heat produced and must be helped out by convection.

But for *red dwarfs*, those stars on the lower part of the MS band, the reverse seems to hold. A red dwarf whose mass is 0.6 and luminosity 0.56 solar units has such a weak self-gravity that its center can reach just about 9×10^6 K, and so only

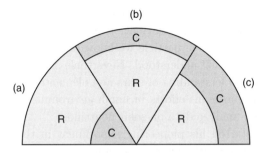

Figure 8.8: Internal structure of stars in terms of prevailing modes of energy transfer, convection (C) or radiation (R): (a) High-mass star; (b) Solar-mass star; (c) Low-mass star.

the p–p chain can be initiated. Since a lower temperature makes stellar matter more opaque, it is difficult for radiation alone to move energy produced at the center all the way to the surface; convection must do the job. A typical red dwarf has a radiative intermediate region enveloping a small hydrogen-burning core; it is surrounded, in turn, by a convective zone, which contains about one tenth of all the star's mass (Fig. 8.8c).

How about the sun? This mid-MS star has a central temperature of about 14×10^6 K. It draws energy from fusion reactions of the p–p cycle and, to a much smaller extent, the CN cycle. It has a structure similar to that of a red dwarf, but its outer convective layer is considerably thinner, containing only about 2% of the sun's mass (Fig. 8.8b). There is a high concentration of mass at the center, which arises from the nuclear 'ashes' accumulated ever since the sun reached the MS some 4.5 billion years ago.

8.3.3 Solar Neutrinos

The description given above of energy production processes in stars is part of the standard model of stellar structure and evolution. How can we test those ideas? The enhanced abundances of the expected products of the p–p and CN cycles over their initial values and the presence of the heavy elements, seen in spectral studies, provide very strong circumstantial evidence that the energy source inside ordinary stars is thermonuclear fusion. But it would be nice to have some more direct proof.

Evidently, the sun is a handy test-bed for the model. Nuclear reactions that take place in the center of this star produce photons and neutrinos, both detectable on earth. The photons interact with the solar material and take several hundred-thousands of years to emerge into the surface layers, so that the light that reaches us from the sun comes from the energy released in processes occurring in the photosphere and is characteristic of the sun's surface, not of its interior. Neutrinos, on the other hand, are weakly-interacting particles that rarely scatter off anything. A neutrino produced in the sun would go directly, according to the standard model, to the solar surface. Thus, when we detect solar neutrinos, we 'see' nuclear processes deep inside the sun and gain direct information about the conditions there now. But the neutrino is the most elusive particle we know of, very hard to detect, and its physics is not completely well understood. Nevertheless, because of the crucial role it plays in testing our understanding of stars and elementary particles, hundreds of physicists have spent enormous efforts in harsh environments during the last four decades to collect and analyze data on solar neutrinos.

Raymond Davis started his pioneering experiment in the mid 1960s, using a chlorine detector inside a South Dakota gold mine. In the 1990s, three other groups of physicists joined the solar neutrino watch. Two of them — Gallex (and its successor, the GNO) in the Gran Sasso east of Rome and SAGE in the Caucasus — use gallium as the detecting medium. The other, Kamiokande in the mountains west of Tokyo and led by Masatoshi Koshiba, records neutrino events with a water Čerenkov detector. Since each detector type has its own energy threshold, and each neutrino-producing solar process has its own energy spectrum (Fig. 8.9), the neutrinos observed by different groups do not originate from the same reactions. Neutrinos from the dominant proton–proton main branch (Eq. (8.2)) can be seen

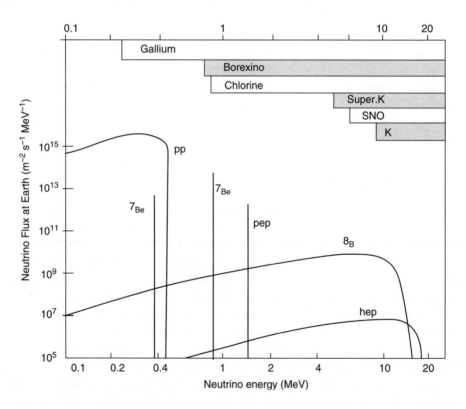

Figure 8.9: Solar neutrino spectra for different fusion reactions predicted by the standard solar model. The regions of neutrino energy accessible to different experiments are also indicated.

only by the gallium detectors, while the Kamiokande detector, with the highest threshold (7.5 MeV), can register only the most energetic neutrinos from the decay of ^{8}B of the side branch (Eq. (8.4)). Davis's chlorine detector, on the other hand, is sensitive to the ^{8}B neutrinos as well as to monoenergetic neutrinos from electron-assisted proton fusion $(p + p + e^- \rightarrow {}^2H + \nu_e)$ and from electron capture by ^{7}Be as in Eq. (8.5).

All four groups have detected neutrinos, and Kamiokande has shown that the observed neutrinos do indeed come from the direction of the sun. However, in all cases, the rates measured are substantially smaller than the rate predicted by the standard solar model. The persistent disagreement between the calculated and the observed fluxes — the famous solar neutrino problem — was troubling, because it could mean that we understood solar physics or neutrino physics less well than we thought.

Neutrinos come in three *flavors*, associated with the three existing charged leptons (the electron, the muon and the tau). The solar p–p chain produces only electron neutrinos; they are *the only ones assumed to emerge from the sun* and, in fact, *the only ones detected by the radiochemical detectors*. But already in 1967, Bruno Pontecorvo suggested the possibility that neutrino flavors could become mixed in vacuo; he and V. Gribov interpreted the solar neutrino deficit first reported by Davis in 1968 as evidence for *neutrino oscillation* (the sinusoidal dependence on path length of the probability of the metamorphosis between two flavors; see Chapter 9 for further details). In 1985 S. Mikheyev and A. Smirnov suggested that such mixing might be sufficiently amplified by interactions with matter on the way out of the sun to explain the missing neutrinos.

Recently, three new solar neutrino detectors, with vastly increased capacities, have come into operation, and are expected to clarify the situation. They are Super-Kamiokande, the successor to Kamiokande; the Sudbury Neutrino Observatory (SNO) in a northern Ontario mine; and Borexino in the Gran Sasso laboratory. The first two are sensitive to ^{8}B neutrinos, whereas the third will concentrate its efforts on observing 863 keV ^{7}Be neutrinos.

All these new detectors can register individual neutrino collision events and can detect neutrinos of any flavor through their elastic scattering off electrons in the detector's liquid. But SNO, with its heavy-water Čerenkov detector, can also register signals from deuteron break-up:

$$^2H + \nu \rightarrow p + n + \nu \tag{8.6}$$

and electro-production off deuterons:

$$^2H + \nu_e \rightarrow p + p + e^- . \tag{8.7}$$

While the former reaction is open equally to all three flavors, the latter is available only to the electron neutrino ν_e (because ν_μ and ν_τ in creating μ and τ would require energies far beyond the solar neutrino spectrum). Thus, the researchers at

SNO can measure the flux of incoming neutrinos for all flavors from Eq. (8.6) and the flux of the incoming ν_e neutrinos from Eq. (8.7). In 2002 they reported that the ν_e component of the ^{8}B solar neutrino flux is 1.76×10^{10} m^{-2} s^{-1} for a kinetic energy threshold of 5 MeV and the total flux, for all flavors, is 5.09×10^{10} m^{-2}s^{-1}, consistent with the standard solar model's flux prediction for the electron neutrinos produced by ^{8}B decays in the solar core (5.05×10^{10} m^{-2} s^{-1}). The non-ν_e component measured, 3.41×10^{10} m^{-2} s^{-1}, constitutes a strong evidence for solar neutrino oscillation.

8.3.4 Post-Main-Sequence Evolution

Although a star remains on the main sequence for a very long time with hardly any changes in its basic properties — deriving its exceptional stability from the abundant energy produced by fusing hydrogen into helium — this is only a temporary pause. When the hydrogen fuel in the central core (about 10% of the star's mass) runs out, all transformed into helium, the star evolves off the main sequence, beginning a march towards its inevitable end. At this time, outside the core lies a thin shell in which hydrogen is still burning and, just above it, a thick envelope in which temperatures have never risen high enough to ignite hydrogen. What exactly happens next depends on whether we are talking about low-mass stars or high-mass stars, but the issues common to both revolve around the burning of successive rounds of nuclear fuels and the flow of energy to the outside.

8.3.4.1 Evolution of Low-Mass Stars

Let us take a low-mass star, like our sun. The core, now depleted of its hydrogen fuel and filled with inert helium, continues to lose its heat and, in the absence of an adequate compensating nuclear energy production, contracts gravitationally, heating itself up as well as the surrounding shell. The energy generated in the hydrogen-burning shell contributes in part to the energy of the envelope above it and in part to the star's surface luminosity. The star expands and its radius increases. As the luminosity remains constant or increases slightly, the effective temperature decreases a little. Thus, immediately after leaving the main sequence, the star moves more or less horizontally to the right in the H–R diagram, turning itself into a *subgiant* (Fig. 8.10).

As the core keeps on contracting, heat continues to rush out into the overlying shell. The burning shell, with its ever increasing temperature, produces energy at rates so high that the pressure exerted on the outer layers causes them to expand greatly and the surface to shine brightly. As the outer envelope expands and cools, the surface temperature falls. But it cannot fall indefinitely. In regions just below the stellar surface, the temperature remains low and hydrogen exists as neutral atoms. But as we go deeper in, we find zones where temperatures are high enough to ionize hydrogen. Bubbles will form, rising quickly through the cooler material to

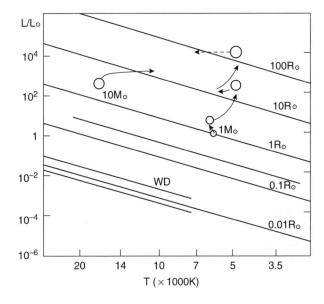

Figure 8.10:   Post-main-sequence evolution of a 1 $M_\odot$ star and a 10 $M_\odot$ star.

the surface, and when the temperature at the surface drops below a certain limit, the whole envelope becomes convective. Hot gases can now flow rapidly upward, and the luminosity increases a thousandfold in about half a million years. As new energy now arrives at the surface, the temperature which has been falling there lately stabilizes around 3 500 K. This temperature barrier forces the evolutionary track to turn almost vertically upward for all but the most massive stars. We now have a very big red star, a *red giant* — 160 times bigger than the sun, 2 000 times brighter and somewhat cooler. This giant star has at its center a tiny core, a few thousand times smaller than the star itself but considerably heavier than initially, having gained mass from the helium ash of the surrounding shell. Meanwhile, outside, a strong hot wind has blown off a large fraction of the stellar material into space.

As the core continues to collapse, it becomes hotter and denser. The electrons that have long been pulled out of their atomic orbits are now so tightly packed that they evolve into a *degenerate quantum state*,[13] while the more massive helium nuclei remain nondegenerate. When the central temperature reaches 2×10^8 K, *triple-α fusion* begins:

$$^4\text{He} + {}^4\text{He} \rightarrow {}^8\text{Be} + \gamma, \quad {}^8\text{Be} + {}^4\text{He} \rightarrow {}^{12}\text{C}^* + \gamma.$$

(An asterisk is used to indicate that ^{12}C is formed in an excited nuclear state). Carbon combines with helium to form oxygen, $^{12}\text{C} + {}^4\text{He} \rightarrow {}^{16}\text{O} + \gamma$. Or it can react with a proton to give ^{13}N, an unstable isotope which quickly decays to ^{13}C to

---

[13] See Appendix C. The key point is that the pressure of a degenerate gas depends on its particle density, not its temperature, so that compression does not lead to heating.

undergo in turn the process $^{13}C + \alpha \rightarrow \, ^{16}O + n$. This reaction is an abundant source of neutrons and will be called on to play a key role in nucleosynthesis in high-mass stars.

Triple-α fusion produces energy copiously at a rate that depends strongly on temperature ($\approx T^{40}$). Whereas in a normal star, the slightest local increase in the core temperature and pressure provokes a local expansion and a concomitant reduction of temperature, here the degenerate pressure is completely decoupled from thermal motions. So any increase in the core temperature leads to an accelerated energy production without the compensating effects of a pressure increase and a stellar expansion. A runaway production of nuclear energy ensues: after ignition, helium burns in a *flash*. This sudden massive release of energy generates enough internal power to inflate the core, lowering both its density and temperature. The potential explosion, which would have catastrophically disruptive effects, is thereby contained, and the electrons return to a normal, nondegenerate state. With a lowered temperature, the triple-α process shuts off, and even hydrogen burning in the outer shell slows down. Both temperature and pressure gradients drop throughout the star and, as the core itself expands, the outer envelope will contract. This period corresponds to the *horizontal branch* in the H–R diagram.

Eventually, the core expansion damps out and, abetted by the infall of the upper layers, the core starts contracting again, raising its temperature back to above 10^8 K. Helium burns anew but now in a nondegenerate environment. When the helium fuel in the core finally runs out and the nuclear energy production there comes to an end, a familiar sequence of events takes place. The outer layers collapse, releasing enough gravitational energy to ignite helium in a shell outside the now dormant carbon core. And in due course, convection sets in, sending the star up to the *supergiant* status, a red globe 180 $R_\odot$ in radius. Now fed by the double-shell sources of hydrogen and helium nuclear energy, the star shines brighter than 3 000 suns. Meanwhile, the core is compressed, and the electrons again become degenerate. The degenerate-electron pressure should be high enough to support the envelope of a low-mass star, so that the core, now a solid carbon–oxygen sphere, remains inert at a temperature below the next fusion point.

During its ascent on the *asymptotic giant branch*, as this period is called, the star loses so much mass in stellar winds that its outer envelope becomes very thin (assuming, as always, a low-mass star). At the top of the climb, successive sharp thermal pulses of extra energy production, interrupted by periods of quiet evolution, start in the thin, unstable helium-burning shell. Each thermal runaway ejects more matter into space until the last burst blows off what is left of the envelope, creating one of the most beautiful sights in the heavens: a *planetary nebula*, a diaphanous ring of ionized gases surrounding a very hot and very bright star, the size of the earth with half the mass of the sun.

The exposed hot core sheds its extended envelope, burns out its hydrogen and helium shells and, finally, with all inner fires extinguished, evolves into an inert carbon–oxygen star that slowly cools and dims to become a *white dwarf*.

Stars with masses of 1–6 $M_\odot$ evolve in a similar way. In the H–R diagram, a typical low-mass star leaves the main sequence horizontally when all hydrogen in the core has been fused into helium, becoming a *subgiant* at the base of the *giant branch*. As hydrogen is burning in the shell surrounding the helium core, the star begins its ascent, and when it reaches the top of the climb, the core helium ignites in a *flash*, leading to its descent along the *horizontal branch*. When the core, now made up of carbon and oxygen, starts contracting, and while the helium shell and the hydrogen shell burn, the star begins its abrupt ascent on the *asymptotic giant branch*. Meanwhile, it is losing a considerable amount of material in stellar winds, making its outer shells spatially so thin that they become unstable to the nuclear inferno. When the last of the outlying layers have been blown off to form a *planetary nebula*, a bare core with a *mass less than* 1.4 $M_\odot$ is exposed, and the star descends the H–R diagram to enter the region inhabited by the *white dwarfs*.

8.3.4.2   Evolution of High-Mass Stars

We turn now to stars with masses greater than six suns. The evolution of these stars differs from that of the low-mass stars we just described on two points. First, they evolve much more quickly because they lose energy through radiation at a much higher rate (remember the mass dependence of luminosity, $L \sim M^4$). Second, their large mass can keep the central core at a sufficiently low density to allow the electron gas to remain essentially nondegenerate (recall $\rho \sim M^{-2}$), but still at a pressure and temperature high enough to ignite nuclear fuels all the way to iron.

In a high-mass star, as helium burning occurs under nondegenerate conditions, there is no helium flash, and the star moves leftward in the H–R diagram, towards higher temperatures (Fig. 8.10). Once all the hot helium fuel has been fused into carbon and oxygen, the newly formed inert core contracts by its own self-gravity and, with this released energy, ignites a ring of helium just outside. The evolutionary curve turns back to the right. There is little time for radiation to reach the surface, so the luminosity changes little and the evolutionary track progresses more or less horizontally in the H–R plot. When the central temperature reaches 5×10^8 K, carbon ignites, starting new chains of reactions, fusing two carbons into magnesium or neon. Although these reactions are extremely sensitive to temperature, they give rise to no disruptive events because carbon ignition has occurred before the electron degeneracy becomes important. The nondegenerate core can adapt itself to the rapidly changing conditions, efficiently releasing its extra pressure to the overlying shells until the carbon fuel is exhausted. So, very massive stars evolve quickly but smoothly — burning each successive round of fuel to exhaustion under normal-gas conditions, undergoing gravitational contraction, reaching quickly the temperature needed to ignite the newly-formed nuclear ash, and starting another fusion cycle. Carbon fuses into neon and magnesium; oxygen into silicon and sulfur; and silicon into nickel and iron. Each successive cycle proceeds at a faster pace: hydrogen burns for 10 Myr, helium for 1 Myr; the silicon layer is built in a few years,

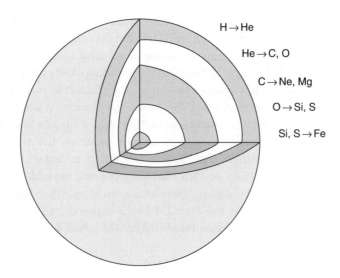

H → He

He → C, O

C → Ne, Mg

O → Si, S

Si, S → Fe

Figure 8.11:   Structure of a high-mass star at the end of the nuclear-fusion phase of its evolution.

and the iron core in a few days. Iron ^{56}Fe is the most strongly bound of all nuclei, so no further energy can be extracted by nuclear fusion. The fusion process shuts off at this point, having played out its role in normal stellar evolution (Fig. 8.11).

When the star reaches the final phase, it has become a *red supergiant*, a thousand times larger than the sun. Its bloated envelope of unburned hydrogen is held tenuously by a central onion-like layered structure: a small inner core of hot iron surrounded by successive burning shells of silicon, carbon–oxygen, helium and hydrogen. Throughout all its post-main-sequence life and specially during the periods of evolutionary changes, the star loses gradually vast quantities of mass from its outer layers in hot winds and ejecta. Whether it will now die the quiet death of a white dwarf, or suffer more violent convulsions before it goes depends on how much matter is left in the central core.

8.3.4.3 Supernovae

Let us now consider a massive star on the edge, a thousand times larger than the sun, consisting of successive layers of burning hydrogen, helium, carbon–oxygen, neon–magnesium and silicon, and a central core of quiescent iron. The core may be some 1 000 km in radius and may hold 1.2–1.5 $M_\odot$ at 10 billion K. As the source of nuclear energy at the center has dried up, the core is held up against its own weight only by the internal energy of a hot matter and the pressure of a degenerate-electron gas. But the support against the gravitational pull is being fast eroded, on the one hand, by energy-consuming processes in the interior (photodisintegration of nuclei, neutrino and neutron-rich nuclei production by electron capture) and, on the other hand, by the added weight of the infalling fusion products from the still-active upper layers.

When the core mass is nudged over 1.4 $M_\odot$, this fragile balance is broken, and the core collapses within a second to a structure 100 km in radius, squeezing matter within it to a density nearing that of nuclear matter. The *implosion* proper involves only the tiny inner core, while the bulk of the star shrinks at a relatively slower pace. Although nuclear matter is hard to compress because of the nuclear repulsive force at short distances (Fig. 8.4), it is not completely incompressible. The implosion, carried by momentum beyond the point of equilibrium, squeezes matter to a density slightly beyond that found in large atomic nuclei, before it finally grinds to a halt and vigorously bounces back. A pressure wave, built up at the center of the core, turns into a *shock wave*, carrying with it matter as it blasts its way out.

However, the shock wave fails to go all the way to the core limits — the material falling from the upper layers is crashing down on it while nuclear reactions saps its energy. In these processes, iron and other nuclei disintegrate into nucleons, costing 9 MeV per nucleon; or nucleons capture electrons and positrons to produce neutrinos and anti-neutrinos which carry away more energy. The shock wave stalls at some distance from the surface of the core.

What gets it started again is neutrinos. In the few seconds after the catastrophic collapse, the core is very dense and very hot, at several times 10^{11} K near its center. The only way it can cool down is by emitting a *prodigious amount of neutrinos* (while electromagnetic radiation is hopelessly trapped) and, thereby, transforming itself into a neutron-rich star. When the neutrinos move out through the star and are captured, they heat up the material there. Heat builds up a violent *convection* that breaks the traffic jam, re-energizes the shock wave and keeps it moving out of the core and into the surrounding envelope. Not only does the shock wave carry with it energy and matter at high speeds and violently heat up the outer layers, it also produces in its wake more heavy elements from the pre-existing materials.

In a few hours, the shock breaks out of the surface, moving at a speed of one tenth the speed of light. The star brightens within minutes to 10 billion times the solar luminosity, out-shining an entire galaxy: a *supernova* is born. Light appears first in the far ultraviolet, then shifts after a day to the visible; after a few months, most of the radiation is in the infrared, and the star fades out. Most of the outer envelope — made up of hydrogen, helium and heavy elements — is blown off, and, accelerated to a few 10 000 km/s, it expands while heating and stirring up the surrounding medium. The metal-enriched ejecta eventually cool off to mix with the interstellar gas and contribute to the composition of the next generation of stars. But it is not in the expanding gas nor in the electromagnetic radiation that the star releases its stupendous energy; it is rather the invisible and almost undetectable neutrinos that carry off 99% of the energy produced — 10^{46} joules, or a hundred times the energy radiated by the sun during its entire main-sequence lifetime.

The explosion throws off all the material of the parent star beyond a certain radius, enriching the interstellar medium with heavy elements, while the rest of the star settles down as a hot, neutron-rich ball, a hundred kilometers across, that quickly shrinks to one tenth of this size and slowly cools off. During this final period of contraction, the remaining electrons are annihilated, and the matter that is left is primarily composed of tightly packed, degenerate neutrons: we have a *neutron star*. If the original star has a mass high enough, the relic core might exceed 3 $M_\odot$, and gravity will overcome the degenerate-neutron pressure to form, probably, a *black hole*.

Supernovae are spectacular and rare events. Among them, four have historical significance: SN1006 (supernova seen in the year 1006), described by medieval Chinese astronomers to be as bright as the half-moon; SN1054, also recorded by them and identified now as the event leading to the formation of the Crab Nebula; SN1572, sighted by Tycho Brahe, whose book about the event, '*De Nova Stella*,' gave these objects a name; and finally SN1604, reported by Johannes Kepler, whose observations and explanations contributed to the final break with the Aristotelian traditional view that the realm of fixed stars was immutable.

But, of course, supernovae occur in other galaxies as well, as shown by Walter Baade and Fritz Zwicky, who initiated a systematic search in the 1930s. Since then, close to 150 supernova remnants have been detected in the Milky Way, and more than a hundred are being discovered every year in distant galaxies.

The more recent SN1987a is the nearest naked-eyed visible supernova seen since 1604. This event provides astrophysicists with a unique opportunity to test their current understanding of the most spectacular phenomenon in the heavens. Up to now, their knowledge about supernovae has been largely gained through theoretical modelings and computer simulations, but hard information about both historical supernovae and extragalactic supernovae is lacking. So it is fortunate that SN1987a is located only some 160 thousand light years away in one of the two nearest galaxies, the Large Magellanic Cloud. Its progenitor was a small blue supergiant having 18 $M_\odot$, and for the first time we detected the burst of high-energy neutrinos from the collapse of a massive star. Its evolution is practically unfolding before our eyes, and has provided us with a wealth of information about supernova physics, which helps to confirm, in particular, the basic predictions of the core-bounce picture.

8.3.5 *Summary*

When a star heats up to ten million degrees at its center, the nuclear fusion of hydrogen (the most abundant of all elements) is ignited. Thermonuclear fusion is the only energy source that can sustain the radiation of stars for very long time-spans. This process provides the most coherent explanation for a vast body of data, from the basic properties of stars on the main sequence to various phenomena marking the late stages of stellar evolution, in passing by the relative abundances of chemical elements observed in the universe.

According to the current stellar models, after 10% or so of hydrogen has been burned into helium, the star becomes structurally unstable. The relationships between its basic properties change drastically, and the star moves away from its position on the main sequence. The subsequent collapse of the core releases enough energy to heat up the layers around the core, igniting a shell of hydrogen. The luminosity generated by the shell source pushes layers lying just above, causing them to expand. The bloated envelope, fed in radiation by convective currents, gives the star the appearance of a red giant.

In a low-mass star, the next cycle of fusion involving helium burning is accompanied by a fast and copious production of energy, which might force the swollen star to shed its hydrogen-rich envelope and give birth to a planetary nebula together with a remnant core. This exposed hot core goes on to become a white dwarf.

High-mass stars evolve much more quickly, because they radiate at much higher rates, yet can manage to adapt themselves to the rapid internal changes. They evolve smoothly, undergoing contraction, raising the temperature to the level needed to ignite the ash of the previous fusion cycle, and starting a new cycle. But when iron is finally formed at the center, the nuclear energy source dries up, and the core has no recourse but to contract. The supernova explosion that ensues leaves behind a compressed neutron-rich core, which evolves into a neutron star or a black hole.

8.4   White Dwarfs

The story of the evolution of stars is the story of a death foretold: pressed by the relentless pull of gravity and surrounded on all sides by a cold and dark space, stars are fighting a battle they cannot win. Once drained of their energies, they will inevitably lose their struggle and die, either in a destructive explosion in which their ashes turn to dust, or in a lingering death in which, as a white dwarf, a neutron star or a black hole, they finally come to terms with both gravity and the universe.

8.4.1   Observations of White Dwarfs

In 1844 astronomer Friedrich Bessel noticed, while examining astronomical records over many years, that Sirius, the brightest star in the northern sky, showed a wobbly motion. He interpreted the irregular motion of Sirius as being caused by the pull of an unseen massive star that gravitated around it. This invisible companion star turned out to be the first white dwarf star ever discovered, now known as Sirius B. Although it is the brightest white dwarf detected in our Galaxy, its faint glow is overwhelmed by the light of Sirius, a star ten thousand times more luminous. Just as for many other stars, the radius of Sirius B is deduced from its light spectrum. From spectral analysis, astronomers first obtain its surface temperature (26 000 K), and from its apparent brightness and an independent measurement of its distance from earth (8.6 ly), they next find its true luminosity. With the temperature and luminosity known, the star radius is derived from the familiar radiation law.

White dwarfs are common. Those that have been identified locally range in radii from 0.01 to 0.02 $R_\odot$ and in temperatures from 5 000 to 50 000 K. In the

H–R diagram, they occupy a narrow band below the main sequence. The first few discovered and still the easiest to discover have temperatures near 10 000 K, and so look 'white,' but there is no particular temperature associated with these stars. Blue white dwarfs are rare because they cool quickly, and red ones are dim and hard to detect. But if we look hard, we will find them: the Hubble Space Telescope in the last years has found white dwarfs 30 times hotter than the sun (at 2×10^5 K) and many faint and cool ones (the brightest no more luminous than a 100 W light bulb seen at the moon's distance).

Their masses are very difficult to determine except when they belong to visual binaries. In these cases, the Newton gravitational law, applied to the observed orbital motions, can give stellar masses, and radii are derived from knowledge of effective temperatures and distances. The recent Hipparcos space astrometry mission, dedicated to precise parallax measurements, yields data from which masses and radii can be estimated even more reliably than before. For the four best known such stars — Sirius B, Procyon B, 40 Eri B and Stein 2051 B — with respective radii of 5846, 6680, 9465 and 7725 km, their masses have been established at 1.0, 0.60, 0.50 and 0.48 solar masses, respectively. White dwarfs therefore have average mass densities of some 1×10^9 kg/m^3, one million times greater than the density of the sun $(1.4 \times 10^3$ kg/m$^3)$ or the earth $(5.4 \times 10^3$ kg/m$^3)$.

8.4.2 *Electron Degeneracy Pressure*

A mass density of 1×10^9 kg/m^3 implies a *number density* for electrons of $n_e = 10^{36}$/m^3 (for comparison, $n_e = 10^{27}$/m^3 in metals). This tells us that the electrons of the white dwarf's atoms evolve considerably closer to one another here than they do in ordinary atoms but are still farther apart than the mean nucleon–nucleon distance. So they are effectively squeezed together outside the atomic nuclei, which remain intact but stripped of their electrons. White dwarfs are composed of a dilute ion gas co-existing with a very dense electron gas.

For a gas that dense — when the distance between two nearest neighbors becomes comparable to or smaller than their de Broglie's wavelength — quantum-mechanical effects become important.[14] This means, in particular, that the electrons must obey the *Pauli exclusion principle*: no two electrons can occupy the same single-particle quantum state. In a normal gas (*e.g.*, metals), this is not a problem since there are not enough electrons around to completely fill up all the energy levels. But in a white dwarf, all the electrons are forced close together, and soon all the available energy levels are occupied, from bottom to top, two electrons in each. We say that the white dwarf has become *degenerate*. Once a star is degenerate, gravity cannot compress it anymore, because there is no more available space to be taken up — electrons cannot go where quantum mechanics forbids.

In a degenerate gas, the mean separation Δx between an electron and its nearest neighbor is very small, and their momenta must differ by at least $\Delta p_x = h/\Delta x$ (by

---

[14]See Appendix C.

Heisenberg's uncertainty relation), otherwise they would violate the Pauli exclusion principle. As Δx is small, Δp_x is correspondingly large; so electrons squeezed to high densities must have high random speeds. These large random motions, which are of quantum-mechanical, not thermal, origins, give rise to a *degeneracy pressure* that depends only on the electron density, not on the temperature. This is because when the electrons fill up all the energy levels up to a certain maximum, the energy of the last occupied level (the *Fermi energy*) is well defined and depends only on the number density. So all energy-related quantities must also have that property. It turns out that the electron degeneracy pressure goes as $P \propto n_e^{5/3}$ when the electrons are non-relativistic, and $P \propto n_e^{4/3}$ when they are ultra-relativistic. (Such relations between pressure and density are known as 'equations of state.') It is important to note that the proportionality factors in both cases are completely defined in terms of fundamental physical constants. The reason that the relativistic pressure increases *less quickly* with increasing density than the non-relativistic pressure is that, as Δx becomes smaller, Δp_x becomes larger and, at some point, the increase of the velocity v_x must saturate, because v_x approaches the speed of light for very large momenta p_x.

Thus, in high densities, the degeneracy pressure dominates over the thermal pressure (which varies only linearly with n_e) and can become strong enough to support a white dwarf against its self-gravity. Degenerate stars survive, not by internal combustion, but by the quantum-mechanical degeneracy pressure.

8.4.3 Mass and Size

A white dwarf can hold up as a stable structure because the outward pressure due to its electron degeneracy, $P_e \propto n_e^{5/3}$, can grow powerful enough to balance out the inward pressure arising from its self-gravity, $P_g \propto M^2 R^{-4}$ (where M and R stand for the star's mass and radius). Since $n_e \propto M R^{-3}$, we see that white dwarfs must satisfy a *mass–radius relation*[15]:

$$M = \text{constant} \times R^{-3}.$$

It is a startling result: the more massive a white dwarf is, the smaller it is. In our ordinary experience, we expect, for instance, a marble twice as massive as another should have twice the volume. Not so with white dwarfs. A white dwarf twice as massive as another occupies only half the volume, because it squeezes four times more electrons together to maintain enough outward pressure to support the extra mass. Thus, an equilibrium state always exists for any reasonable mass.

But there is a limit to how much matter a white dwarf can hold. As its mass increases well above $1\,M_\odot$, its central density increases sharply, so that more electrons become relativistic. When all the electrons become ultra-relativistic, all reaching the speed of light, the degeneracy pressure will also attain its relativistic limit,

---

[15]The 'constant' actually is $2.2 \times 10^{-6}\,M_\odot\,R_\odot^3\,(2Y_e)^5$, where Y_e is the ratio of the electron number to the nucleon number. Numerically, it is $1.5 \times 10^{51}\,\text{m}^3\,\text{kg}$ for $Y_e = 0.5$.

$P_e \propto n_e^{4/3}$, or equivalently $P_e \propto M^{4/3}R^{-4}$. Since the gravitational and internal energies depend on R in the same way, R^{-4}, and hence go up or down in the same ratio, the star cannot seek stability by collapsing to a higher density. On the other hand, as the mass of the star is increased, its gravitational pressure increases as M^2, whereas its internal pressure goes up more slowly as $M^{4/3}$. So, there must be an upper limit to the mass of stable degenerate stars, which is the *Chandrasekhar mass limit*:

$$M_{Ch} = 1.4\,(2Y_e)^2\,M_\odot\,,$$

so called in honor of its discoverer, Subrahmanyan Chandrasekhar (winner of the Nobel Prize for Physics in 1983). This equation, one of the most important in astrophysics, is remarkable in its simplicity: apart from a factor sensitive to the star's composition (Y_e being the ratio of the number of electrons to the number of nucleons), it contains only a numerical factor calculated from universal physical constants. For white dwarfs made up of C, O or Si, $Y_e = 0.5$ and the mass limit is $M_{Ch} = 1.4\,M_\odot$, but it is smaller for an iron-rich white dwarf, because then $Y_e = 0.464$ and $M_{Ch} = 1.2\,M_\odot$. (An iron white dwarf would form when a main-sequence star was massive enough, say 6–8 $M_\odot$, to create iron nuclei in its core but was able to lose mass and angular momentum effectively to collapse into a white dwarf.) Sirius B and Eri B belong to the former group, whereas Procyon B and, perhaps, Stein 2051 appear to belong to the latter, as shown in Fig. 8.12.

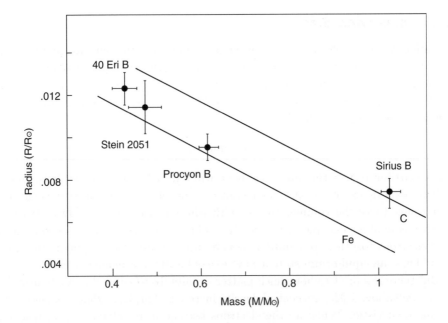

Figure 8.12:   Mass–radius relation for white dwarfs and data for Sirius B, Eri B, Procyon B and Stein 2051 (based on J. Provencal, H. Shipman, E. Høg and P. Thejll, 1998).

8.4.4 Luminosity

With a surface gravity 100 000 times that of the earth, the atmosphere of a white dwarf should have unusual properties. It consists of a very thin envelope, containing ordinary hydrogen and helium at the top and heavier elements at the bottom, wrapped around a 50 km thick rock-like crust. Underneath lies a central solid core made of carbon and oxygen atoms packed in extremely high densities.

The white dwarfs we know of have higher effective temperatures than the sun, but, with much smaller radii, they are 1 000 times less luminous. As their outer layers remain non-degenerate, they radiate energy into space by photon diffusion. But the situation changes completely in the interior. In general, absorption of light by matter takes place when the absorbing electrons can make a transition to a higher-energy state. But if matter is degenerate, as is the case of the white dwarf's core, the Pauli exclusion principle makes such transitions impossible for all but those few electrons near the Fermi level, which alone are free to absorb photons and jump to higher empty levels. And so, most electrons whiz around for a long time before making any collisions and losing energy.

With the electron thermal energy essentially locked up, the only significant source of heat to be radiated is the *ion* thermal energy left over from the original gravitational collapse. This energy store is quite substantial (like 10^{40} J for $T = 10^7$ K), comparable to the radiant output of a supernova. White dwarfs emit radiation at the expense of their heat reserves, glow ever more faintly, cool down and, finally, come into thermodynamic equilibrium with the cold universe. In the process, as the internal temperature falls, the electric forces between the nuclei will eventually dominate the thermal motions, and the nuclei will begin to 'stiffen up' under their mutual Coulomb repulsion. Thus, the white dwarf's core *crystallizes*, consuming more heat from the star's dwindling energy store.

As white dwarfs keep radiating weakly at decreasing rates through small constant surface areas, they will take billions of years to cool off and finally become *black dwarfs*, cold compact objects forever supported by the electron degeneracy pressure.[16]

8.4.5 Summary

When a typical low-mass star reaches the end of its normal nuclear-powered evolution, it has lost most of its outer layers of material and becomes reduced to a

---

[16]Here, we are assuming isolated white dwarfs. When the white dwarf belongs to a binary or multiple system, it can accrete matter from a main-sequence companion that is going through the giant stage, and undergo recurrent outbursts (novae) which blow away its surface layers. If accretion is massive enough or lasts for a long time, the white dwarf may accumulate so much matter that it goes over the Chandrasekhar limit. The electron degeneracy pressure then gives way, and the star blows itself apart entirely in a thermonuclear blast, leaving no remnant core behind. This kind of event is called a Type I supernova, to distinguish it from Type II supernovae, the end state of a massive star described in the preceding section.

compact core containing still-hot but inert carbon. The core is supported against self-gravity by the degenerate-electron pressure in its interior, so that its mass and radius are related. However, for the degeneracy pressure to be effective, its mass must not exceed 1.4 $M_\odot$. Although its residual internal energy is still substantial, an isolated white dwarf continuously radiates away this energy and becomes 'black' after billions of years.

8.5 Neutron Stars

Neutron stars are the extremely dense compact objects left over from initially massive stars at the end of their normal evolution amid the violence of supernova explosions. Predicted by Walter Baade and Fritz Zwicky in 1934 and first studied by J. R. Oppenheimer and G. M. Volkoff in 1939, their existence was confirmed only in 1967, when Jocelyn Bell and Anthony Hewish detected a *radio pulsar*. Many others have been discovered since then; some are solitary, others have companions. Because of their extreme and complex physical properties, neutron stars manifest themselves in unusual and often unexpected guises, of which we only begin to have a cohesive picture, thanks largely to intensive computer simulations and abundant data sent back by a new fleet of orbiting astronomical satellites.

8.5.1 *Formation and Structure*

Once a star with an initial mass between 6 and 20 $M_\odot$ has gone through the successive stages of nuclear burning and reached the final fusion phase, it keeps on accumulating iron in its center (see Section 8.3). Matter builds up so quickly that within a day or two, the iron central core goes over the Chandrasekhar mass limit, and the electron degeneracy pressure, which has so far successfully opposed the increasingly strong gravitational pull, suddenly fails, and the core begins to contract sharply.

As the internal pressure increases, it becomes energetically more favorable for the free degenerate electrons to be captured by *nuclei*, according to the basic equation $e^- + p \rightarrow n + \nu$. The *neutrinos* thus produced scatter a bit, but quickly escape the interior and help the supernova happen. The reacting nuclei which stay behind get one of their protons replaced by a *neutron* and, after a few more captures, become neutron-rich, having more neutrons than they need for their stability. The newly created neutrons must go to higher and higher unfilled nucleon energy levels (because, having spin one-half, they must obey the Pauli exclusion principle), and so are less tightly bound than the neutrons already there. As the capture adds more and more neutrons to the nucleus, they will come closer and closer to the continuum (the region of unbound states), and some eventually begin to *drip out* (float out) of the nucleus as 'free' neutrons. Above a density of $\rho \approx 4 \times 10^{14}$ kg m^{-3}, the core consists of degenerate electrons and partially degenerate nucleons whizzing around the sluggish nuclei.

As the density reaches 5×10^{16} kg m^{-3}, the neutrons have an average kinetic energy of about 8 MeV (the average nuclear binding energy) and can no longer be recaptured by nuclei. Matter is then composed of increasingly abundant free neutrons in addition to electrons and neutron-rich nuclei. The neutrons, which now are supplying a larger and larger part of the density and pressure, have become fully degenerate: their Fermi energy amply exceeds the thermal energy, even at the high temperatures ($T \approx 10^9$ K) now prevailing in the star. When matter in the core becomes as dense as in a typical nucleus — that is, $\rho \approx 2.3 \times 10^{17}$ kg m^{-3} — the nuclei begin to merge together, creating a unique state of nuclear matter composed chiefly of neutrons, with only a few percent of charged particles (protons and electrons). Each cm-cube of matter now contains 10^{41} neutrons, and the *neutron degeneracy pressure* becomes powerful enough to stop the collapse and support the core against its own weight. A violent shock-wave created by the sudden halt of the implosion travels outward, carrying with it the kinetic energy of the infalling material. It blows away the outer layers of the star, leaving behind a bared core packing a typical mass of 1.4 $M_\odot$ into a sphere of only 10 km in radius: a *neutron star* is born.

Fascinating though a neutron star might be, you are well advised not to venture there. On the surface of the star, you would be crushed by a powerful gravitational force (100 billion times stronger than it is on earth) and vaporized by the intense heat (up to 10^6 K). The reduced size of the star also means that it rotates very fast (like 100 rotations per second) and is strongly magnetized (up to 10^8 tesla (1 T = 10^4 G), compared to 10^{-4} T for the sun). Most stars rotate and are magnetized. When such a star collapses spherically, it spins up (by a factor $\propto R^{-1}$ that increases with decreasing radius) to conserve angular momentum, and its magnetic field strength grows as R^{-2} to conserve the magnetic flux (which is 'frozen' by the ionized plasma).[17] The intense magnetic lines which thread the resulting neutron star can influence its internal structure and the medium surrounding it.

So, what is a neutron star's *structure* like? Imagine we start at the surface and burrow our way down, we will find the following zones (see Fig. 8.13):

(1) A thin *surface layer*, at densities of 10^9 kg m^{-3}, which consists of atomic polymers of ^{56}Fe plus a small number lighter elements in the form of a close packed solid. (Neutron stars are the only stars with a solid surface.) Atoms immersed in strong magnetic fields change drastically their energy states and take up cylindrical shapes aligned parallel to the field lines.

(2) An *outer crust* ($10^{10} \leqslant \rho \leqslant 10^{14}$ kg m^{-3}), which is a solid region composed of separate heavy nuclei forming a Coulomb lattice embedded in a relativistic degenerate electron gas, similar to the material found in white dwarfs. As the electrons can

---

[17]See Chapter 3.

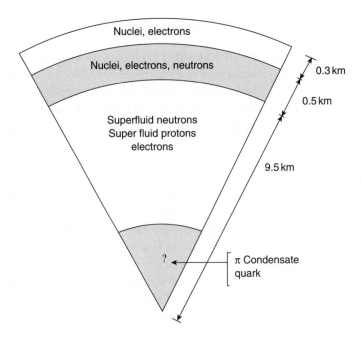

Figure 8.13:   Internal structure of a typical neutron star.

travel great distances before interacting, the electrical and thermal conductivities in this zone are enormous.

(3) When the density reaches 4×10^{14} kg m^{-3}, it becomes energetically more favorable for neutrons to *drip out* of the nuclei and move around as free particles. As the density increases further, more and more of the nuclei dissolve, and the degenerate neutron gas provides most of the pressure.

(4) Deeper down, in a region that occupies 91% of the star's volume, the *neutron liquid* phase occurs at densities above the nuclear density, and matter consists mainly of neutrons, with a 5–10% sprinkling of protons and electrons. The long-range attractive nuclear forces between neutrons grow sufficiently strong to promote the formation of coherent pairs, and the neutron gas becomes *superfluid*. The protons undergo a similar phase transition to a *superfluid* and *superconducting state*. However, we expect the degenerate electrons to remain in the normal phase, their long-range coupling too ineffective to create favorable conditions for pairing.

(5) Finally, in the very center of the star, where the density may exceed 10^{18} kg m^{-3}, all sorts of subnuclear particles will appear as products of reactions between highly energetic electrons, protons and neutrons. Here, we may have really *exotic matter*, like hyperons and isobars, pion and kaon condensates. Or the mesons and hyperons may dissolve into quarks and form quark–gluon plasmas.

At least, that's how theorists see it. The way to test their model is to examine their predictions on observable properties of neutron stars.

First, the *mass* and *radius*, two fundamental parameters of any stellar objects. If we adapt the treatment for the degenerate electrons in Sec. 8.4 to the case of degenerate neutrons we have here, we will find that the maximum possible mass for a degenerate neutron star is 5 $M_\odot$. However, this estimate does not take into account general relativity, which would tend to lower the maximum mass because general-relativistic gravitational forces inside compact stars are stronger than Newtonian gravitational forces, and so can overwhelm the quantum-mechanical degeneracy pressure at lower masses. General-relativistic calculations have led to the conclusion that the mass of a neutron star cannot exceed 3.2 $M_\odot$, independent of assumptions on the details of internal structure.

The actual value of the maximum mass is sensitive to the star's assumed internal structure; calculations have it ranging between 1.5 and 2.7 solar masses, with the lower value corresponding to a core that contains hyperons and strange mesons. Now, whereas we cannot measure the radii of neutron stars with great certainty (although we expect them to be in the 10–13.5 km range), the masses of neutron stars (at least in binary systems) are the best known masses in astronomy: numerous recent data show that they all cluster around 1.4 $M_\odot$, with none above 2 $M_\odot$. This supports a lower value for the maximum possible mass (although its exact value is uncertain) and provides some evidence for the existence of a complex internal structure for the neutron stars.

Another way to verify the model is to examine the neutron star's *thermal history*: how does it cool off?

At the moment of the neutron star's birth, its temperature may reach 10^{12} K, but it falls below 10^{11} K within seconds, and keeps falling fast. In this early epoch (say, within the first 10 years), neutrinos are produced in abundance in the reactions $n \to p + e^- + \bar{\nu}_e$, $n + n \to n + p + e^- + \bar{\nu}_e$ and $n + p + e^- \to n + n + \nu_e$. They escape with little hindrance into space, carrying with them a huge amount of energy and thereby helping the star to cool down. When the temperature has dropped far enough (between 10 and 10 000 years after birth), processes less sensitive to temperature (such as thermal photon emission, thermal pair bremsstrahlung) take over, and the cooling process slows down. In this 'standard cooling' picture, a thousand-year old neutron star would have a surface temperature of a few 10^6 K.

But if the center of the star contains exotic stuff — like strange matter, meson condensates or quark–gluon plasmas — then the neutrino cooling processes, which go with temperature as T^6, can operate for longer than we thought, which would cool things down very fast. In other words, as long as the center is made of exotic subnuclear particles, the interior heat drains off much faster than predicted by the standard cooling model.

The 'observed' thermal evolution of the neutron stars is obtained from the general assumption that all neutron stars are basically the same, and by making the correspondence of the star ages and core temperatures (the latter by extrapolating the surface temperatures, assuming isothermal interiors). Unfortunately, we have few cases where such data are available, and even in these, the experimental errors are large. Nevertheless, the available data show deviations from the standard cooling, and provide some evidence of rapid cooling and, hence, the possible presence of an exotic core.

8.5.2 Solitary Neutron Stars

How can you see a very dim object, twenty kilometers across and thousands of light years away? Nobody knew the answer until 1967, when it came in an unexpected way. Anthony Hewish and his team of astronomers in Cambridge, England, had just set up a radio telescope and were planning to observe the *scintillations*, or the rapid, irregular variations in the flux of radio waves from distant sources. The twinkling arises from irregular diffraction of radio waves when the radiation passes through regions of inhomogeneous density in the plasma of the solar corona before reaching the earth. It can be observed only if the angular sizes of the sources are small enough, less than half a second. Hewish therefore decided to use the scintillation technique to study *quasars*, those extremely distant extragalactic radio sources whose angular sizes are measured in thousandths of a second. Up until then no detection devices had been built to measure short-time changes of radiation for wavelengths in the three-meter region of the radio spectrum. A young graduate student on his team, Jocelyn Bell, noticed something rather odd on the charts produced daily by the antenna. She found an unfamiliar point source that was scintillating even at night. More remarkably still, the flux did not fluctuate at random, as is the case with ordinary scintillations, but varied in a strictly periodic way. Very short pulses of radio waves, lasting about 50 milliseconds, were detected at very regular time intervals of approximately 1 second; the first pulsating emission source, or *pulsar*, was thus discovered.

Since then, more than 1 000 pulsars have been detected. Most exhibit broadband *radio* emission, a few are also seen in the X-ray or visible parts of the spectrum. The pulse intensities vary over a wide range, but the basic pulse is periodic. *Pulsar periods* are remarkably stable, except for a very slight *overall increase over time*. They lie in the 1.55 ms–5 s range; this means a pulsar with the shortest period is seen to flash strongly 660 times per second, whereas one with the longest period only once every five seconds.

But, how do we know that *pulsars are neutron stars*? As a bit of circumstantial evidence, some pulsars are in the middle of supernova remnants, where you would expect to find a neutron star. It is the case of the young pulsars in the Crab and Vela supernova remnants. (But you could not detect old pulsars in this way, because the remains of supernovas more than 10^5 years old would have entirely dissipated.) Further evidence comes from the brevity of the pulses. If a pulse of light lasts, say, 30 ms, it cannot come from an object more than 30 light-milliseconds (or 9 000 km) across. This limits us to white dwarfs, neutron stars, and black holes, which all could send periodic signals via *vibrations*, *rotations* or *binary orbits*. White dwarfs are ruled out because they are large enough to have their maximum vibrational, rotational, or orbital periods more than 1 s (and we have many millisecond-pulsars). Black holes do not have a solid surface to which to attach a light beacon, and so cannot produce rotational or vibrational signals. Black holes or neutron stars in a binary could produce the required range of orbital periods, but they would move closer to their companions very quickly, causing their own orbital periods to decrease, not increase. Finally, vibrations of 10 km objects produce periods of milliseconds, not seconds (and we have many second-pulsars). So, the only possibility left is *rotating neutron stars* as compact objects capable of producing the kind of clockwork mechanism associated with pulsars.

How fast can a neutron star rotate? Clearly there must be a limit to the angular velocity, just as there is a limit to the linear velocity. Even before this limit is reached, however, if the star rotates more rapidly than some critical value, it induces an unstable non-axisymmetric mode of oscillation. This oscillation emits gravitational radiation which takes angular momentum away from the star, thereby pulling it back below the critical value. So there should be a lower bound on the rotation period; calculations show that the minimum possible period of neutron stars ranges from 0.4 to 1.5 ms (or 2500 to 660 rotations per second). We may have already detected one of the fastest possible spins in the pulsar PSR 1937+214, whose period is 1.55 ms. (The numbers in the pulsar's identification label indicate its position in the sky.)

As we have already mentioned, when a supergiant's core collapses, the resulting neutron star inherits a sharply increased spin and magnetization. This combination of a rapid rotation and strong magnetic fields induce enormous electric fields near the surface of the neutron star. The induced electric fields force the electrons and other charged particles to flow off the surface, especially at the two magnetic poles of the star (where the magnetic field density is at its greatest), into space along trajectories parallel to the magnetic field lines. As the electrons are accelerated to extremely high speeds along curved trajectories, they radiate high-energy photons in their directions of motion. These photons interact with the strong magnetic fields and produce jets of electron–positron and other high-energy particle pairs, which proceed to produce more photons, which generate in turn more particles, and so on. The intense light thus emitted covers the whole electromagnetic spectrum, but most of the radiated energy is in the form of gamma and X-rays (with the highest energies exceeding 100 MeV), and only a small fraction (about 10^{-5}) goes into radio emission.

If the star's magnetic fields *are aligned* with the spin axis, light is emitted in two thin beams parallel to this axis through the north and south magnetic poles. But usually they *are not* (for a similar reason that the true north and the magnetic north are at different locations on earth), so radiation streams out in two hollow cones centered about the rotation axis, as if two narrow *rotating* beams of light exiting at the magnetic poles made circles through their sky. As the neutron star rotates and one of its beams points in our direction, the beam will swing away from

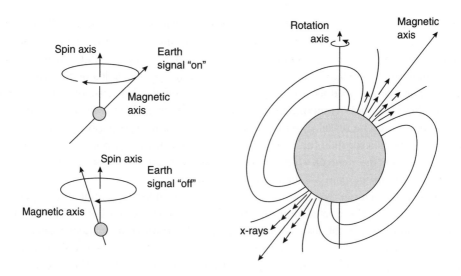

Figure 8.14:   Light-house effect in pulsar radiation.

us and not return until the star has completed a rotation. So the light *appears* to flash on and off at a regular interval as the beam comes in and out of our line of sight. We will not see a neutron star 'pulsing' unless one of its beams happens to sweep across us (see Fig. 8.14). Pulsars are most easily observed using radio telescopes, because their emission is brightest in radio waves — no-one knows why.

The large mass of a neutron star means that it takes a lot of energy to speed up or slow down the star's rotation. So, the timing of its flashes must have exceptional stability, which makes a radio pulsar an extremely accurate time-keeping device in space, as accurate as an atomic clock is on earth. Space travelers could use distant pulsars for navigation.

Any systematic changes in this precise timing are caused by unforeseen external forces, and their observations have led to significant astronomical discoveries. For instance, in 1974 Russell Hulse and Joseph Taylor observed a systematic variation in the arrival times of the pulses from a distant radio source; they reasoned that these smooth periodic variations must be caused by the changing strength of the gravitational field of two compact objects of nearly equal masses in tight orbits around each other. In other words, the radio source must be a pulsar associated with a compact companion, which is either a white dwarf or another neutron star, forming a binary system, now labeled PSR 1913+16. Hulse and Taylor analyzed the star's motion by examining the arrival times of individual pulses. Their careful analysis not only yielded precise values of the parameters of the double-star system, but also revealed the presence of significant effects of gravitational time dilatation. Because a binary system loses energy with time, as orbital energy is converted to gravitational radiation, the orbital period of PSR 1913+16 is decreasing, by 2.4×10^{-12} s/s, in perfect accord with theoretical calculations. This agreement provides us with good circumstancial evidence for the existence of gravitational radiation. A more direct proof could be provided, in the near future, by results gathered by instruments like the Laser Interferometer Gravitational-Wave Observatory (LIGO) and other ground-based observatories under construction worldwide, or sensitive detectors aboard space missions like the Laser Interferometer Space Array (LISA).

Another example based on observations of this kind is the astonishing discovery by Alexander Wolszczan and Dale Frail in 1992 that the radio pulsar PSR 1257+12 has two earth-sized *planets* as well as a moon-sized satellite. Their existence has been confirmed by the detection that their mutual interaction perturbs their Keplerian orbital motion, resulting in subtle changes in the pulsar timing. At present, this is the only confirmed existence of planets around a neutron star.[18]

Neutron stars may be rotating very fast at birth, but they *spin down* afterwards. This is why the youngest pulsars, such as the Crab pulsar (33 ms) and the Vela pulsar (80 ms), have unusually short periods, whereas older neutron stars have longer periods. The neutron star spin-down is due to *magnetic braking*, a mechanism by which the star's magnetic field exerts a strong torque on its surroundings, which are then forced to corotate with the star. This causes the star to decelerate and transfer its rotational energy to the radiation of electromagnetic energy into space at a rate that we can calculate.[19] Take the Crab pulsar as an example of such rotation-powered neutron stars: the rate at which it loses rotational energy is 6.4×10^{31} W, which is similar to the energy requirements of the surrounding supernova remnant in non-thermal radiation and bulk kinetic energy of expansion, 5×10^{31} W.

The spin-down rate is a fundamental observational parameter and is known with great precision (typically 10^{-15} second per second in most cases, but 10^{-13} s/s for a young pulsar like the Crab). From such measurements, researchers have estimated the magnetic fields at the surfaces of radio pulsars to be 10^6–10^8 T. If this sounds unbelievably high,[20] wait until you hear about the *magnetars*!

8.5.3 The Missing Pulsars

Of the billion or so neutron stars thought to exist in our Galaxy, only a thousand have actually been observed. Where have all the others gone?

Until recently, the main way to find a newly formed neutron star is to detect its radio pulsations. But with sophisticated instruments now at their disposal, astronomers are realizing that neutron stars need not only be radio pulsars. For example, some of the missing ones might be *quiet, isolated, cooling* compact stars that are

---

[18]However, we know of at least 100 planets around normal stars outside our solar system. Most have masses in the 1–10 Jupiter mass range, most have highly eccentric orbits and are extremely close to their parent star. For the latest developments, go to http://exoplanets.org/science.html

[19]A star of radius R rotating at angular frequency $\Omega = 2\pi/P$, with its magnetic moment misaligned from the spin axis by an angle θ, radiates energy at the rate of $dE/dt = B^2 R^6 \Omega^4 \sin^2 \theta/6c^3$, where c is the speed of light and B is the magnetic field strength at a magnetic pole.

[20]The highest known field in ordinary stars is 100 T. On earth, we can produce sustained fields of 50 T in a small volume, and fields of 1000 T for milliseconds.

not seen pulsing, because, for unknown reasons, they never 'turned on.' However, their hot surfaces glow quietly in X-rays, and they can be seen only in this part of the spectrum.

Other missing neutron stars might be the unpredictable *soft gamma-ray repeaters* (SGRs), the recurrent soft X-ray transients discovered in the 1980s. Although there are only a few examples known at present,[21] SGRs have attracted a lot of attention lately because of their unique characteristics. They emit frequent, but randomly spaced in time, outbursts of low-energy gamma rays of very short duration, usually tenths of seconds. Each outburst sends out 10^{33}–10^{35} W in luminosity at its peak. In between the bursts, they produce persistent softer radiation, with a clear period of 5–8 s consistent with rotational modulation; and the period is lengthening with each successive eruption at a rapid rate of about 10^{-12} s/s.

Analyses of the data collected by several satellite-borne instruments (such as ASCA, BATSE and RXTE) show unambiguously that SGRs possess the characteristics of neutron stars. In particular, the locations of SGR 0525–66 and SGR 1806–20 are shown to coincide with the faint X-ray sources found in the young supernova remnants N49 in the Large Magellanic Cloud and G 10.0–0.3 in our Galaxy, respectively. Thus, if SGRs are associated with supernova remnants, then they must be neutron stars some tens of thousands of years old — although neutron stars that young with 8 s periods are rather puzzling. The question is, what causes the gamma-ray bursts?

There is an even more intriguing class of objects called *anomalous X-ray pulsars* (AXPs) — 'anomalous' because we do not understand why they shine. We have to date five confirmed objects of this class,[22] all in the galactic plane; they pulse with periods of 6–12 s, which increase with time at the rates of about 10^{-12} s/s; and, finally, they radiate X-rays at modest luminosities. Most oddly of all, at least two are associated with supernova remnants. They exhibit many similarities with SGRs, and might even erupt in sudden emissions, although they show clearly less activity than their cousins. Again, the question is, where does their energy come from?

The two most obvious possibilities — rotation and accretion of material from a nearby star — can be ruled out, both being inadequate to explain the huge power needed for the brief outbursts observed. In the 1990s Robert Duncan and Christopher Thompson advocated another energy source, that of very strong magnetic fields, greater than 10^{10} T. Such powerful fields in neutron stars, dubbed *magnetars*, could provide the energy source for SGR bursts and also provide a mechanism that would quickly spin the stars down to their characteristically long rotational periods. At the quantum level, these strong fields would also modify the atomic energy structure so drastically that the stellar matter becomes essentially transparent to electrons, thereby allowing the huge luminosities observed in SGR outbursts.

---

[21] SGR 0525–66, SGR 1806–20, SGR 1900+14 and SGR 1627–41 are known in 2003.

[22] As of 2003, the AXP catalog consists of 4U 0142+615, 1E 1048–5937, RXS 1708–4009, 1E 1841–045 and 1E 2259+586.

When we apply these ideas to SGR 1806–20 — a specially prolific SGR that exhibits 7.5 s pulsations with a long-term increase of 8×10^{-11} s/s — we find that the neutron star has a spin-down age of 8 000 years and a surface magnetic field strength of 2×10^{10} T. Estimates of the number and ages of SGRs in the Galaxy suggest that magnetars may form one tenth of the entire galactic neutron-star population. Similarly, the spin-down rates of AXPs imply magnetizations in the range of $(1–7)\times10^{10}$ T, which would place them among the strongest fields inferred for neutron stars.

So, are SGRs and AXPs magnetars? The evidence is convincing for SGRs, but less strong for AXPs. The situation will certainly be clarified in the near future with the new data coming back from space missions, such as the Rossi X-Ray Timing Explorer (RXTE) and the Chandra X-Ray Observatory.

8.5.4 Neutron Stars in Binaries

Not all neutron stars lead a solitary life; many are found as the optically invisible partners of normal stars in X-ray binary sources. Some of these neutron stars are born in binaries that survive the supernova events; others may have captured (or have been captured by) ordinary stars in dense stellar regions such as globular clusters. But there is no reason to suppose that they are different in nature from the isolated variety most often observed as radio pulsars, and, therefore, any distinctive phenomena they might exhibit must derive simply from their close association with another star.

Binary X-ray sources, in which the neutron star components are always optically invisible, come in two broad classes. The first is associated with very luminous and massive (late O or early B type) stars, which are among the youngest stars and have rather short main-sequence lifetimes. These binaries are called HMXBs, or *high-mass X-ray binaries*. The second class is formed by the LMXBs, or *low-mass X-ray binaries*, which contain cooler low-mass main-sequence stars, having masses, luminosities, and temperatures similar to those of the sun. In either case, the companion star provides the prime additional source of energy through *accretion*. This mechanism of mass transfer to the compact star proceeds in different ways depending on the mass of the donor star — hence the two-class distinction.

Why do we believe that accretion of matter from the normal (primary) star onto the neutron star (or black hole) is responsible for the X-ray emission from these binary X-ray sources?

First, because when matter falls from the primary star on the secondary star, the kinetic energy it gains is equivalent to the gravitational binding energy of matter on the compact star. Most of this energy is converted into thermal energy when the infalling matter hits the surface, releasing up to 15% of the rest mass of the infalling material as radiation. This represents an excellent source of energy, 20 times more efficient than nuclear energy sources. If we assume that the neutron star gains additional mass in this way at a rate of 10^{-10} $M_\odot$ per year, it can radiate energy with a power of 10^{30} W, which is consistent with the observed luminosity of a strong X-ray binary. If this energy is emitted as black-body radiation, then the star's effective temperature should be no less than 10^7 K, and the star should emit in the X-ray waveband.

Another argument in favor of accretion as the mechanism of powering X-ray binaries concerns the existence of an upper limit to their luminosities: what prevents them from accreting matter at an inordinate rate? The answer is, when the luminosity of the accreting object is too great, the radiation pressure acting on the infalling matter (*e.g.*, scattering of the emergent high-energy photons by infalling electrons) increases and can reach a level such that it can effectively prevent any further matter falling onto the surface of the compact object. Thus, there exists a critical luminosity above which accretion shuts off, known as the *Eddington limit*, given by

$$L_{\mathrm{Edd}} = 1.3 \times 10^{31} \, (M/M_\odot) \, \mathrm{W} \, .$$

This result agrees quite well with the observed cut-off in the X-ray luminosity distributions of the X-ray binary sources in the Galaxy and the nearby Magellanic Clouds.

The *neutron stars in HMXBs* are intense X-ray pulsars rotating with periods of 1–1000 s. The accretion mechanism operating here must also explain why their emission is pulsed. All stars emit *stellar winds*, quiescent mass loss in sun-like stars and more powerful winds in the cases of the more massive stars. In HMXBs associated with luminous, massive O and B stars, the orbiting magnetized compact star is embedded in the strong outflow of the primary (Fig. 8.15). As the infalling matter approaches the surface of the neutron star, the two masses spinning at different speeds will try to come into equilibrium with each other, such that the star's rotation period nearly matches the orbital period of the circling matter just outside the *magnetosphere*, the region where the magnetic field dominates the accretion flow. From the physics of this spin equilibrium situation, one can infer that these neutron stars must have strong magnetic fields, in excess of 10^8 T (similar to values found in typical isolated pulsars). Accreting matter is forced to flow along the strong magnetic field lines and can only funnel down onto the star's surface through the magnetic poles, where the field density is the greatest. So the X-ray emission is concentrated in these two 'hot spots' where the accreted particles hit the surface. If the magnetic field axis and the star's spin axis are misaligned, as is usually the case, then the radiation in our direction sweeps past us once every rotation and we see X-ray pulsations.

From recent observations by orbiting X-ray telescopes, astronomers have detected another remarkable property of these systems: bucking the overall tendency to spin down, some of them occasionally *speed up* while others display abrupt *accelerations and decelerations*. These features can be understood as being associated with variations in the mass flow about the magnetosphere of the neutron star

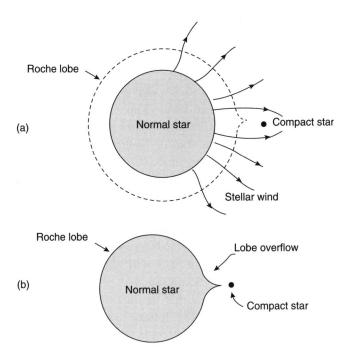

Figure 8.15: Two possible modes of mass transfer in a binary X-ray source. In (a) the normal primary lies inside the Roche lobe (a surface beyond which the primary will begin to shed gas onto its compact companion). The mass transfer is effected via a stellar wind. In (b) the primary begins to expand to and overflow its Roche lobe. (Based on Shapiro and Teukolsky, p. 399.)

and the subsequent transfer of angular momentum from matter to the star. If there is no accretion, there is no acceleration and the star slows down. But, at times, a disk of swirling matter may form briefly around the star and, because the stellar wind is unstable, it may dissipate to reform later, going the other way.

The *neutron stars in LMXBs* have companions with masses less than 1 $M_\odot$. They are older than those in HMXBs — 10^9 years compared to 10^7 years — and have accreted more matter. They show no apparent evidence of dipolar accretion or pulsations in their radiative fluxes, and so must have relatively weak magnetic fields of about 10^5 T or less, being reduced naturally to these values by eons of ohmic decay.

The neutron star in an LMXB usually has a gravitational field stronger than the field of its less massive companion. If the outer envelope of the latter is close enough, the compact star pulls the closer parts of the envelope towards itself, creating a narrow stream of gas rushing out of the envelope. As the compact star is very small compared to its companion, the flowing gas has too much angular momentum to fall directly onto the star, and so must orbit around it, forming a thin *accretion disk* of hot matter. Within the disk, the gas will spiral in swiftly towards the center until it reaches the magnetosphere, where the magnetic field at the neutron star's surface takes control and forces it to flow along the field lines to the magnetic poles. As it swirls in towards the neutron star and collides against itself, the gas is heated to millions of degrees, dissipating its gravitational binding energy into radiation which we observe as a continuous X-ray emission.

Since a typical neutron star in this class is old, its magnetic field is expected to be relatively weak and its magnetosphere small. This means that the swirling gas can orbit very close to the star and acquire high orbital frequencies before being picked up by the magnetic field. Also, as the star has been accreting long enough to gather a great amount of additional material (up to 0.1 $M_\odot$), the angular momentum it acquires from this much high-spin accretion can *spin it up* to very rapid rotation when the stellar magnetic field is weaker than about 10^5 T. However, this spin is notoriously difficult to detect, and it was only in 1998 that the first accretion-powered millisecond pulsar was discovered with RXTE, when the neutron star SAX J1808.4–3658 was revealed to have 2.5 ms pulsations in its persistent X-ray flux and an inferred surface field in the range of $(2–8)\times10^4$ T, comparable to the magnetic strengths observed in millisecond radio pulsars.

Neutron stars in LMXBs show another distinctive phenomenon: they display repeated *X-ray bursts*, which are brief, intense emissions of X-rays caused by unstable thermonuclear reactions in the accreted layer of hydrogen and helium. Due to thermal instability on their surfaces, the stars are unable to burn the accreted matter as fast as it is gathered. Instead, they accumulate hydrogen and helium for hours or days until instability sets in, then burn the fuel in a few seconds when the local temperature exceeds 10^9 K. The runaway thermonuclear process starts at a localized site, probably near a magnetic pole, then spreads around the compact star, eventually burning all the nuclear fuel. When astronomers examine closely the power spectra of these outbursts, they discover pulsations in a range (1.7–3 ms) very similar to the spin periods of millisecond radio stars. This provides strong evidence for the existence of *rotating* neutron stars in LMXBs, supporting the lone direct observation mentioned above.

In summary, the accreting neutron stars in many LMXBs are rotating with millisecond periods; some of them may be magnetized enough to become millisecond radio pulsars once accretion shuts off; but most others, for unknown reasons, do not show persistent pulsations at all.

8.5.5 Summary

Neutron stars are end states of massive stars. They owe their stability to the degenerate-neutron pressure of a very dense neutron-rich matter that makes up its interior. They appear to have the maximum possible mass of 3 $M_\odot$, and are characterized by a rapid rotation and a very strong magnetic field, two properties that determine to a large extent their interactions with the surrounding medium. Whether neutron stars are isolated or associated with a companion star also gives rise to distinctive phenomena. Thus, we find isolated neutron stars in the form of radio pulsars, soft-gamma-ray repeaters or anomalous X-ray pulsars; and neutron stars in high-mass X-ray binaries or in low-mass X-ray binaries, which are strong X-ray sources.

8.6 Black Holes

When a massive star runs out of nuclear fuel, it contracts into a compact inert core. If the core has a mass greater than 3 $M_\odot$, no known forces can prevent it from collapsing relentlessly under the inward pull of gravity until it falls into a region of space from which no light, matter or signal of any kind can escape. This warp in spacetime is called a *black hole*. Although black holes were first studied in 1939, by Robert Oppenheimer and Hartland Snyder, evidence of their occurrences did not exist until recently, thanks especially to advances in X-ray astronomy. Nowadays, it is generally believed that black holes occur commonly; they may be found not only in the stellar-mass range, as end states of normal stellar evolution, but also as supermassive objects produced by the coalescence of many dense stars or even of an entire conglomerate of stars and holes. In fact, they may have any mass.

Gravitation becomes so overwhelming in black holes that it changes the nature of spacetime itself, giving it curvature and causing it to produce disturbances that can propagate in waves. These phenomena, unknown to Newtonian physics, can be studied in the framework of Einstein's general theory of relativity, the most satisfactory theory of gravitation available to us at present.

8.6.1 Gravitational Collapse

Let us consider first the collapse of a massive *spherical star*. When the star, having exhausted its nuclear fuel and having contracted slowly inward, begins to squeeze its electrons onto the atomic nuclei, weakening thereby its source of pressure, it becomes unstable. As the instability develops quickly into a full-scale implosion, the stellar core falls inward on itself. If the degenerate-neutron pressure, which must appear at appropriately high densities in the interior, fails to stop the fall, the collapse keeps pressing on, pulling the star's surface through its *gravitational radius*, where the star's gravitational binding energy becomes comparable to its total mass energy. Once this critical surface is passed, the implosion will not stop until the compressed star has reached a 'point' of zero volume and infinite density — a *singularity*. Within a short lapse of time, about $10^{-2} \, (M/M_\odot)$ s, the structure of total mass M has matured, and the dynamical behavior of spacetime has settled down into a stationary situation described by an exact solution to Einstein's field equations called the *Schwarzschild spacetime* (Figs. 8.16 and 8.17).

This solution represents the geometry of curved spacetime exterior to a spherical static source of total energy Mc^2 and which is also asymptotically flat. It depends

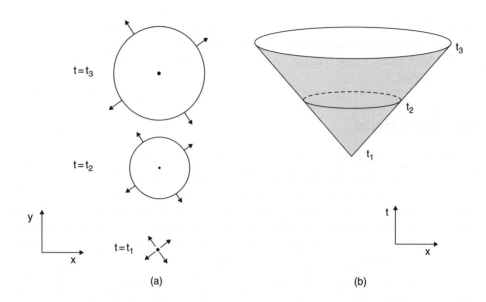

Figure 8.16: Representations of the light fronts at successive times: (a) Spatial representation; (b) Spacetime representation.

only on this one parameter, but is otherwise universal, independent of the nature of the source that has produced it. And it even covers situations more general than stated, for it can be applied to any spherical mass distribution, whether static, collapsing, expanding or pulsating. The gravitational field surrounding the sun, a neutron star, and that of a black hole right up to its singularity are identical, provided their masses are the same. The Schwarzschild solution forms the basis of our understanding of the physics of spherical black holes, which we now summarize.

A spherically symmetric black hole of mass M possesses a characteristic radius known as the *gravitational* (or *Schwarzschild*) *radius* $r_\mathrm{g} = 2GM/c^2$, which represents the effective radius of the black hole.[23] The non-material spherical surface of radius r_g is called the *event horizon*; it has the defining property that signals emitted inside cannot escape, whereas signals emitted outside can escape to infinity. So it acts as an effective physical boundary of the black hole. Putting in the values of the constants G and c, we find $r_\mathrm{g} = 3\,(M/M_\odot)$ km, so that solar-mass black holes have $r_\mathrm{g} = 3$ km.

The gravitational field inside the event horizon becomes so powerful that even light is pulled ineluctably inward toward the center regardless the direction in which it is emitted. The paths of all motions terminate at the central singularity. Outside the horizon, light may escape, but as the emitter approaches the surface from a distance, light rays find it harder and harder to move away. The rays that can

[23]In the following, we will use units in which $G = c = 1$, so that M designates not only mass, but also energy, length and time. The corresponding expressions in conventional units are $Mc^2 = 2 \times 10^{47}(M/M_\odot)$ J, $GM/c^2 = 1.5\,(M/M_\odot)$ km and $GM/c^3 = 5\,(M/M_\odot)$ µs.

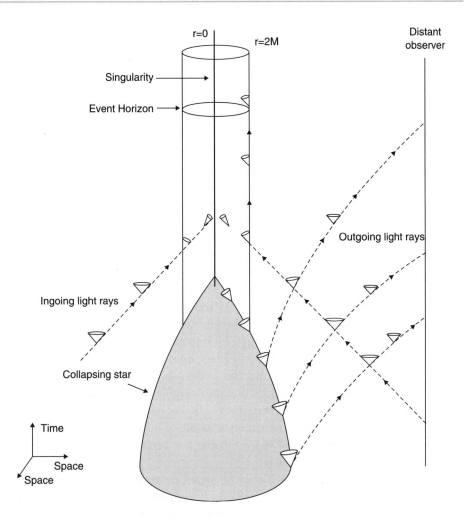

Figure 8.17: Gravitational collapse of a spherical star into a Schwarzschild black hole. Light rays in various directions of propagation are shown as successions of light cones.

get away are restricted to a cone aimed radially outward whose opening becomes narrower and narrower until only a pencil of vertical rays emerge free. On the horizon itself, it is possible for light to remain stationary, sandwiched between those signals just outside of the horizon and those just inside, and hovering in place forever. This is why the surface $r = r_g$ of a Schwarzschild black hole is also called a *static limit*.

Imagine you send toward a black hole a space observatory, programmed to transmit signals at equally spaced intervals (by its clock) at some fixed frequency (by its standard). So long as the space probe is at large distances from the compact star, we receive these signals at nearly equal intervals. But as the spacecraft approaches its target, we observe that the signals are arriving at more and more widely spaced intervals and at wavelengths that become longer and longer. When the probe is close to the black-hole surface, the time dilatation and signal redshift thus observed increase exponentially with time by the earth-bound observer's clock. The signal sent out just above the event horizon itself will take an almost infinite time to reach us, and will be infinitely redshifted when arrived. From our vantage, the space probe takes an infinitely long time to cross the Schwarzschild surface, and once it has crossed it, there is no way for us to learn directly what happens to it afterward. But for the spacecraft, nothing unusual happens as it passes the boundary, which it does smoothly and at a finite time by its clock. But once inside, it is pulled inexorably toward the central singularity, which it reaches in a few microseconds and where it disappears forever.

The most direct manifestation of gravitation and hence of spacetime curvature is the *gravitational tidal effect*, which results from the non-uniformity of the gravitational field in spacetime. A familiar example is the tidal force of the moon, which causes the earth's ocean tides. If an astrophysicist goes with the space mission we described earlier, tidal gravitational forces will stretch him from head to foot and, even worse, compress him on all sides. At some point, when the tension reaches 10 MPa (or 100 atm), his body will be broken. In a mission to a solar-mass black hole, this happens at 200 km from target, far outside the 3 km event horizon, but in an exploration of a supermassive 10^9 $M_\odot$ black hole, the astrophysicist can survive up to a distance of 2×10^5 km from the center, far inside the critical radius of 3×10^9 km. As the spacecraft and its baryonic cargo cross the surface of no return and are propelled toward the center, they will be ripped apart and crushed by rapidly rising tidal forces. By the time the point of infinite curvature is reached, the mash of quarks to which the space mission has been reduced will be squeezed to zero volume and infinite density.

Let us turn now to real stars with asymmetries, magnetic moments, and possibly electric charges. The *non-spherical collapse* is a more complex process which we now understand well from results of a number of calculated examples and precise theorems. The key elements of spherical collapse — instability, implosion, the formation of an event horizon, the rapid decay to a 'black' stationary state, and the presence of some kind of singularity at the center — are believed to occur also in general non-spherical collapse. The main additional phenomenon is the emission of gravitational and electromagnetic radiation during the collapse.

The first local sign that a black hole is forming is the appearance of a *trapped surface* within the collapsing star; by this term, we mean a closed spatial surface (spherical or not) such that light rays emitted perpendicularly from the surface encounter enough matter or gravitation to converge, regardless of whether the rays are emitted in the outward or the inward direction. These converging rays will never reach infinity, and so the trapped surface lies inside an event horizon. For instance, any spherical surface inside the event horizon in the Schwarzschild geometry is a trapped surface; we know then that the convergence of light rays can be attributed to gravity, which pulls the photons to the central singularity. Stephen Hawking and

Roger Penrose proved in the 1970s that a symmetric or an asymmetric spacetime that contains a trapped surface (and satisfies some general properties) must have a singularity, a region of spacetime where physical theory breaks down. In black holes, we interpret this singularity as a region of infinitely strong tidal forces. If one assumes (the Cosmic Censorship Hypothesis) that singularities resulting from gravitational collapse are hidden from view and hence causally cut off from the external universe, as is true in the spherically symmetric situation, then an *event horizon* will arise.

Once formed and left undisturbed, a black hole will settle down rapidly into a final stationary state characterized uniquely by the values of the mass M, spin J and charge Q of the hole (and no other attributes whatsoever: 'a black hole has no hair'). This is the conclusion that suggests itself from a set of powerful theorems due to Hawking, Werner Israel and Brandon Carter. Why should M, J and Q determine completely the final state of the black hole? Because of all quantities intrinsic to an isolated source of gravity and electromagnetism, only the mass, spin and charge are locked to the distant external fields of the source (mass and spin to the gravitational field, charge to the electromagnetic field) by their relations to the conserved flux integrals over surfaces surrounding the hole and far from it. While the collapsing star quickly settles down to the unique configuration corresponding to the given M, J and Q, the fields undergo dynamic changes: all the higher-multipole asphericity in the gravitational field not compatible with M and J, and all electromagnetic moments, except the electric charge, detach themselves and blow off to infinity as radiation. The outcome of this process is a black hole with the specified final values of M, J and Q, and the external fields determined uniquely by those values. All other properties of matter, such as composition, baryon number and lepton number, are swallowed by the black hole, cut off from outside observations.

Black holes are unlikely to have a substantial charge. If a hole has any finite charge, the electric forces it exerts on a distant plasma would overwhelm gravitational forces, causing charge separation and pulling enough particles of opposite charge to the hole to neutralize it. By contrast, most stars that collapse rotate so fast that the black holes they produce have an angular momentum equal or nearly equal to the maximum allowed, $J_{\max} = M^2$. Therefore, to all intents and purposes, all black holes in the universe have just two characteristics, mass and spin. Such a black hole is described by an exact solution to Einstein's field equations, called the *Kerr geometry* (for Roy Kerr who created it). Listen to Chandrasekhar as he comments on this remarkable result: "(Black holes) are, thus, almost by definition, the most perfect macroscopic objects there are in the universe. And since the general theory of relativity provides a single unique two-parameter family of solutions for their description, they are the simplest objects as well."

The Kerr black hole's event horizon (the closed surface that lets no photons escape from inside) has a radius that depends on the angular momentum:

$$r_{\mathrm{g}} = [M + (M^2 - a^2)^{1/2}],$$

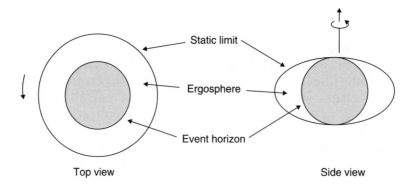

Figure 8.18:   A Kerr black hole has a static limit and an event horizon; the region between these two surfaces is an ergosphere.

where $a = J/M$. It is smaller than the corresponding value for the Schwarzschild horizon and reduces to it when $a = 0$. (We are assuming that $|a| \leq M$; should this not be the case, there will be no horizon and we shall have a naked singularity, in violation of the Cosmic Censorship Hypothesis.) The Kerr black hole is not spherically symmetric, only axially symmetric, because its static limit

$$r_0 = [M + (M^2 - a^2 \cos^2 \theta)^{1/2}] \, ,$$

depends on a certain angle θ defined such that the 'poles' correspond to $\theta = 0$, π, and the 'equator' to $\theta = \pi/2$. This surface touches the horizon at the poles and intersects the equatorial plane on a circle whose radius is larger than that of the horizon (Fig. 8.18). On this surface, no particles can stay in the same place; only a light signal emitted in the radial direction can be at rest, and so must appear as infinitely redshifted to a far-away observer. For radii smaller than r_0, but larger than r_g, all matter, regardless of the forces acting on it, is dragged irresistibly around by the rotation of the black hole, although it can still, if powered properly, escape again to infinity. Only outside the static limit is it possible to surmount this dragging with a finite acceleration and remain fixed with respect to the distant universe.

8.6.2   *Black Hole Dynamics*

The space between the horizon and the static limit of a rotating black hole is called the *ergosphere*. One can in principle extract energy from the hole by having particles move in and out of this region. The key to this mechanism is that there are orbits in the ergosphere that support particles of negative total energy, meaning their gravitational binding energy exceeds the sum of their mass and kinetic energies. Imagine you send in from infinity a particle that splits into two parts in the ergosphere in such a way that one part, as judged by a far-away observer, follows a negative-energy trajectory into the horizon, while the other part escapes to infinity with an energy that is in excess of that of the original particle, by conservation of the total energy. An energy has thus been effectively extracted from the rotational energy. This process causes the black hole to spin down slightly and reduces its mass by an amount equivalent to the extracted energy. There is a limit to the amount of energy that can be extracted; it is given by $\Delta M = M - M_{\text{irr}}$, where M_{irr} is the *irreducible mass* of the hole, defined such that

$$M^2 = M_{\text{irr}}^2 + J^2/4M_{\text{irr}}^2 \,. \tag{8.8}$$

When all the rotational energy has been extracted, we are left with a spherical black hole of mass M_{irr}, whose value cannot be further reduced by any interaction whatsoever. By the same token, a black hole can accrete matter selectively, in the direction of its own rotation, so that it can spin up to the maximum allowed spin, $J_{\text{max}} = M^2 = 2M_{\text{irr}}^2$, at which the rotational energy attains also its upper limit, consisting of 29% of the total energy.

One of the few known general properties of non-stationary black holes is *the area theorem*, which states that in any process involving black holes, the total surface area A of the horizons can never decrease; it can at best remain unchanged, when the conditions are stationary. We write the corresponding equation as $dA \geq 0$, which is equivalent to $dM_{\text{irr}} \geq 0$ for arbitrary small changes in M and J. Equation (8.8) tells us that any decrease in M must come from a decrease in J and, furthermore, M has M_{irr} as the lower limit.

The area theorem leads to a remarkable unification of thermodynamics and gravity, which we will now discuss. The condition $dA \geq 0$ points to a formal analogy between black hole area and entropy of a closed system: both tend to increase with time. The *entropy of a black hole*, S_{BH}, must be a function of A with dimensions of Boltzmann's constant k. So we write $S_{\text{BH}} = kf(A)$. The simplest choice is a linear function, that is, $f(A) = bA/L_{\text{P}}^2$, where b is a pure number, and L_{P} any universal length introduced to make $f(A)$ dimensionless. If we take L_{P} as Planck's length $L_{\text{P}} = \sqrt{\hbar} = 1.61 \times 10^{-35}$ m, then a calculation shows that $b = 1/4$. So, we have a very simple formula, $S_{\text{BH}} = kA/4\hbar$, for the entropy of any black hole of boundary area A. This result was first obtained by Jacob Bekenstein and Hawking. (It was re-derived in 1996 using string theory.)

What does this area–entropy relation mean? In statistical physics as in many other fields, entropy represents missing information. In black holes, it must represent all the information (hair) lost after the collapsing stars have settled down into stationary objects known to us only by the parameters M, J and Q. It represents the external observer's deep ignorance about the star interior, about all the possible microstates of different temperature, shape, color, composition, etc. that contribute to making up a macrostate specified only by M, J and Q. The entropy quantifies this deep ignorance of ours about the physics under the horizon: $\exp(S_{\text{BH}}/k)$ gives precisely the number of those possible microstates. By doing a quick calculation for a solar-mass black hole, you will see that S_{BH}/k is a huge number, larger than the sun's thermal entropy by a factor of 10^{20}.

Briefly, the area theorem looks very much like a version of the *second law* of thermodynamics. We can also obtain an analog of the *first law* of thermodynamics, which gives the energy change of a closed system as it evolves from one state to a nearby state. It is, in fact, an energy conservation condition, which says that the energy change in the transformation, dE, is heat plus any work done on the system: $dE = T\,dS + W$. To simplify, we restrict ourselves to Schwarzschild black holes, in which case $W = 0$, $A = 16\pi\,M^2$, and $M = M_{\text{irr}}$. (Generalization to include charge and rotation is straightforward.) Since we already know $E = M$ and $S_{\text{BH}} = kA/4\hbar$, we easily obtain $T = T_{\text{H}} = \hbar/(8\pi kM)$. If we follow the thermodynamic analogy further, we must consider T_{H} the *black hole temperature*. Note that, just as S_{BH} is proportional to the black hole surface area A, so T_{H} is proportional to its surface gravity, $g = 1/(4M)$ (which is the acceleration, as measured at spatial infinity, of a free-falling test particle near the horizon); the proportionality factors in both cases contain only fundamental physical constants.

That a black hole behaves as a thermodynamic black body with entropy $kA/4\hbar$ and temperature $\hbar g/(2\pi k)$ is rather puzzling, since we know that any object with a finite temperature must emit radiation, and black holes are absorbers rather than emitters. The missing piece of the puzzle was found by Hawking in 1974 when he discovered the thermal radiance of black holes.

The presence of the Planck constant $\hbar$ in black-hole entropy and temperature indicates that these concepts must have quantum origins. According to quantum theory, a vacuum is not an absolute void, but a sort of reservoir of pairs of *virtual* particles and antiparticles of every kind (also known as 'vacuum fluctuations'). These particles and antiparticles are considered 'virtual' because they can only appear transiently, for time intervals too short to measure. Such a pair, with energy ΔE, may be created out of the vacuum, in violation of the conservation of energy, provided it exists only for a short enough lapse of time, $\Delta t < \hbar/\Delta E$, in accordance with the energy–time uncertainty relation.

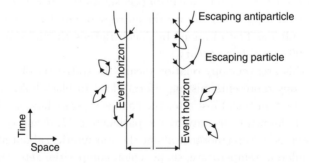

Figure 8.19: Hawking's radiation. In this space-time representation, the black hole is propagating vertically in time at a fixed point in space. In the course of pair creation processes near the event horizon, some pairs are split up, one member falling into the black hole, the other escaping to infinity.

Vacuum fluctuations may take place in any spacetime, including the vicinity of a black hole (Fig. 8.19). Consider, for instance, a fluctuation that produces two photons, one of energy E, the other with energy $-E$. In flat spacetime the negative-energy photon would not be able to propagate freely for long; it would recombine with the positive-energy photon within the time $\hbar/E$. But if produced just outside the horizon, it may cross the horizon before the time $\hbar/E$ elapses; once inside the horizon, it can propagate freely along one of the negative-energy trajectories that exist inside the hole due to the large gravitational binding effects, to vanish eventually at the singularity. Meanwhile, the positive-energy photon has a chance to escape to infinity as a free, real particle, carrying with it an energy E. According to quantum mechanics, we may view an antiparticle falling into the hole as a particle coming out of it but traveling backward in time. At the point where the pair was originally created, the particle is deflected by the gravitational field and moves forward in time. The particle that is escaping appears to a distant observer to have emerged from inside the black hole by tunneling through the gravitational barrier on quantum-mechanical principles.

Hawking found that the particles emitted in these processes have a *thermal distribution* of energies, with characteristic temperature given by the same formula as obtained above, $T_{\mathrm{H}} = \hbar g/2\pi k$ (which holds even in the presence of charge and rotation, with appropriate g). This form of T_{H} can be understood qualitatively by setting a thermal wavelength $\hbar/kT_{\mathrm{H}}$ equal to the gravitational radius $2M$. The numbers work out that $T_{\mathrm{H}} \approx 0.6 \times 10^{-7}(M_{\odot}/M)$ K: the bigger a black hole is, the cooler it is.

When a black hole emits particles, the outside world gains energy. By conservation of energy, there is a corresponding *decrease* in energy of the black hole, and hence a loss of its surface area, or entropy, in violation of the *classical* second law. But because of Hawking's radiation, the entropy of the surrounding medium, S_{ext}, must *increase*. Thus, the second law must be generalized to require that the total entropy, $S = S_{\mathrm{BH}} + S_{\mathrm{ext}}$, be a non-decreasing function of time.

We may readily estimate the power of this radiation: since $r_{\mathrm{g}} = 2M$ and $T_{\mathrm{H}} \sim M^{-1}$, we have $L \sim r_{\mathrm{g}}^2 T_{\mathrm{H}}^4 \sim M^{-2}$. A solar-mass black hole would radiate at the rate of $L \approx 10^{-28}$ W in radio waves. The hole's lifetime is thus limited by evaporation to $\tau \sim M/L$. Putting in all numerical factors, we have

$$\tau = 10^{10} \text{ yr}(M/10^{11} \text{ kg})^3 \,.$$

This makes a solar-mass black hole live almost forever and its radiation unobservable. But black holes with much smaller masses could evaporate at a more perceptible rate. In particular, if small black holes (the primordial black holes) were formed at the beginning of the universe, those with masses greater than 10^{11} kg could have survived to this day; they would have the size of a proton and a temperature of 10^{11} K. At this temperature, they would emit a profusion of photons, neutrinos and gravitons; they would radiate at ever greater rates until they explode violently out of existence. Mini black holes could last for some ten billion years, about the present age of the universe; so they could still be out there in space, sending out strong beams of very energetic gamma rays and waiting to be discovered.

8.6.3 Searching for Black Holes

Once formed, isolated stellar black holes do not emit light and would be hard to see. In principle, if a black hole is located on the line of sight of the earth to a distant star, its presence could be revealed by the gravitational lensing effect, which refers to the large change in the apparent position of the star following the deflection of starlight by the strong gravity of an intervening compact object. But the probability for such an alignment is slim.

However, a black hole is never completely isolated; it is always surrounded by the interstellar medium, which it may capture. When matter falls down toward the horizon, it heats up by friction and emits thermal radiation. The radiation that escapes would be detectable at a distance. This radiation should be intense and more easily observable if there is a nearby rich source of gas or plasma, such as a companion star or a dense interstellar cloud. If the gas falls in radially, most energy goes down the hole in the form of internal energy or radial kinetic energy. But if it has a large spin, the gas will be captured first into a circular orbit, then gradually spiral down into smaller and smaller orbits while it sheds its excess of angular momentum. Matter circulating in these orbits forms a thin *accretion disk* around the black hole. The motion of particles at different speeds in adjacent bands of close orbits produces a viscous flow in the disk. It is this viscosity that transports angular momentum and energy outward, and heats the gas, ionizing it and causing thermal radiation. For a solar-mass black hole, temperatures as high as 10^8–10^9 K may be attained, and the disk radiates 10–100 keV X-rays.

The accreting process in black holes is not dissimilar to that occurring in neutron stars we described earlier, but two new features lead to distinctive observable effects. The first is that the innermost edge of the accretion disk comes closer to the center, at the last stable orbit $r \gtrsim r_g$. After passing this radius, matter spirals quickly inward with negligible further radiation. The total thermal energy liberated by matter as it traverses the width of the disk is just the gravitational binding energy in the innermost stable orbit. The second distinction is that the gravitational pull of very massive black holes may be so strong that the overflow of matter into the disk from the companion star might exceed the Eddington limiting value; only then will the maximum allowed amount manage to get down to the hole, while the rest will be forcefully expelled along the axis of rotation of the black hole in the form of two symmetrical powerful jets of hot gas, which can produce observable emission lines (Fig. 8.20). Needless-to-say, accretion disks around black holes are not well understood, being governed by the complexities of magnetohydrodynamics in general-relativistic settings. Calculations suggest that a thin disk of very hot

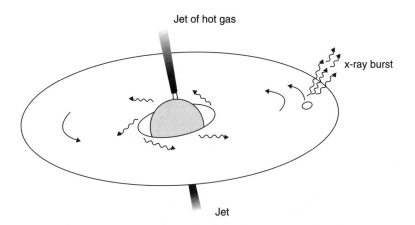

Figure 8.20: Emission of high-energy particles and radiation by an accretion disk around a black hole.

gas is prone to local thermal instabilities. 'Bubbles' form from time to time at distances of several r_g from the center, make several complete revolutions at a velocity approaching that of light, in millisecond periods, then dissolve. To a distant observer, they appear as highly variable bursts of nearly periodic fluctuations in the X-ray flux on millisecond time scales. The detection of these flux variations would provide strong evidence for the presence of a black hole and give the astronomer a good chance to measure the hole's rotation.

So, how would you go about searching for stellar black holes? First, you would look for strong X-ray binary sources that are neither periodic nor recurrent. Then you would examine all the data available from spectroscopic observations of the companion star, assuming it to be optical; they may include the orbital period, amplitude of periodic Doppler shifts that characterize the spectroscopic lines of binary systems, reddening of light by interstellar dust, effective temperature, and brightness. These parameters will allow you to determine a range of values for the mass of the compact star. If the lowest value turns out to be well above 3 $M_\odot$, a conservative estimate for the upper limit of the neutron-star mass, then you might have a black hole. And you would have a stronger case if you can show that the X-ray flux exhibits rapid variations, or the spectrum contains emission lines, which are the signature of ultra-relativistic jets.

The strongest existing case for a stellar black hole is Cygnus X-1, which has been intensively studied ever since its discovery in 1965. It is a powerful X-ray source that varies so rapidly that it has to be a compact object. Its companion is a hot blue giant with a surface temperature of 31 000 K and a mass in the 24–42 $M_\odot$ range (most probably 33 $M_\odot$). Spectral analyses yield an orbital period of 5.6 days and orbital size of 30 million km. From the star's apparent magnitude, corrected by interstellar reddening, its distance should be 2.3 kpc. The absence of eclipses means that the inclination angle between the orbital plane and the direction of observation exceeds 55°. A careful analysis of these data sets a lower limit of 7 $M_\odot$ and assigns a most probable mass of 16 $M_\odot$ for Cyg X-1. Since this is well above the maximum allowed mass of neutron stars, Cyg X-1 must be a black hole.

We now know of many black-hole candidates in binary systems and several isolated black-hole candidates with masses in the 3–30 $M_\odot$ range.[24]

As already mentioned, the theory of gravitational collapse sets no limits on the final masses, and one could imagine black holes of a thousand, a million or even several billion solar masses. In fact, it has been known for some time that supermassive black holes of 10^6–10^9 $M_\odot$ exist in the centers of many galaxies, including our own. They could have developed by condensation of lumps in the early universe, or by gravitational collapse of a star-rich globular cluster, or gradual growth of a seed black hole in a very dense and confined environment. Black holes in the 400–10^5 $M_\odot$ mass range have been found in galactic cores or star clusters; they could act as the seeds for future supermassive black holes.

Our understanding of black hole astrophysics has progressed at a rapid pace, matched by parallel technological advances, which allow construction of large X-ray telescopes, more sensitive gamma-ray detectors and very long baseline radio observatories capable of excellent angular resolution. These technologies will be harnessed in a network of ground-based observatories under construction and a new array of orbiting astronomical satellites, which can make detailed studies of black holes over a large energy range. They will give astronomers tools of unprecedented sophistication to study not only electromagnetic radiation, but also gravitational waves, which are the unambiguous signature of black holes.

8.6.4 Summary

When massive stars become so compact and the gravitational force of gravity becomes so strong that no physical force can oppose total collapse, the final stationary state is a black hole. It is 'black' because light sent out from a limiting surface, called the *static limit*, will take an infinite time to reach a distant observer and will be infinitely redshifted when received. It is a 'hole' because light or matter falling within another surface, called the *event horizon*, cannot escape outside and, once inside, will be inexorably pulled to a central singularity, a point where the gravitational force is infinitely strong.

Astronomers have observed many compact objects in the 3–30 $M_\odot$ mass range thought to be the final black-hole states of massive stars. But there are also compact, supermassive conglomerates in the 400–10^5 $M_\odot$, and even 10^6–10^9 $M_\odot$, mass range. These are thought to have developed by condensation of lumps in the early universe, or by gravitational collapse of star-rich globular clusters, or gradual growth of seed black holes in a very dense and confined environment.

---

[24]The website http://www.johnstonarchive.net/relativity/bhtable.html gives complete lists of black-hole candidates.

8.7   Further Reading

Books

- Charles Misner, Kip Thorne and John Wheeler, *Gravitation* (W. H. Freeman, San Francisco, 1973).
- Stuart Shapiro and Saul Teukolsky, *Black Holes, White Dwarfs and Neutron Stars* (John Wiley and Sons, New York, 1983).
- Edwin Taylor and John Wheeler, *Exploring Black Holes: Introduction to General Relativity* (Addison Wesley Longman, San Francisco, 2000).
- Stephen Hawking, *A Brief History of Time* (Bantam Books, New York, 1988).
- K.S. Thorne, *Black Holes and Time Warps: Einstein's Outrageous Legacy* (Norton, New York, 1994).

The sun and other normal stars

- John Bahcall, M.H. Pinsonneault and S. Basu, *Solar Models: Current Epoch and Time Dependences, Neutrinos and Helioseismological Properties*, Astrophys. J. **555** (2001) 990–1012.
- John Bahcall, Frank Calaprice, Arthur MacDonald and Yoji Totsuka, *Solar Neutrino Experiments: The Next Generation*, Physics Today (July 1996), pp. 30–36.
- Hans Bethe, *Supernovae*, Physics Today (September 1990), pp. 24–27.
- Virginia Tremble, *1987A: The Greatest Supernova since Kepler*, Rev. Mod. Phys. **60** (1988) 859–871.

White Dwarfs, Neutron Stars and Black Holes

- J.L. Provencal, H. Shipman, E. Hog and P. Thejll, *Testing the White Dwarf Mass–Radius Relation with Hipparcos*, Astrophys. J. **494** (1998) 759–762.
- Lars Bilden and Tod Strohmayer, *New View of Neutron Stars*, Physics Today (February 1999), pp. 40–46.
- Roger Blandford and Neil Gehrels, *Revisiting the Black Holes*, Physics Today (June 1999), 40–45.
- F. Melia and H. Falcke, *The Supermassive Black Hole at the Galactic Center*, Ann. Rev. Astron. Astrophys. **39** (2001) 309–352.

Web-sites

- http://www.stsci.edu/resources/ (*Space Telescope Science Inst.*).
- http://imagine.gsfc.nasa.gov/ (*Imagine the Universe*).
- http://heasarc.gsfc.nasa.gov/ (*Astrophysics Archive Resource Center*).
- http://math.ucr.edu/home/baez/relativity.html (*Links to selected websites on relativity*).

8.8 Problems

8.1   Given the temperatures $T = 300$ K (earth), 5800 K (sun), and 10 000 K (class A star), find the corresponding dominant wavelengths, or colors, of the radiations.

8.2   (a) Find the luminosities of the earth and the sun using the following data. For the earth: radius $R = 6.4 \times 10^6$ m, temperature $T = 300$ K; for the sun: radius $R = 7 \times 10^8$ m, temperature $T = 5780$ K. (b) Sirius A has luminosity $L = 23\ L_\odot$ and radius $R = 1.8\ R_\odot$. Calculate its temperature.

8.3   (a) Calculate the apparent luminosity of the sun when you know its absolute luminosity and the earth–sun distance. (b) Given the absolute luminosity $L = 125\ L_\odot$ and the apparent luminosity $\mathcal{L} = 0.022\ L_\odot/(\mathrm{pc})^2$ for Aldebaran A, what is its distance from us?

8.4   (a) How does L change with color for stars with the same radius? (b) Arcturus and δ Orionis have about the same radius ($20\ R_\odot$). What is the color of their radiation if we know that $L = 115\ L_\odot$ for the first star and $L = 10^4\ L_\odot$ for the second.

8.5   (a) On the main sequence, a star has luminosity $L \propto M^a$, where $a = 3$–4. Show how its lifetime on the MS varies with mass. (b) A star on the main sequence has radius $R \propto M^b$, where $b = 0.6$–1. Show how its effective temperature varies with mass.

8.6   Find the energy released in the following reactions:

(a) $4\mathrm{p} \to {}^4\mathrm{He} + 2\mathrm{e}^+ + 2\nu$;
(b) ${}^8\mathrm{B} \to {}^8\mathrm{Be}^* + \mathrm{e}^+ + \nu$, where ${}^8\mathrm{Be}^*$ is a nuclear excited state at 3.95 MeV;
(c) ${}^7\mathrm{Be} + \mathrm{e}^- \to {}^7\mathrm{Li} + \nu$.

Use the following *atomic masses*: 1.007825 u (${}^1\mathrm{H}$), 4.0026 u (${}^4\mathrm{He}$), 8.0246 u (${}^8\mathrm{B}$), 8.0053 u (${}^8\mathrm{Be}$), 7.0169 u (${}^7\mathrm{Be}$), and 7.0160 u (${}^7\mathrm{Li}$), where the mass unit is 1 u $= 931.502$ MeV.

8.7   Prove that the gravitational pressure at the center of a star of mass M and radius R is given approximately by $P_\mathrm{g} = GM^2/R^4$.

8.8   The pressure of a gas is conventionally defined as the momentum transfer per unit time per unit area. It is given by $P = nvp$, where n is the particle-number density, v and p are the velocity and momentum of a particle. Now you can obtain P for a degenerate gas in terms of n only by using the fact that the nearest distance between two particles in a degenerate gas is $\Delta x \propto n^{-1/3}$ and the uncertainty relation $\Delta x\, \Delta p_x > h$. You should have $P_\mathrm{e} \propto h^2 n_\mathrm{e}^{5/3}/m_\mathrm{e}$ for a degenerate gas. From the equilibrium requirement, $P_\mathrm{g} = P_\mathrm{e}$, derive the mass–radius relation for white dwarfs.

8.9   Prove that for ultra-relativistic electrons ($v = c$), the degenerate pressure is $P_\mathrm{e} = hcn^{4/3}$. From this, derive the Chandrasekhar mass limit.

8.10 White dwarfs can still draw their thermal energy from its hot ions. They radiate at the rate given by $L = 0.2(M/M_\odot)T^{3.5}$ W, where T is the uniform internal temperature. What is the cooling time for a solar-mass white dwarf at temperature T?

8.11 (a) A measure of the strength of the gravitational field of a star is its *surface potential*, defined as $\gamma = GM/Rc^2$, a dimensionless parameter. Taking $M = M_\odot$, calculate γ for $R = R_\odot$, $0.01\ R_\odot$, 10 km, and 3 km. (b) Alternatively, one may also define the *surface gravity* of a star (the Newtonian gravitational acceleration at the surface) by $g = GM/R^2$. Calculate g for the values of R given in (a) and compare them with the surface gravity on earth, $g_E = 9.81$ m/s^2.

8.12 The condition for a gas of electrons (or neutrons) to be degenerate is $\varepsilon_F > kT$, where $\varepsilon_F = 4.78\hbar^2 n^{2/3}/m$ is the Fermi energy of a particle (see Appendix C), m its mass; the gas has number density n and temperature T. Find the critical temperature at which the gas becomes degenerate for the cases: (a) Electron gas in metals: $n = 10^{27}$/m^3; (b) Electron gas in white dwarfs: $n = 10^{36}$/m^3; (c) Neutron gas in neutron stars: $n = 10^{44}$/m^3. The mass energy of an electron is $mc^2 = 0.5$ MeV; that of a neutron, $mc^2 = 1000$ MeV.

8.13 Assume that a compact star accretes matter at the rate of $10^{-10}\ M_\odot$ per year. Give this rate in watts. Assume now that all this is radiated back into space as black-body radiation. Use the relation $L \propto R^2 T^4$ to calculate the effective temperature of a neutron star and, from this, its color (wavelength).

8.14 When a proton hits the surface of a neutron star and is absorbed by it, its kinetic energy is converted into gravitational binding energy. Calculate the rate of this conversion.

8.15 How would you measure the mass, spin, and charge of a black hole?

8.16 The gravitational tension exerted by a black hole of mass M on a body at distance r is given by the formula $T = 10^8$ MPa $(M/M_\odot)(r/1\ \text{km})^{-3}$. The human body can withstand a tension less than 10 MPa. Find the distance r at which a human explorer could no longer survive by gravitational pressure from black holes of masses 1, 100, and $10^9\ M_\odot$.

8.17 A Kerr black hole, with a mass equal initially to its irreducible mass M_{irr}, is spun up to its maximum angular momentum given by $J = M^2$, where M is the total mass it has acquired then. (a) Show that $M = \sqrt{2}M_{\text{irr}}$. (b) What is the energy gain?

8.18 (a) Calculate the thermal entropy of the sun, assuming it consists entirely of ionized hydrogen. Suggestion: use $S = E/T$, where E is the total thermal energy. (b) Calculate the quantum-mechanical entropy of a Schwarzschild black hole of one solar mass.

Elementary Particles and Forces

9

9.1 Elementary Constituents of Matter

Ever since the dawn of civilization, one of the most cherished ambitions of human-kind has been to establish order and regularity in our complex surroundings. The ancient Greeks thought that all matter was made up of four elements: air, fire, water, and earth. The ancient Chinese had similar ideas, though they insisted that there were five elements: metal, wood, water, fire, and earth, and they named the five classical planets accordingly. Years of work in chemistry showed that both of these views are, to say the least, somewhat oversimplified. Matter is now known to consist of some one hundred different elements, instead of just four or five. In 1808, John Dalton proposed that there is a most elementary constituent within each of these elements, which is itself unalterable and indestructible. He called it the *atom*. Further progress came in the middle of the 19th century when Dmitry Mendeleyev discovered the regularity of chemical elements in his famous *periodic table*, though the reason behind these regularities was not understood until the end of the first quarter of the 20th century, when quantum mechanics was established.

Ironically, though by no means uncommon in the annals of scientific discoveries, this understanding of the periodic table began with the destruction of the cherished atomic concept of Dalton. *Electrons* were discovered by J. J. Thomson in 1897, and the atomic *nucleus* was discovered by Ernest Rutherford in 1911. These discoveries led to the planetary model of atoms, in which atoms are made out of electrons circu-lating about an atomic nucleus, like planets revolving around the sun. In this picture atoms are no longer the most elementary entities as Dalton envisaged. Electrons and nuclei now share that honor. Yet this destruction of Dalton's atomic concept actually opens the door to an understanding of the periodic table. With electrons we can relate different chemical elements. Elements are different merely because they have a different number of electrons in the atom. The similarity of chemical properties is now translated into the similarity of electronic arrangements, and the latter can be understood from the laws of quantum mechanics (see Appendix B for a discussion of quantum mechanics).

A chemical element may have several *isotopes*. They have identical chemical properties, the same number of electrons in each atom, but they differ from one another in the mass of their nuclei. This simply suggests that it is the electrons of an atom, and not the nuclei, that are responsible for its chemical properties. But why should there be different nuclei for the same element? The mystery was finally solved when James Chadwick discovered the *neutron* in 1932. This is an electrically neutral particle that resides in the nucleus. The other particle in the nucleus is the *proton* which carries one unit of positive charge and is nothing but the nucleus of the hydrogen atom. All nuclei are made up of protons and neutrons, and these two particles are collectively referred to as *nucleons*.

The charge of the electron is equal and opposite to that of the proton, so it carries one unit of negative charge. Now, a given chemical element has a definite number of electrons. Since atoms are electrically neutral, there must be an equal number of protons as electrons in each atom. However, this argument does not limit the number of neutrons in the nucleus because neutrons are electrically neutral. Atoms which differ only by the number of neutrons in their nuclei are isotopes of one another. This is why a chemical element may have several isotopes.

From this reasoning one might conclude that one could obtain an infinite number of isotopes just by adding more and more neutrons. This turns out to be false because a nucleus with too many neutrons becomes unstable by undergoing *beta decay*. We will come back later to discuss what that is.

Now the laurel of elementarity is passed on to the electrons, protons, and neutrons.

Since the 1930's many other new particles have been discovered in cosmic rays, or produced and detected in experiments using high energy accelerators. (A complete listing of them can be found on the Particle Data Group website http://pdg.lbl.gov/.) Most of these are *hadrons*, a name reserved for particles which respond to nuclear forces. Protons and neutrons are definitely hadrons, for what holds them together in a nucleus is by definition the *nuclear force*. In contrast, electrons are not hadrons; what keeps them held to the nucleus in an atom are the *electric and magnetic forces*.

With the presence of tens of hadrons, the stage is set for another shakedown. There were many hadrons and they all seemed to be unrelated. It was just like the old days with many unrelated chemical elements before the discovery by Mendeleyev. For that reason the rush was on to find a regularity, or if you like, some sort of a periodic table for them. Murray Gell-Mann and Yuval Ne'eman finally succeeded in the early 1960's, and the regularity goes under the name '*Eightfold Way*,' or more technically, an $SU(3)$ symmetry. Just like Mendeleyev's periodic table, which opened the way to the discovery of electrons, protons, and neutrons, and the inner structure of an atom, the Eightfold Way led to the discovery of an inner structure of the hadrons. In 1964, Murray Gell-Mann and George Zweig proposed that protons and neutrons, and all the other hadrons, were themselves made out of three fundamental particles called *quarks*. The electron, however, is not a hadron and is still believed to be elementary.

All of this, so to speak, is ancient history. Nevertheless, it serves to illustrate how discoveries are made through an intimate interplay between theory and experiment. Puzzles lead to new ideas, and regularities lead to new structures. This is the same in all fields. What distinguishes this field of elementary particles from others, is the small size of the elementary constituents. As the substructures become smaller and smaller, we need a larger and larger 'microscope' to 'see' them. The only difference is that these large 'microscopes' are called *accelerators*, and they occupy miles and miles of land. An optical microscope yields to an electron microscope when higher resolution is called for, because the beam of electrons in the latter has a smaller wavelength than the beam of light in the former, and better resolution can only be achieved by an illuminating beam with a smaller wavelength.

It is a fundamental fact of the quantum nature of matter that all particles in the microscopic world behave like waves. Moreover, the wavelength of a particle gets shorter as its energy gets higher. This is where an accelerator comes in: to produce a higher energy beam with enough resolution to probe the tiny world of elementary particles.

One never sees a cell directly from a beam of light without the complicated array of lenses in a microscope. Similarly, one cannot see what is illuminated by the beam emerging from an accelerator without a complicated bank of detectors and computers. What is different between the two, however, is the destructive nature of the latter. One can illuminate a sample under an ordinary microscope without destroying it, but it is virtually impossible to have a very energetic beam impinging on a target particle without shattering it to pieces. This complication adds substantially to the difficulty of reconstructing the original image before it broke apart, but the other side of the coin is that such pulverization and disassembly allows us to look into the guts of the target particle. With the help of theory, the reconstruction can be achieved.

In what follows, we will adopt a pedagogical approach. We will skip all these reconstructions, and hence most of the direct experimental data, to concentrate on the physics that finally emerges. We will make free use of the license of hindsight to discuss the beauty and the intricacies of the physics in as logical a manner as possible without mathematics. Let us be forewarned however that this approach also contains an inherent danger of misconception that physics is purely a logical and theoretical science. It is definitely not; experiments are needed to confirm a conjectural theory and to guide our thinking in future developments. The final judge as to whether something is correct or not is experiment.

Accordingly, we will now skip over the intervening history in the second half of the 20th century, and jump directly to a summary of the elementary constituents of matter as we know them today.

These constituents fall into two categories: *quarks* and *leptons*. They are distinct because quarks experience nuclear forces but leptons do not.

There are three *generations* (sometimes known as *families*) each of quarks and leptons. Constituents in different generations are almost clones of each other, except that those in higher generations have larger masses. Constituents of the first generation make up the matter we are familiar with. Constituents of the second and third generations are unstable and cannot be found naturally on earth, but they can be produced by high energy collisions of ordinary matter.

Recent experiments show that there are probably no more than three generations of matter. But why three? Why do we even have a second and a third generation anyway? This nobody really knows. When the *muon*, a second generation lepton, was discovered in 1936, it was so astonishing, and so uncalled for, that it prompted I. I. Rabi to utter his famous remark: 'Who ordered that?' Almost seventy years later, we still do not know the answer.

We will now proceed to give a more detailed description of these constituent particles.

There are two quarks and two leptons in each generation. The quarks of the first generation are called the *up quark* (u) and the *down quark* (d), and the leptons of the first generation are the *electron* ($e = e^-$) and the *electron-neutrino* (ν_e). Symbols in the parentheses are used to designate the particles; a superscript, when it is present, indicates the electric charge carried by that particle.

The quarks of the second generation are the *charmed quark* (c) and the *strange quark* (s), and the leptons are the *muon* (μ^-) and the *muon-neutrino* (ν_μ). Finally, the quarks of the third generation are the *top quark* (t) and the *bottom quark* (b) (sometimes also known, respectively, as the *truth* and the *beauty* quarks), and the leptons are the *tau* (τ^-) and the *tau-neutrino* (ν_τ). All these particles have been experimentally observed.

If the electric charge carried by an electron e is counted as -1, then the electric charges of μ and τ are also -1, but those for the neutrinos ν_e, ν_μ, and ν_τ are zero (hence the name neutrino, which means 'little neutral one' in Italian). The electric charges of u, c, and t are $+\frac{2}{3}$, and the electric charges of d, s, and b are $-\frac{1}{3}$.

The proton (p), which has a unit ($+1$) of electric charge, is made up of two up quarks and a down quark. The neutron (n), which is electrically neutral, is composed of two down quarks and an up quark: $p = uud$, $n = udd$. Other hadrons are made up of three quarks, called *baryons*, or a quark and an antiquark, called *mesons* (see the following for a discussion of antiquarks).

There is an asymmetry between quarks and leptons in that leptons have integral units of electric charges while the electric charges of quarks are multiples of $\frac{1}{3}$. This factor of 3 is actually accompanied and 'compensated' by another factor of 3: each quark comes in three *colors*. We will explain more fully what 'color' is in Secs. 9.4.3 and 9.5.4, but whatever it is, it has nothing to do with the color we see with our own eyes. It is just a name invented to label the different varieties that exist for each of the quarks described above.

To this list of constituents of *matter particles* we must add the *anti-matter particles*, or simply *antiparticles*, for each of them. Antiparticles will be denoted by a bar on top of the symbol for the particle. For example, the antiparticle for the u quark is called the anti-u quark and is denoted by $\bar{u}$. The only exception to this notation is the antiparticle of the electron, called the *positron*, which is usually denoted as e^+. Similarly, the antiparticles of μ^- and τ^- are denoted by μ^+ and τ^+, respectively. Antiparticles have the same mass but opposite electric charge as the particle. They also have all the opposite additive quantum numbers (see Sec. 9.4). When they encounter particles of the same kind, they annihilate each other. Since the earth is made out of particles (by definition), no antiparticles can live for long in our vicinity, so all the antiparticles observed to date are produced either by cosmic ray bombardments or by high energy accelerators.

The existence of antiparticles for matter was first predicted theoretically in 1929 by Paul Dirac long before any of them were found experimentally. This prediction was based on the theory of special relativity and quantum mechanics, and represents one of the greatest triumphs of theoretical physics in the twentieth century. Nowadays the existence of antiparticles is very firmly established experimentally.

Quarks and leptons are known as *fermions*, as they obey Fermi–Dirac statistics and the Pauli exclusion principle, forbidding two of them to occupy the same quantum state.

Beyond the quarks and leptons, present theory (called the '*Standard Model*,' or SM for short) would also like to have — for technical reasons to be discussed in Sec. 9.5.3 — another electrically neutral particle, called the *Higgs boson* (H^0). An intense search for the Higgs is underway, but as yet it has not been found. In fact, we are not completely sure that it has to exist as an elementary (rather than a composite) particle, and we do not know whether there should not be more than one of them. The form and shape of the future theory (which is more fundamental than the existing Standard Model) will depend quite critically on such details, which is why it is such an important matter to discover the Higgs particle or to rule it out experimentally.

The Higgs boson is certainly not a matter particle in the usual sense. Matter generally cannot disappear unless it is annihilated by its own antiparticle. In contrast, H^0 can be created and absorbed singly.

The Standard Model also requires a set of *gauge particles* to transmit the forces (see Secs. 9.2 and 9.3). These particles are again *not* indestructible; in fact, forces are transmitted only because these particles can be freely absorbed and created (see Sec. 3). There are a number of gauge particles, each responsible for transmitting a different kind of fundamental force. As we shall see in Sec. 2, the fundamental forces are the *strong* or nuclear force, the *electroweak* (electromagnetic and weak) force, and the *gravitational* force. The corresponding gauge particles are the *gluons* (g) for the strong force, the *photons* (γ), $W^\pm$ and Z^0 particles for the electroweak force, and the *graviton* (G) for the gravitational force. With the exception of the graviton, all the other gauge particles have already been found.

Gauge and Higgs particles are *bosons*. They obey Bose–Einstein statistics and they always carry integer spins.

Table 9.1: Particles in the Standard Model.

Matter/anti-matter particles				Gauge particles		
Quarks		Leptons		Strong	Electroweak	Gravitational
(u,d)	$(\bar{d},\bar{u})$	(ν_e,e^-)	$(e^+,\bar{\nu}_e)$	g	$\gamma,\, W^{\pm},\, Z^0$	G
(c,s)	$(\bar{s},\bar{c})$	(ν_μ,μ^-)	$(\mu^+,\bar{\nu}_\mu)$		**Higgs particle**	
(t,b)	$(\bar{b},\bar{t})$	(ν_τ,τ^-)	$(\tau^+,\bar{\nu}_\tau)$		H^0	

All in all, there are several kinds of elementary particles. Fermions are responsible for giving us the material universe (quarks and leptons), and bosons are used to transmit the forces (gauge particles) and provide symmetry breaking (the Higgs particle — see Sec. 9.5.3). These particles are summarized in Table 9.1.

9.1.1 Summary

Matter is composed of two kinds of particles, quarks and leptons. Quarks experience the nuclear force but leptons do not. Each quark has three possible colors while leptons are colorless. Quarks and leptons come in three generations each. The first generation quarks are the (u,d), the second generation quarks are the (c,s), and the third generation quarks are the (t,b). The corresponding three generations of leptons are (e^-,ν_e), (μ^-,ν_μ), and (τ^-,ν_τ).

For each quark or lepton there exists an antiparticle with the same mass and opposite electric charge.

On top of these, there may also be one or several Higgs bosons.

Forces are mediated by the exchange of gauge particles. The gauge particles responsible for strong, electromagnetic, weak, and gravitational forces are respectively g, γ, $(W^{\pm},Z^0)$, and G.

9.2 Fundamental Forces

Of equal importance to the identity of elementary particles is the nature of fundamental forces. What holds protons and neutrons together to form a nucleus? What binds electrons and a nucleus together into an atom, and atoms together into a molecule? For that matter, what holds molecules together in ourselves, and what holds us on the surface of this planet? Can we study the forces of Nature and classify them systematically?

At first glance this seems to be quite an impossible task. There are simply too many varieties of forces. After all, you can tie things together with a rope, hold them together with your hands, or bind them together with crazy glue. All these correspond to different forces. How can we possibly classify them all?

Fortunately, things are not so bleak! We could have said the same thing about the classification of matter, for certainly there is a great variety there also. But we now know that all bulk matter is made up of electrons and quarks. Similarly, it turns out that all these complicated forces are nothing but very complex manifestations of four fundamental forces! In order of decreasing strength, these are: the *nuclear force*, the *electromagnetic force*, the *weak force*, and the *gravitational force*.

Before we go into the details of these four fundamental forces, it might be interesting to reflect on what kind of a world we would live in should these forces be very different from what they are.

If the electromagnetic force holding the electrons and the nucleus together had a short range and was very weak, then there could be no atoms and no molecules, because electrons could no longer be bound to the nuclei. The whole world would be a soup of negatively charged electrons and positively charged nuclei, in what is known as a *plasma*. Needless to say we would no longer be present. Now think of the other extreme: if the electric force were much stronger than the nuclear force. Then, the electrostatic repulsion between the protons in the nucleus would overcome the nuclear attractive forces between them, and a nucleus containing more than one proton would be broken up. The world would be left with only one chemical element, hydrogen, though hydrogen isotopes could exist and it is conceivable that very large molecules of hydrogen could be formed. In any case this world would also look very different from ours.

At both these extremes, when the electromagnetic force is very weak or very strong, it is almost impossible to *see* anything. In the former case, when the world is a plasma, light is constantly emitted and absorbed by the charged particles in the plasma. This means that light can never travel very far, and we cannot really see clearly like we do in an electrically neutral surrounding when this does not happen. At the other extreme, when the electromagnetic force is very strong, ironically we cannot see either. The world is now neutral, but another effect takes over to block our vision. In the present world, light can travel great distances as long as there is no matter blocking its way. It is of no importance whether there are other light beams traveling in different directions crossing our light beams — they just go through each other. There are no traffic police in the sky, no traffic lights, but there are never collisions between the light beams in the sky. (Otherwise whatever image light brings would be all shattered by the collisions from cross beams and you would never be able to see anything.) Alas, this happy phenomenon would no longer be a reality if the electromagnetic force were very strong. In that case, it would be easy for the light beam to create 'virtually' an electron-positron pair, which is charged and can thus absorb and emit light. In this way an effective plasma would be present in the vacuum, which as we have seen, would severely limit our vision. The difference between this 'virtual' plasma and the real one of the previous case, is that none of the pairs in this virtual plasma can live very long. But as one dies, others are born, and there are always some pairs there to obstruct the passage of the light beam.

If the nuclear force holding the nucleons together were too weak, we would not have complex nuclei. On the other hand, if the nuclear force were very much stronger than it is, nuclei could be much larger and very heavy chemical elements could be formed. We can go on and on with different scenarios. If the gravity were too weak? Then, we would all float away like an astronaut in orbit.

So the kind of world we have around us depends very sensitively on what forces are present and how strong they are. Our mere existence depends on that too. It is therefore intellectually important to understand the origin and the detailed nature of the fundamental forces. We will indeed discuss this very important and fascinating problem, but in later sections. For now, let us first get familiar with these forces.

It is ironic that the first of these four fundamental forces to be discovered is actually the weakest in strength. It is the gravitational force. This discovery was made by Isaac Newton in the seventeenth century while endeavoring to explain Kepler's three laws of planetary motion. He found out that the planetary motions could be explained by postulating a universal attractive force between any two bodies. The force has to be proportional to the product of the masses of the two objects, and inversely proportional to the square of the distance separating them, no matter what the distance is. For this reason this force law is also known as an *inverse-square law*. This force is universal and all pervasive. It exists between any two of us, but we do not feel a force pulling us towards the other person because neither of us is massive enough. In contrast, the earth is heavy and it is this force that holds us to the surface of the earth. Similarly, it is this force that keeps the planets going around the sun, and the satellites going around the earth, like a weight at the end of a string. Without this force they would fly apart in a straight line. This force also exists between two elementary particles, but it is so feeble compared to the rest of the forces present that it can be safely neglected in present-day experiments. However, it may play a very fundamental role in some future theories. We will come back to this aspect in the last section.

Newton's law of gravitation was modified and improved by Albert Einstein in his famous 1915 *theory of general relativity*. The result differs from Newton's law in mathematical details, but this difference produces only very small effects on planetary motions and day-to-day gravitational actions. If you should fall off a ladder, God forbid, Einstein would neither help you nor hurt you any more than Newton.

This is not to say that Einstein's general relativity is trivial or insignificant. On the contrary, it is probably the deepest theory in the 20th century. It turns gravity into a geometrical concept, by interpreting gravity as a revelation of the curvature of spacetime. As such, light must fall under gravity just like particles do, because both of them are subjected to the same curvature. Observations show that light passing near a massive body is indeed pulled towards that body by gravity, just the way Einstein predicted. This property is nowadays used to find dark objects and dark matter in the universe. In fact, wherever there is energy, there is gravity, according to Einstein. From that point of view, light is affected by gravity because it carries energy. The faster a particle is trying to run away from the capture of gravity, the larger is the force so that eventually (this occurs only at unbelievably high energies) it does not win and cannot get away. According to Einstein, gravitational waves can be emitted from an accelerating object just like an accelerating charge can emit electromagnetic waves. This phenomenon has actually been seen in a binary pulsar (see Chapter 8) whose orbit decays because of the loss of energy due to gravitational wave emission.

Quantum gravity is not well understood so we will not discuss it much further. Suffice it to say that it is an important problem to solve, because much of our eventual understanding of particle physics may hinge on our understanding of quantum gravity.

The next fundamental force to be discovered was the *electromagnetic force*. Electricity and magnetism were first thought to be two unrelated forces, but subsequent experimentation showed that they were intimately connected, a connection that was finally codified and formulated into the famous *Maxwell's equations* by James Clerk Maxwell in 1873. Under ordinary circumstances, the electric force is stronger than the magnetic force. Like gravity, the electric force also obeys an inverse-square law, called *Coulomb's law*. The difference is, instead of being proportional to the masses of the two bodies, the electric force is proportional to the product of their electric charges. Since electric charges can be both positive and negative, electric forces can be attractive (between opposite charges) or repulsive (between charges of the same sign). In contrast, masses are always positive and the gravitational forces are always attractive. Antigravity does not exist; even antiparticles have positive masses.

It is this bipolar property of the electric force that allowed the gravitational force to be discovered. Because of this property, there is no electric force between neutral objects, for the electric force on (or from) the positively charged part of the object is canceled by the electric force on (or from) the negatively charged part of the object. If this were not the case, the stronger electric force between the planets and the sun would have masked completely the gravitational force and prevented its discovery by Newton.

The alert reader may have noticed a peculiar feature of these two forces. The gravitational forces are *attractive* between masses of the *same* sign (because masses are necessarily non-negative), while the electrical forces are *repulsive* between charges of the *same* sign. The difference comes about because of the different nature of these two forces, in spite of the fact that both of them obey the inverse-square law. We will come back to this point later in the next section.

The electromagnetic force is the force responsible for the creation of light, radio, and television waves. It is also responsible for the functioning of electronics. It is the force that binds electrons to the nucleus to make an atom, the atoms together to make a molecule, and the molecules or atoms together to form a liquid or a solid. It is the fundamental force, when dressed up in a complicated way by the presence of these atoms and molecules, which gives rise to the strength of a piece of rope, or the adhesion of crazy glue. In short, other than gravity, the electromagnetic force is the only force we encounter in our daily lives.

The remaining two fundamental forces were discovered in the twentieth century. They are the *nuclear* or *strong force*, and the *weak force*. They are not detectable in our daily lives because of their incredibly short ranges. Nuclear forces between two nucleons are effective only when they are within 10^{-15} m (one million-billionth of a meter) of each other. Weak forces have a shorter range still: about 10^{-17} to 10^{-18} m. At distances short compared to their ranges, these two forces also obey the inverse-square law. Beyond these distances, as is the case in our daily lives, these forces can become extremely small.

The great strength of the nuclear force can be seen from the fact that an enormous amount of energy can be derived from a small quantity of fissionable material used in a nuclear power plant or an atomic bomb.

The weak force is responsible for the instability of neutrons, when left alone. Under the influence of this force, an *isolated* neutron is torn into a proton, an electron, an anti-electron-neutrino ($n \to p + e + \bar{\nu}_e$) within a time of some fifteen minutes or so, in a process known as *radioactive beta decay*, (or *β-decay*). Under the influence of the strong or the electromagnetic force, some hadrons also become unstable, but their lifetimes are typically many orders of magnitude shorter than the neutron lifetime. This shows that the strength of the force responsible for beta decay is weak compared to the other two—hence the name 'weak force.'

Neutrons in a nucleus are protected by the nuclear and the electromagnetic forces, so they may remain stable if there are not too many of them. If there are too many, then such protection would not be sufficient for all of them to remain secure, and the nucleus would undergo a radioactive beta decay. If you have seen a watch with glowing dials, you have probably seen beta decay at work. A radioactive material is painted on the dial, whose emitted electron excites the nearby fluorescent material to produce light.

As a result of the beta decay, a neutron inside the nucleus is changed into a proton, while the produced electron and the anti-electron-neutrino leave the nucleus. The new nucleus has one more proton than the old nucleus, and soon it will capture an electron somewhere to neutralize it. Since the chemical nature of an element is determined by the number of electrons or protons, a chemical element undergoing a beta decay changes into another chemical element. With one more proton, the additional electrostatic repulsion causes the new nucleus to have a higher energy, so *β*-decay of the original nucleus can take place only if the energy released from the neutron *β* decay, about 0.8 MeV, is sufficient to supply this energy difference. Otherwise the original nucleus will remain stable in spite of the tendency of its neutron to decay.

When such a chemical change was discovered by Rutherford at the beginning of the 20th century, it amounted to heresy, especially among the chemists. It seemed like modern alchemy; it certainly violated the cherished 'unalterable and indestructible' doctrine of Dalton's atomic theory. For this reason when Rutherford gave a talk on this discovery at the Physical Society of McGill, he was severely attacked and ridiculed by his chemistry colleague Frederick Soddy. Eventually Soddy became a convert and a close collaborator of Rutherford, and both Rutherford and Soddy went on to win separate Nobel Prizes in chemistry (1908 and 1921 respectively).

The strong force is the strongest of the four forces. As remarked above, the force between two hadrons has a range of the order of 10^{-15} m. The force between quarks, however, appears to be quite peculiar. At short distances, the force obeys the inverse-square law like everything else. From a distance of the order of 10^{-15} m onwards, the force is constant, and quite independent of the separation of the two quarks. That means no matter how far apart these two quarks are, there is always a constant force to pull them back. It also means that the work done and the energy required to pull the two quarks apart is proportional to their separation, hence an infinite amount of energy is needed to separate and isolate them. As a result, two quarks cannot be isolated from each other, a property that has come to be known as *confinement*. Note that hadrons such as nucleons are made out of quarks, and there is no doubt that hadrons can be isolated. So a group of quarks is sometimes confined, sometimes not. We will come back in Sec. 9.5 to discuss which is which. We should also note here that the idea of quark confinement is actually forced upon us by the failure to find *isolated* quarks in spite of an intense search.

9.2.1 Summary

All forces are complex manifestations of four fundamental forces. In order of decreasing strength, these are the nuclear (or strong) force, the electromagnetic force, the weak force, and the gravitational force. The range of the gravitational and electromagnetic forces is long, and the range of the nuclear and weak forces is short.

9.3 Theory of Forces

We saw in the previous section that our mere existence, as well as the character of the whole universe, depend critically on the range, strength, and various other details of the fundamental forces. The electromagnetic and gravitational forces have long ranges and they obey the inverse-square law, but the weak and the nuclear forces have only short ranges. What makes them different, and why are some of the forces attractive and others repulsive?

These questions are unanswerable in classical physics, nor in non-relativistic quantum mechanics of atoms and molecules. A force is just there. One can measure it to determine what it is, and one can ask how it affects the motion of particles. But one has absolutely no idea where it comes from, and why it is in a particular form.

However, when special relativity is incorporated into quantum mechanics, these questions become answerable in the resulting *quantum field theory*. On the one hand, Einstein's special relativity tells us that a particle with mass m has an energy $E = mc^2$ at rest, and a larger one in motion. Here $c = 3 \times 10^8$ m/s is the speed of light. If that amount of energy is made available, a particle of that mass can be created as long as no conservation law is violated. On the other hand, Heisenberg's uncertainty relation in quantum mechanics (see Appendix B) tells us that the energy of a closed system may fluctuate by an amount $\Delta E \sim \hbar/\Delta t$ during a time interval Δt, where $\hbar = 6.6 \times 10^{-25}$ GeV s is Planck's constant, and 1 GeV $= 10^9$ eV $= 0.16 \times 10^{-9}$ J is the energy gained by an electron crossing a billion (10^9) volts. Thus, at short time intervals, enough energies are available to create one or more particles. A particle of mass m thus created can last only a time interval $\Delta t \sim \hbar/mc^2$ before it is absorbed or annihilated, or else the uncertainty principle is violated. For that reason such a particle is known as a *virtual particle*.

A virtual particle created by particle A at one location and annihilated by particle B at another transmits a force between A and B. This is how force arises in a relativistic quantum field theory. The virtual particle which transmits the force is sometimes known as the *exchange particle*.

9.3.1   *Range and Mass*

Even moving at the speed of light, the virtual particle can travel only a distance $\Delta x \sim c\Delta t \sim \hbar/mc$ before it is absorbed, so the *range* of the force is of the order $\hbar c/mc^2$. This argument, invented by Hideki Yukawa in 1935, was used by him to predict the existence of a new particle, nowadays called the *pion* (π).

It was already known at that time that the range of the nuclear force is about 1 fm $= 10^{-15}$ m. Using the value $\hbar c = 0.197$ GeV fm, the rest energy mc^2 of the virtual particle is approximately $\hbar c/1$ fm $= 0.197$ GeV. After the discovery of the pion (whose rest energy is actually 0.14 GeV), Yukawa was awarded a Nobel Prize.

We know now that the pion is not an elementary particle, but that does not affect its ability to transmit a force. There are actually three kinds of pions, π^+, π^0, π^-. Their quark compositions are $\pi^+ = u\bar{d}$, $\pi^- = d\bar{u}$, and π^0 is a combination of $u\bar{u}$ and $d\bar{d}$.

Nuclear forces can also be transmitted by heavier mesons. They give rise to forces with a shorter range.

If there is no other particle nearby, the virtual particle created must be re-absorbed by A. Consequently every particle A is surrounded by a *Yukawa cloud* of virtual particles, whose extent is $\sim \hbar c/mc^2$. The density of the cloud depends on how often the virtual particle is created, which in turn depends on the strength of the local interaction. If the cloud is relatively dense, as will be the case in strong interactions, then the effective size of the particle A will be $\sim \hbar c/mc^2$, which is of the order of 1 fm. If the cloud is thin, as is the case for all the other interactions, the cloud is hardly noticeable and the effective size of particle A is not much increased by the presence of the cloud.

The virtual particle transmitting the electromagnetic force is the photon (γ), and the virtual particle transmitting the gravitational force is the graviton (G). Both are massless so the range of these two forces is infinite.

Weak forces come in two varieties, transmitted respectively by $W^\pm$ and Z^0. They carry the name of *charge-current interactions* and *neutral-current interactions*, respectively. The mass of $W^\pm$ and Z^0 are respectively 80 and 91 GeV/c^2. Taking a round number of 100 GeV/c^2, the range of the weak forces comes out to be about 2×10^{-18} m.

We talked about strong forces between two nucleons, by exchanging pions and other heavier mesons. However, these particles are made out of quarks and anti-quarks. How do quarks themselves interact?

They interact by exchanging gluons (g). As mentioned before, quarks and gluons are strange objects that must be confined inside hadrons. As a result, they are not free to propagate, so the Yukawa argument is no longer valid. Nevertheless, as we shall see, the force is independent of distance, so it is surely long-ranged. Since nucleons are made out of quarks, and the nuclear forces have a range of 1 fm, how can that happen? A look at the more familiar atomic world will give us the answer.

The force between neutral atoms, the so-called *van der Waals force*, also has a range much shorter than the range of the Coulomb forces between electrons and/or protons. This is because an electron in one atom sees both a repulsive force from the electrons of another atom, and an attractive force from the protons of the other atom. These two forces are almost equal and opposite and they tend to cancel each other, especially when the separation between the two atoms is much greater than the size of the atoms. This is why the van der Waals force has a range much shorter than the Coulomb force. The short range nuclear force is presumably also a result of cancelation between the long-range forces from the different quarks inside a nucleon. In fact, the cancelation must be so complete that the range of the residue force must have a range shorter than the pion-force range. In that way, the longest range nuclear force between nucleons is governed by the exchange of pions but not gluons.

9.3.2 *Inverse-Square Law versus Confinement*

A force tapers off to zero beyond its range. What is it going to be at a distance r small compared to its range? We asserted in previous discussions that an inverse-square law holds at these short distances. Why?

At a distance well within the range of the virtual particle, the mass of the virtual particle does not matter. The virtual particle simply spreads out from its source like a ripple. At a distance r from the emitting particle A, the surface area of a sphere is $4\pi r^2$ (see Fig. 9.1a), and the exchange particle can be anywhere on this surface.

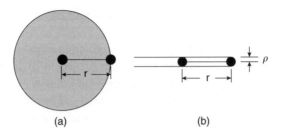

Figure 9.1:   (a) A sphere with radius r, illustrating the origin of the inverse-square law; (b) a cylinder of radius ρ, illustrating the origin of confinement.

The probability of its being along some fixed direction is therefore proportional to $1/r^2$, which is then the origin of the inverse-square law.

This argument may still be valid even if the exchange particle is not emitted isotropically. For example, if it is emitted uniformly within a $30°$ angle, then the area of the surface within this $30°$ cone would still increase like r^2, and the force would still decrease like $1/r^2$. If the exchange particle carries a spin (see Sec. 9.3), anisotropy may indeed occur, but that will not affect the inverse-square law at short distances for those forces.

The argument, however, fails if the emission is confined to a cylinder of constant radius ρ, as illustrated in Fig. 9.1b. For the sake of definiteness let us assume $\rho \ll R$, the range of the force. At a distance $r \ll \rho$, the same argument as before leads to an inverse-square law. For $\rho \ll r \ll R$, the story is different. The exchange particle is now guided down the cylinder, without ever having the chance to spread out into a cone. The cross-sectional area down the cylinder is constant, so the force also remains independent of r. This is what happens to the confinement force between quarks, where $\rho \sim 1$ fm. What is yet to be understood is why the exchange particle is confined to the cylinder. We will return to this in Sec. 9.5.4.

9.3.3   Spin and the Nature of Forces

The spin of the exchange particle also conveys important information about the force it transmits.

Spin is the angular momentum of a particle at rest. It is an intrinsic and unchangeable attribute of the particle, just like its mass. We might think of a particle as a miniscule earth, and its spin is like the rotation of earth about its own axis. While this analogy has some merit, it is somewhat dangerous because it will yield nonsensical results if taken too far. In reality, spin is a quantum mechanical phenomenon with no classical analog.

Angular momentum is described classically by a three-dimensional vector, or equivalently, by its three components (see Appendix A). Quantum mechanically, on account of the uncertainty principle, its three components are not simultaneously measurable. The best we can do is to specify the magnitude of the vector, together with one of the three components. It is traditional to label the component specified to be the z-component.

The length of the spin vector is $\sqrt{s(s+1)}\hbar$, and the z-component is $m_s\hbar$. s must either be a non-negative integer ($s = 0, 1, 2, \ldots$), or a half-integer ($s = \frac{1}{2}, \frac{3}{2}, \frac{5}{2}, \ldots$). For a given s, m_s may take on $2s+1$ discrete values, from $-s, -s+1, \ldots$, to $s-1, s$.

Electrons, nucleons, quarks, and neutrinos are $s = \frac{1}{2}$ objects, and so are their antiparticles. Photons, gluons, $W^{\pm}$ and Z^0 are spin-1 objects, and the graviton is a spin-2 object. The Higgs boson ϕ^0 has a spin 0.

Since the x- and the y-components cannot be fixed once the z-component is specified, we might think of the spin vector as precessing in a cone, as shown in Fig. 9.2.

The nature of the force carried by an exchange particle depends on the spin of that exchange boson.

Quantum field theory dictates that if s is an even integer, the force between identical particles is attractive. If s is an odd integer, the force between identical particles is repulsive, and the force between a particle and its own antiparticle is attractive.

Pions have spin $s = 0$ and gravitons have spin-2. This is the reason why the nuclear and the gravitational forces of two protons are both attractive. Photons have spin-1, so the Coulomb force between two protons is repulsive.

The spin of the exchange particle also determines the complexity of the force. An exchange particle carries with it $2s + 1$ bits of information on its orientation. These different bits can trigger different forces, so the number and complexity of the force increases with the value of the spin. For example, the Yukawa nuclear force between two nucleons is carried by the pion, which has spin-0. So the force is a single attractive one. The electromagnetic force between two protons, however, is

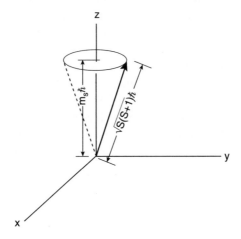

Figure 9.2:   The thick arrow represents a quantum-mechanical angular momentum vector, which can be pictured to be precessing about the z-axis on the surface of the cone.

carried by a photon, which has spin-1. As a result, there is in this case an electric force *and* a magnetic force. Gravitational force is transmitted by a spin-2 graviton, so the force is even more complex than the electromagnetic one. Nevertheless, in all these cases, unless the particles move very fast, the magnetic-like force is much smaller than the electric-like force, so the latter dominates. In the case of electromagnetism, this is the Coulomb force. In the case of gravity, this is the Newtonian force. When we talked about the association between the spin of the exchange particle and the type of force it transmits, we were tacitly referring to these dominant forces.

9.3.4 *Energy and Force*

Many people find it difficult to understand how the exchange of a virtual particle can possibly produce an attractive force. They argue that if A emits a virtual particle towards B, then A must recoil away from B. When B absorbs this virtual particle and its momentum, B must be knocked away from A. Therefore the exchange always causes A and B to move away from each other, which is the hallmark of a repulsive force. How can one possibly get an attractive force?

This argument would have been correct in the absence of quantum mechanics, but then we would not have the virtual particle either. Since the distance between A and B are fixed, the uncertainly principle tells us that the momentum of the virtual particle is undetermined. It could be opposite to what it is classically and intuitively, thereby causing A and B to move towards each other. This gives rise to an attractive force.

How then is one going to determine whether a force is attractive or repulsive? To that end it is better to discuss energy, and not the force directly.

The interaction of A and B, separated by a distance r, is determined by a potential energy $V(r)$. A physical system likes to minimize its potential energy by moving its particles towards a smaller $V(r)$. The sharper the variation of $V(r)$, the faster the particles move. The corresponding force is simply $F = -dV(r)/dr$.

In quantum field theories, the potential energy $V(r)$ consists of three factors, $V(r) = C_A P_s(r) C_B$. The factor $P_s(r) \sim f_s e^{-(mc/\hbar)r}/r$, called the *Yukawa potential*, tells us how the virtual particle propagates. It is the only factor which depends on the distance r. From the exponential factor it is clear that the range is given by $R \simeq \hbar/mc$, as demanded by the Yukawa theory. The factor f_s depends on the spin s of the exchange particle. For the dominant force, it is negative for even s and positive for odd s. This sign determines whether the corresponding force is attractive or repulsive.

The inverse-square law at short range now follows from the formula

$$F = -dV(r)/dr \quad \text{when } r \ll R.$$

The creation and absorption of the exchange particle are given by the *vertex factors* C_A and C_B. For the dominant forces transmitted by neutral gauge particles (γ, Z^0, G), we can take $C_A = gN_A$, $C_B = gN_B$, where the parameter g determines the overall strength of the force and is known as the *coupling constant*. N_A and N_B are respectively the additive quantum numbers of A and B that couples to the gauge particle (see Secs. 9.4 and 9.5). A similar statement can be made for the exchange of $W^\pm$, but it is more complicated in that case because the charge of particles A and B after the emission/absorption of $W^\pm$ is no longer the same as the charge before.

9.3.5 Summary

Forces are transmitted by the exchange of particles. The mass m of the exchange particle determines the range R, and its spin s determines the nature of the resulting forces. For distances small compared to R, all forces obey the inverse-square law. For distance large compared to R, they taper off rather sharply, unless confinement is involved. In that case the force is independent of the distance.

9.4 Quantum Numbers and Symmetry

We have now met a large number of particles: the quarks, the leptons, and the various exchange particles carrying forces. What distinguishes them?

How do we tell two persons apart? By their individual characteristics, of course. John Doe might be 2 meters in height and 100 kg in weight, male, and have black hair and brown eyes. Jane Smith is, say, 1.8 meters tall, weighs 70 kg, female, and has blond hair and blue eyes. They are different and they look different. There is no problem in distinguishing them.

Similarly, we distinguish the different particles also by their special characteristics, except that these characteristics in physics are usually expressed as numbers. We shall refer to these numbers as *quantum numbers*. Other than the mass, they are usually discrete numbers, often integers, but not always. The spin s of the particle is such a quantum number, its electric charge is another.

Quantum numbers are either *additive* or *vectorial*. Unless otherwise stated, all quantum numbers are conserved. This means that the *sum* of these quantum number in an isolated system remains the same at all times. If they did change with time, they would not be permanent features of a particle, and we could not use them to label particles.

The two kinds of quantum numbers differ in how the *sum* is taken. For additive quantum numbers, it is just the arithmetic sum. *Electric charge* is an example of an additive quantum number. Its conservation allows the β-decay reaction $n \rightarrow p + e^- + \bar{\nu}_e$ to occur but not the reaction $n \rightarrow p + e^+ + \bar{\nu}_e$, for example. *Quark number* and the *lepton number* are other additive quantum numbers. Quark number is defined to be the number of quarks minus the number of antiquarks, and the lepton number is defined to be the number of leptons minus the number of anti-leptons. We can check the conservation in the β-decay reaction $n \to p + e^- + \bar{\nu}_e$. The initial state has 3 quarks and zero leptons, and the final state also has 3 quarks and $0 = 1 - 1$ leptons.

One may regard quark-number and lepton-number conservations as the modern version of Dalton's indestructibility of matter.

9.4.1 *Vectorial Quantum Numbers*

A *vectorial quantum number* is the quantum version of a conserved vector, such as angular momentum. Classically, if a vector has d components, then the addition of two vectors is equivalent to the addition of their respective components. If this were true also quantum mechanically, then a vectorial quantum number is no different from a collection of d additive quantum numbers.

Because of the uncertainty principle, a quantum mechanical vector can usually be specified only by $r < d$ components. The other $d - r$ components remain uncertain, just like the momentum of a particle with specified position is uncertain. r is called the *rank* of the vectorial quantum number.

If this were all, then again a vectorial quantum number is no different from a collection of r additive quantum numbers. However, the length and some other characteristics of the vector can also be specified. In general, an additional t discrete numbers $(p_1, p_2, \ldots, p_t)$ are needed to determine the vector as completely as possible. In most of the vectorial quantum numbers we are going to deal with (the so-called $SU(n)$ quantum numbers), the number t is equal to the rank r. These numbers p_i do *not* add arithmetically, and this is what distinguishes a vectorial quantum number from a set of additive quantum numbers. By a vectorial quantum number, we now mean the collection of the r additive numbers and the t non-additive quantum numbers.

Spin is the simplest example of a vectorial quantum number. Let us recall from Sec. 9.3 and Fig. 9.2 what we know about it.

Classically, spin angular momentum is a vector $\vec{s}$, specified by three components. Quantum mechanically, because of the uncertainty principle, only one of the three components (say the z-component) may be specified, although the length of the vector may be simultaneously determined as well. Both of these numbers are *quantized*; they are only allowed to take on certain discrete values. The length of the vector must be of the form $\sqrt{s(s+1)}\hbar$, with s being an integer or a half-integer. For a given s, the z-component must be of the form $m_s\hbar$, with m_s being one of the $2s + 1$ values $-s, -s + 1, \ldots, s - 1, s$. In other words, $r = t = 1$, and $p_1 = 2s$ is a non-negative integer.

Now let us discuss how to find the *sum* of two spins, specified by (s_1, m_1) and (s_2, m_2). If their sum is specified by (s, m_s), then clearly $m_s = m_1 + m_2$. But what about s?

Classically, the length of $\vec{s} = \vec{s}_1 + \vec{s}_2$ depends on the relative angle between the two vectors, $\vec{s}_1$ and $\vec{s}_2$. In particular, if the two vectors are parallel, then the length $|\vec{s}|$ of $\vec{s}$ is simply the sum $|\vec{s}_1| + |\vec{s}_2|$. If they are anti-parallel and if $|\vec{s}_1| \geq |\vec{s}_2|$, then $|\vec{s}| = |\vec{s}_1| - |\vec{s}_2|$.

Quantum mechanically, the non-additive numbers s, s_1, s_2 are integers or half integers, which means that the equivalent angle between $\vec{s}_1$ and $\vec{s}_2$ is not arbitrary. The best s can do is to take on the discrete values between $s_1 + s_2$ and $|s_1 - s_2|$, in integer steps. This is precisely what happens. For example, the sum of $s_1 = s_2 = \frac{1}{2}$ can yield $s = 0$ or $s = 1$. The sum of $s_1 = s_2 = 1$ can yield $s = 2$, 1, or 0.

We can verify these rules by counting the number of components. Take $s_1 = s_2 = \frac{1}{2}$. Each of $\vec{s}_1$ and $\vec{s}_2$ has two allowed orientations, so there should be $2 \times 2 = 4$ states altogether when they are vectorially added. Indeed, $s = 0$ has one state, and $s = 1$ has three states, making a total of four. Similarly, $s_1 = s_2 = 1$ each has three states, so altogether there should be nine states when they are vectorially added. Indeed, $s = 2$ has five states, $s = 1$ has three states, and $s = 0$ has one state, making a total of nine, as it should.

For more complicated vectorial quantum numbers, a branch of mathematics known as *group theory* is needed to tell us what the non-additive quantum numbers p_i are, and how they are combined. See the next subsection for further discussions.

9.4.2 Symmetry and Conservation Laws

Physics is the same today as a billion years ago. It is also the same here on earth as in other parts of the universe. The former, known as time-translational invariance, or time-translational symmetry, can be shown to lead to energy conservation. The latter, known as space-translational invariance, can be shown to lead to momentum conservation. We also get the same physics whether the laboratory is facing east or south. This rotational symmetry leads to angular momentum conservation. These examples illustrate the universal correspondence between symmetries and conservation laws. See Chapter 1 for more discussions on symmetry.

Symmetry is systematically studied in a branch of mathematics called *group theory*. A symmetry is specified by a *group*, which tells us the quantum number that is conserved, and how to add them. Additive quantum numbers typically correspond to the $U(1)$ group. It is an *abelian group*, meaning that the order of symmetry operation is immaterial. Vectorial quantum numbers correspond to *non-abelian groups*, in which a different symmetry results if the order of two symmetry operations is interchanged. The spin angular momentum corresponds to the $SU(2)$ group, which is the first of a series of groups known as $SU(n)$, for $n \geq 2$. It is useful to get acquainted with these groups, as they will come up later in several places.

Let us start from $SU(2)$ which we already know. Its conserved quantum number is specified by a three-dimensional vector, which is the spin vector in the previous application. The length of this vector is discrete, given by a non-negative integer

$p_1 = 2s$. For a fixed p_1, the allowed orientation of the vector is determined by an additive quantum number m_s, which takes on discrete values between s and $-s$ in integer steps. These $p_1 + 1$ possible states are said to form a *multiplet*; the size $(p_1 + 1)$ of the multiplet is called its *multiplicity*, or *dimension*. Note that the sum of the additive quantum number m_s over the multiplet is always zero. This actually is what the 'S' in $SU(2)$ stands for.

The smallest non-singlet multiplet, given by $s_1 = \frac{1}{2}$, is called a *fundamental* multiplet. The multiplet $s_1 = 1$ has a multiplicity three equal to the dimension of the spin vector. It is called the *adjoint* multiplet. The adjoint multiplet can be obtained by vectorially adding two fundamental multiplets, because the vectorial sum of $s_1 = \frac{1}{2}$ and $s_2 = \frac{1}{2}$ yields both $s = 1$ and $s = 0$.

These properties can be generalized to $SU(n)$. The $SU(n)$ quantum number is specified by an $(n^2 - 1)$-dimensional vector, whose allowed length and multiplicity are determined by $n - 1$ non-negative integers $(p_1 p_2 \cdots p_{n-1})$. Each $SU(n)$ vectorial quantum number contains $n - 1$ additive quantum numbers, which are the generalization of m_s. As pointed out before, the number of additive quantum numbers in a group is known as the *rank* of the group, so $SU(n)$ has a rank of $n - 1$. The sum of each of these additive quantum numbers in a multiplet is always zero. We shall refer to this relation as the *traceless constraint*; this is actually what the 'S' in $SU(n)$ stands for.

The multiplet $(p_1 p_2 \cdots p_{n-2} p_{n-1})$ and the multiplet $(p_{n-1} p_{n-2} \cdots p_2 p_1)$ are said to be *conjugate* to each other. They have the same multiplicity. In physics, they often represent particles and antiparticles, respectively, especially for $n > 2$.

One often denotes a multiplet by its multiplicity. Conjugate multiplets are distinguished by putting an asterisk on one of them. For example, in $SU(2)$, multiplets with $s = 0, \frac{1}{2}, 1, \frac{3}{2}$ are denoted by **1**, **2**, **3**, **4** respectively.

For $n > 2$, there are two *fundamental* multiplets of dimension n. They are $\mathbf{n} = (10 \cdots 0)$ and $\mathbf{n}^* = (0 \cdots 01)$.

The *adjoint* multiplet $\mathbf{n^2 - 1} = (10 \cdots 01)$ has the same dimension as the conserved vector. It can be obtained by vectorially adding $\mathbf{n}$ and $\mathbf{n}^*$, for such an addition yields a singlet **1** and an adjoint multiplet $\mathbf{n^2 - 1}$. The reason why the multiplicity of the adjoint is $n^2 - 1$ but not n^2 is intimately related to the traceless constraint of $SU(n)$.

As mentioned before, these general properties of $SU(n)$ are already present in the familiar case $n = 2$. Readers who do not follow these semi-quantitative details should have little problem understanding most of the remaining chapter.

The same symmetry may be applicable to different physical quantities. For example, charge conservation, quark number conservation, and lepton number conservation all correspond to a $U(1)$ symmetry. Unlike energy, momentum, and angular momentum, these symmetries have nothing to do with ordinary spacetime. For that reason they will be referred to as *internal symmetries*. Different members of an internal multiplet correspond to different elementary particles.

We shall now proceed to discuss an internal $SU(2)$ symmetry known as isospin.

9.4.3 Isospin

Isospin is an internal quantum number, mathematically (though not physically) identical to a spin. In other words, it is an $SU(2)$ quantum number, specified by two numbers, (I, I_3), with I being an integer or a half integer (like s in spin), and I_3 varying in integer steps between I and $-I$ (like m_s in spin). They add like spins as well.

Each isospin state is a particle. The $2I+1$ particles that have the same I are said to form an isospin *multiplet*. Multiplets with isospins $I = 0, \frac{1}{2}, 1$ are respectively known as isosinglets, isodoublets, and isotriplets.

Isospin was invented by Werner Heisenberg to describe the similarity between the proton and the neutron. These two have slightly different masses, but that difference is attributed to the proton having an electric charge and the neutron not. If we were able to switch off the electromagnetic interaction, then the neutron and the proton are believed to have identical masses and identical strong interactions. For that reason, we might regard both as different manifestations, or different states, of the same *nucleon*. The nucleon multiplet has isospin $I = \frac{1}{2}$, with the proton having $I_3 = +\frac{1}{2}$ and the neutron having $I_3 = -\frac{1}{2}$.

Similarly, the three pions (π^+, π^0, π^-) form an isotriplet with $m_s = +1, 0, -1$ respectively.

Isospin conservation allows us to relate pp, pn and nn interactions. According to the Yukawa mechanism, one nucleon emits a virtual pion, and the other nucleon absorbs it. A proton can emit a π^+ to become a neutron, or a π^0 to remain a proton. A neutron can emit a π^- to become a proton, or a π^0 to remain a neutron. Similar statements can be made about the absorption site. The different emission and absorption processes are related by isospin conservation. Hence, the different interactions are also related.

Nucleons and pions are made up of u, d quarks and their antiquarks. Correspondingly, the (u, d) quarks form an isodoublet, with $I_3 = \frac{1}{2}$ for u and $I_3 = -\frac{1}{2}$ for d. The $(\bar{d}, \bar{u})$ quarks also form an isodoublet, with $I_3 = -\frac{1}{2}$ for $\bar{u}$ and $I_3 = \frac{1}{2}$ for $\bar{d}$.

The isospin discussed above is sometimes known as the *strong isospin*, because it relates to the strong interaction of hadrons. It does not apply to leptons, which have no strong interactions.

There is another $SU(2)$ quantum number called *weak isospin*, which plays an important role in weak interactions (see the next section). The six quarks form three weak isodoublets, (u, d), (c, s), (t, b), one for each generation. The six leptons also form three weak isodoublets, $(\nu_e, e^-), (\nu_\mu, \mu^-)$, and (ν_τ, τ^-), also one for each generation.

9.4.4 Color

The other important internal quantum number is *color*. It is a vectorial quantum number related to the group $SU(3)$. In spite of the name, it has absolutely nothing to do with the colors we see in daily life.

Classically, color is an 8-dimensional vector. Quantum mechanically, these vectors are discrete and are labeled by two non-negative *integers*, $(p_1 p_2)$, which are the counterparts of I in isospin. The size of the multiplet is determined by p_1 and p_2. For example, $\mathbf{1} = (00)$, $\mathbf{3} = (10)$, $\mathbf{3}^* = (01)$, and $\mathbf{8} = (11)$.

The particles within each multiplet are labeled by *two* additive quantum numbers, which are the counterpart of I_3 in isospin.

Elementary particles carrying non-zero color have strong interactions (see the next section). Those with no color do not. Hence, only quarks and gluons carry a non-singlet color. *Each* of the six quarks u, d, c, s, t, b constitutes a color triplet $\mathbf{3} = (10)$, and each of the six antiquarks constitutes the conjugate color triplet $\mathbf{3}^* = (01)$. Gluons belong to the adjoint multiplet $\mathbf{8} = (11)$. All in all, there are eight gluons, three u-quarks, three $\bar{u}$-quarks, etc.,

The color quantum number comes about in the following way. There is an excited state of the nucleon by the name of Δ^{++}, made up of three u quarks ($\Delta^{++} = uuu$). Its spin is $\frac{3}{2}$, so the spins of the three quarks are all lined up along the same direction. It has no orbital angular momentum between the quarks. Its wave function is therefore symmetric under the interchange of any two quarks, because each of the orbital, spin, and isospin parts of the wave function is symmetric. That, however, is against the Pauli exclusion principle, which demands the wave function to be completely antisymmetric under an interchange. The simplest way out is to assume there are three varieties (three different 'colors') of u quarks, $u = (u_1, u_2, u_3)$, and that Δ^{++} is composed of $u_1 u_2 u_3$. Then, there is no difficulty in constructing an antisymmetric wave function in color so that the exclusion principle is satisfied.

This then is the origin of color, and the color triplet for each of the six quarks.

9.4.5 Vacuum Condensates

At the beginning of the universe, temperature is high and energy is plentiful. Elementary particles too massive to be observed at the present may appear at that time. At high temperatures, particles might possess additional conserved quantum numbers which are no longer conserved at the present temperature of 2.7 K (see Sec. 10.7). This actually includes the weak isospin. In this subsection we will discuss how that happens.

The ground state of the universe is called the *vacuum*. It is the state with no particles, as the presence of each particle invariably brings along an additional amount of energy. As the temperature decreases from its initial value, the vacuum is believed to undergo a series of *phase transitions*. Let us see what they are and how it affects the conservation of quantum numbers.

We are familiar with the phase transitions of water. For temperatures above $100°C$, it is in the vapor or gaseous phase. Between 0 and $100°C$, it is in the liquid phase, and below $0°C$, it is in the solid or ice phase.

Phase transitions are caused by the desire of a physical system to seek the lowest possible energy. Since thermal energy is part of the total energy budget, the ground state configuration may depend on the temperature of the environment. The change of one configuration in to another is a phase transition.

Because of the hydrogen bonding between water molecules, a lower energy is obtained if these molecules can stick close together. This is not possible at temperatures above 100°C because they are being torn apart by the thermal motion. Below 0°C, thermal energy is so small that water molecules can arrange in a regular crystalline configuration that makes best use of the hydrogen bonding. Between 0 and 100°C, thermal energy is large enough to disrupt the regular crystalline pattern, but still not large enough to tear the water molecules from the grip of one another. This is the liquid phase.

The same is true for the vacuum, but this time the bonding comes from the attraction of some *scalar* (*i.e.*, *spin-0*) particles. The attraction between the scalar particles must be strong enough to overcome the burden of their individual masses, and the thermal energy that tends to tear them apart. If that happens, a phase transition occurs in which the vacuum is modified. It is now energetically more favorable to have lots of scalar particles, whose mutual attraction makes the vacuum with their presence to have a lower energy. A vacuum endowed with these scalar particles is said to have a *vacuum condensate*.

In order for the new vacuum to be stable, the attraction must turn to repulsion when the density of scalar particles in the condensate reaches a certain value. Otherwise, there is nothing to stop the continuing creation of scalar particles to cause the vacuum to collapse.

The condensate no longer looks like individual scalar particles, because they would have all merged into one giant continuum. Different scalar particles may have different binding energies and therefore different phase-transition temperatures. The larger the binding is, the higher the phase-transition temperature would be.

The vacuum condensate must not contain higher-spin particles. These particles are expected to have spin interactions to line up their spins along some direction. Since our vacuum is isotropic, this must not occur.

The scalar particles in the condensate might contain certain quantum numbers. If that happens those quantum numbers will no longer be conserved. This is so because the vacuum now has an infinite supply of these quantum numbers, which can be used to alter these quantum numbers in any particle reaction. Since we know that electric charge, quark number, lepton number, and color are conserved, the vacuum must be electrically and color neutral, and it must contain no quarks nor leptons except in particle-antiparticle pairs. Weak isospin turns out not to be conserved at the present temperature, though it did at a higher temperature. Hence, the vacuum condensate should consist of scalar particles carrying a weak isospin.

9.4.6 Space Reflection and Charge Conjugation

It was believed before the mid-1950's that physics is the same under spatial reflection, otherwise known as *parity transformation*. This means that if something can happen, then the process viewed in a mirror can happen as well. This symmetry leads to a conserved quantum number known as *parity*. In 1957, Tsung Dao Lee and Chen Ning Yang discovered that while this is true for strong and electromagnetic interactions, it is not so for weak interactions. In other words, parity is not a conserved quantum number in the presence of weak interactions. This discovery of Lee and Yang has rather far-reaching consequences, so let us discuss it in some detail.

We need to introduce two concepts: *helicity* and *chirality*.

The spin component along the direction of motion is called *helicity*. In the case of quarks and leptons, whose spin is $s = \frac{1}{2}$, these components are either $m_s = \frac{1}{2}$ or $-\frac{1}{2}$. The former is known as a *right-handed helicity*, because when you stand behind the particle, the angular motion of its spin turns like a right-handed screw. The latter is called a *left-handed helicity*. In relativistic quantum mechanics, helicity is a more important concept than spin. Spin-orbit coupling is present in a relativistic theory, so neither spin nor orbital angular momentum is conserved, though the total angular momentum still is. Since orbital angular momentum has no projection along the direction of motion, the projection of total angular momentum is equal to the projection of spin alone, so in spite of the spin-orbit coupling, helicity is still conserved.

Chirality is defined to be twice the helicity for a massless spin-$\frac{1}{2}$ particle. If the particle is massive, the concept of chirality is mathematically well-defined, but it is hard to describe it in physical language.

A right-handed screw looks like a left-handed screw in a mirror. Thus parity transformation reverses helicity and chirality. Since the late 1950's, the weak interaction has been known to be *left-handed*. This means that only quarks and leptons of left-handed *chirality*, and only antiquarks and antileptons of right-handed chirality, participate in the charged-current interactions mediated by the exchange of $W^{\pm}$. This preference of one handedness over another clearly violates parity ($\mathcal{P}$) symmetry. It also violates *charge conjugation symmetry* ($\mathcal{C}$) as we will explain, though it preserves the combined $\mathcal{CP}$ symmetry.

Charge conjugation symmetry means that if we change all particles to their anti-particles (γ, Z^0, H^0 are their own anti-particles, and W^+ and W^- are anti-particles of each other), without altering any other quantum number, then the new process is still an allowed one occurring with the same probability. The weak interaction violates charge conjugation invariance because interactions occur for left-handed quarks and leptons but not for left-handed anti-quarks and anti-leptons. It is the right-handed anti-quarks and anti-leptons that participate in the interactions. So $\mathcal{C}$ symmetry is violated. Nevertheless, the weak interaction is still invariant under a combined $\mathcal{CP}$ operation, for this changes the left-handed quarks and leptons to right-handed anti-quarks and anti-leptons.

Strictly speaking, the $\mathcal{CP}$ symmetry is violated at a small level. Such violations are very important, for we probably owe our existence to it. We will come back to discuss this in a later section.

9.4.7 Summary

Conserved quantum numbers may be additive, or vectorial. The former type add arithmetically, but the latter also contains non-additive quantum numbers which do 'add,' but in a peculiar way.

Examples of additive quantum numbers are electric charge, quark number, and lepton number. Examples of vectorial quantum numbers are isospin and color.

Strong and electromagnetic interactions are also invariant under charge conjugation ($\mathcal{C}$) and parity transformation ($\mathcal{P}$), but the weak interaction is not. $\mathcal{CP}$ is almost conserved in all weak interaction processes, but a small violation does exist.

9.5 Standard Model

The current theory of strong, electromagnetic, and weak interactions is known as the *Standard Model*, or SM for short. It has so far passed all the experimental tests thrown at it, so it is a very reliable theory. The only exception is neutrino oscillations, which will be discussed in the next section.

As for gravity, Einstein's theory of general relativity works for macroscopic bodies, but owing to the weakness of its strength, quantum gravity has not yet been tested. In any case, we do not have a very good understanding of the theory of quantum gravity. We will discuss that problem further in the last section of this chapter.

The three fundamental forces in the SM are all described by *gauge theories*. The fourth, Einstein's theory of gravity, is also a kind of gauge theory, though technically sometimes not regarded as such.

A gauge theory is a very special theory of force, probably the most elegant theory at present. The simplest and the oldest of these gauge theories is the Maxwell theory of electromagnetism. It will serve as a template for the other gauge theories, so let us study it in some detail.

9.5.1 Electromagnetic Theory

Classical electricity and magnetism were unified and completed by James Clerk Maxwell in the late 19th century. Its quantum version was invented in the late 1920's. This quantum theory is known as *quantum electrodynamics*, or QED for short. From then on it was just a matter of studying its properties and obtaining numbers from it to compare with experiments. Calculations of this sort actually ran into a snag which did not get resolved until the late 1940's. The solution goes by the name of *renormalization*, a topic which we shall discuss at the end of this section.

In order to expose the elegance of the theory, and to motivate how similar considerations can be applied to other gauge theories, let us now pretend not to know the Maxwell theory, and see how far a few simple facts can lead us.

From the long range nature of the inverse-square law, we deduce that the exchange particle must be massless. It must also be electrically neutral, for otherwise the charge of the electron will be altered after a photon is emitted. The fact that it transmits both an electric and a magnetic field shows that it must carry a non-zero spin. In fact, the photon spin is known to be 1.

We will now argue that this spin-1 nature is extremely important in keeping its mass zero, and hence the infinite range of the Coulomb force.

A particle with a mass m carries a minimum amount of energy equal to mc^2. According to quantum mechanics, energy generally shifts with interactions, which leads to a shift in the mass m. For the photon, this shift must be positive, because there is no such thing as a negative mass. A positive mass will cause the Coulomb force to have a finite range, contrary to experimental observation.

There are two ways out. It is logically possible for the photon to start with a finite mass, and the forces of the universe are so finely tuned to make the final photon mass zero. Such an accidental zero is highly contrived and seems to be highly unlikely.

It is also conceivable to have some sort of a mechanism to protect the photon mass, so that no shift of it can occur. One such mechanism is the *spin trick*.

A particle with spin s has $2s + 1$ possible spin orientations *at rest*. According to the principle of special relativity, this should continue to be the case when the particle is in motion.

Since a *massless* particle must move with the speed of light, and cannot be put at rest, this argument does not apply to them. In fact, the photon has spin 1 but it has only two helicities, corresponding to the *transverse* polarizations (right-handed and left-handed circular polarizations) of light. The longitudinal polarization is absent.

Conversely, if somehow we can keep the number of physically realizable helicities of a particle to be fewer than $2s + 1$, then it must be massless. This is the spin trick.

A *gauge theory* is a theory in which the longitudinal polarization of the spin-1 boson (called the *gauge particle*) is decoupled from the rest. In other words, it is invariant under a *gauge transformation* in which the magnitude of the longitudinal polarization is changed. The Maxwell theory of electromagnetism is a gauge theory, which is why the photon has a zero mass even in the presence of interactions.

One can envisage a situation in which every point in space has one or more photons. Since none of them may have a longitudinal polarization, gauge invariance must be true at every point in space. Thus, the symmetry of gauge transformation is a *local* symmetry, which is a much more powerful symmetry than the *global symmetries* that lead to energy, momentum, angular momentum, charge, isospin, and color conservations.

In the presence of matter, it is possible to maintain this local symmetry only when the gauge particle couples *universally* to all matter particles. This means a $U(1)$ symmetry must be present in the matter sector to produce an additive conserved quantum number, Q. For Maxwell theory, Q is just the electric charge. By universality, we mean that the coupling to matter is proportional to Q, whatever the other quantum numbers are. It is the democratic principle, if you like. Everybody with the same ability (Q) has the same rights, whatever his/her sex, religion, or ethnicity is.

This kind of gauge theory, coupled universally to an *additive* quantum number, is known as an *abelian gauge theory*, or a $U(1)$ gauge theory.

9.5.2 Yang–Mills Theory

One might wonder whether there are gauge particles coupled universally to *vectorial* quantum numbers. Such a theory was constructed by Chen Ning Yang and Robert Mills in 1954, and is nowadays known as a *non-abelian gauge theory*, or a *Yang–Mills theory*.

As in the abelian gauge theory, the masslessness of the gauge particle is enforced by the spin trick, which in turn is realized by demanding a local gauge invariance. For that to be fulfilled, the gauge particle once again has to couple universally to the conserved vectorial quantum number.

The vectorial nature of the quantum number however brings about some complications.

The gauge particles always belong to the adjoint multiplet, so there are $n^2 - 1$ of them in $SU(n)$. Unlike the photon which carries no electric charge, and therefore cannot emit other photons, the non-abelian gauge particle carries the adjoint quantum number and can therefore emit other gauge particles, on accoant of universality. This affects the nature of short-distance interactions, as we shall see in Sec. 9.5.5.

9.5.3 Electroweak Theory

Weak interaction was discovered at the turn of the century in beta decay. The first theory to describe it was written down by Enrico Fermi in 1934 shortly after Pauli postulated the existence of the neutrino. Indeed it was Fermi who christened this elusive particle emerging from the beta decay $n \to p + e^- + \bar{\nu}_e$ to be the (anti-)neutrino.

Fermi's theory was very successful in predicting the energy distribution of the electron, but when parity violation was discovered, it had to be modified to take into account parity and charge-conjugation violations. This was carried out successfully in the late 1950's.

The resulting interaction resembles an isospin gauge theory. However, on the one hand, the mass of the would-be gauge particle has to be very, very large to explain the extremely short range of the weak forces; on the other hand, the mass of any gauge particle has to be zero to allow for gauge invariance. These two requirements do not seem to be compatible with each other.

This dilemma was solved using the Higgs mechanism, which we proceed to explain.

9.5.3.1 Higgs Mechanism

The Higgs mechanism enables a gauge particle to become massive.

To make it work we need to have a scalar (spin-0) particle present to form a vacuum condensate (see Sec. 9.4.5).

The gauge particle W, which couples universally to any isospin object, will interact with this condensate if the latter carries an isospin. As a result, W receives a drag as it travels through the vacuum, causing it to slow down from the speed of light. This is analogous to light traveling in a dielectric material. Its interaction with the molecules in the material causes its speed to be $c/n < c$, where $n > 1$ is the dielectric constant of the medium.

To travel slower than the speed of light, W must have a mass. In order to have a mass, each spin-1 W must acquire from somewhere a longitudinal polarization. In the Higgs mechanism, these extra degrees of freedom are supplied by three extra scalars, one for each of the three gauge particles W^+, W^0, and W^-.

With the Higgs mechanism to make W massive, the weak interaction may now be described by an (weak) isospin gauge theory. To that end we need to introduce four scalar particles: one for the vacuum condensate, and three to supply the longitudinal polarizations. In the SM, these four form an isodoublet (ϕ^+, ϕ^0), and its antiparticles $(\bar{\phi}^0, \phi^-)$. The extra degrees of freedom for W^+, W^-, W^0 are provided respectively by ϕ^+, $(\phi^0 - \bar{\phi}^0)/\sqrt{2}i$, ϕ^-, so these scalar particles disappear as a separate entity, to be counted as part of the W's. The remaining degree of freedom, $(\phi^0 + \bar{\phi}^0)/\sqrt{2}$, condenses in the vacuum, but the excitation of this condensate also gives rise to a new particle, which is the Higgs particle H^0 listed in Table 9.1.

Actually, a refinement is needed to describe Nature. In this refinement, weak and electromagnetic interactions are combined into a single electroweak theory. The W^0 boson mixes with another neutral gauge boson to form the Z^0 boson and the photon γ of Table 9.1. We will now proceed to discuss this refinement.

9.5.3.2 The Electroweak Interaction

The electromagnetic and weak interactions in the SM are combined into a single electroweak theory by Sheldon Glashow, Abdus Salam, and Steven Weinberg. It is a combined abelian and non-abelian gauge theory. The non-abelian gauge boson W is coupled to the weak isospin, as described before. The abelian gauge boson B^0 is coupled to an additive quantum number known as the (weak) *hypercharge*, defined to be $Y = 2(Q - I_3)$, where I_3 is the third component of the weak isospin and Q is the electric charge of a particle. Table 9.2 below lists the elementary particles discussed in Table 9.1, together with their Q, I, I_3, Y, and Y_Z quantum numbers.

Table 9.2: Electroweak quantum numbers of the Standard Model particles.

Particle	I	I_3	Q	Y	Y_Z
(u_L, d_L)	$\frac{1}{2}$	$\left(\frac{1}{2}, -\frac{1}{2}\right)$	$\left(\frac{2}{3}, -\frac{1}{3}\right)$	$\frac{1}{3}$	$(0.35, -0.42)$
(u_R, d_R)	0	$(0, 0)$	$\left(\frac{2}{3}, -\frac{1}{3}\right)$	$\left(\frac{4}{3}, -\frac{2}{3}\right)$	$(-0.15, 0.08)$
(ν_L, e^-)	$\frac{1}{2}$	$\left(\frac{1}{2}, -\frac{1}{2}\right)$	$(0, -1)$	-1	$(0.5, -0.27)$
e_R^-	0	0	-1	-2	0.23
(W^+, W^0, W^-)	1	$(1, 0, -1)$	$(1, 0, -1)$	0	
B^0	0	0	0	0	
(ϕ^+, ϕ^0)	$\frac{1}{2}$	$\left(\frac{1}{2}, -\frac{1}{2}\right)$	$(1, 0)$	1	$(0.27, -0.5)$

The Y_Z quantum number is defined to be $I_3 - (0.23)Q$, whose significance will be discussed in the subsection 'Mixing of Neutral Gauge Bosons.' Only fermions in the first generation are listed. Those in the second and third generations have identical quantum numbers as the first. Anti-fermions have the same I and opposite Q and I_3, and are not listed either. Gravitons (G) and gluons (g) are absent in the table because they do not participate in electroweak interactions.

It is important to note that particles with left-handed chirality (L) do not have the same isospin and hypercharge as particles with right-handed chirality (R), so parity is explicitly broken in this theory. However, it can be shown that parity is still conserved in the electromagnetic part of this theory.

Here are some items in the table worth noting:

- The $2I+1$ members inside an isospin multiplet all have the same hypercharge Y.

- The gauge boson B and the gauge bosons W are decoupled, because the isospin of B is zero and the hypercharge of W is zero. Since B couples only to hypercharge Y and W couples only to isospin (I, I_3), B is not subject to the isospin force and W is not subject to the hypercharge force.

- Only left-handed (L) components of a fermion are affected by the charged-current weak interaction mediated by W. The right-handed (R) components have no isospin so they are not affected.

- The right-handed neutrino ν_R is assumed to be absent. If it were there, it should have $I = I_3 = 0$ like all other right-handed fermions. It is electrically neutral, so $Q = 0$. As a result, $Y = 0$ as well. Like all leptons, it does not carry a color. Hence, except for gravity, it is not subject to any force at all. Such a particle is extremely difficult to detect, even if it were present. For that reason it is called a *sterile neutrino*. We will return to ν_R later when lepton mixing is discussed.

The strength of the isospin and hypercharge couplings will be denoted by the coupling constants g_I and g_Y, just like the strength of the electromagnetic coupling is determined by e (with $\alpha = e^2/\hbar c \simeq 1/137$). More precisely, the coupling of W^0 is proportional to $g_I I_3$, the coupling of B^0 is proportional to $g_Y(Y/2)$, and the coupling of γ is proportional to eQ. The *Weinberg angle* θ_W, defined by $\tan\theta_W = g_Y/g_I$, has a value $\sin^2\theta_W \simeq 0.23$ at the Z mass.

The electroweak theory is an $SU(2)_L \times U(1)_Y$ gauge theory. $SU(2)_L$ refers to the Yang–Mills theory in which W is coupled to the isospin quantum number. The subscript L is there to remind us that only the left-handed fermions get coupled. $U(1)_Y$ refers to the abelian gauge theory in which B is coupled to the weak hypercharge Y.

After the Higg's condensate is formed, isospin I and hypercharge Y are no longer conserved, because the condensate carries these quantum numbers. However, electric charge Q is still conserved, because the condensate is neutral. The surviving gauge theory is therefore QED, or $U(1)_{\text{EM}}$, where the subscript denotes 'electromagnetism.' This is described by saying that the vacuum condensate *breaks* (the symmetry) $SU(2)_L \times U(1)_Y$ down to (the symmetry) $U(1)_{\text{EM}}$.

9.5.3.3 Mixing of Neutral Gauge Bosons

Since the condensate carries both isospin and hypercharge, one might expect B and W both to become massive via the Higgs mechanism. This is not the case.

What happens is that the coupling of B^0 and W^0 to the condensate induces a coupling between the two, which makes neither of them have a definite mass. Instead, one gets a 2×2 mass matrix that has to be diagonalized. This situation is familiar in quantum mechanics. When an interaction is introduced, states originally having definite energies will no longer be so. Instead, states having the same quantum numbers will mix to form new states of definite energies. In relativistic theories, since mass is equivalent to energy, the same will happen to states with definite masses.

As a result of the mixing, one of the two resulting normal modes is massive; it is the Z^0 boson. The other one is massless; it is the photon γ. Since these are mixtures of B^0 and W^0, they couple to mixtures of I_3 and Y. It turns out that the photon couples only to the mixture $Q = I_3 + Y/2$, as expected. The electromagnetic coupling constant is $e = g_I \sin\theta_W$, from which we can determine g_I. The Z^0 boson couples to $Y_Z \equiv I_3 - \sin^2\theta_W Q$, with a coupling strength $g_I/\cos\theta_W$. The value of Y_Z is also listed in Table 9.2 above, assuming a value $\sin^2\theta_W = 0.23$. This quantity tells us the relative coupling strength of different particle to the Z^0 boson.

9.5.3.4 W and Z Masses

As mentioned before, W and Z gain a mass through the Higgs mechanism. The magnitude of the mass depends on the amount of vacuum condensate, hereafter parameterized by a constant v of energy dimension. This parameter can be determined from β-decay to be $v \simeq 246$ GeV. In terms of this parameter, the W and Z masses turn out to be $M_W = g_I v/2$ and $M_Z = M_W / \cos \theta_W$. From the values of g_I, v, and θ_W we can determine the masses to be $M_W \simeq 80$ GeV and $M_Z \simeq 91$ GeV. More accurate computations can be made and the result agrees completely with the measured values: $M_W = 80.423 \pm 0.039$ GeV, and $M_Z = 91.1876 \pm 0.0021$ GeV.

9.5.3.5 Fermion Masses

In quantum field theories, fermion masses may be considered to be determined by the energy it takes to produce a fermion-anti-fermion pair from the vacuum. If m is the mass of the fermion, an energy $2mc^2$ or more is required. In this production process, momentum and angular momentum projection along the direction of movement must be kept at the vacuum value, namely 0, since both of these quantities are exactly conserved.

Figure 9.3 depicts such a process. The fermion f and the anti-fermion $\bar{f}$ must be produced back-to-back as shown, to conserve momentum. Spins are shown by double arrows in the picture. They must point in opposite directions to maintain a zero angular momentum. That means, the fermion and the anti-fermion so produced must both be left-handed, or they must both be right-handed.

In light of these requirements, the electroweak theory forbids any fermion to have a mass before the condensate is formed. To see that, suppose the fermion f ($f = u$, d, e^-, ν_e, etc.) is left-handed. To remind us of this we shall henceforth denote it as $f_{\rm L}$. To have a mass, the anti-fermion $\bar{f}$ in Fig. 9.3 must also be left-handed, namely, $(\bar{f})_{\rm L}$. However, as remarked at the end of Sec. 9.4.6, with $\mathcal{P}$ violation and $\mathcal{CP}$ conservation, the antiparticle of $f_{\rm L}$ is right-handed, not left-handed. Namely, $\overline{f_{\rm L}} = (\bar{f})_{\rm R}$. In order to have a mass, we need $(\bar{f})_{\rm L} = \overline{f_{\rm R}}$ instead. Now, in the electroweak theory, $f_{\rm L}$ is an isodoublet but $f_{\rm R}$ (and hence $\overline{f_{\rm R}} = (\bar{f})_{\rm L}$) is an isosinglet, so the produced pair is again an isodoublet. Since the vacuum must be an isosinglet, the process in Fig. 9.3 does not conserve isospin, and is hence forbidden. This is why all fermions must be massless before the vacuum condensate is formed. When it forms, isospin conservation is no longer valid, then the process in Fig. 9.3 may proceed and a fermion mass may be produced. The magnitude of the mass depends on the strength of the coupling between the fermion and the condensate. This cannot be computed in the electroweak theory so these masses are left as parameters that can be determined only by measurements.

Figure 9.3: A pair of left-handed fermions created from the vacuum.

Fermion masses produced this way are known as *Dirac masses*. Neutrinos cannot have a Dirac mass even in the presence of a vacuum condensate because right-handed neutrinos ν_R are absent. However, there is a way to get a neutrino mass just with the left-handed neutrinos ν_L. We can do it by taking f in Fig. 9.3 to be ν_L, and $\bar{f}$ also to be ν_L. This kind of a mass is known as a *Majorana mass*.

To do so, we identify ν_L with $(\bar{f})_L$, so fermion number is no longer conserved. Another way of saying this is that the pair in Fig. 9.3 now has a fermionic number 2, and the vacuum has a fermionic number 0, so this process does not conserve the fermionic number.

All the other fermions carry an electric charge, so none of them may have a Majorana mass. Otherwise, electric charge is not conserved, contrary to observations.

9.5.4 Strong Interactions

The nuclear force between nucleons was explained by H. Yukawa to be due to the exchange of pions (see Sec. 9.1). We now know that both nucleons and pions are made up of quarks and anti-quarks. How do the quarks/anti-quarks interact to bind them into the nucleons and pions?

If the interaction is given by a gauge theory, the only thing we have to know is what conserved quantum number it couples to; not isospin, nor electric charge, for those have been taken up by the electroweak theory already. The only thing left is color, and that turns out to be correct. The theory of strong interaction is therefore known as *quantum chromodynamics*, or QCD for short.

We shall now discuss what led to that conclusion.

9.5.4.1 Asymptotic Freedom

In the late 1960's, an experiment performed by Jerome Friedman, Henry Kendall, Richard Taylor, and collaborators at the Stanford Linear Accelerator (SLAC), exposed an important property of quark interactions that was hitherto unknown.

In that experiment, high energy electrons thrown at a nucleon target were detected at *large angles* (a more precise technical description is *large 'momentum transfers'*). As a result of the collision, energy is taken from the electron to create other particles. Friedman and his collaborators were interested in those electrons which suffered a great loss in energy. Collisions of this type are known as *deep inelastic scattering*.

The probability of detecting an electron depends on the energy loss of the electron, and the angle at which it is detected. James D. Bjorken had predicted earlier that in the deep inelastic scattering region, the probability depends only on one *independent* combination of these two variables. When the experiment was analyzed in the deep inelastic region, this *Bjorken scaling* was indeed obeyed. Richard Feynman then realized that this scaling phenomenon could be understood *physically* if the nucleon was composed of many constituents, which he called *partons*, provided that the partons do not interact among themselves, nor radiate other particles when being knocked around in the deep inelastic scattering process. Since nucleons are made up of quarks bound together by gluons, these partons must be nothing but quarks, anti-quarks, and gluons.

There are two puzzling features in the parton picture. Firstly, a nucleon is supposed to contain three quarks, but the parton idea of Feynman requires that the nucleon contain very many quark partons. So how many quarks are there in a nucleon? Secondly, quarks and gluons interact strongly, but for Feynman's idea to work, they must either have no interaction, or at best a very weak one. How can that be?

To understand the first puzzle, let us be reminded (Sec. 9.3) that every particle has a Yukawa cloud around it. The cloud surrounding a quark or a gluon is made up of quark-anti-quark pairs and other gluons. If we examine the particle with a large probe (one with a long wavelength), we will see the particle as a whole, endowed with its full Yukawa cloud. Such a quark is known as a *constituent quark*, and a nucleon is made up of three constituent quarks.

On the other hand, if the probe is small (short wavelength), it can only see a small fraction of the cloud. We may then pick up a quark (or an anti-quark) hidden in the cloud, without picking up the rest of the cloud. These are Feynman's partons, since a high-energy electron scattered at large angles qualifies as a short wavelength probe. This argument shows that there can indeed be many quark partons.

The second puzzle is more difficult to understand. In fact, its resolution led to the discovery that the dynamical theory of the strong interaction is QCD.

The only way to reconcile the second dilemma is for a strongly interacting quark to become weak when bombarded in a deep inelastic scattering experiment. Such a property is known as *asymptotic freedom*. To find out whether there are theories with such a property, an intense theoretical effort was undertaken. After many unsuccessful attempts, it was finally discovered by David Gross and Frank Wilczek, and by David Politzer, that this is possible if quarks interact through a non-abelian gauge theory. The reason for that will be explained in Sec. 9.5.5. As discussed in Sec. 9.4.4, we are forced by the Pauli exclusion principle to introduce this new quantum number 'color' for the quarks, and not for the leptons. Since the other quantum numbers 'isospin' and 'electric charge' have already been used by weak and electromagnetic interactions, it is reasonable to assume that 'color' is the quantum number that governs the new Yang–Mills gauge theory for strong interactions. This $SU(3)_C$ non-abelian gauge theory is now accepted as the correct theory of strong interaction between quarks. The subscript C is there to remind us that the $SU(3)$ vectorial quantum number used is color.

Recall that the electroweak theory is an $SU(2)_L \times U(1)_Y$ gauge theory. Hence, the SM of strong, electromagnetic, and weak interactions is given by an $SU(3)_C \times SU(2)_L \times U(1)_Y$ gauge theory.

9.5.4.2 Confinement

Quarks have never been seen in isolation, so QCD must carry with it the confinement property (see Sec. 9.3.2). Ordinary hadrons like nucleons and mesons are not subject to confinement because the quarks and anti-quarks within them conspire to form color singlets. As such they no longer interact directly with the gluon, which is why the dominant nuclear force between two nucleons is carried by the pion and not by the gluon.

The mechanism behind confinement is still not clear. This ignorance lies in our inability to calculate reliably strong interaction field theories, except when the momentum transfer is large and the effective coupling small, as is the case in the deep inelastic scattering region. There are, however, approximate numerical calculations (known as *lattice gauge theory* calculations) which indicate that confinement indeed occurs in QCD.

We will now describe a qualitative model of confinement which has some intuitive appeal, but whose quantitative validity is not yet known. Nevertheless, it will serve to illustrate what kind of mechanisms might induce confinement.

According to this model, at temperatures below about 0.2 GeV, a *colorless gluon condensate* is formed in the vacuum. Like the Higgs condensate, this will happen only when the vacuum with the condensate is more stable than the one without. In order for color to remain conserved, the gluons in the condensate must pair up to form colorless objects. Ideally color electric and magnetic fields should be absent from this condensate, for otherwise they will break up the colorless condensate and increase its energy. However, if two quarks are present, a color field connecting them is inevitable. In that case, to minimize the energetic damage, this field should be concentrated in as small a volume as possible, and that is in a tube between the two quarks with a cross-sectional area like that of the quarks. As discussed in Sec. 9.3.2, the color force between the two quarks is then a constant, and confinement results.

If this scenario is correct, one might be able to achieve *deconfinement* by high energy heavy ion collisions, if a temperature of ~0.2 GeV can be created and trapped by these ions to melt away the condensate. Experiments of this kind are underway in RHIC at the Brookhaven National Laboratory.

9.5.5 Renormalization

Has it ever occurred to you that it is a miracle we can discover any physical law at all?

Newton managed to discover his three famous laws of mechanics and many others without ever knowing any atomic or molecular physics, because he did not have to. Microscopic details are really not important to macroscopic physics. This is not to say that there is absolutely no connection between the microscopic and the macroscopic worlds. The mass of an object, its thermal conductivity, its tensile strength, etc., are controlled by the details of atomic and nuclear physics. Macroscopic theories, however, treat these quantities as experimentally measured parameters. If we want to *compute* these parameters, then a detailed knowledge of atomic and/or nuclear physics will be needed.

This is the normal world. The theory in such a normal world is known as a *renormalizable theory*.

We could imagine a completely different world in which the *details* of atomic and nuclear physics are required to describe *macroscopic* physics. We can even write down theories with these characteristics. Such a world would be so complex that physical laws could never be discovered.

In our normal world, physics at different scales is described by different equations. In the macroscopic world we have Newtonian mechanics; in the microscopic world there is quantum mechanics. The motion of an object on a frictionless table can easily be described by Newtonian mechanics. In principle, it can also be described by the Schrödinger equations of its molecules, but this latter description is so horribly complicated that it is virtually useless. Similarly, The motion of water in the macroscopic scale is described by the 'Navier–Stokes equation,' but on the atomic and molecular scale, the motion of the water molecules would be given by the quantum-mechanical Schrödinger equation. It would be foolhardy to treat the waves in the ocean with more than 10^{26} coupled Schrödinger equations for the water molecules; we will never get anywhere with such complexity. The Navier–Stokes equation bears very little resemblance to the Schrödinger equation, though the former must be derivable from the latter. The Navier–Stokes equation contains a number of parameters, viscosity being one of them. In the classical regime, viscosity is measured, but it can also be calculated if the microscopic physics is known.

Renormalization is the procedure whereby we relate physics of one scale to physics of another scale.

In the water example above, the equations in the two scales are different: Navier–Stokes in one, and Schroedinger in the other. In other systems it is also possible for the two equations to be essentially the same, but the parameters in one scale are different from the parameters in another. In such examples the task of renormalization is simply to determine how these parameters change with scale. SM equations are of this type.

It is not difficult to understand why the SM parameters change with scale. Take for example, an electron in QED. Let $r = \hbar c/\mu$ be the size of the probe, measured with a parameter μ, which has the dimension of energy. Different μ see different amounts of the Yukawa cloud around the electron, hence a different mass. For that reason the effective mass $m(\mu)$ depends on the probe scale μ. Since the cloud contains e^+e^- pairs, these pairs will shield the electric field of the bare electron at the center. This is like an electron placed in a dielectric material. The electric field it generates polarizes the medium which in turn shields the electric field to make it weaker. The larger μ is, the deeper we penetrate into the cloud, and the less is the shielding. Hence, the effective electric charge $e(\mu)$ is[1] also a function of μ. We see from this example that the effective parameters in quantum field theories generally depend on the scale μ of the probe. This is renormalization at work. Such parameters are known as *running parameters* (running mass, running coupling constants, etc.).

There is a fundamental difference between the running coupling constant of an abelian gauge theory, and a non-abelian gauge theory. In an abelian gauge theory, as we saw above, $e(\mu)$ grows with increasing μ. In a non-abelian gauge theory, say QCD, the cloud is no longer color neutral. In fact, gluons can emit more gluons to spread the color throughout the cloud. When we increase μ to penetrate deeper into the Yukawa cloud surrounding a bare quark, on the one hand we are seeing less $q\bar{q}$ shielding than in QED. This tends to increase the effective color charge, or equivalently the effective coupling constant g_S. On the other hand, because of the spreading, we are seeing less net color charge in the cloud, and that tends to decrease the coupling strength g_S. The final effective coupling constant $g_S(\mu)$ receives corrections from these two competing effects. Calculation shows that the gluon spreading wins out, so that $g_S(\mu)$ becomes a decreasing function of μ, opposite to the behavior in QED. For large energy or momentum scale, $g_S(\mu)$ can become quite small. This is the origin of asymptotic freedom, observed in the deep inelastic scattering experiment in SLAC (see Sec. 9.5.4).

Similarly, the $SU(2)_L$ gauge theory is also asymptotically free.

Historically, renormalization was discovered in the late 1940's, in QED. Quantum field theory with interaction is such a difficult theory that it can be solved only approximately, by making use of the smallness of the coupling constant e in QED to obtain an iterative solution. Upon the first iteration, every thing works well and the result agrees approximately with experimental results. If we iterate once more, instead of getting a small correction to the first iteration, one gets something which is infinite.

After the second World War, microwave equipments developed for radar became available for precise spectroscopic measurements. Using them Willis Lamb and Robert Retherford found an unexpected energy shift in an atomic hydrogen line, a shift that is very small but nevertheless cannot be accounted for using the first iterative solution. In response to the challenge to explain this *Lamb shift*, people were forced to tackle the infinities brought on by the second iteration.

The infinity comes from very short-range contributions, a range much shorter than the resolution of any existing probe. From the result of the first iteration, we know our QED theory to be a good description of Nature at the present energies, or probe resolution. The infinity occuring at the second iteration tells us that the QED theory is *not* a good description of reality at a much smaller distance scale, or else infinities should not occur. This is analogous to the Navier–Stokes equation,

---

[1]The actual electron mass with $mc^2 = 0.51$ MeV is the parameter when $\mu = mc^2$. The coupling constant e with the measured value $\alpha = e^2/\hbar c = 1/137$ is also the value at $\mu = mc^2$.

which is true at the macroscopic scale, but not true on the quantum scale. Without knowing the physics at the quantum scale, parameters like viscosity cannot reliably be computed, so it must be measured. Similarly, in QED, since its description at the very small distance scale is wrong, we must take parameters like the electron mass and the coupling constant at their measured values, at least at one scale μ, with no attempt to compute them. The necessary machinery to accomplish this was developed by Richard Feynman, Julian Schwinger, and Sin-Itiro Tomonaga. The result agrees very well with the Lamb shift, as well as all other precision measurements of QED.

With renormalization, the SM becomes a precision theory for particle physics at the presently available energies.

9.5.6 Summary

Gauge theories are theories whose forces are transmitted by massless spin-1 particles. To keep them massless, a gauge invariance is enforced to decouple the longitudinal polarization. Gauge invariance is a local invariance; it can be achieved only when the gauge boson couples universally and democratically to a conserved quantum number.

Abelian gauge theory is just the Maxwell theory. Non-abelian, or Yang–Mills gauge theories are more complicated. The former couples to an additive quantum number, and the latter couples to a vectorial quantum number. The gauge bosons in abelian theories do not couple directly among themselves, but those in a non-abelian gauge theory do.

The electroweak interaction is given by the gauge theory $SU(2)_L \times U(1)_Y$. To make it work, a Higgs condensate is required.

The strong interaction is given by the gauge theory $SU(3)_C$, also known as QCD. Asymptotic freedom exists in QCD; confinement may be present there as well.

The SM of the strong, electromagnetic, and weak interaction is given by a gauge theory based on the gauge group $SU(3)_C \times SU(2)_L \times U(1)_Y$.

Parameters like masses and coupling constants run with the probing scale μ. In particular, the coupling $e(\mu)$ in QED increases with μ, and the coupling $g_S(\mu)$ in QCD decreases with μ, giving rise to asymptotic freedom.

9.6 Fermion Mixing

According to the electroweak theory, W couples universally to all isospin doublets. Among the left-handed fermions, there are three quark doublets, (u,d), (c,s), and (t,b), and three lepton doublets, (ν_e, e), (ν_μ, μ), and (ν_τ, τ). However, these fermions may not have a definite mass, so they may not be the fermions detected in experiments.

To understand why these fermions may not carry a definite mass, let us recall the mass generating mechanism discussed at the end of Sec. 9.5.3, and Fig. 9.3. Two virtual fermions $(f_1)_L$ and $\overline{(f_2)_R} = (\overline{f_2})_L$, or $(f_1)_R$ and $\overline{(f_2)_L} = (\overline{f_2})_R$, are created from the vacuum, with opposite momentum, opposite electric charge and color, and opposite quark and lepton numbers. In Sec. 9.5.3, we concentrated on the situation when $f_1 = f_2$, but there is nothing to prevent $f_1 \neq f_2$.

To conserve quantum numbers, if f_1 is any one of (u, c, t), then f_2 must be one of (u, c, t). If f_1 is any one of (d, s, b), then f_2 must be one of (d, s, b). If f_1 is any one of (e, μ, τ), then f_2 must be one of (e, μ, τ). The 3×3 combinations in each of these three categories can be arranged into three 3×3 matrices, called the *mass matrices*. Unless the matrix is diagonal, *i.e.*, non-zero only when $f_1 = f_2$, the fermions f mix with one another and they will not have a definite mass. The states f_m that have definite masses, known as *mass eigenstates*, are linear combinations of these fermions f, known as *flavor states*. This is similar to the mixing discussed in Sec. 9.5.3.3 to form the Z^0 and the photon. The combinations can be computed mathematically once the mass matrix is known, but unfortunately we do not know what it should be. For that reason, we simply have to parameterize the combinations and determine the parameters by experiment.

We will discuss the quark sector below. The lepton sector is similar, and will be discussed in a subsequent subsection. The mixing parameters can be measured experimentally, and they may shed light on the origin of the interaction that causes the mixing.

9.6.1 Quarks

There is a 3×3 mass matrix for (u, c, t), and another one for (d, s, b). In principle, neither of them is diagonal. In practice, as we shall explain, we can always assume the (u, c, t) matrix to be diagonal, so only linear combinations of the flavor states (d, s, b) in terms of the mass eigenstates (d_m, s_m, b_m) have to be specified. The subscript m denotes quarks with definite masses.

The (u, c, t) mass matrix can be chosen to be diagonal because of the universality of charge-current electroweak interactions. W couples to each of the three left-handed quark isodoublets in exactly the same way. Hence it also couples the same way to any linear combination of (u, d), (c, s), and (t, b). We will choose the linear combination that makes the (u, c, t) mass matrix diagonal. With this new combination, quarks in the flavor basis will change from (u, d), (c, s), (t, b) to something else, say (u', d'), (c', s'), (t', b'). By construction, $(u', c', t') = (u_m, c_m, t_m)$ have definite masses. As remarked above, W still couples universally to (u', d'), (c', s'), and (t', b'). We will now drop all the primes in the notation. In this way, we see that we may indeed assume the mass matrix of (u, c, t) to be already diagonal, as claimed.

The linear combinations of (d_m, s_m, b_m) are specified by a 3×3 unitary matrix V, known as the *Cabibbo-Kobayashi-Maskawa* (CKM) matrix:

$$\begin{pmatrix} d \\ s \\ b \end{pmatrix} = V \begin{pmatrix} d_m \\ s_m \\ b_m \end{pmatrix} = \begin{pmatrix} V_{ud} & V_{us} & V_{ub} \\ V_{cd} & V_{cs} & V_{tb} \\ V_{td} & V_{ts} & V_{tb} \end{pmatrix} \begin{pmatrix} d_m \\ s_m \\ b_m \end{pmatrix}. \tag{9.1}$$

With this mixing, the decay $W^- \to d\bar{u}$ can turn into the decay $W^- \to d_m \bar{s}_m$, for example, with a probability amplitude given by $V_{ud}V_{us}^*$. From the decay rates of various combinations of particles, and other experimental inputs, we can measure the various CKM matrix elements.

A 3×3 unitary matrix is specified by nine real parameters. However, the five relative phases of the six states u, d, s, u_m, d_m, s_m are arbitrary, so they can be used to eliminate five parameters. This leaves four parameters, but their choice is not unique. One set, due to Chau and Keung,[2] uses three rotation angles $\theta_{12}, \theta_{13}, \theta_{23}$, and one phase angle δ. Rotational angles give rise to real numbers but phase angle gives rise to complex numbers. $\mathcal{CP}$ violation is present only when there are complex numbers around, namely, only when $\delta \neq 0$.

Another set of parameterization, more suitable for quark mixing, was proposed by Wolfenstein,[3] where the four parameters are λ, A, ρ, and $i\eta$, with $i\eta$ being an imaginary number controlling $\mathcal{CP}$ violation, and $\lambda \simeq 0.22$. From experimental data, we also know that A, ρ, η are all of order 1. With that, we can see that[3] the mixings given by the non-diagonal matrix elements of V are small.

Note that if we had only two generations of quarks, V would be a 2×2 unitary matrix, parameterized by $4 - 3 = 1$ parameter. It is a rotational angle, not a phase angle. For that reason, $\mathcal{CP}$ violation cannot occur from quark mixing if we had only two generations.

$\mathcal{CP}$ violations have been detected experimentally. The manner and amount of the violations are so far consistent with a quark-mixing origin. $\mathcal{CP}$ violation is important cosmologically, for without it we should have the same number of protons as antiprotons, which our universe does not have. However, we need a stronger $\mathcal{CP}$

---

[2] The exact form in the parameterization suggested by Chau and Keung is

$$V = \begin{pmatrix} c_{12}c_{13} & s_{12}c_{13} & s_{13}e^{-i\delta} \\ -s_{12}c_{23} - c_{12}s_{23}s_{13}e^{i\delta} & c_{12}c_{23} - s_{12}s_{23}s_{13}e^{i\delta} & s_{23}c_{13} \\ s_{12}s_{23} - c_{12}c_{23}s_{13}e^{i\delta} & -c_{12}s_{23} - s_{12}c_{23}s_{13}e^{i\delta} & c_{23}c_{13} \end{pmatrix},$$

where $c_{13} = \cos\theta_{13}$, $s_13 = \sin\theta_{13}$, etc.

[3] The exact form in the parameterization suggested by Wolfenstein is

$$V = \begin{pmatrix} 1 - \frac{1}{2}\lambda^2 & \lambda & A\lambda^3(\rho - i\eta) \\ -\lambda & 1 - \frac{1}{2}\lambda^2 & A\lambda^2 \\ A\lambda^3(1 - \rho - i\eta) & -A\lambda^2 & 1 \end{pmatrix}.$$

There are $O(\lambda^4)$ corrections left out in this expression, but that is reasonable because $\lambda \simeq 0.22$ is a small number.

violation than whatever can be provided by quark mixing alone to account for the present amount of matter (over antimatter) in the universe, so another stronger source of $\mathcal{CP}$ violation must occur somewhere, presumably at higher energies than we can reach right now.

9.6.2 Leptons

As in the quark sector, we may assume the charged-lepton mass matrix to be already diagonal, $(e, \mu, \tau) = (e_m, \mu_m, \tau_m)$. The neutrino mass matrix and the neutrino mixings will now be discussed.

In what follows we shall denote the neutrino states with definite masses by ν_1, ν_2, ν_3, rather than $(\nu_e)_m$, $(\nu_\mu)_m$, $(\nu_\tau)_m$. The latter notation is not only more cumbersome, it is also not justified if the mixing is large. In the case of quarks, the mixings are small, so we can uniquely associate the three mass eigenstates with the three flavor states. In the case of neutrinos, the mixings are large, so the former notation is better.

The corresponding neutrino masses will be denoted by m_1, m_2, m_3.

In the SM, $m_1 = m_2 = m_3 = 0$. It is so because there is no right-handed neutrino, so there can be no Dirac mass. Lepton number is conserved, so there can be no Majorana mass. A direct measurement of the $\bar{\nu}_e$ mass shows it to be small, being bounded by 3 eV/c^2 or less.

The latest fit to available cosmological and astrophysical data by the WMAP collaboration suggests an upper bound of 0.23 eV/c^2.

However, from the observation of solar and atmospheric neutrino deficits, to be discussed later, one deduces that at least two of the three masses must be non-zero. So $m_1 = m_2 = m_3 = 0$ is ruled out by experiment.

In light of that, something must be missing in the SM. What then can we say about the neutrino mass matrix and neutrino mixing?

A neutrino mass may be a Majorana mass, or a Dirac mass (see Sec. 9.5.3). In the former case no right-handed neutrino is needed, but lepton number is violated because the vacuum state turns into a two-neutrino state (instead of one neutrino and one anti-neutrino). In the latter case the presence of a right-handed neutrino is required. If such a particle did exist, it is presumably not like the other right-handed fermions in the SM, because the neutrino mass is so much smaller than the other fermion masses.[4] There are speculations of the existence of such a right-handed neutrino, either at high energy or in an extra dimension (see Sec. 9.7.3), or both. Since they are not directly visible right now, we can 'integrate them out' to get an effective mass using only left-handed neutrinos. In doing so we are back to the Majorana mass again.

---

[4] The smallest charged fermion mass is that of the electron, which is 0.5 MeV/c^2, some 200 000 times larger than the largest conceivable neutrino mass.

A Majorana-neutrino mass matrix is a complex symmetric matrix, because f and $\bar{f}$ in Fig. 9.3 are both equal to ν_L. The mathematics of diagonalizing it is slightly different from that used to diagonalize the (d, s, b) mass matrix. Nevertheless, the flavor states $(\nu_e, \nu_\mu, \nu_\tau)$ are still unitary linear combinations of the mass eigenstates (ν_1, ν_2, ν_3) as in Eq. (9.1):

$$\begin{pmatrix} \nu_e \\ \nu_\mu \\ \nu_\tau \end{pmatrix} = V \begin{pmatrix} \nu_1 \\ \nu_2 \\ \nu_3 \end{pmatrix}. \tag{9.2}$$

This time the matrix V is called the *Pontecorvo–Maki–Sakata* (PMNS) matrix. Since the mixings turn out to be large, it is not useful to parameterize it in the Wolfenstein manner. It is the Chau–Keung form[2] of parameterization that is usually used.

In addition, since the mass matrix is complex, the resulting masses m_i are complex as well. If we want to write them as positive numbers, we should replace m_i by $m_i e^{i\delta_i}$. An overall phase does not matter, so we may set $\delta_1 = 0$, but we are left with δ_2 and δ_3, which have no counterpart in the mixing of quarks. These extra phase angles can be traced back to the Majorana nature of the neutrino masses.

We have assumed the neutrinos to be Majorana in order to get a mass. If so, there may be important consequences. Lepton number violation might turn into quark number violation, and possibly a mechanism to generate more protons than antiprotons, and more electrons than positrons in the universe, as required by observation. It is therefore important to ascertain directly whether the neutrino is Majorana or not. The smoking gun is the discovery of neutrinoless double β-decay. Two neutrons in a nucleus may decide to undergo the β-decay $n \to p + e^- + \bar{\nu}_e$ simultaneously. In such a decay, two $\bar{\nu}_e$ are produced. However, if the neutrino is Majorana, the two antineutrinos may combine into the vacuum, thereby producing a double β-decay without any neutrino. Measurements are being carried out to find such a reaction, but so far none have been observed.

9.6.3 Neutrino Oscillations

We will explain in this subsection the evidence for non-zero neutrino masses. We will also discuss the qualitative difference between neutrino and quark parameters, and its possible implication.

The energy from the sun is primarily powered by the fusion reaction $4p \to {}^4\text{He} + 2e^+ + 2\nu_e$, in which two protons are turned into two neutrons through the inverse β-decay reaction $p \to n + e^+ + \nu_e$. In addition to this nuclear reaction, there are several others which also produce neutrinos in the sun.

Since the early 1960's, Raymond Davis and his group had been trying to detect these solar neutrinos. The experiment is a very difficult one because neutrinos interact very weakly, so a large chunk of matter is needed to capture some of them.

Moreover, cosmic rays overwhelm the solar neutrino events, so the detector must be heavily shielded, usually by putting it deep underground in a tunnel or a mine. Even so, natural radioactivities contained in the detector and the nearby rocks must be largely removed to give a chance for the solar neutrino events to show through. What Davis used as a target was a large 615 tons tank of dry cleaning fluid. Most of the neutrinos pass straight through, but an occasional one strikes a chlorine atom ^{37}Cl in the fluid to turn it into a radioactive argon atom ^{37}Ar. Periodically, the few argon atoms in the fluid are extracted and measured by counting its radioactivity. For over twenty years, Davis' experiment was the only one set up to detect solar neutrinos. It was a difficult job and data accumulated very slowly.

What he found was quite astonishing. Only about 1/3 of the expected neutrinos were found. What happened to the other 2/3?

In the 1980's, another large detector was built in Japan using water as the target, which the Čerenkov light emitted by the struck electrons as the signal for a neutrino. This group is led by Masatoshi Koshiba; the upgraded version of the detector is called the Super-Kamiokande. It uses 50 thousand tons of pure water as the target. Because of the large size of the tank, most solar neutrino data to date have come from this source. They found about 1/2 of the expected neutrinos.

By now there are also two gallium detectors in Europe, and a heavy water detector called SNO in Canada. They all find a deficit in solar neutrinos. These deficits are consistent with a similar deficit from reactor antineutrinos, measured after they travel some 180 km to reach a liquid scintillator detector known as KamLAND.

Different detectors are sensitive to different neutrinos and/or different energies. They are all needed to analyze the outcome.

The explanation for the deficit is *neutrino oscillation*. Unlike quarks which are detected in their mass states, neutrinos are produced by weak interaction, and also detected through weak interaction, so at both ends we are dealing with the flavor neutrino ν_e. In between the sun and the earth, it is the mass eigenstates ν_i that propagate with a definite frequency, when treated as a quantum mechanical wave. On account of relativistic kinematics, the frequency is proportional to m_i^2. To compute the deficit, we must use Eq. (9.2) to decompose the initial ν_e into the mass eigenstates ν_i, and at the detector use Eq. (9.2) again to convert each ν_i back to ν_e. If all the masses m_i are the same, their propagating frequencies are identical, therefore all of them have the same phase at the detector. In that case, the mass eigenstates ν_i simply recombine into ν_e, and no deficit is present. If, however, the masses are different, the phases of the different ν_i's will be different at the detector, so the ν_i's will no longer recombine back into ν_e, for some of them will now combine into other kinds of neutrinos. In that case a deficit in ν_e is observed.

Conversely, the fact that deficits are observed tells us that we cannot have $m_1 = m_2 = m_3 = 0$.

Super-Kamiokande also discovered a deficit in atmospheric neutrinos. Cosmic rays hitting the upper atmosphere generate many charged pions which eventually

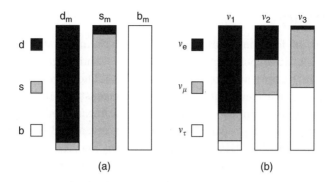

Figure 9.4: (a) Quark mixing, and (b) neutrino mixing. Mass states appear at different columns, and flavor states appear at different rows.

decay, mostly via $\pi^+ \to \mu^+ + \nu_\mu$ and $\pi^- \to \mu^- + \bar{\nu}_\mu$. The muons are themselves unstable; they decay via $\mu^+ \to e^+ + \nu_e + \bar{\nu}_\mu$ and $\mu^- \to e^- + \bar{\nu}_e + \nu_\mu$. Combining these two decays, we expect to find two μ-type neutrinos and antineutrinos for every e-type neutrinos and antineutrinos. What is observed is more like one-to-one, not two-to-one.

The explanation for a deficit in the μ-like neutrinos is again neutrino oscillation.

The solar neutrino deficit tells us about the oscillation of the e-type neutrino, and the atmospheric neutrino deficit tells us about the oscillation of the μ-type neutrino. There are other accelerator and reactor experiments which provide us with additional information. By analyzing these experiments carefully, one finds that neutrino mixings are large, and neutrino mass differences are small. In fact, $m_2^2 - m_1^2 \simeq (7.5 \text{ meV})^2$, $|m_3^2 - m_2^2| \simeq (50 \text{ meV})^2$ ($1 \text{ meV} = 10^{-3} \text{ eV}$), $\tan^2 \theta_{12} \simeq 0.4$, $\tan^2 \theta_{13} < 0.026$, $\tan^2 \theta_{23} \simeq 1$.

Given the small mass differences and the mass upper bound for $\bar{\nu}_e$, it follows that all neutrino masses must be smaller than 3 eV or so. Even at this upper limit, the mass is a factor of 2×10^5 smaller than the electron mass, the lightest of the charged fermions. Neutrino mixings are also very different from quark mixings. For quarks, all three mixing angles are small. For neutrinos, there are two large angles and one small one. These two different mixing patterns are illustrated in Fig. 9.4.

Many experiments are being set up or planned to confirm the oscillation hypothesis, and to obtain better values for the neutrino parameters. Neutrino oscillation is the only confirmed phenomenon not explained by the SM, and as such it might provide a stepping stone to discover new physics beyond the SM. The fact that neutrinos behave so differently from charged fermions also suggests the presence of new physics that only neutrinos can see. Neutrinos are also important in astrophysics and in cosmology. They are ubiquitous, and promise to be an important tool in astronomy if only we can detect them more efficiently. For all these reasons, neutrino physics promises to be a very active field in the near future.

For their important discoveries, Davis and Koshiba were awarded the Nobel Prize for Physics in 2002.

9.6.4 Summary

The isodoublets of fermions coupled to $W^{\pm}$ do not automatically have definite masses. However, it is possible to choose (u, c, t) and (e, μ, τ) to have definite masses, but then (d, s, b) and $(\nu_e, \nu_\mu, \nu_\tau)$ will not. The flavor states (d, s, b) can be expressed as a linear combination of the mass eigenstates (d_m, s_m, b_m), and the flavor states $(\nu_e, \nu_\mu, \nu_\tau)$ can also be expressed as a linear combination of the mass eigenstates (ν_1, ν_2, ν_3). The linear combinations are given by 3×3 unitary matrices. It is known as the CKM matrix in the quark sector, and the PMNS matrix in the neutrino sector. These matrices are parameterized by three rotation angles and one phase angle. The rotation angles measure the amount of mixing, and the phase angle gives rise to $\mathcal{CP}$ violation. The mixing in the quark sector is relatively small, but the mixing in the neutrino sector determined by neutrino oscillations is large.

Neutrino masses are much smaller than the mass of any charged fermion. This fact, together with their different mixing pattern from the quarks, may hopefully lead us to some new physics beyond the Standard Model.

9.7 Outlook

We have reviewed in the last seven sections some of the exciting developments in particle physics. The importance of this field can be gauged by the number of Nobel Prizes awarded: close to thirty people mentioned in the previous sections are Nobel Laureates. Moreover, in order to keep this chapter to a reasonable length, we have been forced to leave out the important work along with the names of some other Nobel Laureates in the field.

As mentioned before, the SM for strong and electroweak interactions has been very successful in explaining the precision data now available to us. The only outstanding task is to find the Higgs boson H^0 experimentally, a challenge that hopefully will be met in the Large Hadron Collider (LHC) under construction at CERN.

Is there a life beyond the SM, is there any more fundamental physics we can learn?

The answer is definitely yes, though what the new physics will be is not clear at the moment. To be sure, there is no shortage of suggestions, but we must await experimental data to guide us in making a choice.

In this section we will discuss some of the more popular proposals.

9.7.1 Grand Unified Theories

There are three different forces in the SM, one for each of the gauge groups $SU(3)_C$, $SU(2)_L$, and $U(1)_Y$. These forces have different ranges, and different strengths. At a distance r much less than the range, all forces obey the inverse square law, but still,

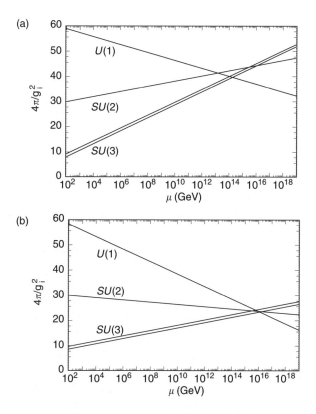

Figure 9.5: A schematic plot for the energy variation of the three inverse coupling constants $4\pi/g_i^2$ as a function of energy. (a) Without supersymmetry. (b) With supersymmetry.

they appear to be different because $g_Y < g_I < g_S$ at the present energy. However, we learned in Sec. 9.5.5 that coupling constants run with the energy scale μ, so it is conceivable that all coupling strengths become equal at some high energy $\mu_{\rm GUT}$. If that happens, the three different forces may simply be three different manifestations of a single force. A theory of that kind, uniting the different Standard Model forces and particles, is known as a *grand unified theory*, or GUT for short. The GUT group must contain the SM group $SU(3)_C \times SU(2)_L \times U(1)_Y$ as a subgroup.

Let us first look at the possible unification of $g_S(\mu)$ and $g_I(\mu)$, which for later convenience we will re-label as $g_3(\mu)$ and $g_2(\mu)$, respectively. Both of them decrease with increasing μ, but since there are eight gluons and only three W's, $g_3(\mu)$ decreases faster than $g_2(\mu)$. With $g_3 > g_2$ at the present energy, these two are getting closer together at higher μ's (see Fig. 9.5, but note that it is plotted in $1/g^2$, not g), so they will intersect at some large $\mu_{\rm GUT}$. The strengths of the electroweak and the strong forces are then the same for $r \sim \hbar c/\mu_{\rm GUT}$.

What about the hypercharge force? g_Y is smaller than g_2 and g_3 at current energies, and being an abelian theory, it increases with μ. So at some high μ it will catch up with $g_2(\mu)$ and/or $g_3(\mu)$. If it should intersect with both of them at μ_{GUT}, then all the three strengths become the same at that energy. Does it do that?

It actually does not, but there is a problem of normalization. In a gauge theory, the gauge particle couples universally to a conserved quantum number, with some strength g. However, there is an arbitrariness in the normalization of the quantum number, which is reflectd in an arbitrariness in the strength g. Let us explain what we mean by looking at the coupling of W^0.

W^0 couples to the third component of isospin through a term proportional to $g_I I_3$. However, we could have used $I_3' = a I_3$ as the additive quantum number instead, in which case the coupling strength would have to be $g_I' = g_I/a$ in order to preserve $g_I' I_3' = g_I I_3$, and hence the same interaction. Without specifying the normalization of the quantum number, there is no way that different coupling strengths can be meaningfully compared.

For $SU(2)$, one uses I_3 by convention, not I_3'. Note that the sum of I_3^2 over the two components of the (fundamental) doublet is $2 \times (\frac{1}{2})^2 = \frac{1}{2}$.

The conventional normalization for $SU(n)$ is a generalization of this. Let Q_i be any of the $n-1$ additive quantum numbers of $SU(n)$. We require each of them to be normalized like I_3 of $SU(2)$, namely, the sum of Q_i^2 over the n members of a fundamental multiplet be equal to $\frac{1}{2}$.

We shall see below that the correct normalization of the hypercharge quantum number in a GUT is not $Y/2$, but $(Y/2)\sqrt{3/5}$. This means that the correct coupling strength for which we seek a unification is $g_1 = \sqrt{5/3} \, g_Y$. The question now is whether $g_1(\mu), g_2(\mu)$, and $g_3(\mu)$ all intersect at some common point μ_{GUT}.

When this idea of unification was first proposed, data were not very precise, and indeed these three coupling constants seem to meet at a common point. With the precise data available nowadays, they no longer do, as seen in the sketch in Fig. 9.5a. However, with the insertion of *supersymmetry*, they once again meet at a common $\mu_{\text{GUT}} \simeq 10^{16}$ GeV, as sketched in Fig. 9.5b. This gives an impetus to the possible correctness of supersymmetric grand unified theories.

We shall discuss supersymmetry in the next subsection. Briefly, it proposes a doubling of the existing particles, with opposite statistics. The new 'supersymmetric particles' are heavy, with a mass believed to be of order of 1 TeV (10^{12} eV). As far as the running of the coupling constants is concerned, these new particles provide additional shielding that serves to decrease the slope of every line, as can be seen by comparing Fig. 9.5b with 9.5a.

In the rest of this subsection, we will discuss GUT by ignoring supersymmetry. What the latter does is to add additional particles to the spectrum, but it does not invalidate the discussion in the rest of this subsection.

The SM group $SU(3)_C \times SU(2)_L \times U(1)_Y$ has rank 4. Namely, it contains four additive quantum numbers (I_3, Y, and two additive quantum numbers from color). Since a GUT group must contain the SM group, its rank must be four or more. There is a rank 4 candidate group, $SU(5)$, and a rank 5 candidate group, $SO(10)$, which contains $SU(5)$ as a subgroup. We will discuss these two cases separately below.

There are $8 + 3 + 1 = 12$ gauge bosons in the SM group $SU(3)_C \times SU(2)_L \times U(1)_Y$. A GUT contains all those gauge bosons, and more. Below μ_{GUT}, the GUT group breaks down into the SM group, presumably triggered by the formation of a vacuum condensate that contains a GUT quantum number but no SM quantum number. The gauge bosons not in the SM will then interact with the condensate to get a mass, of the order of μ_{GUT} (possibly multiplied by some coupling constants).

There are 15 *left-handed* fermions per generation in the SM. Those in the first generation are the u and d quarks, and the left-handed $\bar{u}$ and $\bar{d}$ anti-quarks, each coming in three colors, making a total of 12. Then, there are two left-handed leptons, e^- and ν_e, as well as *one* left-handed anti-electron, e^+, making a grand total of 15. There is no anti-neutrino with a left-handed helicity in the SM. These 15 fermions must fit into one or more multiplets in the GUT group. Since they are massless in the SM at high temperature, they are expected to be massless in the GUT theory at high temperature as well. Neutrino oscillation experiments (Sec. 9.6.3) indicate that right-handed neutrinos ν_R might exist. If so, we should include $\overline{\nu_R} = (\bar{\nu})_L$ in the GUT multiplets as well.

9.7.1.1 SU(5)

The multiplets of $SU(5)$ are specified by four non-negative integers (see Sec. 9.4.2): $(p_1 p_2 p_3 p_4)$. In particular, the fundamental multiplets are $\mathbf{5} = (1000)$ and $\mathbf{5}^* = (0001)$, and the adjoint (gauge-boson) multiplet is $\mathbf{24} = (1001)$.

In addition to the 12 SM bosons, there are 12 additional ones which we will call X and $\overline{X}$. X is a color triplet ($\mathbf{3}$) and an isospin doublet. $\overline{X}$, its antiparticle, is a conjugate color triplet ($\mathbf{3}^*$) and an isospin doublet. The emission of X can change a quark into a lepton, and that will cause the proton to decay, say into $e^+ \pi^0$. The decay rate is severely limited by the ultra short range of the X-exchange force, because its mass is very large. Experimentally, the proton has not been found to decay; its lifetime into $e^+ \pi^0$ is larger than 10^{32} years. This experimental limit places considerable constraint on the details of the $SU(5)$ theory.

The 15 left-handed SM fermions are fitted into $\mathbf{5}^*$ and $\mathbf{10}$ as follows:

$$\mathbf{5}^* = (\bar{d}_1 \bar{d}_2 \bar{d}_3 \nu_e e^-)$$

$$\mathbf{10} = (\bar{u}_1 \bar{u}_2 \bar{u}_3 u_1 u_2 u_3 d_1 d_2 d_3 e^+). \tag{9.3}$$

Their $\mathcal{CP}$ partners are right-handed fermions, and they are fitted into the multiplets $\mathbf{5}$ and $\mathbf{10}^*$.

We can use Table 9.2 to compute the sum of Y^2 for the members of $\mathbf{5}$, or $\mathbf{5}^*$. It is $3 \times (2/3)^2 + 2 \times (1)^2 = 10/3$. The corresponding additive quantum number $\mathcal{Q}_1$, which is correctly normalized, namely, whose square sum is $\frac{1}{2}$, is then $\mathcal{Q}_1 = \sqrt{3/20}\, Y$. The coupling constant g_1 so that $g = g_1 = g_2 = g_3$ at $\mu = \mu_{GUT}$ is defined by $g_1 \mathcal{Q}_1 = g_Y (Y/2)$. Hence, $g_1 = \sqrt{5/3}\, g_Y$, as previously claimed.

Remember from Sec. 9.5.3 that the Weinberg angle is defined by $\tan \theta_W = g_Y/g_I$, and its present value is given by $\sin^2 \theta_W \simeq 0.23$. At the GUT scale, $g_1 = g_2$, so $\tan \theta_W = \sqrt{3/5}$. This translates into a value of $\sin^2 \theta_W$ to be $3/8 = 0.375$. The running of coupling constants between these two energy scales takes care of this difference.

$SU(5)$ breaks down into the SM group by forming a vacuum condensate. There are many ways of doing that, each giving different results for masses and other relations. Unfortunately, none of them seems to be both successful and simple.

9.7.1.2 SO(10)

The vector defining the quantum number of $SO(10)$ is 45 dimensional, which is also the dimension of its gauge multiplet. The (color, isospin) content of this multiplet is given by

$$\begin{aligned} \mathbf{45} = &(\mathbf{3}, \mathbf{1}) + (\mathbf{1}, \mathbf{2}) + (\mathbf{1}, \mathbf{1}) + (\mathbf{3}, \mathbf{2}) + (\mathbf{3}^*, \mathbf{2}) + (\mathbf{3}^*, \mathbf{1}) + (\mathbf{3}, \mathbf{2}) \\ &+ (\mathbf{1}, \mathbf{1}) + (\mathbf{3}, \mathbf{1}) + (\mathbf{3}^*, \mathbf{2}) + (\mathbf{1}, \mathbf{1}) + (\mathbf{1}, \mathbf{1}) \,. \end{aligned} \tag{9.4}$$

The first line is the $\mathbf{24}$ of $SU(5)$, so it contains the 12 SM gauge bosons and the 12 X, $\bar{X}$ bosons. The second line contains 20 more bosons in the $\mathbf{10}$ and $\mathbf{10}^*$ multiplet of $SU(5)$. The last line is an additional singlet in $SU(5)$. After the vacuum condensate is formed, the SM bosons remain massless but the other 33 bosons must gain a large mass.

Since $SO(10)$ contains $SU(5)$ as a subgroup, it is also possible that $SO(10)$ is conserved at an even higher energy scale, and then it breaks down to $SU(5)$ before μ_{GUT}. However, it does not have to do it that way.

There is a $\mathbf{16}$ multiplet in which all the 15 left-handed fermions and $\overline{\nu_R} = (\bar{\nu})_L$ are fitted in. This is an appealing feature of $SO(10)$ over $SU(5)$, in that the right-handed neutrinos suggested by neutrino oscillations are automatically included, and that all the fermions are contained in a *single* multiplet, not two as is the case in $SU(5)$.

The details of the outcome again depend on the scalar particle contents, and how the vacuum condensate is formed.

9.7.2 Supersymmetry

The symmetries in SM and GUT produce multiplets of particles of the same spin and statistics. *Supersymmetry* (SUSY) is a hypothetical symmetry whose multiplets contain an equal number of bosons and fermions.

Since bosons have integral spins and fermions have half integral spins, a SUSY multiplet necessarily involves particles of different spins and statistics.

With this symmetry, every SM particle is doubled up with a partner of the opposite statistics. The partner of a lepton of a given chirality is a spin-0 particle called a *slepton*, the partner of a gauge particle is a spin-$\frac{1}{2}$ particle called a *gaugino*, the partner of a Higgs boson is a spin-$\frac{1}{2}$ particle called a *Higgsino*, and the partner of the graviton is a spin-$\frac{3}{2}$ particle called the *gravitino*.

None of these SUSY partners have yet been found. If SUSY is correct, these partners must have high masses, which means that SUSY is broken. There are many suggestions as to how SUSY breaking takes place, but none have been universally accepted.

There are theoretical reasons to believe these new particles to have masses in the TeV (1,000 GeV) region, so there is hope that they may be found in the Large Hadron Collider (LHC), under construction at CERN, or some other high-energy accelerators in the future.

With no SUSY partners found, why do people think that SUSY might exist? The reason is primarily because exact SUSY brings with it many nice properties, and some of those are preserved at least qualitatively even when SUSY is broken.

What then, are the nice properties of SUSY that warrant the introduction of this whole zoo of unobserved particles?

Recall that quantum mechanics allows the appearance and disappearance of virtual particles. In a SUSY theory, the probability *amplitudes* of creating and annihilating a virtual boson-antiboson pair is opposite to that of a fermion-antifermion pair, because of their opposing statistics. As a result, when the two amplitudes are added together, they cancel each other as if no virtual pairs were present, at least for some purposes.

The mass of a particle is affected by its surrounding Yukawa cloud, as discussed in Sec. 9.5.5. Without virtual pairs, there is no cloud, the mass will not be affected by interactions and it will not be renormalized.

Similarly, the energy of the vacuum is not shifted in the presence of a SUSY interaction. In fact, it can be shown that the energy of the vacuum must be zero in that case.

However, SUSY is broken, so the cancelation between the boson-antiboson pairs and the fermion-antifermion pairs is not complete. The Yukawa cloud is back, though reduced in strength compared to the one without SUSY. Mass shifts are back, but in reduced amounts. Vacuum energy is shifted, but not by as much as without SUSY. Above the SUSY breaking scale μ_{SUSY}, exact SUSY is restored, so the shifts should somehow depend on this scale.

The mass shift of a fermion is fairly insensitive to this scale, but not so for a spin-0 boson. If this scale is too high, the mass of the Higgs boson H^0 will be shifted too much for us to understand why we can still keep it among the SM particles. This is known as the *hierarchy problem*. For the H^0 mass to be tolerable, this scale cannot be too high, and that implies the SUSY particles will make their appearances in the TeV region.

Broken SUSY also helps the unification of coupling constants. As mentioned in the last subsection, the three lines in Fig. 9.5a do not intersect at a single point in light of the precision data now available. The introduction of SUSY partners around 1 TeV increases the number of virtual pairs, and reduces the slopes of the lines from 1 TeV on. With this change, the three lines now do meet at a point, as seen in Fig. 9.5*b*, so the possibility of grand unification is restored.

There is another very attractive feature of SUSY, broken or not. We know that the universe is full of *cold dark matter*, which is some five times more abundant than the usual matter (see Sec. 10.3.4). If the lightest SUSY partner is stable, then it qualifies quite well as a cold dark matter candidate. They are therefore being searched for in the dark matter around us, though so far none have been detected.

9.7.3 Extra Dimensions

A cherished dream of Albert Einstein was to unite the theory of gravity and electromagnetism. Although he never succeeded in his lifetime, a novel attempt in that direction by Theodor Kaluza in 1919, later improved by Oskar Klein in 1926, pioneered an important idea that is still being explored today.

Mathematically, Einstein's gravity is described by a two-index object $g_{\mu\nu}$, which is symmetric in μ and ν (a symmetric tensor). Electromagnetism is described by a one-index object A_{μ} (a vector). The indices μ and ν are either 1, 2, 3, or 4, indicating in that order the component along the x-axis, the y-axis, the z-axis, and the time-axis. Kaluza suggested that if there was an extra spatial dimension, then five-dimensional gravity would automatically contain a four-dimensional gravity and a four-dimensional electromagnetism. The reason is simple. Five-dimensional gravity is described by g_{MN}, where M, N now range over 1, 2, 3, 4, and 5. Namely, μ and 5. The $g_{\mu\nu}$ components of g_{MN} describe four-dimensional gravity, the $g_{\mu 5} = g_{5\mu}$ components describe four-dimensional electromagnetism. In addition, it also contains a four-dimensional scalar field g_{55}.

Klein modified this idea by assuming the extra dimension to be a circle, rather than an infinite line used in Euclidean geometry.

Klein's theory possesses a rotational symmetry in the fifth dimension. Since the extra dimension gives rise to electromagnetism, the conserved quantum number for which symmetry is the electric charge. We may now regard the electric charge to be on the same footing as energy, momentum, and angular momentum, since all of them are symmetries of the five-dimensional spacetime.

More spatial dimensions have since been introduced. For example, the so-called *superstring* theory requires six extra spatial dimensions, and the M theory requires seven. If the internal space (extra dimensions) possesses a larger symmetry, one might imagine the other SM quantum numbers also to be a consequence of the symmetry. Whether that is the true origin of internal quantum number is presently unclear.

What is the size R of these extra dimensions? We have never seen it in daily life, so it must be small. We can actually do better than that by giving an estimate of how small it must be. Let $\vec{p} = (\vec{p}_e, \vec{p}_i)$ be the momentum of a particle, with $\vec{p}_e$ its components in the usual four-dimensional space, and $\vec{p}_i$ its components in the extra dimensions. The uncertainty principle allows internal momentum p_i to be either 0, or some discrete multiples of $\hbar/R$. According to the theory of special relativity, the energy of this particle of mass m is $E = c\sqrt{\vec{p}^2 + (mc)^2} = c\sqrt{\vec{p}_e^2 + \vec{p}_i^2 + (mc)^2}$. Not knowing the presence of the extra dimensions, we would interpret $\vec{p}_i^2/c^2 + m^2 \equiv M^2$ as the mass square of an excited state, separated from m^2 by an amount $(p_i/c)^2 = (n\hbar/Rc)^2$, for some integer n. Current accelerators can produce and detect masses of the order of 1 TeV, and none of these excited particles have been seen. From that we can estimate the size of the extra dimensions to be smaller than $\hbar c/(1 \text{ TeV}) \sim 10^{-19}$ m.

In the last few years, inspired by the discovery of branes in superstring theory, another possibility was proposed in which the extra dimension could be as large as 0.1 mm.

In this *braneworld scenario*, it is assumed that none of the SM particles can escape the confines of our four-dimensional spacetime, known as a 3-brane. Only SM singlets like the graviton and the right-handed neutrino ν_R may roam freely in the extra dimensions. Since the SM particles are confined to four dimensions, they have no Kaluza–Klein excited states, so their absence cannot be used to limit the size of the extra dimensions. We must somehow use gravity, or the right-handed neutrino to do it.

The gravitational force between two particles of masses m_1 and m_2 follows an inverse square law, Gm_1m_2/r^2, where G is the Newtonian gravitational coupling constant. As explained in Sec. 9.3.2, this inverse square law originates from the surface area $4\pi r^2$ for a sphere of radius r. In a spatial dimension of $3 + n$, the surface area of the sphere is proportional to r^{2+n}. The gravitational force in that case is $G'm_1m_2/r^{2+n}$, with G' being the gravitational coupling constant for the higher-dimensional theory. This r dependence is correct for $r < R$, the size of the extra dimension. For $r > R$, the sphere with radius r will be squashed to a size R along the extra dimensions, so the corresponding surface area will be proportional to $\sim R^n r^2$. The force law for $r > R$ is then $\kappa G'm_1m_2/(R^n r^2)$, where κ is a calculable geometrical factor. So even in the presence of extra dimensions, the inverse square law is restored for distances $r > R$. This property can be used to measure R, simply by reducing r until a deviation from the inverse square law is found. Current experiments detect no such deviation down to about 0.1 mm, so R cannot be larger than that. However, a size almost as large as 0.1 mm is not yet ruled out.

By comparing the inverse square law in four dimensions, and in higher dimensions in the region $r > R$, we see that $G' = GR^n/\kappa$. So G' can become fairly large for large R or n. If $R = 0.1$ mm and $n = 2$, G' will become sufficiently large to render gravity strong at the TeV scale, or an equivalent distance of $\hbar c/(1 \text{ TeV}) \sim 10^{-19}$ m, making gravitational radiations and other effects possibly detectable in the next round of accelerators.

Another way to probe such extra dimensions is through neutrino oscillations. As discussed in Sec. 9.6.3, neutrino oscillations require the neutrinos to have a finite mass, which in turn suggests the possible presence of a right-handed neutrino ν_R, which is an SM singlet. If so, this neutrino can roam in the extra dimensions, and be detected through neutrino oscillation experiments. The unusual properties of the neutrinos, with tiny masses and large mixings, might be attributed to the fact that ν_R can see the extra dimensions but other fermions cannot. Unfortunately, there are as yet an insufficient amount of data either to support or to refute this idea.

9.7.4 *Superstring and M Theories*

The fundamental entity in a superstring theory is not a particle, but a string. Its various vibrational and rotational modes give rise to particles and their excited states. Supersymmetry is incorporated into the formulation. Without it, a *tachyon* appears, which is a particle with a negative mass square. This is not allowed in a stable theory, hence supersymmetry.

A supersymmetric theory of particle physics contains an equal number of bosons and fermions. Similarly, a superstring theory brings in a fermionic string as the partner of an ordinary (bosonic) string. A bosonic string is just like a violin string, but it is a rather harder to get an intuitive understanding of what a fermionic string is, because there are no analogs in daily life. But it can be written down mathematically.

The lowest superstring modes have zero mass. The excited modes have higher masses and possibly higher spins. The mass squares are separated by constant gaps, given by a parameter usually designated as $1/\alpha'$. In the limit $\alpha' \to 0$, only massless states remain, and string theory reduces to the usual quantum field theory.

There are two kinds of fundamental strings, open or closed. An open string has two ends, which are free to flap around. A closed string does not. Later on, we shall discuss strings whose two ends are tied down, but that necessitates the introduction of objects which the string is tied down to. These objects are called *D-branes*.

There is a simple property of a vibrating string which is not shared by a vibrating membrane (a drum head), or a vibrating solid. This property is responsible for the magical features of the string theory, which we will describe.

A pulse propagating along a (closed or infinite) string does *not* change its shape, whereas a pulse propagating in a membrane or a solid does. Imagine now an additive quantum number, like the electric charge, carried by the pulse. Since the shape of the pulse does not change, the charge carried in *any segment* of the pulse will still be conserved. In this way a string generates an *infinite* number of conserved quantities, each for a different segment of the pulse. This infinite number of conserved quantities corresponds to a huge symmetry, known as the *conformal symmetry*. It is this symmetry that allows very nice things to happen in the string theory.

For example, it tells us how strings interact, by joining two strings into one, or by splitting one string into two. This symmetry guarantees such interactions to be unique, and specified only by a coupling constant g_s, telling us how often such acts happen, and nothing else.

This description of string interaction is useful only when g_s is not too large. In other words, it is a perturbative description. We do not quite know how the string behaves when the coupling is strong, but hints are available, some of which will be discussed later.

This conformal symmetry present in the classical string is destroyed by quantum fluctuation unless the underlying spacetime is ten-dimensional. Thus there are six extra spatial dimensions in a consistent quantum superstring theory. To preserve supersymmetry, this compact six-dimensional space must be a special kind of mathematical object known as *Calabi–Yau* manifolds.

We saw in the Kaluza–Klein theory (Sec. 9.7.3) that a single object in a five-dimensional spacetime (the gravitational field) can decompose into several objects in a four-dimensional spacetime (the gravitational, electromagnetic, and scalar fields[5]). This one-to-many correspondence is again there when we view the ten-dimensional superstring in four dimensions at the present energy. The detailed correspondence depends on what Calabi–Yau space we choose, and how to break down the internal and super symmetries to get to the Standard Model. There is a great deal of arbitrariness in such a process, which we do not know how to fix in perturbative string theory. It is partly for that reason that superstring theory has yielded no experimentally verifiable predictions so far. It is also why we must find a way to deal with the non-perturbative aspects of the theory.

The internal symmetry group of superstrings is also restricted by our desire to preserve quantum conformal symmetry. A careful analysis shows that there are only five types of consistent superstrings. They are, an open string which is called Type I, whose internal symmetry is $SO(32)$, two closed strings called Type IIA and Type IIB, and two 'heterotic' strings whose internal symmetry is either $SO(32)$ or $E_8 \times E_8$. We have not explained what the groups $SO(32)$ and $E_8 \times E_8$ are. Suffice it to know for our purpose that they give rise to well-defined vectorial quantum numbers and multiplets, and the multiplicity of the adjoint (gauge) multiplet of both is 496.

Each of these five superstrings has its own spectrum. Since $1/\alpha'$ is likely to be much larger than the available experimental energy, it is the zero mass particles in each case that are of the most interest. In this regard, all particles in the SM are assumed to be massless in the high energy limit at which the string theory is supposed to operate.

The spectra are supersymmetric; every boson has a fermionic partner. In what follows we shall describe only the bosonic part. It should also be pointed out that the Type I string also has a closed string sector, obtained by gluing two open strings together end to end. In other words, a closed string is present in all five superstrings.

---

[5]A classical field may be thought of as the wave function of a single particle, and a quantum field may be thought of as a collection of many particles.

Now onto the massless spectra. The gravitational tensor $g_{\mu\nu}$ ($\mu, \nu = 1, 2, \ldots, 10$) is contained in every superstring. So is an antisymmetric tensor $B_{\mu\nu}$, and a scalar ϕ called the *dilaton*.

In addition, the Type I and the $SO(32)$ heterotic string theory all contain an $SO(32)$ gauge field A_μ, and the $E_8 \times E_8$ heterotic string contains an $E_8 \times E_8$ gauge field A_μ.

Type IIA also contains totally antisymmetric tensors with 1, 3, 5, and 7 indices, and Type IIB also contains totally antisymmetric tensors with 2, 4, 6, and 8 indices. These are known as the *Ramond–Ramond (RR) potentials*. Besides this, these two theories are also different in the fermionic sector. The fermions in a IIA theory have both (ten-dimensional) chiralities, whereas the fermions in a IIB theory have only one.

We know what the gravitational and gauge fields are, albeit only in four-dimensional spacetime. We have also encountered the scalar fields, which are objects needed to form vacuum condensates. What generates these anti-symmetric objects, $B_{\mu\nu}$ and the RR potentials in a Type II theory?

A p-dimensional object, when traced through time, generates a $(p + 1)$-dimensional spacetime manifold called a *world volume*. There is a natural way to couple mathematically an antisymmetric tensor with $(p+1)$ indices to a $(p+1)$-dimensional world volume. With this coupling, we may regard the p-dimensional object to be the source of the antisymmetric tensor with $(p + 1)$ indices. The presence of these antisymmetric tensors therefore suggests the presence of these p-dimensional objects in a superstring theory.

For $B_{\mu\nu}$, the one-dimensional source is just the string itself.

For the RR potentials with $(p + 1)$ indices (p is even for IIA, and odd for IIB), the source must be some p-dimensional object, called a *p-brane*. It can be shown that the 1-brane cannot be the original string, it has to be something else.

We have introduced these p-branes to be the source of the RR potentials. But what are they? They are certainly not present in the original formulation of the perturbative string theory. If they are really there, they must come from the non-perturbative sector of the string theory. We may therefore regard the RR potentials as something that can guide us to an understanding of the elusive non-perturbative string theory.

In what follows we will try to motivate the existence of p-branes in another way: through *T-duality*.

There are certain excited modes in the closed string which are not present in an open string, nor a particle. That happens, for example, when one of the extra dimensions (say, the 9th) is a circle. A closed string can wrap around the circle, one or more times, like a rubber band. To do so, it has to stretch, and that costs it a certain amount of energy. This energy (actually c times the momentum along the 9th dimension) depends on the length of stretch, and the stiffness of the string. It is equal to $wR/\hbar c\alpha'$, where w is the number of times the string is wrapped around the circle of radius R, whose circumference is $2\pi R$. The proportionality constant

$1/2\pi\hbar c\alpha'$, which has the correct dimension of energy/distance (because $1/\alpha'$ has the dimension of square energy) tells us how stiff the string is. We may actually use this relation to define the parameter $1/\alpha'$ used in measuring the mass square gaps of the excited states.

Winding is an excitation mode unique to closed strings. There are excitation modes common to everything, obtained by having a non-zero momentum component along the 9th direction. This Kaluza–Klein momentum is given by the uncertainty principle to be $n\hbar/R$, for some integer n. The mass square of the excited state of a closed string thus receives a contribution $(n\hbar c/R + wR/\hbar c\alpha')^2$ from the 9th dimension. This expression remains the same if we make the swap $n \leftrightarrow w$, $R \leftrightarrow (\hbar c)^2\alpha'/R \equiv R'$. Thus, from an energetic point of view, there is no way to distinguish R from R', provided we also interchange n and w. This is known as *T-duality*.

Note that supersymmetry has not been mentioned in the discussion. T-duality is true for any closed bosonic string. Note also that the duality is between two perturbative strings, with radii R and R', respectively. The non-perturbative effect is not involved.

T-duality can be interpreted as saying that there is a minimum length $\hbar c\sqrt{\alpha'}$ in a closed string theory, because any size less than that is equivalent to a size larger than that, so we never have to consider a closed string theory with $R < \hbar c\sqrt{\alpha'}$.

That leaves the open string, which does not have a winding mode. So does it have a T-duality symmetry? If so, what is dual to the Kaluza–Klein momentum n/R?

An open string always has a closed string sector, which obeys T-duality. So it is very hard to imagine that the duality does not apply to the open string sector as well. In order to see what n/R is dual to, we go back to the closed string for guidance.

In a closed string, a pulse traveling in one direction will do so forever, because there are no ends to reflect it. We are particularly interested in the pulse along the 9th direction. The amplitude for the pulse traveling to the right will be denoted $f_R(\tau - \sigma)$, and the amplitude for the pulse traveling to the left will be denoted by $f_L(\tau + \sigma)$. The parameter σ, with values between 0 and π, is a scale painted on the string to tell us where we are on the string. The parameter τ is just time. The total amplitude is therefore $f(\tau, \sigma) = f_R(\tau - \sigma) + f_L(\tau + \sigma)$.

Under a T-duality transformation, $n \leftrightarrow w$. This can be shown to be equivalent to flipping the sign of the right-moving amplitude. In other words, under this transformation, $f(\tau, \sigma) \rightarrow f'(\tau, \sigma) = -f_R(\tau - \sigma) + f_L(\tau + \sigma)$. This now is a definition of T-duality which we can apply even to the open string.

The result is the following. The T-dual of an open string whose two ends are free to move in space, is an open string whose two ends are separated in the 9th dimension by a fixed distance $2\pi nR'$. Actually, further consideration shows that we can alter this distance, but whatever it is, the 9th coordinates of the two ends are fixed when the string vibrates and moves. There is no restriction on the other coordinates for the ends.

We can now imagine having two parallel 8-branes, each having its 9th coordinate fixed. The dual of the open string is now 'tied' to these two branes as it vibrates. These branes are called *D*-branes, or in this case *D*8-branes, because ends with a fixed (9th) coordinate are said to obey the *Dirichlet* boundary condition.

So far we have talked about the *T*-duality when one extra dimension is compactified into a circle. Similar discussions apply when $a > 1$ dimensions are so compactified. In that case the ends of the dual open string are tied down to two parallel $D(9 - a)$-planes.

These branes turn out to be the same ones that give rise to the *RR* potentials in a Type II theory. Since open strings end on them, you can pull the strings to deform them. They are indeed dynamical objects, though they do not appear originally in a perturbative string theory. They should be present in a full string theory including non-perturbative contributions.

An open string whose two ends are stuck to the same *D*-brane may have the middle off the brane when it is excited. However, such a configuration may be decomposed into a string lying completely in the *D*-brane, and a closed loop above it, as shown in Fig. 9.6. If we associate the open-string vibrations in a *D*3-brane with SM particles, then we are led to the braneworld scenario discussed in Sec. 9.7.3: that SM particles are always confined to our three-dimensional space. The closed

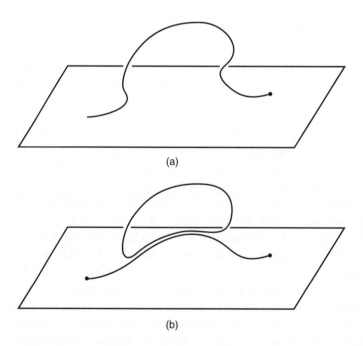

(a)

(b)

Figure 9.6: (a) A string with two ends restricted to a brane; (b) the same string viewed as the composite of two strings: an open string lying completely in the brane, and a closed string above it in the bulk.

loops which may flow into other dimensions, known as the *bulk*, must be SM singlets. In a superstring theory, they contain the graviton, the dilaton, the antisymmetric tensor $B_{\mu\nu}$, and their super-partners. Whether the right-handed neutrino is among them is a matter of conjecture.

There is another kind of duality known as S-duality. It interchanges g_S with $1/g_S$, so weak coupling becomes strong coupling. This is a more difficult subject so we will not venture into it. Suffice to say that with the studies of S-duality and T-duality transformations, one is able to change any of the five superstring theories into any other, so it seems that if these two transformations are taken to be a basic symmetry in our fundamental theory, then the five superstring theories are simply different manifestations of one and the same theory. There is very little known about what this fundamental theory is, except that it is believed to be an 11-dimensional theory, rather than the ten for strings. It is known as the M theory.

9.8 Further Reading

http://pdg.lbl.gov/. This is the Particle Data Group website. It contains a comprehensive listing of all known particles and their properties. It also contains a lot of other useful information on particle physics.

9.9 Problems

9.1 Use http://pdg.lbl.gov/2002/contents_tables.html — the Particle Data Group webpage — to produce a list of particles whose rest energy mc^2 is less than 1 GeV. Construct a table listing the name of the particle, its rest energy, and its spin (called J there), and electric charge. For fermions, list only the particles with fermionic number $+1$.

9.2 Find the range $R = \hbar c/mc^2$ of the forces obtained from the exchange of ρ, ω, η, and η'.

9.3 Let us assume the energy required to pull a quark-antiquark a distance r apart from their equilibrium position is $V = \sigma r$, with $\sigma = (400 \text{ MeV})^2/\hbar c$. At what r is this energy large enough to create a $\pi^+\pi^-$ pair?

9.4 Let q be the quark number, Q the electric charge, and I_3 the z-component of the strong isospin of a u, d quark, or their antiquarks. Show that the relation $Q = I_3 + q/6$ is valid in each case.

9.5 The measured magnitudes of the CKM matrix elements in Eq. (9.1) can be found on the webpage http://pdg.lbl.gov/2002/kmmixrpp.ps. Take in each case the central value in the range of values given. Find the parameters λ, A, ρ, η in the Wolfenstein parameterization of the VKM matrix (footnote 3) to fit approximately the measured values.

Cosmology

10

10.1 Hubble's Law

Modern cosmology is founded on the 1929 discovery of Edwin Hubble: distant galaxies are flying away from us, and the farther the galaxy, the faster is its recession. This discovery can be summarized into *Hubble's law* $v = H_0 s$, where s is the distance to the galaxy and v is its receding velocity. The proportionality constant H_0 is called *Hubble's constant*. Its value is frequently expressed in units of 100 (km/s)/Mpc, namely, $H_0 = h_0 \times 100$ km/s per Mpc. The latest value of h_0, obtained using the Hubble Space Telescope and published in 2001, is 0.72 ± 0.08, but values around $h_0 = 0.65$ obtained from other data have also frequently been used. The unit 'Mpc' stands for *megaparsec*; it is a distance unit which will be discussed below.

To establish the law, both the distance to a galaxy and its receding velocity must be measured. These are discussed separately in the next two subsections.

10.1.1 Velocity Measurements

The velocity of an object along the line of sight can be measured using its *Doppler shift*. This effect is what highway police use to catch a speeding driver. It is also what makes the whistle of an approaching train have a higher pitch, and the whistle of a receding train, a lower pitch.

When a train moves away from us, it stretches the sound wave between the train and our ear, thereby elongating its wavelength and lowering its pitch. An approaching train does the opposite, compresses the wavelength, and increases the pitch. The faster the train, the greater the stretch or compression, and the greater the change in pitch. This change of wavelength, or of frequency, is the Doppler shift.

A similar effect also occurs with light waves, except that this time what we experience is not the musical sensation of pitch, but the visual sensation of color. A shift towards a longer wavelength in the visible spectrum moves light towards the red color, and is thus called a *red-shift*. A shift towards a shorter wavelength moves light towards the blue color and is therefore called a *blue-shift*. Again, the amount of shift indicates the velocity of the object; red-shift for recession, and blue-shift for approach. It is by measuring such shifts in the light reaching us that the velocity of stars and galaxies is determined.

Astronomers use not just visible light, but electromagnetic radiation at other wavelengths as well. Even though this radiation is beyond the visible range, the terms red-shift and blue-shift are still used.

If λ is the wavelength at the source, receding at a velocity v, and if λ' is the wavelength observed on earth, then the quantity $z = (\lambda'/\lambda) - 1$ measures the amount of shift, and is called the *red-shift factor*. It is approximately equal to v/c when v is small, where $c \sim 3 \times 10^8$ m/s is the speed of light. For the galaxies measured by Hubble, v/c is indeed small.

To make red-shift a useful tool, we must know the original color of the light in order to figure out the amount of shift. Since we are unable to go to the star to find out, we must resort to other means. For this we make use of a remarkable property of the atoms: the existence of *spectral lines*.

Just like finger prints and dental records, which vary from individual to individual, and can be used to identify a person, atoms and molecules can also be distinguished by their 'finger prints' — in the form of their characteristic spectral lines. An object, when it is heated, emits two kinds of electromagnetic radiation. One contains a broad spectrum of all wavelengths. It reflects the ambient temperature of the surroundings but not the identity of the atom (see Sec. 10.5.2 and Appendix C). The other kind appears at discrete and fixed wavelengths characteristic of the atom. When analyzed through a spectrograph, this kind of light appears as a series of lines on a photographic plate, hence they are called spectral lines. The spectral lines are caused by the electrons of the atom jumping from one discrete orbit to another. The wavelengths characterize the energy differences of the orbits, which characterize the atom. It is for this reason that spectral lines can be used as a tool to identify atoms.

By looking at the light coming from a star or a galaxy, two interesting points were discovered.

First, except for possible Doppler shifts, the spectral lines of light from distant stars are identical to the spectral lines of the same atoms on earth. This must mean that the physics on a distant star is identical to the physics on earth! This is an extremely important discovery, for it allows us to apply whatever knowledge of physics we learn here to other corners of the universe.

Secondly, the spectral lines from many galaxies are red-shifted (but seldom blue-shifted). This means that these galaxies are flying away from us. The velocity of this flight can be determined by measuring the amount of red-shift.

Between 1912 and 1925, V. M. Slipher measured the red-shifts for a number of galaxies. Using Slipher's compilation of red-shifts of galaxies, the receding velocities can be computed. With his own measurement of distances, Hubble was then able to establish his famous law that the receding velocity increases linearly with the distance of the galaxy.

This law revolutionizes our thinking of the universe, as will be discussed in the rest of this chapter.

10.1.2 *Astronomical Distances*

The second requirement to establish Hubble's law is a way to measure astronomical distances.

Distances between cities are measured in kilometers (km), but this is too small a scale in astronomy. Astronomical distances are huge. It takes light, traveling with a speed $c = 3 \times 10^5$ km/s, a speed which allows it to go around the earth more than seven times in a second, 500 seconds to reach the earth from the sun, and some four years to reach the nearest star. The appropriate unit to measure such vast distances is either a *light-year*, or a *parsec*. A light-year is the *distance* covered by light in one year. The nearest star is four light-years from us, the center of our Milky Way galaxy is some twenty-six thousand light-years away. In ordinary units, a light-year is equal to 9.46×10^{12} km, or in words, almost ten million million kilometers.

One parsec is 3.26 light-years. The name comes from the way astronomical distances are measured. Distances to nearby stars are measured by triangulation. As the earth goes around its orbit, the position in sky of a *nearby star* seems to shift against the background of distant stars, for much the same reason that nearby scenery appears to move against distant mountains when you look out of the window of your car. If the star is at a distance such that the shift from the average position is 1 second of arc each way, then the star is said to have a parallax angle of one second, and the distance of the star is defined to be 1 *parsec*. This distance turns out to be 3.26 light-years. A thousand parsecs is called a kiloparsec (kpc), and a million parsecs is called a megaparsec (Mpc). It is this last unit that has been used in the value of the Hubble constant above.

The more distant the star, the smaller the parallax angle, so after some 20 pc or so, the parallax angle will become too small for this method of measuring distances to be useful. Other methods are required.

One such method was discovered by Henrietta Leavitt and Harlow Shapley at the beginning of the twentieth century. It makes use of the special properties of a class of stars, called the *Cepheid variables*.

During his exploration in 1520, Magellan found two luminous 'clouds' in the sky of the southern hemisphere. These two 'clouds' turn out to be two nearby irregular galaxies, nowadays called the *Large Magellanic Cloud* (LMC) and the *Small Magellanic Cloud* (SMC). In 1912, Leavitt found in the SMC twenty five variable stars of the type known as the *Cepheid variables*. A variable star is a star whose brightness changes with time. A Cepheid variable is a kind of variable star which has a fixed period of brightness variation, and this period usually falls within a range of a few days to a month or more. They are called the Cepheid variables because a typical variable star of this type is found in the constellation Cepheus.

By studying these Cepheid variables in the SMC, Leavitt found that the period of variation of these twenty five Cepheid variables follows a simple law. The period of a star is related to its average brightness observed on earth: the longer the period, the brighter the star.

The brightness of astronomical objects is measured in *magnitudes*. It is a logarithmic scale. If star A has a magnitude x and star B has a magnitude $x - y$, then star B is $(100)^{y/5} \simeq (2.512)^y$ times brighter than star A. The larger the magnitude, the dimmer is the object. The brightest star in the sky, Sirius, has a magnitude -1.47. Our sun has a magnitude -26.7.

These are the *apparent magnitudes*, which measures the brightness of these objects seen on earth. This apparent brightness depends on the intrinsic brightness of the object, as well as its distance R from us. Light emitted from a star is spread over a spherical shell whose area grows with R^2, thus the apparent brightness falls as $1/R^2$. The intrinsic brightness of an object is measured in *absolute magnitudes*. It is defined to be the apparent magnitude if the object is put 10 pc away. The absolute magnitude of Sirius turns out to be $+1.4$. The sun, whose apparent brightness is so much greater than Sirius, turns out to have a much larger absolute magnitude of $+4.8$.

If we know the distance of the star from earth and its apparent brightness as seen here, then we can use this relation to compute its intrinsic brightness. This is important because it must be the intrinsic brightness that is controlled by physics of the star, and not the apparent brightness.

At the time of Leavitt's discovery, the distance to the SMC was not known. In 1917, Harlow Shapley was able to use a statistical method of stellar motions to estimate this distance, as well as the distances to some globular clusters containing Cepheids. The distance to each individual star in a galaxy or a globular cluster is different for different stars, but since a galaxy or a globular cluster is very far away from us, the difference in distances between individual stars within a galaxy or a globular cluster is negligible in comparison, so we may treat them all to be at some average distance to the galaxy or globular cluster.

This distance measurement allowed Harlow Shapley to convert the apparent brightness of these stars to their intrinsic brightness. Like Leavitt, he plotted the period of these Cepheids against their brightness, but this time he used the *intrinsic* brightness. In this way he found that the period of variation of a Cepheid variable bears a *definite* relationship with its *intrinsic brightness*, whether the star is in one galaxy or another globular cluster. In other words, he found this to be a universal physical law independent of the location of the star. This celebrated law is called the *period–luminosity relation* for the Cepheid variables.

Once established, this period–luminosity relation can be used as a yardstick to measure the distance to a new Cepheid variable. For that reason Cepheid variables can be regarded as a *standard candle*. What one has to do is to measure its period as well as its apparent brightness on earth. From the period one gets its intrinsic brightness through this relation. Knowing now both the intrinsic and the apparent brightness, one can calculate its distance R.

In 1923, Hubble found twelve Cepheid variables in a nearby nebula in the Andromeda constellation. This nebula is actually a galaxy and is now called the *Andromeda galaxy*, or by its Messier catalog number M31. He also found twenty-two other Cepheid variables in another nebula, M33. In this way, he determined both of their distances to be several hundreds of kiloparsecs away, placing them well outside of the confines of our Milky Way galaxy, and settling the debate once and for all in favor of these objects being distant galaxies in their own right, rather than some luminous clouds within our own galaxy.

It also allowed him to establish Hubble's law.

The distance to nearby galaxies can be determined by the brightness of their Cepheid variables. The farthest Cepheid variable observed to date was through the Hubble Space Telescope, in 1994, in a galaxy called M100 in the Virgo cluster of galaxies. Its distance is determined to be 56 ± 6 million light-years.

In recent years, a much brighter standard candle has been found to measure greater distances. This is the type Ia supernova. As discussed in Chapter 8, a supernova is an exploding star. It comes in two types. Type II results from the collapse of a single massive star at the end of its life, while Type Ia is always associated with a binary star system. One of the two stars in such a binary system is a white dwarf (see Chapter 8), and the other is a live star which keeps on dumping hydrogen from its atmosphere onto the nearby white dwarf. When the accumulated weight becomes too large for the white dwarf to support, a collapse and a subsequent explosion takes place. Since the two kinds of explosions have different origins and the stars involved have different histories, the spectral lines seen in their explosions also have different chemical compositions. For Type II supernova, the outermost atmosphere consists of hydrogen, so hydrogen lines are usually seen. For Type Ia, hydrogen dumped on the surface of the white dwarf collapses and fuses, so no hydrogen lines are present. The light curves of the two, as the supernova brightens and then dims, also have different characteristics. It is through such differences that one can distinguish a Type Ia supernova from a Type II supernova.

After the explosion, it takes less than a month for a Type Ia supernova to reach its peak brightness. At its peak, it can be several billion times brighter than our sun, though being so much further away than our sun, of course we do not see it on earth as bright as the sun. The time it takes the brightness to decay away is found empirically to depend on the peak brightness: the brighter it is the longer it takes to decay. This is then very much like the Cepheid variables, but instead of using the period of brightness variation to determine the intrinsic brightness we can now use the decay time of a Type Ia supernova to determine its intrinsic brightness, once the intrinsic brightness of some of them is determined by other means. So like the Cepheid variable, it can be used as a standard candle to measure the distance.

Since the supernova is so much brighter than a star, we can use it to reach a much greater distance.

With this method one can reach galaxies much farther than those seen by Hubble. The farthest Type Ia supernova to date has a red-shift factor of $z = 1.7$. The result of such deep sky measurements will be discussed in a later section.

10.1.3 Summary

The velocity of a galaxy can be measured by the amount of red-shift in its spectral lines. Its distance can be measured using either Cepheid variables or Type Ia supernova as standard candles, or some other means not discussed here.

By measuring the velocity and the distance of relatively nearby galaxies, Edwin Hubble established the most important law in modern cosmology, that galaxies are receding from us with velocities proportional to their distances. The latest value of Hubble's constant is $H_0 = 72$ km/s/Mpc.

10.2 The Big Bang

Since galaxies are flying away from us, they must have been closer a million years ago, and closer still a billion years ago. Running this backwards, at some time past all galaxies must come together to a single point. That point can be taken to be the beginning of the universe. The 'explosion' that sets every galaxy flying outward has come to be known as the *Big Bang*.

We can estimate how long ago that happened, if we assume the velocity v of every galaxy to be constant in time. If t_0 is the present time, then the present distance of the galaxy from us is $s = vt_0$. Comparing this with Hubble's law, $s = vH_0^{-1}$, we arrive at the conclusion that the present age of the universe is $t_0 = H_0^{-1}$. If we take $H_0 = 72$ km/sec/Mpc, then H_0^{-1} turns out to be about 14 billion years.

However, v cannot be a constant because there are gravitational pulls from the rest of the universe, and there is also pressure exerted by the other galaxies. Both cause a deceleration, making v today to be smaller than that in the past. Taking this into account, we shall see that the age of the universe is reduced to $(2/3)H_0^{-1}$. With the above value of H_0, the age comes out to be about 9 billion years. This causes a serious problem, because we know of stars in globular clusters which are 11 or 12 billion years old, and it does not make sense to have a star to be older than the universe it lives in! This serious dilemma has been resolved by a recent observation which we will come back to in the next section.

For now, let us discuss some of the implications of Hubble's law, and the Big Bang. Galaxies fly away from us in every direction. Does it mean that we are at the center of the universe? Well, yes, if it helps our ego to think so. However, we had better realize that everyone else in other galaxies can claim the same honor as well. This is because we know from observation that the large-scale structure of our universe is homogeneous and isotropic. There can therefore be no distinction between a galaxy here and a galaxy in another part of the universe. To help us understand this point, we may imagine the universe to be a very large loaf of raisin bread being baked in the oven, with the galaxies being the raisins, and the expansion of the universe simulated by the rise of the bread. Whichever raisin you live on, all the other raisins are moving away from you isotropically as the bread rises, so every raisin can claim to be at the center of the universe.

The idea that the universe started from a Big Bang also brings about many other interesting questions. What caused the explosion, and what happened before the explosion? Are space and time already there when the universe exploded, or are they created together with the universe? Why is the universe created in a three-dimensional space, and not in ten dimensions, for example? If it is ten dimensions, why are three of them big and the other seven so small to be unobservable? How does the universe evolve after creation, and how is it going to end up? There are many more questions of this kind that can be asked. We do not know most of the answers, but we know a few. Those will be discussed in the rest of this chapter.

10.2.1 Summary

The universe as we know it began in an explosion, known as the Big Bang, some ten to twenty billion years ago. This picture of the universe brings up many interesting philosophical and physical problems, some of which will be discussed in the rest of the chapter.

10.3 After the Big Bang

In this section we discuss how the size of the universe evolves under the influence of gravity and pressure. Derivations and quantitative details are relegated to the last subsection.

10.3.1 Gravity and Pressure

The distribution of matter in the universe, averaged over a scale of hundreds of millions of light-years, looks the same everywhere. For that reason we may model the universe as a homogeneous fluid, with the galaxies[1] playing the role of molecules of the fluid.

A homogeneous fluid at any time t is specified by an energy density[2] $\rho(t)$, and a pressure $P(t)$. If r is the present distance of a galaxy (from us), then the uniform expansion of the universe is described by a common scale factor $a(t)$, so that the

---

[1] Note that real galaxies are not formed until much later, but even in the early universe, it is convenient to refer to the chunks of matter that make up the fluid to be 'galaxies.'

[2] According to the theory of relativity, the equivalent mass density is $\rho(t)/c^2$.

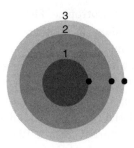

Figure 10.1: The position of a galaxy (black dot) in an expanding universe at three successive times. Darker shading indicates a higher density.

distance of the galaxy at time t is $s(t) = a(t)r$. In particular, this fixes the normalization of $a(t)$ to be $a(t_0) = 1$ at the present time t_0. The present distance r is known as the *comoving distance*.

Such a model of the big-bang universe is known as a Friedmann–Robertson–Walker (FRW) universe.

The black dots in Fig. 10.1 represent a galaxy at three different times, $t_1 < t_2 < t_3$. As the universe expands, the scale factor $a(t)$ gets bigger and the density $\rho(t)$ becomes smaller. This is indicated in the diagram by the different size of the spheres, and the different gray-scale shadings. The black-dot galaxy on the surface of the sphere is surrounded on all sides by other galaxies. Only galaxies inside the sphere exert a net gravitational pull on the black-dot galaxy. This pull can be represented by a mass M placed at the center of the sphere, with M equal to the total mass of all the galaxies inside the sphere. The gravitational pull from galaxies outside the sphere all cancel one another. The isotropic expansion of the universe may now be thought to be the expansion of the spherical surfaces.

The expansion of the sphere is affected by two factors: gravitational pull from the galaxies inside the sphere, and pressure exerted on the sphere by the galaxies outside to resist its expansion. Both tend to slow down the expansion of the sphere and the outward flight of the black-dot galaxy. The total effect of these two is a force proportional to $\rho(t) + 3P(t)$, as we shall demonstrate later. The first term is due to gravity, while the second term comes from pressure.

Right after the Big Bang, the universe is hot and the energy density is high. This causes the galaxies to have a highly relativistic random motion, thereby exerting a large pressure on the sphere and the surrounding galaxies. This pressure is related to the energy density by the equation $P = \rho/3$. Such an equation is known as an *equation of state*. This period is known as the *radiation-dominated era* because the galactic motions are relativistic, like light. In this epoch, $a(t)$ grows like $\sqrt{t}$ and $\rho(t)a^3(t)$ falls like $1/a(t)$, as we shall show. The total energy inside a sphere of comoving radius r is proportional to ρa^3. This energy falls because part of it is used up to do work to counter the pressure exerted by the galaxies as the sphere expands.

As the universe expands, its energy is being shared by a larger volume, so its temperature drops. Eventually, the universe will cool down sufficiently so that

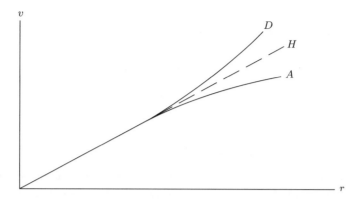

Figure 10.2: Curves H, D, and A represent schematically a universe with a constant velocity, a universe which is slowing down, and a universe which is speeding up, respectively.

the random motion of the galaxies becomes negligible. When that happens the pressure P is practically zero, so only gravity is there to hold back the expansion of the universe. In that period $a(t)$ grows a bit faster, as $t^{2/3}$. If gravity were absent as well, then the velocity would be constant, and $a(t)$ would be proportional to t. The persistent presence of gravity is what makes it grow slower than t, as $t^{2/3}$. This also causes the present age of the universe to be $(2/3)H_0^{-1}$ rather than H_0^{-1}. The absence of pressure means there is no need to expend energy to do work, hence $\rho(t)a^3(t)$ remains a constant. This is the *matter-dominated era*.

The universe decelerates through both of these epochs, so the velocity $v = ds(t)/dt = r(da(t)/dt)$ should be larger in the past than at the present. If no deceleration occured, then $a(t) = t/t_0$. By differentiation, we get back to Hubble's law $v = r/t_0 = H_0 s(t_0)$, with $H_0 = 1/t_0$. This is the straight line labeled H in Fig. 10.2. With deceleration, v should be larger in the past than at the present. Since light reaching us today was emitted some time ago, looking at a distant galaxy is like looking back in time. For example, the light we see today from a galaxy one billion light-years away was emitted one billion years ago. Consequently, the curve in the Hubble plot with deceleration taken into account should lie above the straight line H, as in D in Fig. 10.2.

10.3.2 The Red-Shift Factor in an Expanding Universe

In Sec. 10.1.1 we used the Dopper shift formula $z \simeq v/c$ to measure the velocity v of a nearby galaxy. This formula is derived for a static universe. How should it be modified when the universe is expanding?

For a nearby galaxy whose red-shift is small, there is no time for the size of the universe to change significantly before its light reaches us, so the static universe formula used in the original Hubble plot is valid.

For faraway galaxies when the red-shift is large, this is no longer the case. It will be shown in Sec. 10.3.5 that the correct formula is $z+1 = a(t_0)/a(t)$, the ratio of the scale factors at the time of detection and the time of emission of the electromagnetic radiation. Since the present scale factor is $a(t_0) = 1$, this formula can be written simply as $z+1 = 1/a(t)$. This is a useful formula because a measurement of red-shift tells us the size of the universe at the time of emission.

For nearby galaxies, this formula reduces to the naive Dopper shift formula[3] $z \simeq v/c$.

10.3.3 The Accelerating Universe

Using type Ia supernovae as standard candles to measure distant galaxies, the Hubble plot at large r was found to curve downward, like curve A in the schematic drawing of Fig. 10.2, rather than the curve D we expected. This means the universe is accelerating, rather than decelerating. How can that be?

Whatever the reason, this amazing discovery in 1998 resolves the dilemma regarding the age of the universe mentioned in Sec. 10.2. Recall that the formula $t_0 = (2/3)H_0^{-1}$ gives an age of the universe smaller than the age of some stars. If the universe were to expand at a constant rate, the age would have been H_0^{-1}, and there would be no dilemma. The extra factor $2/3$ comes from deceleration, so it should be replaced by a larger factor when the present-day acceleration is taken into account. The effective factor turns out to be close to 1, making the true age of the universe to be 14 to 15 billion years, longer than the age of all known stars, so the dilemma is no longer present.

To understand the acceleration mechanism, let us recall that the effective force on a galaxy is proportional to $\rho+3P$. The first term is due to gravity, and the second term is due to pressure. The energy density ρ is positive, but the pressure P may be either positive or negative. If it is negative and larger than $\rho/3$ in magnitude, then the effective force is negative, thereby causing acceleration. In other words, the black-dot galaxy in Fig. 10.1 and the surface on which it lies are being sucked out by the negative pressure rather than being pushed in. When the suction is sufficiently strong, it overcomes the gravitational pull as well, in which case the universe accelerates. That, presumably, is what is happening in the present universe.

For relativistic matter, $P = \rho/3 > 0$. For non-relativistic matter, $P = 0$. In both cases the pressure is non-negative. How can a negative pressure possibly be present then?

The answer is *dark energy*. By definition, it is a new form of energy whose pressure is negative. To figure out its equation of state, let us consider the following.

In the radiation-dominated era, ρa^3 decreases as $1/a$, because energy must be spent to counteract the positive pressure. In the matter-dominated era, ρa^3 becomes a constant, for there is no work to be done when the pressure is zero. We might

---

[3]The Hubble constant is $H_0 = \dot{a}(t_0)/a(t_0)$, where a dot indicates the time derivative. For nearby galaxies, $t \sim t_0$, so we can approximate $a(t)$ by $a(t) \simeq a(t_0) + \dot{a}(t_0)(t - t_0)$. Hence, $z = (a(t_0) - a(t))/a(t) \simeq H_0(t_0 - t) = H_0 r/c = v/c$, according to Hubble's law.

therefore suspect that if ρa^3 increases with a, then P might turn negative. This is indeed the case. In fact, if ρa^3 is proportional to a^{-3w}, then $P = w\rho$. The radiation-dominated era corresponds to $w = 1/3$, and the matter-dominated era corresponds to $w = 0$. To have a negative pressure, we must have $w < 0$. This kind of energy with a negative pressure is known as the dark energy. If $w = -1$, ρ is independent of a, and time. The energy is then known as the *vacuum energy*.

Vacuum energy was first introduced into cosmology by Einstein, before the discovery of Hubble's law. At that time the universe was thought to be static. To achieve that, a force opposing the gravitational attraction must be present. This is why Einstein introduced the *cosmological constant*, which is a quantity that is proportional to the vacuum energy. More exactly, if ρ_1 is the energy density of matter and ρ_2 is the energy density of the vacuum, then their corresponding pressures are $P_1 = 0$ and $P_2 = -\rho_2$. The total force is zero if $(\rho_1 + \rho_2) + 3(P_1 + P_2) = 0$, which requires $\rho_2 = \frac{1}{2}\rho_1$.

Later on, when the universe was found to be expanding, Einstein regretted his introduction of the cosmological constant, saying that it was the biggest mistake he had made in his life. Too bad he did not live to see the recent discovery of the accelerating universe, for the cosmological constant may be needed after all. To have acceleration, we must have $\rho_2 > \frac{1}{2}\rho_1$. In fact, from the Hubble plot and other observations to be discussed later, one finds the present dark energy density to be $\rho_2 \simeq 2\rho_1$.

If the dark energy is not the vacuum energy, then $P_2 = w\rho_2$ and the condition for acceleration is then $\rho_2 > \rho_1/(-3w - 1)$. With $\rho_2 \simeq 2\rho_1$ giving rise to the present-day acceleration, this also shows we must have $w < -1/2$.

The universe must have been decelerating until the recent past, when it turned to acceleration. This is so because the vacuum energy density ρ_2 is independent of a, so it remains the same at all times, but the matter energy density ρ_1 is proportional to $1/a^3$ for non-relativistic matter, so it grew bigger in the past. Consequently, at some time back the inequality $\rho_2 > \frac{1}{2}\rho_1$ needed for acceleration could no longer true. A present ratio of $\rho_2/\rho_1 \simeq 2$ would turn into a ratio of $\simeq \frac{1}{2}$ when $a \simeq (1/4)^{1/3} \simeq 0.63$. This occurs at a red-shift factor of $z \simeq 0.59$, a very recent past in the cosmological scale. A similar estimate can be carried out if $0 > w > -1$.

Where does the vacuum energy come from? There are models, but we do not really know the answer. Vacuum energy is believed to be present in the modern theory of elementary particle physics (Chapter 9), but the energy density obtained from particle physics is vastly greater than the vacuum density needed to explain the present acceleration. This is one of the greatest unsolved puzzles, known as the *cosmological constant problem*.

10.3.4 *The Critical Universe*

A crucial parameter governing the evolution of the universe is the *critical density* ρ_c. As will be shown in Sec. 10.3.5, this parameter depends on the Hubble constant H_0.

With the present value of H_0, the critical mass density is $\rho_c/c^2 \simeq 10^{-29}$ g/(cm)^3, which is the density of five protons per cubic meter.

It is conventional to use the ratio $\Omega = \rho/\rho_c$ as a measure of the density ρ of the universe. If this is larger than 1, the universe is said to be *closed*. If it is less than 1, the universe is said to be *open*. If it is equal to 1, the universe is said to be *critical*.

A galaxy in an expanding universe possesses a kinetic energy. Being attracted by other galaxies it also possesses a gravitational potential energy. The kinetic energy is respectively larger, equal, and smaller than its potential energy when $\Omega > 1, = 1, < 1$. A critical universe is spatially flat. Euclidean geometry is applicable there, so parallel lines never meet. A closed universe has a positive spatial curvature like the surface of a sphere. Similar to the longitude lines on earth which meet at the north and south poles, seemingly parallel lines do meet further on in a positively curved universe. An open universe has a negative curvature, seemingly parallel lines will get farther from each other further out.

The critical density is awfully small. The best vacuum that can be achieved in the laboratory contains about 2.5×10^6 molecules per cubic meter, which is at least some 500,000 times bigger than the critical density of the universe. Surely our universe must be closed?

It is not so, because there are lots of empty space in the universe. In fact, for years astronomers believed our universe to be open. Stars account for only about 0.35% of the critical density, but that by itself should not be taken as evidence for an open universe, because lots of matter do not reside in the living stars. By measuring the abundance of light elements made at the beginning of the universe, a subject which will be discussed in Sec. 10.6.4, one deduces that the total amount of ordinary matter, including the bright ones in shining stars, and the dark ones in dead stars, non-luminous clouds, planets, etc., accounts for about 4% to 5% of the critical density. On top of that, it has been known for some time that there must be additional *dark matter* around whose nature we do not know, because no such matter have yet been found through direct observation. We are aware of its presence only from its gravitational influence on stars and galaxies. We do know that most of it has to be non-relativistic in order for galaxies to cluster the way they do. For that reason this kind of dark matter is sometimes referred to as *cold dark matter*.

Since the universe is accelerating, we know that dark energy is present as well. Dark energy and matter of various kinds not only change the Hubble plot, they also affect the cosmic background radiation (see Sec. 10.7), and the clustering of galaxies, so we can obtain their amounts from these observations as well. The result shows that our universe is compatible to being critical. About 70% of the critical density resides in the dark energy, and about 30% is contained in matter. As mentioned before, only about 4% to 5% of the critical density is in ordinary matter, so the cold dark matter outnumbers ordinary matter more than 5 to 1.

More than twenty years ago, when astronomers swore to an open universe, this possibility of a critical universe was already suggested by Alan Guth and others, to explain some of the puzzles encountered in the classical Big Bang theory. We will come back to discuss this inflationary theory of the universe in the next section.

10.3.5 The Dynamics

We study in this subsection how gravity and pressure *quantitatively* affect the expansion of the universe. In particular, formulas used in the previous subsections will be derived. Readers not interested in such details may skip now to the next section.

We are interested in obtaining the equations governing the time variations of $a(t)$, $\rho(t)$, and $P(t)$. To do so we need three equations, which can be obtained as follows.

Let us look at a nearby galaxy, represented by a black dot in Fig. 10.1. Suppose it has a mass m and a distance $s(t) = a(t)r$ from the center.[4] Its kinetic energy at any time is $\frac{1}{2}m(ds/dt)^2$, and its potential energy is $-GMm/s$, where M is the mass inside the sphere and G is the Newtonian gravitational constant. Its total energy is a constant, which we shall denote by $-\frac{1}{2}mr^2k$. The total energy inside the sphere is $Mc^2 = (4\pi/3)s^3\rho$, and hence the equation expressing the total energy of the galaxy can be written as

$$\left(\frac{da}{dt}\right)^2 - \left(\frac{8\pi}{3c^2}\right)G\rho a^2 = -k. \tag{10.1}$$

In this form the equation is independent of the specific galaxy used to derive it. It is an equation governing the temporal change of the scale factor $a(t)$. The ratio $(da/dt)/a$ is the Hubble constant H at time t. This equation is known as the *Friedmann equation*.

A closed universe corresponds to $k > 0$, an open universe corresponds to $k < 0$, and a critical universe corresponds to $k = 0$. In the last case, the critical density ρ_c of the universe is related to the present Hubble constant H_0 by the relation

$$\rho_c = \frac{3c^2 H_0^2}{8\pi G}.$$

Using $H_0 = 72$ km/sec/Mpc, we obtain $\rho_c = 10^{-29}$g/(cm)^3, as reported before.

To solve Eq. (10.1) for $a(t)$ we must have an equation for $\rho(t)$. This comes about by considering the work done by the sphere needed to overcome the surrounding pressure. To expand the sphere of radius s to $s + ds$, the amount of work required is $P(4\pi s^2)ds$. The energy needed to do this work is supplied by the energy stored inside the sphere, and hence the work done must be equal to $-d[(4\pi/3)s^3\rho]$. This can be written as

$$\frac{d(\rho a^3)}{da} = -3Pa^2. \tag{10.2}$$

By applying $d/dt = (da/dt)(d/da)$ to Eq. (10.1) and then using Eq. (10.2), we obtain an equation for the acceleration:

$$\frac{d^2a}{dt^2} = -\left(\frac{4\pi}{3c^2}\right)G(\rho + 3P)a. \tag{10.3}$$

---

[4]You can think of the center as where we are, or just any other fixed point in the universe.

Multiplying both sides by mr, this becomes the equation of motion for the galaxy with mass m, which shows that the effective force on a galaxy is proportional to $\rho + 3P$, as previously claimed.

Since P is introduced in Eq. (10.2), we need a third equation in order to determine the three quantities $a(t)$, $\rho(t)$ and $P(t)$. That equation depends on the structure of the fluid, and is known as the *equation of state*. It has nothing to do with the size of the universe.

At the beginning of the universe when it is hot, the galaxies acquire a highly relativistic random motion whose equation of state is given by $P = \rho/3$. Later on, when most of the matter becomes cool and non-relativistic, pressure is lost so that the equation of state is simply $P \simeq 0$. Dark energy has an equation of state $P = w\rho$ for some negative w.

Given an equation of state, Eq. (10.2) can be used to find out how ρ depends on a. For $P = \rho/3$, we get $\rho \propto 1/a^4$. For $P = 0$, we get $\rho \propto 1/a^3$. And for $P = w\rho$, it is $\rho \propto a^{-3(w+1)}$.

For the critical universe where $k = 0$, this information can be used to obtain an easy solution of $a(t)$ in each of these three cases. Letting $\kappa = \sqrt{8\pi G/3c^2}$ and assuming $\rho = \beta^2 a^{-2p}$, the solution of Eq. (10.1) is

$$a^p(t) = a^p(t_i) + (\beta\kappa/p)(t - t_i), \ (p \neq 0),$$
$$a(t) = a(t_i)\exp[\beta\kappa(t - t_i)], \quad (p = 0),$$

(10.4)

where $a(t_i)$ is the initial size at time t_i. For the three scenarios, relativistic matter, non-relativistic matter, and dark energy, the parameter p is respectively 2, 3/2, and $(3/2)(1 + w)$.

The scale factor $a(t)$ of the universe can be determined by measuring the red-shift of light leaving a galaxy at time t. The relation is

$$z + 1 = \frac{a(t_0)}{a(t)}.$$

(10.5)

The right-hand side simply reduces to $1/a(t)$ if we normalize the magnitude of the present-day scale factor $a(t_0)$ to be 1. This formula can be derived as follows. By definition, $z + 1 = \nu_E/\nu_0$ is the ratio of the frequency of light emitted at the galaxy (ν_E) to the frequency received on earth (ν_0). Imagine now a pulse of light is sent out by the galaxy at the beginning of every period. This means that the first pulse and the second pulse are sent out a time $1/\nu_E$ apart. The time difference of the arrival of these two pulses on earth is the period, or $1/\nu_0$. Now over a time period dt, light travels from the galaxy towards earth a distance $ds(t) = a(t)dr = cdt$. The comoving distance traveled by the first pulse of light between the galaxy and earth is therefore $r = \int_t^{t_0} cdt/a(t)$. The second pulse of light is emitted at time $t + 1/\nu_E$, and received at time $t_0 + 1/\nu_0$. It travels the same comoving distance r. Equating these two equations, we obtain

$$\int_t^{t+\nu_E^{-1}} dt/a(t) = \int_{t_0}^{t_0+\nu_0^{-1}} dt/a(t).$$

Since the integration intervals on both sides are very small, each integrand stays practically constant, and hence $\nu_E^{-1}/a(t) = \nu_0^{-1}/a(t_0)$. This is equivalent to Eq. (10.5).

10.3.6   Summary

Expansion of the universe after the Big Bang is described by a scale factor $a(t)$, whose time dependence is controlled by the energy density of the universe. Our universe is found to be critical, and hence spatially flat. The required energy density is equal to the rest energy of about 5 protons per cubic meter. About 70% of that resides in the form of dark energy, and 30% in the form of matter. Of the latter, about 4 to 5% of the critical energy is composed of ordinary hadronic matter, with the rest mostly in the form of cold dark matter, which has not yet been directly discovered. The origin of dark energy is not yet understood, but we do know that it causes the present universe to accelerate.

10.4   Before the Big Bang

10.4.1   Problems of the Classical Big Bang Theory

What caused the Big Bang?

A possible answer came to Alan Guth around 1980 when he investigated the problems arising from the standard Big Bang theory. He solved these problems by inventing a period of inflation preceeding the Big Bang, during which the universe expands exponentially fast. The end of inflation erupts into an explosion causing the Big Bang.

The inflation theory predicted our universe to be critical some fifteen years before its recent verification. It also predicted a power spectrum consistent with recent measurements of microwave background radiation and galactic distribution. To explain it, let us review the problems Guth was trying to solve.

The most serious is the *horizon problem*. This is the problem of understanding why different parts of the universe have such a uniform temperature, 2.725 K, as revealed by the cosmic microwave radiation (CMB), which will be discussed in Sec. 10.7.

Right after the Big Bang, the expansion rate $da/dt \propto 1/\sqrt{t}$ is almost infinite, so galaxies were flying apart at that time faster than the speed of light. Even so, special relativity is not violated, because that merely asserts no particle and no information may propagate faster than the speed of light. Expansion of the universe comes from the stretching of space, and that cannot be used to propagate information or a particle from one *place* to another. However, that does make it difficult for information sent out from one galaxy at the beginning of the universe to reach another galaxy, because the other galaxy is running away faster than light can catch up. Gradually, as the expansion slows down, the speed of light is able to surpass the expansion rate, so information can begin to reach nearby galaxies. If we give it enough time, then eventually light emitted at the beginning of the universe from one galaxy can reach every other galaxy in the universe. However, calculations show that by the time the CMB is emitted, this has not yet happened, so there are many parts of the universe which are not in causal contact with one another, meaning that light or information sent out from one cannot have reached the other. That makes it very hard to understand how the CMB emitted from different parts of the universe can carry the same temperature, as observed, unless they are already at the same temperature at the very beginning. But it is not in the nature of most explosions to produce the uniform temperature everywhere, so what happens?

This is called the horizon problem because two galaxies that are not causally connected are said to be outside each other's *event horizon.*

The second problem is the *flatness problem.*

Even twenty odd years ago when astronomers believed the universe to be open, it was already known that the present density of the universe is not that different from the critical density ρ_c — a few percent or tens of percent of the critical density, but not orders of magnitude smaller. If we go back towards the beginning of the universe, then the ratio becomes very close to 1. This is so because at the beginning of the universe, the expansion rate is much greater, so both terms on the left-hand side of Eq. (10.1) are much bigger than their difference, *viz.* the right-hand side. This means that the right-hand side becomes more and more negligible as we go back further and further in time. It also means that the actual density is getting closer and closer to the critical density, as the latter is obtained by setting the right-hand side of Eq. (10.1) equal to zero. In other words, the universe would be getting to be almost completely flat at the beginning. This is deemed to be ugly and unnatural, unless the universe is actually critical and completely flat.

Finally, there is the *monopole problem.* Some speculative but widely accepted theories of particle physics suggest that plenty of magnetic monopoles should be produced in the early universe, but none of them have been found. The same is true for many other hypothetical particles.

10.4.2   *The Theory of Inflation*

If for some reason the density ρ of a patch of the universe remains constant for a sufficiently long period of time, then $a(t)$ expands exponentially during that period, giving,[5] $a(t) = a(0)\exp(\lambda t)$.

Calculations show that if the expansion undergoes a factor of 10^{25} or more, then the problems mentioned in the last subsection will disappear. The flatness problem is solved because the curvature becomes almost zero after such a huge inflation. This is like living on a spherical earth without noticing its curvature because it is so large. The horizon problem disappears because the patch that inflated was originally very small, so everywhere inside can easily reach a uniform temperature

---

[5]This follows from Eq. (10.1) if $k = 0$ and ρ is time independent. In that case $\lambda = \sqrt{8\pi G\rho/3c^2}$. If the universe has an initial curvature, then the inflation solution is more complicated, but for large time, this is still the solution, with an effective $k = 0$.

which the subsequent inflation by and large maintains. The monopole problem no longer exists because the original density of monopoles is greatly diluted by the subsequent inflation.

If ρ is the vacuum energy, then it is truly constant, in which case the inflation will go on forever. This is clearly not our universe. What we need is a mechanism for it to stay approximately constant for a long time, and then decay away to end the inflation. There are many models proposed to simulate such a mechanism but there is no agreement as to which, if any, is the correct one. The origin of the initial constant density is also somewhat controversial.

During that period, the volume has grown a factor of 10^{75} or more, and so has the energy, because ρ is a constant. Note that the *total* energy is still conserved because the growth is in both the kinetic and the potential energies, in such a way that the total energy remains constant. This huge amount of kinetic energy is released at the end of inflation, causing a big explosion which we call the Big Bang. This is a special kind of explosion because the uniform temperature in different parts of the universe is maintained.

10.4.3 Summary

The classical Big Bang theory cannot explain the isotropy of the cosmic microwave background radiation. This problem is solved by the theory of inflation, which proposes that the universe expanded exponentially by at least a factor of 10^{25}. This expansion causes a large kinetic energy buildup, whose eventual release causes the Big Bang in classical cosmology. The inflationary theory predicted a critical universe long before it was observed in recent years. It is also supported by the evidence from cosmic microwave background radiation as well as galactic distributions.

10.5   Thermal Physics

Now that we know how fast the universe expands, it is high time we find out what goes on inside. For example, we would like to know what sort of particles the universe contains, and in what proportions. Is the composition static, or does it change with time? Are there traces left of what went on billions of years ago that can still be observed today? If so, what are they?

Temperature is the key variable that controls the answer to these questions, so let us start by reviewing what it is. For further discussions, please consult Appendix C.

10.5.1 Equipartition of Energy

When particles move about and collide, energy is transferred from the energetic participants to others. After many collisions, a thermal equilibrium is reached, where energy is shared equally by all the particles present. If some species of particles were absent at the beginning, they can be created by collisions, provided the thermal energy present is sufficient to allow this to take place. At the end, every particle of every energetically available species will have the same average energy. This basic fact of statistical mechanics is known as the *principle of equipartition of energy*. It means, for example, that at the beginning of the universe when the temperature is high, we have equal numbers of electrons, positrons, protons, antiprotons, neutrons, antineutrons, photons, neutrinos, antineutrinos and other particles,[6] and they all have the same average energy.

The average kinetic energy of a non-relativistic particle[7] is $3k_BT/2$, where the *Boltzmann constant* $k_B = 1$ eV/$(1.16 \times 10^4$ K) is simply the conversion factor[8] between the Kelvin scale of temperature and the electron-volt scale of energy.

This is not to say that every particle has the same kinetic energy $3k_BT/2$. Only the average is the same. Individual particles may have any energy, though the probability of having an energy $E \gg k_BT$ is damped by the *Boltzmann factor* e^{-E/k_BT}. In particular, a particle with mass m may disappear if $mc^2 \ll k_BT$, whenever there is a reaction to change it into something lighter. For example, electrons (e^-) and positrons (e^+) of mass m can annihilate each other and produce two photons (γ) through the reaction $e^+ + e^- \rightarrow \gamma + \gamma$, but the inverse reaction $\gamma + \gamma \rightarrow e^+ + e^-$ cannot take place efficiently for $k_BT \ll mc^2$, when there is not a sufficient amount of thermal energy around. Since $mc^2 = 0.5$ MeV, electrons and positrons in the initial universe will largely disappear when the temperature cools down to $k_BT \ll 1$ MeV. We will come back to this phenomenon in Sec. 10.6.2.

10.5.2 Stefan–Boltzmann and Wien's Laws

Photons are always relativistic and their number can fluctuate. Their average energy density ρ_γ is given by the *Stefan–Boltzmann* law to be[9] $\rho_\gamma = (\pi^2/15)(k_BT)^4/(\hbar c)^3$. The average number density n_γ is equal to $n_\gamma = (2.402/\pi^2)(k_BT/\hbar c)^3$. The average energy of a photon is therefore $\rho_\gamma/n_\gamma = 2.7k_BT$.

There is a spread in photon energies as well. The spread is once again determined by the Boltzmann factor, which when incorporated correctly gives rise

---

[6]Please consult Chapter 9 for information on elementary particles.

[7]A particle of mass m is relativistic if its total amount of energy is far greater than its rest energy mc^2. It is non-relativistic if the opposite is true. A photon is always relativistic, and a massive particle with $mc^2 \gg k_BT$ is non-relativistic.

[8]The absolute temperature T in the Kelvin (K) scale is equal to 273 plus the temperature in the Celsius scale. The eV (electron volt) is an energy unit equal to the amount of energy gained by an electron dropping through 1 volt. It is numerically equal to 1.6×10^{-19} joules.

[9]Other than the numerical factor $\pi^2/15$ which can be obtained only by a detailed calculation, the other factors in ρ_γ can be understood in the following way. The photon has no mass, so its energy density can depend only on temperature T and fundamental physical constants such as the Planck's constant $\hbar = h/2\pi$, the Boltzmann constant k_B, and the speed of light c. $\hbar c$ is equal to 0.2×10^{-6} (eV) m (m = meter). If we measure ρ_γ in eV/m^3, then $(\hbar c)^3\rho_\gamma$ is a quantity measured in (eV)^4. Since this must depend on T, and k_BT is measured in eV, $(\hbar c)^3\rho$ must be proportional to $(k_BT)^4$, or else the units on the two sides of the equation would not be the same. Similarly, $(\hbar c)^3 n_\gamma$ is the number density measured in (eV)^3, so it must be proportional to $(k_BT)^3$.

to *Planck's law of radiation*. The resulting distribution peaks at a wave-length $\lambda_m = (2\pi/4.965)(\hbar c/k_B T)$. This relation is known as *Wien's displacement law*. The photon energy at the peak is $hc/\lambda_m = 4.965 k_B T$.

The present universe has a temperature of $T_0 = 2.725$ K. The corresponding number density is 414 photons per cubic centimeter. In comparison, the critical density ρ_c of the universe is about 5 nucleons per cubic meter, and the nucleon density is about 4% of the critical density, thus the nucleon number density is about 2×10^{-7} nucleons per cubic centimeter. This makes the ratio of nucleons to photons to be $\eta \simeq 5 \times 10^{-10}$, a very small number. This ratio is the same at all times when the nucleons remain non-relativistic, because both the photon number density and the nucleon number density vary with time like $1/a^3(t)$.

10.5.3 Baryogenesis

We might wonder why η is so small. Actually, the right question to ask is why is η so large!

In the beginning, when the temperature of the universe is high, there are as many protons as photons, because thermal energy is equally shared among the particles of all species. For that matter, there are also equal numbers of antiprotons, electrons, positrons, and so on.

As discussed before, when the temperature ($k_B T$) drops far below the rest energy $E = mc^2$ of a particle, the particle may disappear through annihilation with its antiparticle. We therefore expect almost no protons, no neutrons, and no electrons left today, with an η many orders of magnitude smaller than the small number quoted above.

This means that somewhere along the line, a mechanism must have existed to create more protons than antiprotons, more neutrons than antineutrons, and more electrons than positrons. The present value of η reflects these extra nucleons that have been created. Such a mechanism is theoretically possible, though it must occur when the universe is not at thermal equilibrium. The force that creates them must be a special kind,[6] and must violate nucleon number conservation. Although these conditions first pointed out by Sakharov are known, at what epoch it occured and the precise forces involved, are still a matter of debate. It is clearly a very important problem for it is at the very root of our existence. This is known as the *baryogenesis problem*.

10.5.4 Relativistic Electrons and Neutrinos

The average energy and number densities of relativistic electrons are a bit smaller than those of photons, because electrons obey Pauli exclusion principle, forbidding them to get too close together. For the energy density, it is a factor of $7/8$ down, and for the number density, it is a factor of $3/4$ down. The same is true for positrons (anti-electrons).

For neutrinos, which also obey the Pauli exclusion principle, their total energy density is $(6/2) \times (7/8) = 21/8$ times the photon energy density, and their total number density is $(6/2) \times (3/4) = 9/4$ times the photon number density. The additional factor of $6/2 = 3$ comes about because there are three kinds of neutrinos and three kinds of antineutrinos, but each has only one spin orientation rather than two.[10]

In the radiation-dominated era, the pressure is[11] $P = \rho/3$, which as we saw implies that the energy density ρ be proportional to $1/a^4$, where $a(t)$ is the scale factor describing the expansion. We have just seen that at thermal equilibrium ρ is also proportional to T^4 for all relativistic particles. Hence, $a(t)T(t)$ remains the same at all times as long as the number of relativistic species does not change with time.

10.5.5 Summary

Statistical mechanics predicts how matter and radiation are distributed at a given temperature. For example, it tells us that at the present temperature of 2.725 K, the universe should be about 400 photons per cubic centimeter, a factor of 2×10^9 larger than the number of nucleons present. The fact that there are any nucleons present at all is actually surprising; it points to some features of elementary particle physics which we have not yet understood.

10.6 MeV Temperatures

Three important events take place when the temperature of the universe cools down from a few MeV to about 0.1 MeV. We shall discuss them separately below.

---

[10] We have assumed in this estimate three kinds of massless left-handed neutrinos. This cannot be strictly correct because neutrino oscillation experiments have shown that neutrinos have masses. In this chapter we will continue to use the massless estimate because presently we do not know what the neutrino masses are, except that they are all small.

[11] This equation of state comes about in the following way. Imagine a large semi-infinite tube with cross-sectional area A immersed in the relativistic fluid. The tube is capped at one end, and we wish to compute the pressure exerted on the end cap by the relativistic fluid. First, concentrate on all the relativistic particles moving towards the end cap at an angle θ to its normal. These particles all move with a speed c, so their velocity component normal to the end cap is $c\cos\theta$. After a time T, all particles within a length $L = c(\cos\theta)T$ of the end cap would have hit the end cap. The total amount of energy these particles carry is then $E_{\text{tot}} = \rho AL = \rho c(\cos\theta)AT$. Each such particle carries an energy E and a momentum of magnitude $p = E/c$. The normal component of the momentum is $(E/c)\cos\theta$. The total amount of momentum hitting the end cap in a normal direction is then $(E_{\text{tot}}/c)\cos\theta = \rho(\cos^2\theta)AT$. The particles hitting the end cap bounce backward, so the total amount of change of momentum at the end cap is $2\rho(\cos^2\theta)AT$. The pressure P_θ the end cap endures is the force divided by A, and the force is the total change of momentum divided by T. Hence, $P_\theta = 2\rho\cos^2\theta$. This formula is true when $0 \leq \theta \leq \pi/2$. For $\pi/2 < \theta \leq \pi$, the relativistic particles are moving away from the end cap, so the corresponding $P_\theta = 0$. The final pressure is obtained by averaging P_θ over all angles, with a volume factor proportional to $(\sin\theta)d\theta$ (in a spherical coordinate system). Hence, $P = \int_0^\pi P_\theta(\sin\theta)d\theta / \int_0^\pi (\sin\theta)d\theta = \rho/3$. For non-relativistic particles, their momenta are much less than mc, or their total energy divided by c. The pressure P is only a small fraction of ρ, hence it is a good approximation to set $P = 0$.

10.6.1 Neutrino Decoupling

At a temperature well below 100 MeV, the only *relativistic* particles present are electrons, positrons, neutrinos and photons. To be sure, there are also neutrons and protons around, but their rest energy is close to 1000 MeV, so they are non-relativistic. Besides, there are very few of them left by that time.

Thermal equilibrium is established by sharing energy through collisions. The rate of collision increases with temperature, and also with the strength of interaction of the colliding particles. Neutrinos interact weakly. By the time the temperature drops to a few MeV, neutrinos would have fallen out of thermal equilibrium with the rest of the universe. This means that they will no longer share their energies through collisions. Nevertheless, the temperature of the universe T is still proportional to[12] $1/a$. As we shall see in the next subsection, this decoupling makes the present-day temperature of the cosmic neutrinos lower than the temperature of the cosmic photons.

10.6.2 Electron–Positron Annihilation

The electron has a rest energy of 0.5 MeV, which is the same for the positron. At a temperature well below 1 MeV, most of the positrons and electrons have disappeared through the annihilation process into photons.

The annihilation increases the photon temperature by an amount[13] equal to $(11/4)^{1/3} = 1.4$, but it does not increase the neutrino temperature because neutrinos are no longer in thermal equilibrium with the rest of the universe. The present photon temperature is $T_0 = 2.725$ K; this means that the present neutrino temperature is $T_0/1.4 = 1.95$ K.

10.6.3 Temperature and Time

In the radiation era, time and temperature are related by a simple formula:

$$(t/1 \text{ sec}) \simeq \left(\frac{1.71}{\sqrt{g}}\right)(1 \text{ MeV}/T)^2. \qquad (10.6)$$

---

[12] This is because the expansion of the universe stretches all the wavelengths. From Wien's displacement law and Planck's black-body radiation law, this means a decrease of temperature proportional to $1/a$.

[13] In an isotropic universe there is no exchange of heat between different comoving volumes. This fact has already been used in Eq. (10.2). The change of energy inside a comoving sphere of volume $V \propto a^3$ is $-d(\rho V)$, and the work done during the expansion against pressure is PdV. The difference of these two is the heat flowing out of the volume, which by definition is $-T d(sV)$, with s being the entropy density. We therefore have $d\rho = (sT - \rho - P)(dV/V) + Tds$. Since ρ depends only on T but not on V, we must have $s = (\rho + P)/T = 4\rho/3T$, which is proportional to gT^3, where g is the effective number of relativistic species present. With no heat flow $S = sV$ must be the same at all times, which means $g(aT)^3$ must be constant. Discounting the neutrinos because they have already decoupled, we have $g = 1 + 2 \times 7/8 = 11/4$ before annihilation, and $g = 1$ after. Since annihilation takes place quite suddenly, a remains essentially constant during that period, hence $(11/4)T_{\text{before}}^3 = T_{\text{after}}^3$, or $T_{\text{after}} = (11/4)^{1/3} T_{\text{before}}$.

The number $1.71/\sqrt{g}$ is equal to 1.32 after e^+e^- annihilation and 0.74 before,[14] up to a temperature of about 100 MeV.

10.6.4 Big Bang Nucleosynthesis

The chemical element helium is so named because it was first discovered in the sun's atmosphere (by J. N. Lockyer in 1868; *helios* is Greek for *sun*). It is a latecomer in the annals of the discovery of the elements, but it is the second element synthesized in the universe, and also the second most abundant one.

We are so used to the richness of elements around us that sometimes we do not realize most of what we see on earth is made up of extremely rare materials indeed. The universe is predominantly made up of hydrogen, then helium, with about one helium atom for every twelve hydrogen atoms. The rest of the elements occur only in trace amounts. The reason why hydrogen and helium seldom occur on earth in a natural form is because they are too light to be retained by earth's relatively weak gravitational field.

In the beginning, right after the Big Bang, the universe is very hot. The intense heat prevents any compound nucleus from being formed, so the primordial soup consists of single protons and neutrons, as well as other particles. The only chemical element (or rather, the nucleus of a chemical element) present is hydrogen, which is just the proton. As the universe expands and cools down, at some point neutrons and protons can combine to form other nuclei, such as helium and a (very) few others. This formation is called *primordial nucleosynthesis*, or *Big Bang nucleosynthesis* (BBN). Much later when stars appear, hydrogen and other fuels are burned in their interiors to produce thermonuclear energy. This also produces heavier elements as a by-product. The larger the star is, the heavier the elements that can be produced. But no matter how large the star is, elements beyond iron can only be produced in supernova explosions (see Chapter 8 for more details).

Nuclei are formed from two-body collisions. The probability of three or more particles colliding is so small that they can be discounted. Two protons cannot bind into a nucleus because of their electrostatic repulsion, and two neutrons cannot bind into a nucleus because of their tendency to β-decay.[6] A proton and a neutron, however, can form a deuteron (D), which is the nucleus of deuterium, or heavy hydrogen. The binding energy is 2.225 MeV, meaning that it takes that amount of energy to tear them apart. This binding energy is sufficiently large to prevent the neutron inside from undergoing a β-decay, $n \rightarrow p + e^- + \bar{\nu}_e$, for the following reason. The neutron (n) has a rest energy 1.3 MeV above that of the proton (p), and the electron (e^-) itself has a rest energy of 0.5 MeV. The electron antineutrino

---

[14]The total radiation density is $\rho = g\rho_\gamma$, where $g = 1 + (21/8)(4/11)^{4/3} = 1.68$ after e^+e^- annihilation and $g = 1 + (2+3) \times (7/8) = 5.375$ before, up to about 100 MeV. In the radiation era, $a(t)$ is proportional to $\sqrt{t}$, so $[(da/dt)/a]^2 = 1/4t^2$. Moreover, according to Eq. (10.1), for a critical universe this is also equal to $8\pi G\rho/3c^2 = 8\pi GgT^4/3c^2$. Equating these two, we obtain the equation above.

$(\bar{\nu}_e)$ may have a mass, but it is too small to be presently detectable. If the neutron decays, it would disintegrate the deuteron, and that requires 2.225 MeV of energy, of which only 0.8 MeV can be supplied by the β-decay process. For that reason, the neutron cannot decay and the deuteron is stable.

However, a binding energy of 2.225 MeV is very small compared to the typical binding energies of other nuclei, so the deuteron is fragile. When two deuterons come together, they can actually form a much more stable nucleus, helium (^{4}He), which contains two protons and two neutrons. The binding energy of a helium nucleus is 28.31 MeV, a large number. In fact, helium is so stable that when radioactive nuclei decay, sometimes the whole helium nucleus is ejected. In that form it is called an α particle and the decay is called α-decay.

Deuteron is formed from the process $p + n \rightarrow D + \gamma$. At high temperatures, the inverse process $\gamma + D \rightarrow p + n$ also takes place to tear the deuteron apart. With a binding energy of $E_0 = 2.225$ MeV, one might think that the photo-disintegration of deuteron would stop at a temperature approximately equal to E_0. This is not the case because $\eta^{-1} \simeq 2 \times 10^9$ is a very large number, with two billion photons present for each nucleon. It is true that the probability of any photon having an energy E_0 is given by the Boltzmann factor $\exp(-E_0/k_B T)$. This factor is small when $k_B T \ll E_0$, but if it is not smaller than one billionth, there would still be one of the two billion photons that can attain the energy E_0 to disintegrate the deuteron. For that reason the deuteron would not become stable until the temperature is such that $\exp(-E_0/k_B T) \simeq \eta$, which works out to be $k_B T \simeq 0.1$ MeV. According to Eq. (10.6), that occurs about 130 seconds after the Big Bang.

Neutrons and protons have a mass (meaning rest energy) difference of 1.3 MeV. If they were in thermal equilibrium until deuterium formation at 0.1 MeV, there would be too few neutrons left to account for the approximately 25% of ^{4}He (by weight) left over from the early universe. What happens is that proton and neutron numbers are frozen at their values at $k_B T \sim 0.7$ MeV, so their ratio is given by the Boltzmann factor to be $\exp(-1.3/0.7) \simeq 1/7$. This means one ^{4}He (2 protons and 2 neutrons) to every 12 protons, or a weight ratio between ^{4}He and H of 1/3. Consequently, some 25% of the nucleonic mass is locked up in the helium nucleus.

To understand freezing, we must first understand how neutrons and protons turn into each other at thermal equilibrium. That proceeds mainly through the *weak* interaction processes $\nu_e + n \leftrightarrow p + e^-$ and $e^+ + n \leftrightarrow p + \bar{\nu}_e$. As the temperature decreases, the reaction rate goes down. The disappearance of $e^\pm$ through annihilation around a temperature of 1 MeV slows down these processes even further. At some point these reactions effectively stopped freezing out the number of neutrons and protons. Calculations show that this happens at about 0.7 MeV.

As mentioned before, deuterons are fragile so they tend to combine to form the more stable ^{4}He. If we increase η, *i.e.*, the number of nucleons, then $D = {}^2$H would be formed at a higher temperature, at which time more neutrons would be present. With more neutrons around, there would be more deuterons formed. At a higher temperature, it would also be easier for two colliding D's to overcome their electrostatic repulsion to form a ^{4}He, so at the end we would find fewer D's but more ^{4}He's at a larger η, or a higher nuclear (baryon) density. This is what Fig. 10.3 shows.

The fact that deuterons are created at such a low temperature makes it difficult to form most other nuclei. Two lighter nuclei can fuse together into a heavier one only when their electrostatic repulsion can be overcome by the thermal energy, and only when the universe is dense enough for an appreciable fusion rate to occur. Neither happens effectively below 0.1 MeV. As a result, practically nothing beyond ^{3}He and ^{7}Li can be formed, and these only in a very small amounts.

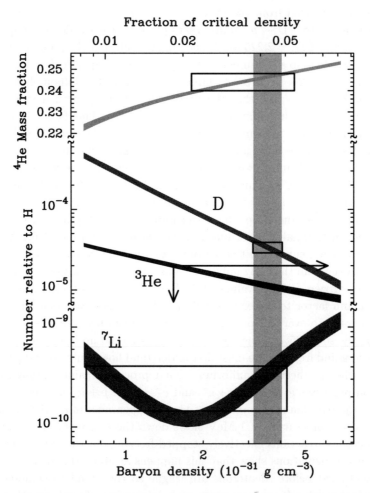

Figure 10.3: Figure taken from D. Tytler, J. M. O'Meara, N. Suzuki, and D. Lubin, *Physica Scripta* **T85**, 12–31 (2000). Note the two breaks in the vertical scale. Smooth curves are theoretical results, with the top curve being ^{4}He. Experimental results with error bars are given by the boxes. The vertical bar indicates the density of the universe that explains the BBN data.

The abundance of these elements all depends on the parameter η, or the nucleon density, as shown in Fig. 10.3. By comparing observation with calculations, one can determine η to be around 5×10^{-9}. This is equivalent to a nucleonic density equal to 4% of the critical density, as quoted before.

It should also be pointed out that it is a difficult task to observe the abundance of these primordial elements, not only because they occur in small quantities, but also because they must be isolated from the elements formed much later in stars.

10.6.5 Summary

The energy density of relativistic particles is proportional to T^4, and the number density is proportional to T^3, where T is the temperature of the universe.

Neutrinos are decoupled from the rest of the universe at a temperature of a few MeV. The permanent annihilation of electron–positron pairs at about 1 MeV reheats the photons, but not the already decoupled neutrinos. As a result, the present neutrino temperature is only 1.95 K, although the photon temperature is 2.725 K.

Deuteron is the first nucleus to be created, at a temperature of about 0.1 MeV. It is fragile; most of it is converted into the much more stable ^{4}He. Besides ^{3}He and ^{7}Li formed in trace amounts, practically no other nuclei are formed at this stage of the universe. By comparing calculation with observation, one finds $\eta \simeq 5 \times 10^{-9}$, which translates into a nucleon density equal to about 4% of the critical density.

Below 1 MeV, the temperature of the universe is inversely proportional to the scale factor $a(t)$. In the radiation era, temperature and time are related by Eq. (10.6).

10.7 The Cosmic Microwave Background (CMB)

10.7.1 Observation

The age of communication satellites brought us an unexpected gift from heaven. It led to the accidental discovery of the cosmic microwave background radiation (CMB) by Arno Penzias and Robert Wilson in 1965.

Penzias and Wilson worked for the Bell Telephone Laboratories in New Jersey. They used a radio horn, originally designed to link up with communication satellites, to make astronomical observations at a microwave wavelength of 7.35 cm. After eliminating all known sources of background and interference, an annoying noise remained. It could not be removed no matter what they tried, and in what direction they pointed the horn. That turned out to be one of the most important discoveries in astronomy, and the unwanted noise became a new and powerful tool for astro-archaeology.

Later observations show that this electromagnetic noise has a black-body distribution with a temperature $T_0 = 2.725$ K.

This was not the kind of accuracy obtained initially. The peak wavelength of this temperature occurs at 1 mm. Electromagnetic waves in that vicinity are absorbed by the water vapor in the atmosphere, so this kind of accuracy can be obtained only when water vapor is greatly reduced or eliminated. This can be achieved by building ground-based observatories at high altitude and/or high latitude, by putting the detectors in a balloon, and better still, on a satellite.

Unknown to Penzias and Wilson, the noise was actually predicted by George Gamow in the late 1940's to be the reverberating radiation left over from the big bang. He even estimated the temperature of this radiation to be around five or six degrees Kelvin; not a bad estimate at such an early stage!

The CMB was discovered at the beginning to be isotropic to an accuracy of 10^{-3} (one part in a thousand). Soon afterwards, using a borrowed U2 spy plane flying high above most of the atmosphere, Luis Alvarez and his group were able to improve the accuracy to 10^{-4}. With that accuracy, they found a 'dipole' deviation from isotropy, namely, one direction of the sky seems to be a bit hotter than the opposite direction. This is attributed to the Doppler shift arising from the motion of the solar system,[15] with a velocity of 371 km/s. The COBE satellite, now retired, then found a complicated temperature fluctuation to set in at the level of 10^{-5}. This fluctuation turns out to yield very rich information on the state of the universe, stretching all the way back to the inflationary era. We will discuss the physics of that in the next subsection.

The COBE satellite had two antennae separated by some $7°$ to measure the temperature difference ΔT in these two directions. Subsequent balloon and ground-based experiments are able to measure down to a much smaller angular difference θ. The present available data of these measurements are shown in Figs. 10.4 and 10.5. The vertical axis can be thought of as the average of $(\Delta T/T)^2$, and the horizontal axis is related to θ roughly by $\ell \simeq 2\pi/\theta$. Figure 10.4 shows the data from the WMAP data released in February, 2003. The curve is a fit to a popular model of cosmology. Figure 10.5 summarizes the earlier data, together with the fitted curve taken from Fig. 10.4. Some of the parameters determined from the fits are, the Hubble constant $h_0 = 0.72 \pm 0.05$, the total matter content $\Omega_m = 0.29 \pm 0.07$, and the baryonic (ordinary) matter content $\Omega_b = 0.047 \pm 0.006$. The latter number translates into a baryon number density $n_b = 0.27 \pm 0.01/\text{m}^3$, and a baryon-to-photon ratio $\eta = (6.5 \pm 0.4) \times 10^{-10}$. The age of the universe is $t_0 = 13.4 \pm 0.3$ billion years, and decoupling (the time when CMB was emitted — see next subsection) occurs at a red-shift of $z_* = 1088 + 1/- 2$, and a time $372\,000 \pm 14\,000$ years after the Big Bang. The matter-dominated era took over from the radiation-dominated era at a red-shift of $z = 3454$.

---

[15] Doppler shift changes the peak and other wavelengths of the black-body radiation, hence the effective temperature.

The universe is flat, with a total $\Omega = 1.02 \pm 0.02$. The equation of state for the dark energy component must have a $w < -0.78$.

It is remarkable that the WMAP parameters, determined from the physics at a red-shift z_* of about 1000 and at a time more than 300 000 years after the Big Bang (see the next subsection), are essentially the same as those determined in a completely different way, and at a much different epoch of the universe. For example, h_0 determined directly from the Hubble plot involves objects with $z < 2$, and the baryonic content determined from BBN concerns the universe at about 3 minutes old.

The European Planck satellite to be launched in a few years will have an angular resolution of $5'$, which enables it to obtain very precise results at large ℓ.

10.7.2 The Physics of CMB

The energy required to tear the electron away from a hydrogen atom is $E_0 = 13.6 \, \text{eV}$. Above that temperature, the universe consists of a plasma of charged electrons and nuclei, with photons trapped in between because they are being continuously absorbed and emitted by the plasma.

Below 13.6 eV, an electron and a proton can combine into a hydrogen atom H. In the process, a photon γ is emitted to carry away the energy E_0. Just like the photo-disintegration of deuteron considered earlier, the large number of photons around makes it possible for the inverse reaction $\gamma + \text{H} \rightarrow n + p$ to take place all the way down to a temperature determined by $E_0 \cdot \ln(\eta^{-1}) \simeq 1$, *viz.*, $k_B T \simeq 0.6 \, \text{eV}$. A more careful calculation, taking distribution and reaction rates into account, shows that the neutralization of plasma into hydrogen atoms takes place at about $(1/4) \, \text{eV}$. This epoch, known as *recombination*, or *decoupling*, will be indicated by an asterisk subscript. It is called decoupling because after the recombination, photons in this neutral environment are hardly absorbed or scattered, so they effectively decouple from the rest of the environment. With $k_B T_* \simeq (1/4) \, \text{eV}$, the corresponding temperature is $T_* \simeq 3000$ K. Since the photon temperature is inversely proportional to the scale factor a, which in turn is inversely proportional to the red-shift factor $z + 1$, we get $z_* \simeq 1000$. This is so because at the present time t_0, the parameters are $T_0 \simeq 3$ K, $z(t_0) = 0$, and $a(t_0) = 1$.

The wiggly structure in Figs. 10.4 and 10.5 is caused by the acoustic waves set up at the beginning of the universe. In the inflationary era when the universe is very small, quantum fluctuations are important. These fluctuations set up an acoustic wave in the cosmic fluid, in much the same way a snap of your finger generates a sound wave in the air. The amplitude of a sound wave in the air, with wave number $k = 2\pi/\lambda$ and generated with no initial velocity, has a time dependence $\cos(\omega t) = \cos(k c_s t)$, where ω is the angular frequency and c_s is the speed of sound. With the equation of state $P = \rho/3$, the speed of sound is $c_s = c/\sqrt{3}$. The amplitude of an acoustic wave in the cosmic fluid is similar, but with one crucial difference.

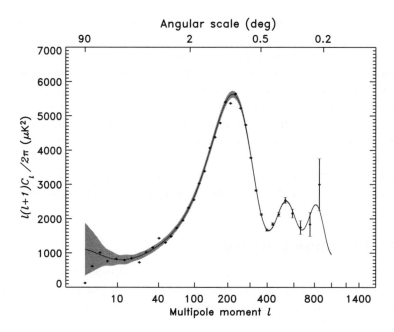

Figure 10.4: Figure taken from the WMAP collaboration. It shows the precision WMAP data with the model fit. (Source: http://arxiv.org/ps/astro-ph/0302207.) (Courtesy of the WMAP Science Team.)

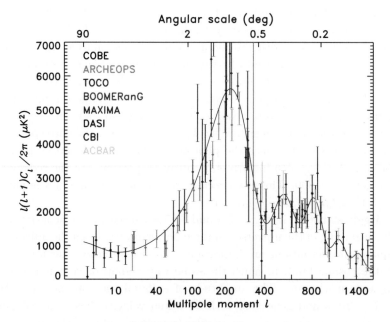

Figure 10.5: Figure taken from the WMAP collaboration. It shows a summary of data before WMAP, and the model fit obtained in the previous figure. (Source: http://arxiv.org/ps/astro-ph/0302207.) (Courtesy of the WMAP Science Team.)

Air is stationary, but the universe is expanding, so to use the expression we must use a distance and time which are not affected by the expansion. We saw before that the appropriate distance is the comoving distance r, related to the expanding distance $s(t)$ and the scale factor $a(t)$ by $r = s/a$. With the comoving distance, the correct time scale to go with it — to ensure that c is the speed of light — is the *conformal time* τ. It is defined by the equations $ds = a(t)dr = c(dt)$ and $dr = c(d\tau)$. In other words, $\tau = \int_0^t dt/a(t)$. The acoustic amplitude in the cosmic fluid should then be modified to read $\cos(kc\tau/\sqrt{3})$. At decoupling time, the dependence of this amplitude on the wavelength λ is then $\cos(2\pi c\tau_*/\lambda\sqrt{3})$. These amplitudes peak at wavelengths for which the argument is $n\pi$, for $n = 1, 2, 3, \ldots$.

A different wavelength at decoupling time subtends a different angle in the sky today. (See Fig. 10.6.) The inner sphere in the picture is the sphere of last scattering at decoupling (conformal) time τ_*. The outer sphere indicates the farthest reach in space, from which light emitted at the beginning of time has just reached us. Its conformal time is 0. The center of the spheres indicate our present position, at a conformal time τ_0. The comoving distance of the inner sphere from us is $c(\tau_0 - \tau_*)$. Since $\tau_0 \gg \tau_*$, we may approximate this by $c\tau_0$. The angle subtended by a wave of length λ is therefore $\theta \simeq \lambda/c\tau_0$. The peaks of the amplitudes, and of $(\Delta T/T)^2$, should therefore occur at $\ell \simeq 2\pi/\theta \simeq 2\pi c\tau_0/\lambda = \sqrt{3}n\pi(\tau_0/\tau_*)$. In the matter-dominated era, $a(t) \sim t^{2/3}$, and hence $\tau(t) \sim t^{1/3} \sim \sqrt{a(t)}$. For that reason, $\tau_0/\tau_* = \sqrt{a(t_0)/a(t_*)} = \sqrt{z_* + 1} \simeq \sqrt{1000}$. The peaks are therefore estimated to occur at ℓ values of $172n$. A more detailed calculation shows that the first peak $(n = 1)$ actually occurs at an ℓ slightly larger than 200, rather than 172.

There is one implicit assumption in this estimate, which can be seen in Fig. 10.6. We are estimating the angle θ using Euclidean geometry. Namely, we are assuming the universe to be flat. This is the case if $\Omega = 1$. On the other hand, if $\Omega < 1$, it is closed like a sphere. Seemingly parallel lines will intersect at a distance, so the observed angle θ is seemingly larger than that in a flat universe, or ℓ smaller. In an open universe with $\Omega > 1$, it is the other way around, with a larger ℓ. The fact

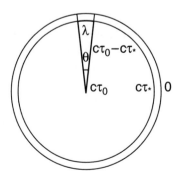

Figure 10.6: Wavelength at conformal decoupling τ_* and the angle θ it subtends today. Flat geometry is assumed.

that the observed first peak in Fig. 10.4 is so close to 200 is further independent support for the flatness, or the criticality, of the universe.

The various peaks in Figs. 10.4 and 10.5 are not of the same height, contrary to the simple argument given above. There are at least two reasons for that. First, gravity compresses the cosmic fluid, and unlike photons, matter provides no pressure to counteract this compression. Hence, the amplitudes of the compression phases of the wave (odd n's) are larger than the decompression phases (even n's). This is borne out by the observation that the height of the $n = 2$ peak is lower than that of the first or third peak. Secondly, electromagnetic coupling is not all that strong, so photons are not completely trapped. The leakage makes the fluid less bouncy and thereby reduces the amplitude of the larger-ℓ peaks.

For very long wavelengths, or very small wave numbers k, $\cos(kc\tau_*/\sqrt{3}) \simeq 1$ at decoupling time. Thus, the quantum fluctuations at these wavelengths are completely frozen between the inflationary era and the decoupling time. This offers the exciting possibility of using the CMB to probe directly into the inflationary era. Most inflationary models predict a near scale-invariant fluctuation, which is verified by the CMB observation, thus providing further support for the inflationary origin of the universe.

At the other end of the wavelength scale, gravitational influence becomes very important and that leads to the formation of clusters and galaxies after the decoupling. The size and distribution of the clusters are sensitive to the amount of light and heavy matter involved. One can vary these amounts and compare the result of computer simulations directly with the observed distributions through galaxy surveys. The resulting content of matter is again consistent with the amount obtained by other methods. So all in all, the general composition of the universe, with about 70% dark energy and 30% dark matter, of which only a small fraction is the ordinary hadronic matter, seems to be well supported by different observations.

10.7.3 Summary

The Cosmic Microwave Background (CMB) radiation provides us with very detailed information about the universe, at decoupling time and before. The observed mean temperature of $T_0 = 2.725$ K today allows us to determine various cosmological parameters. Its dipole component tells us the velocity of the solar system, and its higher multipole fluctuations provides evidence for the acoustic wave in the cosmic fluid generated by quantum fluctuations in the inflationary universe. These fluctuations also provide seeds for cluster and galaxy formation in the later universe.

Detailed cosmological parameters can be obtained from the CMB data alone. They are completely consistent with those obtained in different ways and from a different epoch of the universe.

10.8 Further Reading

Useful websites

- http://lambda.gsfc.nasa.gov/m_uni.html
- http://www.damtp.cam.ac.uk/user/gr/public/cos_home.html
- http://casa.colorado.edu/ (contains links to many other webpages)
- http://www.galacticsurf.com/cosmolGB.htm (contains links to many other webpages)
- http://map.gsfc.nasa.gov/index.html (the home page of CMB satellite WMAP)

10.9 Problems

10.1 A galaxy lying on the Hubble curve has a red-shift $z = 0.1$. The Hubble constant is $H_0 = 72$ km/sec/Mpc.

1. What is its recessional velocity v in km/sec?
2. What is its distance s in light-years?
3. If its absolute magnitude is -20, what is its apparent magnitude?

10.2 Assume the present critical universe to consist of 2/3 dark energy and 1/3 non-relativistic matter. Assume further that the equation of state for the dark energy is $P = -0.8\rho$.

1. What is the scale factor a when the deceleration of the universe turns into acceleration?
2. What is the red-shift z at that time?

10.3 Suppose $k_B T$ right after the Big Bang is 10^{15} GeV, and the number of species at that time is $g = g_U$. Suppose further that the universe has inflated 10^{25} times before it reaches the Big Bang. What is the original size of the inflationary region before the inflation?

10.4 Take the present temperature of the universe to be $T_0 = 3$ K.

1. What is the scale factor $a(t)$ of the universe when $k_B T = 0.2$ MeV?
2. What is the corresponding red-shift factor $a(t)$?
3. How long after the Big Bang does it take to reach this temperature?

10.5 Suppose a darkened room on a summer's day has a temperature of $27°C$.

1. What is the photon number density n_γ in the room?
2. What is the peak wavelength λ_m of these photons?
3. With so many photons in the room, why is the room dark?

10.6 Elements in the universe.

1. What is the most abundant element in the universe?

2. Which of the following elements are produced in Big Bang nucleosynthesis?

   (a) ^{4}He
   (b) D
   (c) ^{12}C
   (d) ^{235}U

3. Which of these four is the most abundant in the universe?

10.7 The solar system moves with a velocity $v = 371$ km/s in the background of the cosmic microwave radiation. This causes the sky in front to appear slightly hotter than the sky behind.

1. Compute the Doppler shifts of the cosmic microwaves in front and behind.
2. Use Wien's displacement formula to estimate the temperature difference in front and behind.

Appendix: General Concepts in Classical Physics

A.1 The Physical Universe

As its name indicates, *physics* in its broadest sense is the study of nature. It is the study of all inanimate forms of matter and energy, encompassing the smallest constituents of matter and the largest bodies in the universe, even the whole universe itself. It is the attempt to observe the formation and evolution of all objects, to probe their inner structure and their interactions with one another. It is also the search for the basic rules, rules with three basic functions: to correlate seemingly disparate facts in a logical structure of thought, to focus thinking along new directions and, finally, to predict future observable events.

The scope of physics is illustrated by the three charts in Fig. A.1, which show the scales of size, mass, and age (or characteristic time) of some physical objects. We express these fundamental quantities in the *International System of Units* (SI): length in meters (m), mass in kilograms (kg), and time in seconds (s). This is one of the two systems of units most often used by physicists, the other being the *cgs system* in which the base units for length, mass, and time are centimeter (cm), gram (g), and second (s).[1] To deal with measurements that may be very small or very large, it is convenient to adopt the exponent notation of numbers — as in 10^n (written as 1 followed by n zeroes) and 10^{-n} (a decimal point followed by $n-1$ zeroes, then 1) — some of which are assigned names with suggestive prefixes, such as *kilo* (k) $= 10^3$, *mega* (M) $= 10^6$, *giga* (G) $= 10^9$, *tera* (T) $= 10^{12}$, *peta* (P) $= 10^{15}$; *milli* (m) $= 10^{-3}$, *micro* (μ) $= 10^{-6}$, *nano* (n) $= 10^{-9}$, *pico* (p) $= 10^{-12}$, *femto* (f) $= 10^{-15}$, etc. It is evident from the figure that the physical universe englobes objects differing widely in nature and separated by huge factors in size, mass and age. Less evident, but no less significant, is the fact that these dry numbers give more than just the

---

[1] Other units can be found. For example, *Planck's units*, which appear in some basic theories, are defined in terms of the speed of light c, the quantum of action $\hbar = h/2\pi$, and the Newtonian gravitational constant G: $L_{\rm P} = \sqrt{\hbar G/c^3} = 1.6 \times 10^{-35}$ m, $m_{\rm P} = \sqrt{\hbar c/G} = 2.18 \times 10^{-8}$ kg, and $t_{\rm P} = \sqrt{\hbar G/c^5} = 5.4 \times 10^{-44}$ s.

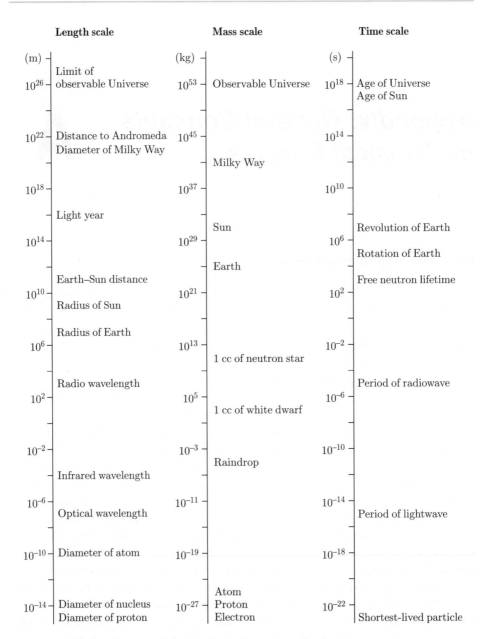

Figure A.1: Scales of length, mass and time of the physical universe.

numerical values of things; they epitomize the fruits of labor by several generations of scientists and represent important signposts for efforts by new generations.

We characterize size, mass and time as 'fundamental' because all other physical quantities can be considered as derived quantities, expressible in units that are various combinations of powers of the base units. Among the more familiar examples, force is measured in *newton* ($N = m \ kg/s^2$), pressure in *pascal* ($Pa = kg/m \ s^2$) and energy in *joule* ($J = kg \ m^2/s^2$) in SI units; or, respectively, in *dyne* (10^{-5} N), *dyne per square-cm* (0.1 Pa) and *erg* (10^{-7} J) in cgs units.

In atomic and subatomic physics, one often prefers a smaller, more appropriate unit of energy, the *electron volt* (eV), defined as 1 eV $= 1.60 \times 10^{-19}$ J. The temperature is a measure of heat, and heat is a form of energy (Appendix C). Thus, the unit of temperature (degree Kelvin or K) may be considered a derived unit; it is related to the unit of energy by a numerical factor which one may identify with a constant called the *Boltzmann constant k*.

The Boltzmann constant is one of the many *fundamental* (or *universal*) *constants* which recur again and again in physics. They are considered universal, *i.e.*, unchanging in any physical circumstances and, of course, constant in time. Their numerical values are the results of many delicate and laborious experiments (and therefore subject to experimental errors) and are as yet unexplainable by theories. Some of the more important physical constants are given in Table A.1.

The popular belief that scientific information doubles every seven years may well be a myth, but it is nevertheless true that a vast body of facts has been accumulated by mankind ever since the ancient Greeks started pestering each other with endless discourses and questions about earth and heaven, thus embarking on a great intellectual adventure that modern scientists are still pursuing. But facts alone do not necessarily give understanding and knowledge by itself is not science. Explanations of specific phenomena in terms of known facts must be generalized into rules that apply to larger classes of phenomena, and rules must further be distilled into a few universal brief statements, or *laws* that cover not only observed cases, here and now, but also unobserved cases, anywhere and at any time. Finally, laws are further organized into a coherent and logical structure, called *theory*.

In physics, existing theories can be divided into three broad categories: *mechanics*, the *theories of matter and interactions* and the *physics of large systems*. Mechanics, the study of motion, plays a unique role central to all of physics. On the basis of the oldest of its three components, *classical mechanics*, modern physicists have constructed *quantum mechanics* and *special relativity*, two superb theories that underpin all contemporary physics and serve as guiding principles for further research. Loosely speaking, the fundamental dynamical theories deal with forces.

Table A.1: Fundamental physical constants (SI units).

Speed of light in a vacuum	$c = 3.0 \times 10^8$ m/s
Gravitational constant	$G = 6.67 \times 10^{-11}$ m^3/s^2 kg
Planck constant	$h = 6.63 \times 10^{-34}$ J s
Boltzmann constant	$k = 1.38 \times 10^{-23}$ J/K
Stefan–Boltzmann constant	$\sigma = 5.67 \times 10^{-8}$ J/s m^2 K^4
Elementary charge	$e = 1.60 \times 10^{-19}$ coulomb
Electron mass	$m_e = 9.11 \times 10^{-31}$ kg
Proton mass	$m_p = 1.67 \times 10^{-27}$ kg

But, in one of the greatest developments of this century, it was discovered that *forces* (or interactions) are in fact intimately related to *matter*. All matter is composed of a few species of fundamental, or elementary, particles, and these particles interact via four fundamental forces: the gravitational force, the electromagnetic force, the strong force and the weak force. Finally, when we deal with macroscopic bodies, the presence of a very large number of component parts requires specialized techniques. The study of the behavior of such systems is the domain of both *thermodynamics* (a theory based on a set of four general rules deduced from observations of controlled experiments on macroscopic bodies) and *statistical mechanics* (a statistical treatment of the mechanics of a very large number of particles).

In this appendix, we give a very brief discussion of some general aspects of classical mechanics. Quantum mechanics will be the subject of Appendix B, while thermal physics and statistical mechanics will be considered in Appendix C. Other more specialized parts of physics are dealt with at appropriate places in the main body of the book.

A.2 Matter and Motion

Space and Time

Whenever physicists describe their observations or formulate their theories, they cannot fail to make use of certain quantities called *space* and *time*, two basic physical concepts that are intuitively evident to us all and yet cannot be formally defined in terms of any simpler entities. Since space and time cannot be defined, ways of measuring them must be devised — a distance in space is measured by comparing it with some unit length and a time interval is measured in terms of the period of some cyclical phenomenon. The concepts of distance and time are ultimately defined by the operations carried out in making the measurement. This measurement then is equivalent to the missing formal definition.[2] Such measurements, absolutely essential to the development of physics, may appear trivial to us now, but had not always been possible in the past. It was perhaps not a coincidence that the first detailed study of the pendulum and the first systematic experiments on motions were carried out by the same person, Galileo Galilei.

It is a fact of experience that the space in which we live is three-dimensional and, to a very good approximation, Euclidean. That our space is *three-dimensional* (*i.e.*, any point can be located relative to another by specifying three and only three coordinates) is intuitively clear. The assertion that the geometry of our space is 'flat,' or *Euclidean* (*i.e.*, the postulates of Euclid are valid for our world) is less evident but nevertheless true. A simple way to check it is to measure the sum of the interior angles of a planar triangle; it is always found to be 180° to within the

---

[2] As P. W. Bridgman, physicist and philosopher, once wrote: 'The true meaning of a term is to be found by observing what a man does with it, not what he says about it.'

measurement error. Euclidean geometry is valid on or near the earth but breaks down in the immediate vicinity of very massive bodies (*e.g.*, the sun or a neutron star).

Day-to-day experience tells us that time flows in one direction only, from past to present to future. This is indeed confirmed by experiment; time reversal never occurs in the macroscopic world. Time has another important property: it is *absolute*, that is, it flows at a rate that does not depend on position or velocity. Suppose we have two identical well-made clocks, initially at rest and perfectly synchronized. Suppose one of them is transported in motion along some path at varying moderate speeds, then brought back to its initial position. If we can verify that the two clocks are still synchronized, we say that time is absolute. Experiments show that time is indeed absolute to a very high precision in ordinary circumstances on earth. But this notion of absolute time is not exact in any situation; in particular, it fails when very rapid motions or gravitational effects are involved: clocks moving at high velocities or exposed to strong gravity are found to lose or gain time relative to stationary clocks.

Classical (or Newtonian) physics is based on the assumption that space is Euclidean and time absolute. Because Newton's theory rests on postulates that do not exactly hold in all situations, we admit that it is not an exact theory. It is only an approximation, but nevertheless an excellent approximation to the real world, quite adequate in most circumstances we encounter on earth or in space. Furthermore, its formulation is restricted to certain special frames of reference, called the *inertial reference frames*, *i.e.*, those frames that are at rest or, at most, moving at a constant velocity relative to distant stars. Thus, a reference frame fixed on the ground is an inertial frame, but one fixed on an accelerating train is not. As any good theory, Newtonian physics has few, simple and plausible assumptions, which are general enough to let it 'grow' to maturity and in various directions, and eventually blend itself with new, improved theories.

Three Laws of Motion

With the notions of space and time understood, one can define *velocity* and *acceleration* in the usual way. The only other concept we need is that of *force*. Again, this concept comes naturally to us, as part of our everyday experience. We can instinctively perceive a force wherever it exists and can gauge its strength. We can feel it as the airplane carrying us on board is taking off or landing; we know the amount of push or pull we must exert to set an object at rest into motion. So, we can consider force as a basic entity and define it by measuring the effect it would have on some standard device (for example, the amount of stretching of a standard spring). With those elementary definitions on hand, we can state Newton's three laws of motion, which together form the basis of classical mechanics and dynamics.

First Law: *A body at rest remains at rest and a body in a state of uniform linear motion continues its uniform motion in a straight line unless acted on by an unbalanced force.*

This statement (often referred to as the law of inertia) simply means that a body persists in its state of rest or motion unless or until there exists an unbalanced force (*i.e.*, when the sum of forces acting on it does not vanish). What happens then is described by the second law:

Second Law: *An unbalanced force (F) applied to a body gives it an acceleration (a) in the direction of the force such that the magnitude of the force divided by the magnitude of the acceleration is a constant (m) independent of the applied force. This constant, m, is identified with the inertial mass of the body.*

Stated algebraically, this law reads: $F = ma$, which is referred to as 'Newton's equation of motion' because it describes, for a given force, the motion of the body via its acceleration. Three points should be noted. First, it is a vectorial equation; it relates the three components of a vector (force) to the three components of another vector (acceleration). Second, it is general, applicable to any force, independently of its nature. On the other hand, to solve the equation for the body's acceleration, one must know the exact algebraic expression of the force, which is a separate problem. Finally, the Second Law plays a dual role: it gives a law of motion and also states a precise *definition of mass* by indicating an operational procedure to measure it. At this point, turning things around, one could choose *mass* as a basic quantity, define a mass unit and use the Second Law to derive the force.

Third Law: *If a body exerts a force of any kind on another body, the latter exerts an exactly equal and opposite force on the former.*

That is, forces always occur in equal and opposite pairs in nature. We walk by pushing backward with our foot on the ground, while the ground pushes the sole of our foot forward. The pull of the sun on the earth is equal to the earth's pull on the sun, and the two forces are along the same line but opposite in direction. Similarly, the positively charged nucleus of the hydrogen atom exerts an attractive force on the orbiting electron, just as the electron exerts an equal and opposite force on the nucleus.

As long as we are describing the *kinematics* of an isolated body, we just need to know how its velocity changes. But when we deal with the *dynamics* of the particle, we must take into account its mass as well. It is then more meaningful to deal with mass and velocity together in a single combination, called the *momentum*, p. The momentum of a body is given by the product of the mass, m, and the velocity, v, of the particle. It is a vector pointing in the same direction as the velocity. Since the acceleration of a particle is the rate at which its velocity changes, the

Second Law simply tells us that the force exerted by the particle is the rate at which its momentum changes. It follows then: (1) the momentum remains constant in the absence of unbalanced forces, and (2) the total momentum of an isolated two-particle system is a constant of motion. We recognize in these results a simple restatement of the First and Third Laws, which, however, has the virtue of making a generalization to systems of more than two particles almost trivial: if a system is isolated so that there are no external forces, the total momentum of the system is constant (*the law of conservation of momentum*). If there are external forces, the rate of change of the momentum is vectorially equal to the total external force. The law of conservation of momentum has a more general validity and a deeper meaning than its derivation given here would indicate; it is a manifestation of the existence of a space symmetry (Chapter 1).

Angular Momentum

So far, we have discussed motion as if it involved only translational motion, *i.e.*, a motion that keeps the relative orientation of different parts of the body unchanged. But pure translation rarely occurs in nature; there is always some degree of revolution or rotation in any motion. In revolution, a particle is acted on by a force that continually changes its direction of action, making it move along some curved path. In rotation, no straight line (except one) connecting any two points in the body remains parallel to itself; the line that does is called the body's axis of rotation. In pure rotation, a body changes its orientation without changing its position and so must be acted on by something other than simple force.

Suppose we nail a long rod at one end loosely to the ground. If we push or pull the rod near its free end, the applied force would generate an equal and opposite reaction of the rod at the fixed end. Two equal and opposite forces, which are *not* on the same line and hence uncompensated, act on the rod and make it rotate about the fixed end. Thus, a correct measure of the effects observed cannot be given by the applied force alone, but rather by that combination of forces, called a *torque* (meaning 'twist'). Clearly the amount of torque depends on where the force is applied. It also depends on the orientation of the force with respect to the rod (or, generally, on the position vector that defines the location of the application point relative to a fixed point); in particular, it vanishes if the force is parallel to the rod. A non-vanishing torque is a vector perpendicular to both the force and the position vector. Just as force produces changes in (linear) momentum, torque produces changes in *angular momentum*; if no torque exists, the angular momentum does not change; it is said to be conserved.

The angular momentum (L) of a particle is a vector perpendicular to both the position vector (r) and the momentum (p) of the particle relative to some origin of coordinates and whose magnitude is $rp\sin\theta$, where θ is the angle between the two vectors r and p. As we have seen, in the absence of forces, a particle moves along a straight line at constant velocity. The direction of its angular momentum relative to an arbitrary reference point is perpendicular to the plane defined by the line of motion and the reference point. Its magnitude is simply the product of its momentum (p) and the shortest distance between the origin of coordinates and the line of motion ($r \sin \theta$). Thus, it is a constant vector for the chosen origin; in the absence of forces, both vectors p and L are constants of motion. If the origin changes, L changes to a new constant vector. In particular, we can make L vanish by locating the origin on the line of motion ($r \sin \theta = 0$). Now, take a particle in uniform circular motion, *i.e.*, a motion with constant velocity along a circle. Its momentum is constant in magnitude but continuously changing in orientation. There is a force (*centripetal force*) directed toward the center of the circle continuously acting on the particle to keep it on its circular orbit. The angular momentum of the particle, defined with respect to the center of the circle taken as the origin of coordinates, is simply $L = rp$, the product of the circle radius and the particle momentum; it is perpendicular to the plane of the circle. So, again, there is no torque, and L is a conserved vector although the momentum is not. But, in contrast to the previous case, there is no way we can make L vanish by some choice of the reference point. Thus, angular momentum is a property characteristic of rotational motion.

Kepler's Second Law (*the radial line segment from the sun to a planet sweeps out equal areas in equal times*) can be seen as an example of application of the principle of conservation of angular momentum. Since the gravitational force of the sun on a planet is along the sun–planet line, there is no torque, and the angular momentum of the planet (which is essentially equal to the area swept out in a unit time) is conserved. Just as the conservation of momentum implies a symmetry in the laws of nature, so does the conservation of angular momentum. This symmetry — the rotational symmetry — implies that space is *isotropic*: the geometry of space is the same in all directions.

Work and Energy

Let us now turn to the concepts of *work* and *energy*. To simplify the discussion, we take the case of one-dimensional motion. If a constant force F acting along the x-axis causes a particle to move some given distance x, then the *work* done is defined by $W = Fx$. This is a reasonable definition, because it corresponds precisely to the effort we would provide and what we would be paid for to push the object that distance, regardless of the nature of the object. If the force applied is not constant over the distance being covered, then a careful summation (integration) over all small segments composing the distance must be made to get the correct result. When we do work on a body, the body changes by acquiring 'energy,' a quantity defined such that the *acquired energy* equals the work done on the body by the force. If we apply a force to a particle of mass m at rest on a smooth (frictionless) surface in a vacuum, the particle will be moving at a definite speed v after the force stops acting on it. The force applied is $F = ma$ and the work done is $W = \frac{1}{2}mv^2$, which we equate with the acquired energy. Similarly, if the particle has initial speed v_1 and acquires a final speed v_2, the work done is $W = \frac{1}{2}mv_1^2 - \frac{1}{2}mv_2^2$. This is the acquired energy. Since the energy appears in the form of motion, we call it the *kinetic energy* of the particle and define it as $K.E. = \frac{1}{2}mv^2$.

But energy may appear in other forms as well. When it represents the capacity of the particle to do work by virtue of its position in space, it is called the *potential energy*. A simple example is that of a particle moving under the influence of a constant gravitational force (*e.g.*, the earth's). To lift vertically a body of mass m a height z above the surface of the earth, the work we must do on the particle is $-mgz$, where g is the gravitational acceleration and the minus sign comes from the downward direction of the force of gravity. We call the function $V(z) = mgz$ the *gravitational potential energy*. It is defined relative to the surface of the earth, that is, the particle's energy at height z is greater by mgz than it is on the earth's surface. If we now release the particle to fall back freely from this height, it will reach a speed v when it hits the ground. If gravity is the only force acting on the particle, the work done by gravity on the particle must be the same in both directions. Thus, the sum $E = K.E. + V(z)$, called the *mechanical energy*, has the same value at any height although the two energy components may vary separately. This is the law of *conservation of mechanical energy*.

So far we have ignored dissipative forces, *e.g.*, friction. If now we attempt to push a heavy object on a rough surface, we must do more work to obtain the same result because not all work done will be expended into changing the body's mechanical energy; some energy will be dissipated in heat, for which we do not usually get paid. In this case, the mechanical energy clearly is not conserved. However, the loss is only apparent because we are taking the macroscopic point of view, *i.e.*, working on the level of the whole body, ignoring its microscopic components. As the body is rubbing against the surface, the energy lost by the body is in fact transmuted into disorderly kinetic and potential energy of the atoms in the body and the surface. This disorderly mechanical energy of atoms is what is meant by *heat*. No energy has been lost, only part of it has been transformed into another kind of energy. To be fair, we should be paid for work done on both the body and the atoms it contains.

Energy can manifest itself in many other ways as well. Some of the basic forms are the following. *Chemical energy* is the term often used for molecular energy or atomic energy, which are, respectively, kinetic and potential energy of atoms in molecules or electrons in atoms. *Nuclear energy* is kinetic and potential energy of particles (neutrons and protons) contained in atomic nuclei. *Electromagnetic energy* is the form of energy carried by all sorts of electromagnetic radiation, such as moonlight and TV signals. Another important form of energy is *mass*, as expressed by the famous equation discovered by Albert Einstein: $E = mc^2$. This equation means that an amount m of mass is equivalent to an amount E of energy given numerically by product of m and the square of the speed of light c. The equation can be read in either direction: energy can be transformed into mass, and mass can be converted into energy — energy and mass are equivalent.

It is a general property of our physical world that the *total amount of energy of any isolated system* never changes in time, although the *form of energy* may change. This is the *general law of conservation of energy*. Again, this conservation law arises from a symmetry of space, the irrelevance of the absolute measure of time.

To summarize the above, let us restate that the laws of classical mechanics depend neither on the origins of space and time coordinates nor on the orientations of the coordinate system. Also, they are insensitive to the state of motion of physical events, as long as this motion is uniform, rectilinear and free of rotation. All these invariance properties are products of experience, not *a priori* truths. It is only because regularities such as these exist in Nature that we can hope to discover its secrets.

As a simple application of the above discussion, consider a particle in uniform circular motion with constant speed v along a circular path of radius r under the influence of some force F. The situation just described can be seen as an idealization of the motion of an artificial satellite around the earth ($F = $ gravitational force), or of an electron around the atomic nucleus ($F = $ electric force). One can show (with the help of calculus) that the magnitude of the centripetal acceleration is $a = v^2/r$. It follows from $F = ma$ that $mv^2 = rF$, which represents work done to move the particle to infinity (where it has zero potential energy by convention). Hence, the kinetic energy of the particle, $K.E. = \frac{1}{2}mv^2$, satisfies the equation $K.E. = -\frac{1}{2}P.E.$, where $P.E.$ stands for the potential energy of the particle. Although the derivation relies on a particular situation, the relation obtained gives the statement of an important general result of classical mechanics, the so-called *virial theorem*. The theorem is statistical in nature in the sense that it involves the time average of various mechanical quantities. It is applicable to a large class of physical systems (those in quasistationary motion, in which coordinates and momenta always remain within finite limits).

As another application, let us consider how the *law of universal gravitation* could be derived. To begin with, we observe that planets must be under the action of some net (attractive) force, because otherwise they would be moving in straight lines instead of curved paths, as required by Newton's First Law. Such a force acting on a planet must be directed at any instant toward the center of motion, otherwise Kepler's empirical Second Law could not be satisfied. For an elliptical orbit, the center of motion is one of the foci of the ellipse — for a circular orbit, the center of the circle. Now, Newton proved mathematically that the centripetal force acting on a body revolving in an ellipse, circle or parabola must be proportional to the inverse square of the distance of the body to the focus.

What is the origin of that force? Newton suggested that the force governing the motion of the planets around the sun, or the revolution of the moon around the earth is of the very same nature as the gravitational attraction that makes an apple fall to the ground. Whatever their makeup, all heavy bodies experience the same basic force. Consider any two objects, for example the earth (mass m_1) and a stone (mass m_2). We know by experience that the stone is pulled toward the earth by its weight, which is proportional to its own mass m_2; but, similarly, there must be a weight of the earth pulling it toward the stone and proportional to m_1. By the symmetry of action and reaction, we see that the gravitational force is proportional to both m_1 and m_2. Assuming this force to depend only on the masses of the two bodies in interaction and the distance between them, we come to the conclusion that the gravitational attraction between any two objects of masses m_1 and m_2 separated by a distance r is given in magnitude by $F = Gm_1m_2/r^2$. It is a vector directed from one body's center-of-mass to the other's. This is Newton's famous *law of gravitation*. The constant of proportionality, G, has come to be known as the universal gravitational constant.

A.3　Waves and Fields

The physical universe of the late nineteenth century was so dominated by Newton's ideas that it must have appeared to be endowed with a superb unity. Not only could the motion of all bodies, whether on earth or in space, be explained by the same laws of motion, but the invisible world of the atoms was also made part of the realm of mechanics when heat was treated as a mere mechanical phenomenon by the kinetic theory of gases. Certainly, there would not be any problems, most physicists of the time must have thought, that could not be ultimately solved within the existing framework.

But problems there were, and not necessarily anodyne. First, *what is the nature of light?* Is light composed of something similar to microscopic particles, as Newton believed, or of something immaterial akin to wave-like impulse, as Christiaan Huygens, Thomas Young, and Augustin Fresnel advocated? Then, how to incorporate into the existing mechanical framework the increasingly large number of new *electric and magnetic phenomena* observed by Hans Oersted and Michael Faraday, among others? Finally, what to do with the clearly unacceptable *action-at-a-distance interpretation* — *i.e.*, that two particles interact even though they are not touching — as suggested by Newton's law of gravitation and the newly discovered Coulomb's law of electric force? It turns out that the answers to these questions are not unrelated, all relying on two new concepts, *waves* and *fields*, which are to play key roles in the physics of the twentieth century.

Waves and Wave Propagation

Before discussing the nature of light, let us pause briefly to describe what turns out to be an analogous phenomenon, the vibrations of a string that is fixed at one end. If we take hold of the free end of the string and move it rhythmically up and down, imparting to it both *motion* and *energy*, successive segments of the string, from the free end on down the line, also move up and down rhythmically. A crest (maximum displacement) is generated, then a valley, and both travel down the string, soon to be followed by another maximum, then another minimum, and so on. Thus, we have a periodic wave traveling along the string, carrying with it energy (due to the motions and positions of the particles) from one end to the other. The speed with which a crest travels may be taken as the speed of propagation of the wave, v; the distance between two successive crests is called the *wavelength*, λ; and the time needed for a crest to cover that distance is called the *period* of vibration, T. The three quantities are related by $v = \lambda/T$ and the reciprocal of T gives the *frequency* of vibration. Here, the wave propagates horizontally while segments of the string move up and down; such a wave is called a *traveling transverse wave*.

Now, if we take a long rubber cord or a row of particles connected by massless springs and if we snap the free end of the string back and forth horizontally, a disturbance is forced on the string and is transmitted from one particle to the next. The disturbance begins traveling along the string and will produce successive variations of particle density — condensations and rarefactions of particles, which oscillate in the same direction as the motion of the wave itself. Such a wave is called a *traveling longitudinal wave*. Let us imagine we have a number of such identical strings, of either kind, fixed to the same wall and forced to vibrate together by the same disturbance. We may then define a *wavefront* as an imaginary surface, generally perpendicular to the direction of propagation of the wave, which represents the disturbance as it travels down all the strings.

Waves have a characteristic property that distinguishes them from *particles*. When two or more waves propagate in the same medium, they fuse together to form a new wave. We call this phenomenon an *interference*; it can be either *constructive* (if the incident waves reinforce each other) or *destructive* (if the incoming waves tend to cancel out). Streams of particles do not, apparently, exhibit such behavior under normal conditions.

Nature of Light

To the question, *Is light particle-like or wave-like?*, experiments have given a clear-cut answer: transmission of light exhibits all the properties characteristic of the propagation of waves. One of the first decisive experiments which helped to establish the wave theory of light is the famous Young's double-slit interference experiment, described in Chapter 2. But we will find it instructive to repeat the historical argument here. It hinges on explaining why a light ray crossing the interface between a rare and a denser medium is *refracted*, that is, deflected toward the normal to the interface. If light is particle-like, one would naturally expect the light ray to be slowed down by the denser medium (more precisely, its velocity's component perpendicular to the interface to be reduced and the component parallel to the interface unchanged) and hence to be deflected *away from the normal*, contrary

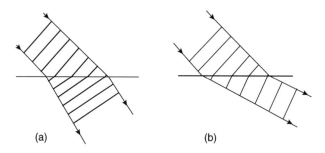

Figure A.2: Refraction of a light pulse going from air to water; wavefronts are at right angles to the light ray and delineate successive maxima. In (a) it is assumed that the speed of light is smaller in water than in air, and so wavefronts travel closer together in water; in (b) light is assumed to move faster in water.

to observations. However, if the 'light particles' are accelerated at the interface by some unknown force, then their path would be bent *toward the normal*, as observed.

On the other hand, if light is wave-like, one would expect that when a light pulse strikes an interface at an oblique angle, the wavefront is split into two, one traveling in the first medium with the old speed, the other in the second medium with a new speed. From Fig. A.2, one sees that light will be bent toward the normal if its speed is *smaller* in the new medium and will be deflected away from the normal if its speed is *greater*. This means that if light is wave-like, it is slowed down in going from air to water, just the opposite of the behavior of a light particle. The predictions of two competing theories could not be more unambiguous. To decide between the two theories, it suffices to compare the speed of light in air with that in water or glass. It was not until the mid-nineteenth century that such a delicate experiment could be carried out, with the result by now well-known: light travels faster in air than in water.

Electric and Magnetic Fields

The *electric force* between two stationary charged particles was experimentally discovered by Charles-Augustin de Coulomb in 1785. Its magnitude is given by $F = Kqq'/r^2$, where q and q' are the two interacting charges, r is their separation distance and K a numerical constant; its direction is along the line joining the two particles. The electric force is repulsive if the charges are of like signs and attractive if the charges are of opposite signs. When the particles are set *in motion*, there is an extra force acting on these particles; this extra force is called the *magnetic force*. The first concrete evidence that moving charges induce a magnetic force was given by Hans Oersted in 1820 when he showed that a strong electric current (flow of charges) sent through a wire aligned along the north–south direction caused a magnetic compass needle, originally set parallel to the wire, to rotate by 90° and settle in the east–west direction. When the direction of flow of the current was reversed, the needle immediately rotated by 180° and aligned itself perpendicular to the wire. The behavior of the needle indicated that it was acted on by a force quite unlike the electric force, emanating from the current and perpendicular to it.

Like every physics student, anybody who has seen the regular alignment of small bits of thread between a pair of charged plates or the pattern of iron filings near a magnet, is convinced that space is modified by the presence of charges or magnets. Something new must have appeared in the intervening space.

To discover what it is, let us perform this experiment. Let us attach a small body to the end of a non-conducting massless string and give it a small charge q, so small that it will not affect any system of charges we want to study (we assume the charges to be at rest for the moment). As we place this device at every point in space, we realize that our probe experiences an electric force which depends on the location of the body, the source charges that set up the action and the probe charge q itself. To obtain a quantity intrinsic to the system, independent of q, we define at any given point P the *electric field strength* (call it E) as the net force F on charge q at point P divided by q. E is a vector, as is F. We could fill the whole three-dimensional space with imaginary arrows to a given scale to represent all such vectors. We call the full set of the E-vectors in a given region the *electric field* in that region. As an aid to visualization, we may represent the electric field by *electric field lines*. At every point along such a line, the electric field is tangential to the line. Also, the density of the lines is proportional to the field strength. When lines crowd together, we have a relatively stronger field; when they spread out, we have a relatively weaker field (Fig. A.3).

The *magnetic field* (B) can be similarly represented by *magnetic field lines*. Repeating Oersted's experiment by placing a compass needle at various points around a current-carrying wire, we discover that the force exerted by the current on the magnet is circular. We can mentally picture the wire as surrounded by concentric lines of force. The tangent to a field line indicates the direction of the field B and the density of field lines, the strength of the field.

What Oersted showed was that a *steady* electric current produced a constant magnetic field around the current-carrying circuit. Inversely, one may ask, could

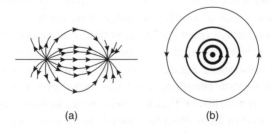

(a)   (b)

Figure A.3:   (a) Representative electric field lines generated by two charges equal in magnitude and opposite in sign. (b) Representative magnetic field lines around a straight wire carrying an electric current. The current emerges perpendicular to the plane of the page.

a constant magnetic field or a steady current flowing in one circuit generate a current in another circuit nearby? The answer was given by Michael Faraday, who discovered the general principle of *electromagnetic induction*: (1) a constant magnetic field or a steady current in one wire cannot induce a current in another wire; (2) only *changing* lines of magnetic force can induce a current in a loop; the change of the force lines can be caused either by moving a magnet relative to the loop, or by suddenly varying the current in a wire nearby.

What now transpires from these results is that: (1) the electric and magnetic fields are sensitive to both space and time; and (2) they are interdependent; if one changes in any way, so does the other. It remained for James Clerk Maxwell to discover the exact physical laws governing the electric and magnetic fields, which he stated in the form of a set of differential equations for E and B. These equations, which form the basis of his electromagnetic theory, describe how the space and time variations of the electric and magnetic fields can be determined for a given electromagnetic source, or distribution of charge and current.

At points far removed from a localized source, Maxwell's equations are reduced to two separate equations, one for E and one for B. They are called the wave equations because they show that the two fields propagate together as periodic oscillations, both perpendicular to each other and perpendicular to the direction of propagation of the waves. Thus, the electric and magnetic fields behave as transverse waves. They cause any electric charges or magnetic poles found anywhere in their region of action to oscillate with a characteristic frequency, in the same manner as a wave transmitted along a string would force particles on its path to fluctuate.

Maxwell's theory predicts, remarkably, that *electromagnetic waves propagate with the same speed as light*. Generally, a wave equation relates the spatial variations of a field amplitude, A, to its time variations. The spatial variations of A are essentially A divided by the square of a small distance, symbolically A/d^2, and the time variations are essentially A divided by the square of a small time interval, or A/t^2. These two quantities are dimensionally different, one has the dimension of the reciprocal of a squared length, the other the dimension of the reciprocal of a squared time. They cannot enter as two terms in the same equation unless the denominator of the second term is multiplied by the square of a speed. This speed is precisely the speed of propagation of the wave, which Maxwell showed to be numerically equal to the speed of light. It was a most remarkable result with far-reaching implications. His theory also predicted that a beam of electromagnetic waves would be reflected off metallic surfaces and that it would be refracted on entering a layer of glass. In other words, electromagnetic radiation is expected on theoretical grounds to behave in every way like light.

Maxwell had thus unified not only electricity and magnetism but also light in a single theory, the electromagnetic theory of radiation, which ultimately encompasses the whole electromagnetic spectrum, from the longest radio waves to the shortest gamma rays. The correctness of his views was eventually confirmed by Heinrich

Hertz, who demonstrated by experiment that an oscillating current indeed sent out electromagnetic waves of the same frequency as the emitting oscillations, and that these waves carried momentum and energy that could induce a fluctuating current in a wire nearby. Maxwell gave the electromagnetic theory mathematical rigor, Hertz gave it physical reality.

To summarize, the concept of *force* can be conveniently replaced by the concept of *field* of force. Instead of saying that a particle exerts a force on another, we may say that a particle creates a field all around itself which acts on any other particles placed in its zone of action. While in non-relativistic physics, the field is merely another mode of describing interactions of particles, in relativistic mechanics, where the speed of light is considered finite, the concept of field takes on a fundamental importance and, in fact, acquires a physical reality of its own. The picture of particles interacting at a distance gives place to a picture of interaction by contact, in which a particle interacts with a field, and the field in turn acts on another particle, such that there is overall conservation of energy and momentum. Applied to electricity and magnetism, the concept of field is essential to a unified treatment of these two phenomena. The unified field thus introduced — the *electromagnetic field* — generates *electromagnetic radiation*, which behaves in every way like *transverse waves* in regions far from the source that produces it. Light is but one form of such radiation. The field concept, introduced by physicists of the nineteenth century, will blossom to full significance in the physics of the twentieth century.

Appendix: General Concepts in Quantum Physics

B

Quantum mechanics is now known to be the correct framework theory for describing all physical phenomena in detail, that is at the microscopic level. Its predictions generally disagree with those of classical mechanics (see Appendix A). The numerical disagreement is, however, pronounced only in the domain of the very small, which is roughly the case for objects having very small inertia, such as electrons and many other atomic and sub-atomic particles. For the commonly experienced objects of sensible magnitudes, the deviations are practically negligible. It is this circumstance that tends to hide from us the otherwise profoundly different character of quantum mechanics. This appendix briefly acquaints you with some of the remarkable features of Quantum Mechanics, which is undoubtedly the most profound of all contemporary scientific thought.

B.1 Heisenberg's Uncertainty Principle

In Classical Mechanics, the state of a particle at any given instant of time is specified completely by giving simultaneously its position and momentum (or velocity) at that instant of time. (The statement can, of course, be generalized to a system of particles.) This, together with the knowledge of all the forces acting on the particle, will completely determine its trajectory, which is then calculable through *Newton's laws of motion*. In general the accuracy of our prediction will be limited by the accuracy of the initial data which is always beset with the ubiquitous errors of measurement, by our incomplete knowledge of the forces acting on the particle, and not a little by our limited computational capability. But nothing in principle prevents our predicting the trajectory of the particle with absolute precision. Besides, unknown and unobserved as it may be to us (the observers), the particle is assumed to have a well defined trajectory, existing as an unobserved objective reality. Now, all this has to be renounced in Quantum Mechanics — the position and the momentum cannot be determined simultaneously with arbitrary high degrees of precision, even in principle. More specifically, if the instantaneous particle position, its x-coordinate say, is determined to within an interval of uncertainty Δx and if the corresponding x-component of its momentum p is determined simultaneously to within an interval of uncertainty Δp, then the *Heisenberg uncertainty principle* asserts that the product of these two uncertainties will be greater than or equal to an irreducible minimum: $\Delta x \Delta p \geq \hbar/2$, where $\hbar$ is Planck's constant divided by 2π ($\hbar \simeq 10^{-34}$ Joule-second). The very act of measurement involved in determining any one of them perturbs the other sufficiently so as to satisfy the above inequality, often by a wide margin, no matter how carefully and cleverly we design the measuring apparatus. This reciprocal latitude of fixation, or shall we say the 'frustration of errors' has observable consequences. For a particle of mass 'm' localized within a box measuring Δx on the side, the uncertainty of momentum would be of the order of $\hbar/\Delta x$. The associated 'zero-point' kinetic energy would be about $(\hbar/\Delta x)^2/2m$. Thus, for $\Delta x \sim 1$ Å, the 'zero-point' energy for an electron is about 3 eV (1 electron-Volt $= 1.6 \times 10^{-19}$ Joule). This will exert a pressure on the walls of the cube of several million times atmospheric pressure. The lighter the particle, the greater is the zero-point energy. It is precisely this zero-point energy that prevents the inert gases of *light* atoms from solidifying even at the absolute zero of temperature. Thus, helium remains liquid down to the lowest temperatures known. True that hydrogen, the *lightest* of all elements, does form a solid at low enough temperatures, but the reason is that the attractive potential between the two hydrogen atoms is sufficiently strong to 'contain' this zero-point motion. For the inert atoms like those of helium, the attraction is relatively much too weak. Again, it is this zero-point energy that prevents the electron in a hydrogen atom from collapsing onto the proton and staying stuck there. The uncertainty principle holds for other pairs of dynamical variables too, for example, energy and the life-time, or angle and the angular momentum.

B.2 Wave Function and Probability: State of the System

Having renounced the classical mechanical idea of sharply defined trajectories, the state of the particle is now given by a quantum mechanical *wave function* $\psi(r,t)$ which is a complex function of position and time, *i.e.*, $\psi = |\psi|e^{i\theta}$. The wave function has a statistical interpretation. Thus, $|\psi(r,t)|^2$ gives the probability (density) of finding the particle at point r at time t, *if* such a measurement is made. The associated probability current (density) is given by $(\hbar/m) \cdot (|\psi|^2) \cdot$ (gradient of phase θ). The wave function is to be determined in any specific case by solving a differential equation, the Schrödinger equation that now replaces Newton's equations of motion. The point is that ψ contains all the observable information about the particle motion. Thus, the classical observables like energy, momentum, angular momentum, etc., retain their usual meaning, but are now obtained from the wave function through certain unusual set of rules prescribed by *quantum mechanics*.

For the simplest case of a freely moving particle, the wave function is a plane wave, $\psi \propto e^{i\vec{p}\cdot\vec{r}/\hbar}$. It has a well-defined momentum $\vec{p}$ and therefore, as demanded by *quantum mechanics*, its position is completely uncertain, *i.e.*, $|\psi|^2$ is the same for all points. This 'matter wave,' or the *de Broglie wave*, carries an energy $p^2/2m$ as in classical particle mechanics. It has a wavelength h/p. In point of fact, except for its probabilistic significance, this wave is like any other wave motion. Thus, for a free electron of energy 1 eV, the de Broglie wavelength is about 12 Å, much shorter than the wavelength of ordinary light. A shorter wavelength means a higher spatial resolution. This is the idea underlying the high resolution electron-microscope where light is replaced by high energy electrons.

For the not-so-simple case of a hydrogen atom, where an electron is bound to the proton by Coulomb attraction, much the same way as planets are gravitationally bound to the sun, the solution of the Schrödinger equation gives stationary states that roughly correspond to the elliptical orbits of the planets around the sun. The set of these stationary states is, however, discrete corresponding to only certain allowed values of energy, angular momentum and its component along a chosen direction. Accordingly, these states are conveniently labeled by certain integers, called *quantum numbers*. Thus, we write ψ_{nlm}. Here n $(= 1, 2, \ldots)$ is the principal quantum number giving the energy $E_n = -1/n^2$ in Rydberg units (1 Rydberg = 13.6 eV); ℓ $(= 0, 1, 2, \ldots, n-1)$ is the angular momentum quantum number and gives the angular momentum in units of $\hbar$; m $(-\ell, -\ell+1, \ldots, \ell-1, \ell)$ is the component of angular momentum along any arbitrarily chosen axis. In addition to these negative-energy (*i.e.*, bound) states, we also have the positive-energy 'scattering' states having a continuous range of energy from zero to infinity — these would correspond to the case of some comets that are merely deflected by the sun into open hyperbolic orbits, and not bound by it.

The same principles, of course, apply to more complex atoms, molecules, bulk matter and sub-nuclear matter. Quantum mechanics has unrestricted validity. Thus, the electromagnetic waves (light), classically described by Maxwell's equations, must also be 'quantized.' The resulting quantum is a 'packet' of energy called a *photon*. If λ is the wavelength (with frequency ν) of light, then the photon carries energy hc/λ, or $h\nu$ where c is the speed of light. It also carries a momentum h/λ. Interaction of quantized radiation with matter involves absorption (emission) of a radiation quantum ($h\nu$) accompanied by an electronic transition from a lower (higher) energy state 1 (2) to a higher (lower) energy state 2 (1) such that $h\nu = E_2 - E_1$ ($E_1 - E_2$). This is the origin of energy 'spectrum' characteristic of atoms, molecules, etc. Similarly, the sound wave-like oscillatory motion of atoms in a solid also must be quantized — the resulting quantum is the *phonon*.

Certain processes, forbidden classically, can take place quantum mechanically. Thus, a particle having an energy which is less than a potential barrier will be classically reflected back by it — it cannot escape over the barrier. The all-pervasive waviness (or fuzziness) of the quantum wave function enables the particle to take the barrier, so to speak, in its stride — it can '*tunnel*' through it even at sub-barrier energies. This is what enables an electron to jump across thin insulating layers separating metallic or superconducting electrodes.

B.3 Superposition of States and Wave Interference

This is a characteristic feature of quantum mechanical waves. If ψ_1 and ψ_2 are the two possible states, then their linear combination (superposition) $a_1\psi_1 + a_2\psi_2$ is also a possible state. It is clear that $|\psi|^2$ will then contain a cross-term (interference term) whose sign will depend on the relative phases of the two complex components, leading to constructive or destructive interference. Thus, these probability waves interfere and diffract just as any other wave; hence, the phenomenon of electron or neutron diffraction which is of great practical use in studying crystal structures — the latter act as diffraction gratings. It is important to note that a particle-wave interferes with *itself* — in a Young's double-slit experiment the electron can propagate through the two slits as alternatives, and the interference is between these two alternatives.

B.4 Indistinguishability of Identical Particles

Classically, identical particles remain distinguishable even if only in virtue of their being spatially separated. For, in principle, we can keep track of which is which by following their *trajectories continuously*. This is obviously *not* allowed in quantum mechanics. Thus, if $\psi(r_1, r_2)$ is the wave function of two identical particles then $|\psi(r_1, r_2)|^2$ only tells us the probability density for finding one of them at r_1 and the other at r_2 without specifying which one. It follows then that under interchange of the labels 1 and 2, the wave function either remains the same (symmetrical) or changes sign (anti-symmetrical). The particles obeying the 'symmetrical statistics' are called *bosons*, and those obeying the 'anti-symmetrical statistics' are called the *fermions*. For identical *fermions*, not more than one particle can occupy a given one-particle state. For *bosons*, any occupation number is permitted. It turns out that particles having intrinsic spin angular momentum half-integral $(1/2, 3/2, \ldots)$ in units of $\hbar$ are fermions, *e.g.*, the electron, the isotope ^{3}He, etc. Particles with integral spin are bosons, *e.g.*, the photon, ^{4}He, etc. Here spin refers to an intrinsic angular momentum that a particle may have. This may be likened to a particle spinning about its axis just as the earth spins about its axis.

This indistinguishability (symmetry or anti-symmetry) under interchange gives rise to purely quantum effects which are pronounced when the amplitude of 'zero-point' motion is comparable with the inter-particle spacing. This happens for electrons in metals, or for helium atoms in liquid helium. We call these *quantum fluids*. Superconductivity or superfluidity of these fluids is an important consequence of *quantum statistics.* The same indistinguishability also gives rise to the 'exchange interaction' responsible for magnetic ordering.

B.5 Quantum Mechanics and Naive Realism

Quantum mechanics not only forbids the simultaneous measurement of position and momentum (and, therefore, trajectories), it even *denies* their objective existence, independent of our measurement, as physically meaningless. Thus, prior to measurement, an electron in a room remains in a state of 'potentiality' (the probabilistic wave function), which 'collapses' to a certain point when it is detected on a photographic plate, say. All this is philosophically very disturbing. Attempts have been made to introduce variables 'hidden' from our reckoning that make the observed particles behave apparently probabilistically while the entire system of particle-plus-hidden variables is *deterministic* in the spirit of classical statistical mechanics. But crucial experiments have ruled out all such 'hidden variable' theories so far. There are paradoxes. There is philosophical uneasiness. But the detailed agreement of quantum predictions with experiments has muted much of the criticism so far.

Appendix: Thermal Physics and Statistical Mechanics

C

C.1 Thermodynamics

Heat is a form of energy. It flows from a hot spot to a cold one. Heat energy in a macroscopic body is carried by the *random motion* of its microscopic constituents, which may be molecules, atoms, elementary particles, or whatever. The constituents at a higher temperature possess higher energies than those at a lower temperature. These constituents collide with one another as they move about, as a result of which energy is constantly being exchanged, thus causing those possessing more energy to share the extra amount with those with less, and this transfer of energy is interpreted as heat flowing from hot to cold. In this way energy is eventually shared equally by every part of the system, the temperature of the whole body becomes identical, and the system is said to reach a state of *thermal equilibrium*. These collisions also change the direction of motion of the constituents, so that any constituent could be moving in any direction at any given time. For that reason these motions carrying heat are called 'random motions.'

Temperature can be measured on any of the familiar scales, but for scientific purposes it is usually measured on the Kelvin scale (K). The *Kelvin scale* differs from the familiar Celsius scale (C) by 273.15°: the freezing point of water at 0°C is 273.15 K; the room temperature of 26.85°C is 300 K. The temperature measured in the Kelvin scale is called the *absolute temperature*. It is 'absolute,' because it can be shown that no temperature can ever be lower than 0 K. At 0 K, all random motions stop and there is not the tiniest amount of heat energy left in the body, hence it is not possible to lower the temperature any further.

So a system tends to have less random motion at a lower temperature, but the ordered nature of it is not completely determined by its temperature. Look at H_2O at 0°C, where water and ice can both exist. The hydrogen and oxygen atoms inside a piece of ice are arranged in an orderly crystal structure, whereas those in liquid water, because water can flow and move freely, clearly are not. So although water and ice are at the same temperature, ice has a more orderly structure than water, and therefore there must be another thermodynamic quantity which signifies the orderliness (or the lack of it) of the constituents of matter. This is called *entropy*, and is usually denoted by S: a substance in a state with a larger amount of entropy is *less* orderly than when it is in a state with less entropy. Looking at the example of ice and water, we can also understand entropy in another (more quantitative) way. Fix the temperature T, in this case at $0°C$. To freeze water into ice, you must put it in a freezer to extract out the *latent heat* in the water. As a result, the final product (ice) becomes more orderly with less entropy, so there must be a connection between the decrease of entropy S at this fixed temperature T (and a fixed volume) and the heat that is extracted from the liquid water to turn it into ice. In fact this amount of heat is precisely TS, and this can be used as the definition of entropy: at a fixed temperature T, the *decrease* of entropy (S) is given by the amount of heat extracted from the substance divided by the temperature T.

If we ignore the influence of heat, as we usually do in studying mechanics, an object becomes stable when it settles down at the bottom of the potential well, where its potential energy U is minimal. When heat is taken into account, then we must require the total energy in a given system, viz., the sum of the potential energy U and the *inflow* of heat energy $-TS$, to be minimal for thermodynamic stability. This sum $F = U - TS$ is called the *free energy*; it replaces the mechanical energy U in the absence of heat.

This example of water and ice illustrates another important aspect of thermodynamics, that a substance can have different *phases*. For H_2O at normal atmospheric pressure, it is in the steam phase above $100°C$, in the water phase between $0°C$ and $100°C$, and in the ice phase below $0°C$. The temperatures $100°C$ and $0°C$ at which a phase changes are called the *transition temperatures*. As a rule, the phase at a lower temperature is usually more orderly than one at a higher temperature. Another way of saying the same thing is that the phase at a lower temperature usually has less *symmetry* (see Chapter 1) than one at a higher temperature. This last statement may seem contradictory, but it is not. A drop of water is spherical and it has spherical symmetry — or symmetry upon rotation for any amount about any direction. A piece of ice is a crystal, so we must *confine* the rotation to some specific amounts about some specific directions in order to move one atom to the position of another atom, and hence its symmetry is smaller.

This phenomenon of *phase transition* occurs not only in water, but also in most other substances and systems. For example, the change from a normal metal to a superconductor at a low temperature is a phase transition (see Chapter 3). As the universe cooled down from its Big Bang beginning towards its present temperature of 2.7 K (see Chapter 10), various phase transitions are believed to have taken place, so that the universe we see today is not in the same phase as the universe that was in the very beginning; at the early epoch the universe was much more symmetrical. Phase transitions are important not only in superconductivity (Chapter 3) and in cosmology (Chapter 10), but also in the modern theory of elementary particle physics (Chapter 9).

C.2 Statistical Mechanics

At thermal equilibrium, each constituent carries the same amount of energy *on average*, but this does not mean that the energy of every constituent is identical at any given time — only the average over time is the same. To obtain the energy spread about the mean, or the *energy distribution*, we must employ the full apparatus of *statistical mechanics*, which is beyond the scope of this book. Nevertheless, it is not difficult to describe and to understand the outcome of these calculations.

Energy distribution depends on whether quantum effects (see Appendix B) are important or not. According to quantum mechanics, a particle in a given volume with energy lying within a specific range can occupy only one of a finite number of *quantum states*. How these quantum states may be occupied depends on what type of particles they are. Particles of this universe are of two types: they are either *bosons* or *fermions*. Two bosons of the same kind (*e.g.*, two photons) have an additional tendency to occupy the same quantum state. It is this remarkable property that leads to the feasibility of a laser (see Chapter 2) and the remarkable phenomena of superfluidity and superconductivity (see Chapter 3). On the other hand, fermions of the same kind (*e.g.*, two electrons) are absolutely forbidden from occupying the same quantum state. It is this special property that leads to the distinction between a conductor and an insulator, and prevents the collapse of a white dwarf or a neutron star (see Chapter 8).

If the total number of particles in the macroscopic system is small compared to the number of quantum states available, then it would be very unlikely for two particles to occupy the same quantum state anyway, in which case whether these particles are bosons or fermions is immaterial. The energy distribution then follows what is known as the *Boltzmann* (or *Maxwell–Boltzmann*) *distribution*. At a given (absolute) temperature T, the *average energy* per *degree of freedom* for a non-relativistic particle is $\frac{1}{2}kT$, where $k = 1.38 \times 10^{-23}$ J/K (Joules per Kelvin) is known as the *Boltzmann constant*. What this says is that the average energy per degree of freedom at $T = 1$ K is 0.69×10^{-23} J; at $T = 300$ K it is 2.07×10^{-21} J. Sometimes units are chosen so that $k = 1$. In that case we simply identify each degree Kelvin as an energy unit of 1.38×10^{-23} J, and the energy per degree of freedom becomes simply $\frac{1}{2}T$. Since one Joule is equal to 6.24×10^{18} eV (electron-Volt), each degree of temperature can also be thought of as carrying the energy of 8.617×10^{-5} eV, or 86μ eV (micro-electron-Volt, or 10^{-6} eV). Thus at a room temperature of $27°$C, or $T = 300$ K, the energy per degree of freedom will be $\frac{1}{2}T = 150$ K $= 0.013$ eV.

So far we have not explained what a 'degree of freedom' is. A point particle can move in any one of the three spatial directions and is counted to have three degrees of freedom; a rigid body of finite size can do so as well as rotating about each of the three axes pointing at the three possible directions, so it has six degrees of freedom. The average energy per point particle is therefore $\frac{3}{2}kT$, and that for a rigid body is therefore $3kT$.

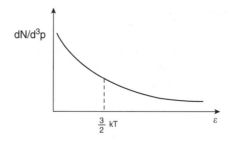

Figure C.1: Boltzmann distribution, depicting the number of particles per unit momentum volume (dN/d^3p) with an energy ϵ at a given temperature T. The average energy per point particle is $(3/2)kT$.

We have also not explained what a 'Boltzmann distribution' is. This distribution is shown in Fig. C.1, in which the number of particles per unit momentum volume is plotted as a function of the particle energy.[1] The average energy of each non-relativistic point particle, as mentioned before, is $\frac{1}{2}kT$. The Boltzmann distribution of other objects is qualitatively similar.

We have so far considered only the Boltzmann distribution, which is valid when the number of particles N in the system of volume V is so *small* that it is improbable for any two of them to occupy the same quantum state. If this is not so, then quantum effects become important and we must distinguish bosons from fermions.

In order to decide whether quantum effects are important or not, we must first ask: 'how small is *small*'? That depends on the temperature T of the system, because the number of quantum states N_0 present depends on it. To see that, recall that the energy of the particle depends on the temperature. Energy is related to momentum, and according to quantum mechanics, momentum is related to the wavelength λ of the particle,[2] which measures the effective size of that particle. Thus, the effective volume λ^3 occupied by one such particle is temperature-dependent, and so is the number of quantum states $N_0 = V/\lambda^3$.

Thus, whether quantum effects are important or not depends on whether N is large or small compared to N_0, and this in turn depends on the temperature T of the macroscopic body. At high temperatures, the wavelength λ is small, the number of quantum states N_0 is large compared to N, and quantum effects are relatively unimportant. At low temperatures, the opposite is true.

The temperature below which quantum effects become important is called the *degeneracy temperature*, and will be denoted by T_0. The energy distribution below

---

[1] The mathematical formula is $dN/d^3p = C\exp(-\epsilon/kT)$, where dN is the number of particles in the momentum volume d^3p whose energy is ϵ. C is a normalization constant determined by the total number of particles in the system.

[2] The formula for λ is: $\lambda = \hbar(2\pi/mkT)^{1/2}$, where m is the mass of the particle. This formula may be understood as follows. The average energy of a particle at temperature T is of order kT, so its average momentum is of order $(2mkT)^{1/2}$, and its average wavelength according to quantum mechanics is of order λ given above. The exact numerical coefficient appearing in the expression for λ can be obtained only through detailed calculations.

the degenerate temperature will be said to be *degenerate*. From the discussion above, T_0 must increase with increasing N/N_0.[3] As a result, it increases with the number density N/V of these particles and decreases with the particle mass m. For example, a *proton* (hydrogen nucleus) or a hydrogen atom has a mass of $m \simeq 2 \times 10^{-24}$ g. At a normal density of $N/V \sim 10^{22}$ cm^{-3}, the degeneracy temperature is $T_0 \simeq 1$ K. A *photon* has $m = 0$ so a photon gas is always degenerate, as $T_0 = \infty$ in this case. For *electrons* in metals, typically $N/V \simeq 10^{22}$–10^{23} cm^{-3}. With the electron mass being $m \simeq 10^{-27}$ g, the degeneracy temperature is $T_0 \simeq (16$–$20) \times 10^3$ K, and hence the electronic distribution in a metal at room temperature is degenerate.

We will now proceed to discuss the distributions when the quantum effects are important. For bosons, the distribution is known as the *Bose–Einstein distribution*; for fermions, the distribution is known as the *Fermi–Dirac distribution*. These two distributions are different, but for $T \gg T_0$, both approach the Boltzmann distribution discussed before.

Electrons, protons, and neutrons are examples of fermions. They obey the Fermi–Dirac distribution, and two of these particles are not allowed to occupy the same quantum state. Therefore, at absolute zero temperature, the N particles in the system must simply fill out the N quantum states with the lowest energy. The distribution is therefore given by the solid line of Fig. C.2a; the energy of the last occupied state is called the *Fermi energy* and is denoted by ϵ_F.[4] When T rises above absolute zero, the distribution smears out a bit as given by the dashed line in Fig. C.2a, but the amount of smearing is still relatively small as long as $T \ll T_0$. Finally, as $T \gg T_0$, the amount of smearing becomes so large that eventually it approaches the Boltzmann distribution given by Fig. C.1.

Helium atoms (^{4}He) are bosons. They obey the Bose–Einstein distribution and these particles have the tendency to occupy the same quantum state. Thus, at absolute zero temperature, these particles *all* occupy the lowest possible quantum state, as indicated by the heavy vertical bar in Fig. C.2b. This is the complete opposite of the fermions, no two of which are allowed to be in the same state. As T rises from zero, the available thermal energy pumps more and more of these bosons to quantum states of higher energy, as shown by the dashed line in Fig. C.2b. However, as long as T is lower than some *condensation temperature* T_c, a finite percentage of the particles will remain in the lowest energy state. This unusual phenomenon, where a macroscopic number of particles occupy a single quantum state and therefore exhibit quantum mechanical properties such as phase coherence, is known as *Bose–Einstein condensation*. The condensation temperature T_c is of the order of the degenerate temperature T_0.[5] For liquid helium, this turns out to be 3.1 K, if all the inter-particle interactions are ignored as we have done so up to the present. The actual value for T_c is 2.19 K, because of the presence of these interactions. Below this

---

[3]The exact formula is $T_0 = T(N/N_0)^{2/3} = 2\pi(N/V)^{2/3}\hbar^2/mk$.
[4]$\epsilon_F = (\hbar^2/2m)(3\pi^2 N/V)^{2/3} \equiv kT_F = 1.31 \, kT_0$.
[5]$T_c = 0.527T_0$.

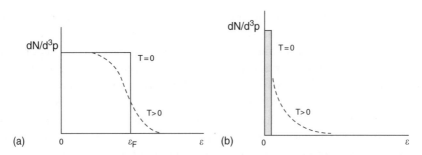

Figure C.2: The number of particles per unit momentum volume (dN/d^3p) with an energy ϵ at temperature T. (a) Fermi–Dirac distribution, with Fermi energy ϵ_F; (b) Bose–Einstein distribution, where the heavy vertical line shows Bose condensation.

condensation temperature, all sorts of interesting quantum mechanical behaviors (*superfluidity*) occur. Another interesting example of Bose–Einstein condensation is the phenomenon of *superconductivity*. This happens because a *pair* of electrons also behaves like a boson. See Chapter 3 for detailed discussions of these topics.

Photons are also bosons, but they differ from helium atoms in two respects: photons have mass $m = 0$, and they can be created and absorbed by the walls of the container or any other material body in the box. As a result of their masslessness, the degenerate temperature T_0 is infinite. However, photons of zero energy cannot exist, so there is no Bose–Einstein condensation of the photons — these photons must disappear and be absorbed by the walls of the container or other material bodies.

The Bose–Einstein distribution of photons is known as *Planck's distribution*. This distribution law is also called the *black-body radiation law*. It was discovered in 1900 by Max Planck a long time before the general Bose–Einstein distribution was worked out; it is through this law that Planck's constant was first introduced, and the seeds for quantum mechanics sowed.

Planck's distribution is sketched in Fig. C.3 for two temperatures. The peaks of the curves at various temperatures are connected by a dashed line.

Note that this differs from Fig. C.2(b), because we are now plotting the *energy density* of photons per unit *wavelength* (which we denote by u) against the *wavelength* λ of the photon, rather than the photon *number* per unit *momentum volume*.[6] Since the distribution is fixed for a given temperature, a measure of this distribution can determine the temperature of a body even when this body is unreachable — as long as its light reaches us. This is the case if we want to know the temperature inside a steel furnace, or the temperature of a distant star (see Chapter 8). It is useful to learn more about two special aspects of this distribution, expressed in *Wien's law* and *Stefan–Boltzmann's law*.

---

[6]Momentum p is related to wavelength λ by $p = 2\pi\hbar/\lambda$, and frequency ν is related to the wavelength by $\nu = c/\lambda$. The energy of a photon is $\epsilon = 2\pi\hbar\nu$.

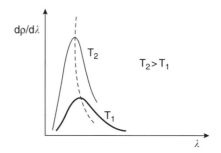

Figure C.3:   Planck distribution, showing the photon energy density per unit wavelength $(d\rho/d\lambda)$ as a function of the wavelength λ at a given temperature T. The dotted line joins the peaks at different temperatures and is the curve for Wien's law.

Wien's law states that the wavelength $\lambda_{\max}$ at the peak of Fig. C.3 is inversely proportional to the temperature: $\lambda_{\max}T = 0.2898$ cm K. This is so because, quantum mechanically, the energy carried by a photon of wavelength λ is inversely proportional to λ (energy $= 2\pi\hbar c/\lambda$). Thermodynamically, this energy is derived from heat and so is proportional to T. Consequently, $\lambda_{\max} \propto T^{-1}$ as required by Wien's law. The proportionality constant ($\simeq 0.3$), however, cannot be obtained without a detailed calculation.

To illustrate this relation, let us just approximate the number 0.2898 by 0.3, and consider the radiation at a room temperature of $T = 300$ K. The peak wavelength works out to be $\lambda_{\max} = 10^{-3}$ cm $= 10$ micron (1 micron $=$ one millionth of a meter). The wavelengths of the visible light being from 0.4 to 0.7 microns, this radiation at the room temperature is too long to be visible. Nevertheless, it is there in the form of the infrared. To emit visible light the body would have to have a higher temperature. The surface of the sun, for example, has a temperature of 6000 K. Its peak wavelength of 0.5 micron is right in the middle of the visible light region. That is why sunlight is visible, and why it contains all these beautiful colors of the rainbow visible at sunrise and sunset.

The *Stefan–Boltzmann* law states that the total amount of power density, or the radiation energy of all wavelengths emanating from a body per unit surface area and per unit time, is given by $f = \sigma T^4$, where $\sigma = 5.67 \times 10^{-5}$ erg/(cm^2 s K^4) is called the *Stefan–Boltzmann constant*. Wien's law tells us what 'color' the radiation has; this law tells us how bright the radiation is. The fact that it is proportional to the fourth power of the absolute temperature is the reason why hot objects are so bright.

A related and a more useful quantity for our purpose is the energy density of the radiation. If one imagines putting the heated body and the radiation it emits in a big box, then the radiation and the body will eventually reach an equilibrium state. The energy density ρ_γ of the radiation in the box is then given by $\rho_\gamma = aT^4$, where $a = 7.56 \times 10^{-15}$ erg/(cm^3 K^4) $= 4.73 \times 10^{-3}$ eV/(cm^3 K^4).

Another version of the Stefan–Boltzmann law concerns the number of photons per unit volume, n_γ. It is proportional to T^3: $n_\gamma = bT^3$, with $b = 3.2 \times 10^{13} T^3 / \text{eV}^3 \text{ cm}^3 = 20.4 T^3 / \text{K}^3 \text{ cm}^3$.

C.3 Dimensional Analysis

To understand the various versions of the Stefan–Boltzmann law, let us resort to a very useful tool known as *dimensional analysis*.

Every quantity in physics carries a *dimension* (*i.e.*, unit) made up of three fundamental units: second (s) for time, centimeter (cm) for length, and gram (g) for mass (in the cgs system of units). Dimensional analysis is simply the statement that whatever quantity we want to consider must have the right units (dimensions). This by itself is of course trivial, and is not going to get us anywhere. However, if we also know on physical grounds what that quantity may depend on, then quite often this will yield very valuable information as we shall illustrate below with the 'derivation' of the Stefan–Boltzmann law.

To begin the analysis for the photon number density n_γ, we must first decide what it can depend on. It will clearly depend on the temperature T. It must also depend on the dynamics and the kinematics governing the emission and the absorption of photons, the balance of which generates the eventual photon density in a volume. The details of these do not concern us, as long as we realize that they are controlled by two fundamental physical constants: the speed of light $c = 3 \times 10^{10}$ cm/s, and the Planck's constant $\hbar = 1.054 \times 10^{-27}$ erg s. It depends on c because anything that has to do with electromagnetic radiation involves c, and it depends on $\hbar$ because we are in a degenerate regime where quantum effects are important. The numerical magnitudes of these two constants do not even matter for our purpose, but their units do.

So the photon number density n_γ is a function of T, c, and $\hbar$: $n_\gamma = f(T, c, \hbar)$, where f is some appropriate function which we seek to determine. The unit for n_γ is number per unit volume, say cm^{-3}. We must now choose the function f above so that $f(T, c, \hbar)$ comes out to have this unit cm^{-3}. The unit of T is energy (with units chosen so that $k = 1$), say erg $= \text{g (cm/s)}^2$; the unit of c is cm/s, and the unit of $\hbar$ is erg s. The only combination of these three variables, $T, c, \hbar$, to yield something of a unit cm^{-3} is clearly $c^{-3}\hbar^{-3}T^3$, which gives a unit $(\text{cm/s})^{-3}(\text{erg s})^{-3} \text{erg}^3 = \text{cm}^{-3}$. Thus, we must have $n_\gamma = b_0 T^3 / (\hbar c)^3$, and the proportionality constant b_0 is dimensionless (*i.e.*, without any unit, or is a pure number). To calculate this number, we must know the technical details of quantum electrodynamics and statistical mechanics. Nevertheless, as we have just seen, the very fact that n_γ is proportional to T^3 can be deduced simply from dimensional analysis.

The fact that the energy density ρ_γ is given by $\rho_\gamma = a_0 T^4 / (\hbar c)^3$ can be obtained in exactly the same way through dimensional analysis, once we note that the dimension of energy density is erg/cm^3.

The beauty of this argument is not only that it is simple, but that it is also quite general. We used the argument above to obtain n_γ and ρ_γ, but nothing in it says that we are dealing with photons γ rather than some other relativistic particles. The appearance of the speed of light c, which is the speed of travel of a massless particle, however suggests that this argument is not valid for massive particles unless they are at such a high temperature to render the mass m negligible $(mc^2 \ll kT)$ and their velocity approaches that of light. Indeed, it is easy to see that if we had to assume f to be a function of m as well as T, c, and $\hbar$, then the dimensional argument would fail because there are infinitely many combinations of these four variables to yield the right dimension. No useful conclusion can then be drawn. However, as indicated before, the argument for photons is equally valid for any *relativistic particle* (meaning $m \ll kT/c^2$), in which case its number density is still given by $n = b_0 T^3/(\hbar c)^3$, and its energy density is given by $\rho = a_0 T^4/(\hbar c)^3$.

The proportionality constants a_0 and b_0 cannot be obtained without detailed dynamical inputs. With them, they are calculable and we shall quote the results. For bosons, the values are $a_{0B} = (\pi^2/30)g = 0.33g$, and $b_{0B} = (1.202/\pi^2)g = 0.12g$, where g is the *spin degeneracy factor*, or the number of spin orientations the particle can have. This number is $g = 1$ for pions and $g = 2$ for photons. For fermions, the two proportionality constants are $a_{0F} = (7/8)a_{0B}$ and $b_{0F} = (3/4)b_{0B}$. That in both a_0 and b_0 the values for fermions are smaller than the values for bosons is understandable. It is due to the fact that fermions tend to avoid one another while bosons tend to stick with their own clans. As a result, there must be more bosons per unit volume than fermions, everything else being equal. Hence, $b_{0B} > b_{0F}$, and similarly $a_{0B} > a_{0F}$.

If we write the energy density for the photon as $\rho_\gamma = aT^4 = a_{0B}T^4/(\hbar c)^3$, and substitute into it the known values of a_{0B}, $\hbar$, and c, we get $a = 4.73 \times 10^{-3}$ eV/cm^3/K^4. Similarly we can calculate the constant b defined by $n_\gamma = bT^3$ from the value of b_{0B} and get $b = 20.4$/cm^3/K^3.

Note that the photon density is fixed at a given temperature. This could not possibly be the case for the *electron* density at low temperatures because the electron density depends on how many electrons we put in the volume, so it is a number which could in principle be anything. This is not so for photons, because photons can be *created* and *absorbed*. If we force more photons in the volume at a given temperature, they will simply be absorbed by the walls. If we take out photons, then the walls will radiate enough photons so that the Stefan–Boltzmann law remains valid.

We also mention that Stefan–Boltzmann's law is valid for any relativistic particle, and that the proportionality constants a_0 and b_0 are completely fixed as well. That this is so is again because all these particles can also be created and absorbed *if the temperature is high enough*. For electrons, if the temperature is considerably larger than 1 MeV (million eV), then there is sufficient heat energy present to create electron–positron pairs. At low temperatures, such creations are energetically forbidden (electrons and positrons both have a mass of 0.5 MeV), and such automatic adjustments of their numbers cannot occur.

It should also be mentioned that the pressure P these particles exert on one another and on the walls of the box is given by $P = \rho/3$, both for bosons and for fermions. The fact that P is proportional to ρ can be obtained from dimensional analysis, but the factor three in the denominator can be obtained only through detailed calculations.

Appendix: Solutions to Problems

Chapter 2. Lasers and Physics

2.1 We know that E/ν, $\lambda\nu$ and λE are universal constants, which can be obtained from the stated equivalences. So that, with λ given, we can calculate the corresponding ν and E by the relations $\nu = (\lambda_1\nu_1)/\lambda$ and $E = (\lambda_1 E_1)/\lambda$, where $\lambda_1\nu_1 = (1240 \text{ nm})(2.4 \times 10^{14} \text{ Hz}) = 3 \times 10^{17} \text{ nm Hz}$ and $\lambda_1 E_1 = (1240 \text{ nm})(1 \text{ eV}) = 1240 \text{ nm eV}$.

So 1 nm is equivalent to 3×10^{17} Hz or 1.24 keV; 500 nm is equivalent to 6×10^{14} Hz or 2.48 eV; and 10^4 nm is equivalent to 3×10^{13} Hz or 0.12 eV.

2.2 (a) 10.2 eV. (b) An energy of 1.9 eV is released. (c) No, because absorption would excite the atom to an energy of 3.6 eV, where no atomic state exists.

2.3 Since $\tau = 1/\gamma$, we have $\tau_2/\tau_3 = \gamma_3/\gamma_2 = 1/100$. Level 3 lives 100 times longer than level 2.

2.4 We know that $\Delta\nu/\nu = v_x/c$. Now $v_x^2/c^2 = kT/Mc^2$. We have $Mc^2 = 10^9$ eV (mass of the hydrogen atom) and $kT = 25 \times 10^{-3}$ eV (room temperature). It follows that $\Delta\nu/\nu = 5 \times 10^{-6}$.

2.5 $E/T = k$ is a universal constant, so that $E = T(E_1/T_1)$, where $E_1/T_1 = 1/12000$ (eV/K). So 300 K is equivalent to 25 meV, and 6500 K to 550 meV.

2.6 In thermal equilibrium, excited states have very small atomic populations. The stimulated emission rate depends on the population of the upper transition level, and so must be much smaller than the absorption rate from the ground state because of the large population of the latter.

2.7 When the first spontaneous radiations occur, they run into all directions, and so the stimulated photons that follow also go into all directions, since no cavity is there to guide them in a particular direction. No amplification is possible without a cavity. So directionality and the high intensity of laser light are the qualities determined by the resonant cavity.

2.8 (a) The irradiance is $10^{-3} \text{ W}/(\pi(10^{-3} \text{ m})^2/4) = 1273 \text{ W/m}^2$. A photon at a wavelength of 700 nm has energy 1.8 eV or 2.88×10^{-19} J. So a power of

1 mW or 10^{-3} J/s corresponds to a flux of 3.5×10^{15} photons per second. Given that the speed of light is 3×10^8 m/s, a cavity 1 m long should contain 10^7 photons at all times to sustain such a flux.

(b) NOVA produces 100 kJ of infrared light in 3 ns pulse lengths. This is equivalent to a power of 100 kJ/3 ns, or 34×10^{12} W. Assuming a diameter of 10 mm for the beam, the irradiance is 430×10^{12} W/m².

2.9 (a) With the laser wavelength $\lambda_R = 500$ nm, we must have $L = 250$ nm to guarantee a monochromatic beam at the resonant wavelength $\lambda_1 = 2L = 500$ nm. If $L = 5$ cm, the resonant mode corresponds to $n = 2 \times 10^5$, so there will be at least 4×10^5 modes. The most likely are $\lambda_R = 500$ nm and a few more adjacent ones separated from it (roughly) by multiples of 0.0025 nm.

(b) Wavelength spacing $\Delta\lambda_n = |\lambda_{n+1} - \lambda_n| = 2L/[n(n+1)]$; frequency spacing $\Delta\nu_n = |\nu_{n+1} - \nu_n| = c/(2L)$. The latter is independent of n, whereas the former decreases rapidly with increasing n.

2.10 (a) 30 μm. (b) 300 m. (c) 2.5 cm.

2.11 (a) 3.1×10^{-4} rad. (b) 3.5×10^{-3} rad.

2.12 $D = 2R = 2\theta x = 2(6.94 \times 10^{-7})(400 \times 10^6) \approx 550$ m.

2.13 (a) When a slit is covered, the photons pass through the uncovered slit and go on to strike the screen: we will see a bright stripe on the screen opposite the uncovered slit. There might be faint dark lines near the edges of the stripe due to the bending of light around the edge of the slit (diffraction).

(b) When a detector is placed behind each slit, we will hear a clicking sound from the detector through which a photon passes, and not from both detectors at the same time. So a photon must pass through either slit, not both.

(c) When both slits are left uncovered, we will *not* see two separate bright bands, as we would expect from observations in (b), but rather a pattern of overlapping bright and dark stripes (with a white ripple at the center) opposite the two slits. This indicates that the light waves have constructively and destructively interfered.

2.14 The condition $T \geqslant 1/\Delta\nu$ also says that $\Delta\nu \geqslant 1/T$. (a) To have $T = 100$ fs, we must have $\Delta\nu \geqslant 1/(100 \text{ fs}) = 10^{13}$ Hz. For a 1 m long cavity, adjacent modes are separated by $\Delta\nu_0 = c/2L = (3 \times 10^8 \text{ m})/(2 \text{ m} \cdot \text{s}) = 1.5 \times 10^8$ Hz. So, to have $T = 100$ fs, we must lock in at least $\Delta\nu/\Delta\nu_0$, or 6.7×10^4 modes.

Chapter 5. Nanostructures

5.1 If the film thickness tends to zero, the particle position uncertainty also tends to zero, which implies that the conjugate momentum uncertainty and the momentum itself tend to infinity: no confinement is possible.

5.2 $\Delta E_{21} = 3h^2/8m^*d^2 = 1.185(m_e/m^*)/d^2$ in eV if d is in nm. For $d = 10$ nm and $m^* = 0.07m_e$ one has $\Delta E_{21} = 0.170$ eV. Since $\Delta E_{32} = 5h^2/8m^*d^2$ is $5\Delta E_{21}/3$, we have $\Delta E_{32} = 0.283$ eV. Converting to wavelength $\lambda = hc/\Delta E$, we have respectively $\lambda_{21} = 7290$ nm and $\lambda_{32} = 4380$ nm in the infrared. When d is reduced by a factor of 2, the energy is increased by a factor of 4 and the wavelength reduced by the same factor.

5.3 Electrons in a grain scatter photons whose energy is less than a minimum determined by the size of the grain, and absorb those with higher energy. The color we see is determined by the scattered light. The larger crystallites can absorb lower energy photons and so appear red, whereas the smaller grains absorb higher-energy quanta and so appear yellow.

5.4 The condition $\Delta E_{21} > kT$ translates into $d^2 < 1.2 m_e/m^* kT$, with d in nm and kT in eV. At room temperature, $kT = 0.026$ eV, so that $d < 6.8$ nm for $m^* = m_e$ and $d < 25$ nm for $m^* = 0.07 m_e$.

5.5 The Pauli exclusion principle dictates that only two electrons can occupy any wave-guide mode. The detection of a current ($I = e/\Delta t$) means an electron has scattered inelastically inside the output reservoir (so that the energy changes by $\Delta E = eV$), thereby vacating a waveguide mode and enabling an incoming electron to occupy it. The uncertainty relation $\Delta E \, \Delta t \sim h$ then becomes $e^2 V/I \sim h$, hence $I \sim e^2 V/h$ for each electron. With two electrons filling a mode, we must have $I \sim 2e^2 V/h$. So the conductance for each mode in the wire is $I/V = 2e^2/h$.

5.6 The condition $E_c > kT$ is equivalent to $R < e^2/\varepsilon kT$. Hence, the limiting value for R is 4.6 nm, which corresponds to a diameter of 9.2 nm.

5.7 As the atomic size changes, there are two opposing effects. First, the Coulomb energy arising from repulsion between electrons orbiting around the nucleus *decreases* when the average distance between electrons increases. But there is another energy scale in the problem: the energy spacings of the orbits of electrons. As the atomic size increases, the differences in the orbital energies *decrease faster* than the Coulomb energy. It follows that the effects of the electron–electron interaction are *relatively* more important in artificial atoms than in natural atoms.

5.8 (a) $E(N, \delta) = -N(N + 2\delta)/2C$; $\Delta_+ = (1 - 2\delta)/2C$; $\Delta_- = (1 + 2\delta)/2C$; and $\Delta_+ + \Delta_- = 1/C$. We note that the first three depend on the capacitance and the voltage, but the last depends only on the capacitance. The last three do not depend on N. (b) The results follow directly from the cancelation of Δ_+ for $\delta = 1/2$ at $N = 0, 1, N$, and Δ_- for $\delta = -1/2$ at $N = 1, 2, N + 1$.

5.9 Just one kind, since all carbons in C_{60} are exactly equivalent. There are 12 pentagons and 20 hexagons. A simple way is to calculate first the total surface area of 12 pentagons and 20 hexagons, all of equal sides a, which is $72.6 \, a^2$. Equating this to $4\pi R^2$ gives $R = 2.4 \, a$. For $a = 0.142$ nm, we have $R = 0.34$ nm, *i.e.*, a diameter of 0.68 nm.

5.10 The width is equal to the diameter of C_{60}, that is 0.68 nm, and the length is the same diameter increased by 0.142 nm, that is 0.82 nm. One could also use the method of the preceding exercise; the answer is the same.

5.11 By the same reasoning as given in the text, but now including also heptagons and octagons, one has the relation $2e = 3v = 5p+6h+7s+8g$, $v-e+f-2 = 0$, $f = p + h + s + g$, where s is the number of heptagons, and g the number of octagons. It follows that $p = s + 2g + 12$.

5.12 This is a purely geometrical problem. One should draw a hexagon of edges with length a, and apply trigonometry to obtain $|a_1| = |a_2| = 2a\cos 30° = \sqrt{3}a$. Next, $C^2 = n^2a_1^2 + m^2a_2^2 + 2nma_1 \cdot a_2$. Given $60°$ as the angle between a_1 and a_2, one obtains the answer. Similarly, with $C \cdot a_1 = |C||a_1|\cos\theta = na_1^2 + ma_1 \cdot a_2$, one obtains θ, as stated.

5.13 Since C_{60} has a radius $R = 0.34$ nm, the circumference is 2.14 nm. Then, $C^2 = 4.58$ nm^2, which is also equal to $0.058(n^2 + nm + m^2)$. It follows that the possible values of (n, m) are $(9,0)$, $(8,1)$, $(8,2)$, $(7,3)$, $(6,4)$ and $(5,5)$. Of these, $(9,0)$, $(8,2)$ and $(5,5)$ are metallic.

5.14 In the copper atom, there is a single electron in the external orbital, which happens to be an s-state (which allows only two spin states). When the atoms get together in a bulk volume, another electron of the same kind will fill up the only empty spin state to complete the orbital, corresponding to an anti-aligned spin configuration. Since there is a single possibility, there is no exchange interaction and no splitting.

5.15 Using the conversion 1 in$^2 = 6.45 \times 10^{14}$ nm^2, we see that 1 bit occupies an area of 6.45×10^5 nm$^2/x$. One bit also occupies an area equal to NL^2. Hence, $x = 6.45 \times 10^5$ nm$^2/NL^2$(nm^2). Now assume $L = 15$ nm and $x = 1$, we have $N = 2900$.

5.16 With $L = 10$ nm, we have $x = 6450/N$, which amounts to 2.15 for $N = 3000$, and 6450 for $N = 1$. A CD, 12.8 square inches in area, can hold 27.50 Gb in the first case, and 82560 Gb (or 82.50 Tb) in the second. Now, 1 Pb is 10^6 Gb, hence storing 1 Pb requires 36400 CDs in the first case, and 13 CDs in the second. If $L = 3.5$ nm, the grain area size is ten times smaller. Hence, the density is correspondingly larger, $x = 64500$ for $N = 1$. In this case, one CD can hold 825 Tb, and so two CDs are enough to store 1 Pb, with plenty of space to spare.

Chapter 6. Quantum Computation

6.1 Two theorems and one game:

1. 10000000, 1001100.

2. (a) When we decrease any binary number s_j, some of its bits must change. The sum d'_a at that bit must become odd.

   (b) Suppose among the n sums, $d_a, d_b, \ldots$ are odd, and a is the leftmost bit among these odd sums. To make the sums all even, all we have to do is to pick a pile whose ath bit is 1, and change its ath, bth, $\ldots$ bits.

6.2 The NAND gate is a combination of the AND and NOT gates. The NOT gate, which interchanges 0 and 1, can be obtained from the XOR ($\oplus$) gate with the help of an auxiliary bit 1, because $0 \oplus 1 = 1$ and $1 \oplus 1 = 0$.

6.3 $d \oplus c = d$ if $c = 0$, and $d \oplus c = \bar{d}$ if $c = 1$.

6.4 $1000 = 13 + 47 \times 21$, hence $1000 \bmod 21 = 13$.

6.5 Different gates:

1. $(1d) \to \boxed{\text{CN}} \to (1, 1 \oplus d) = (1, \bar{d})$.

2. $(cc'0) \to \boxed{\text{CCN}} \to (c, c', c \times c')$.

3. $(cd) \to \boxed{\text{CN}} \to (c, c \oplus d)$.

The last bit of these three operations give the desired result.

4. $(ab) \to \boxed{\text{FANOUT}_1} \to (aab) \to \boxed{\text{FANOUT}_3} \to (aabb)$

        $\to \boxed{\text{AND}_{13}} \to \;\; (a \times b, a, b) \to \boxed{\text{XOR}_{23}} \to (a \times b, a \oplus b)$

        $\to \boxed{\text{XOR}_{12}} \to \;\; (a \times b) \oplus (a \oplus b)$.

The final result is 1 if either a or b is 1, and is 0 if neither of them is 1. Hence, it is a OR b.

6.6 Hadamard gate:

$$|0\rangle \to \boxed{\text{H}} \to \frac{1}{\sqrt{2}}(|0\rangle + |1\rangle) \to \boxed{\text{R}_X} \to e^{-i\theta}\frac{1}{\sqrt{2}}(|0\rangle + |1\rangle) \to \boxed{\text{H}} \to e^{-i\theta}|0\rangle,$$

$$|1\rangle \to \boxed{\text{H}} \to \frac{1}{\sqrt{2}}(|0\rangle - |1\rangle) \to \boxed{\text{R}_X} \to e^{i\theta}\frac{1}{\sqrt{2}}(|0\rangle - |1\rangle) \to \boxed{\text{H}} \to e^{i\theta}|1\rangle.$$

6.7 Consider $\text{H}_2(\text{CU})\text{H}_2$ operating on $|cd\rangle$, where c is the control bit and d the data bit. If CU were absent, then nothing changes because $\text{H}_2\,\text{H}_2 = \mathbf{1}$. With CU sandwiched in-between, the intermediate state $|11\rangle$ gets a minus sign, thereby interchanging the two intermediate states $|1\rangle\frac{1}{\sqrt{2}}(|0\rangle + |1\rangle)$ and $|1\rangle\frac{1}{\sqrt{2}}(|0\rangle - |1\rangle)$. After the operation by the final H_2, this has the effect of interchanging the initial $|10\rangle$ with $|11\rangle$, while leaving $|00\rangle$ and $|01\rangle$ untouched. This is therefore a CN gate.

6.8 The state $|0\rangle$ is given by $\mathcal{H}|\overline{\Upsilon}\rangle$, where $\mathcal{H}$ is the 4-bit HADAMAD gate, and

$$|\overline{\Upsilon}\rangle = \frac{1}{4}\sum_{x=0}^{15}(-1)^{f(x)}|x\rangle = \frac{1}{4}\left(\sum_{x=0}^{7}|x\rangle - \sum_{x=8}^{15}|x\rangle\right) = \frac{1}{4}\sum_{a,b,c=0,1}(|0abc\rangle - |1abc\rangle).$$

Now pass this state through the 4-bit H gate. Then

$$\mathcal{H}|\overline{\Upsilon}\rangle = \text{H}_1\text{H}_2\text{H}_3\text{H}_4|\overline{\Upsilon}\rangle = \text{H}_1\frac{1}{\sqrt{2}}(|0000\rangle - |1000\rangle) = |1000\rangle = |a\rangle.$$

6.9 Finding the greatest common divisor of two numbers:

1. $R_2 = 19$, $R_3 = 2$, $R_4 = 1$, $R_5 = 0$. Hence, $\gcd(124, 21) = 1$.

2. $R_2 = 0$. Hence, $\gcd(126, 21) = 21$.

3. $R_2 = 0$. Hence, $\gcd(21, 7) = 7$.

4. $R_2 = 3$, $R_3 = 0$. Hence, $\gcd(21, 9) = 3$.

6.10 Factorizing a number:

1. $r = 6$ because $5^6 = 15625 = 1 + 744 \times 21$. Therefore, $b = 5^3 - 1 = 124$ and $c = 5^3 + 1 = 126$. From Problem 6.9, we know that $\gcd(124, 21) = 1$ and $\gcd(126, 21) = 21$. Thus, $a = 5$ and $r = 6$ cannot be used to find the prime factors p and q.

2. $r = 6$ because $2^6 = 64 = 1 + 3 \times 21$. Therefore, $b = 2^3 - 1 = 7$ and $c = 2^3 + 1 = 9$. From Problem 6.9, we know that $\gcd(21, 7) = 7$ and $\gcd(21, 9) = 3$. Thus, the prime factors of $N = 21$ are 7 and 3.

Chapter 8. Bright Stars and Black Holes

8.1 Wien's law, $\lambda = 0.0029$ m K$/T$, gives for the earth $\lambda = 10$ μm (infrared), the sun $\lambda = 500$ nm (yellow) and a class A star $\lambda = 290$ nm (blue).

8.2 (a) Use $L = 7.14 \times 10^{-7} R^2 T^4$ (in SI units) to get for the earth $L = 2.37 \times 10^{17}$ W, and for the sun $L_\odot = 0.39 \times 10^{27}$ W.

(b) $(T/T_\odot)^4 = 23/(1.8)^2 = 7.1$, hence $T = 1.63 T_\odot = 9450$ K.

8.3 (a) Using $d = 1$ AU $= 1.496 \times 10^{11}$ m and $L_\odot = 0.39 \times 10^{27}$ W, we get the solar apparent luminosity $\mathcal{L}_\odot = 1.4$ kW/m^2.

(b) Use the formula $d^2 = L/(4\pi\mathcal{L})$ and the data to get $d = 21$ pc.

8.4 (a) For stars of the same radius, $L\lambda^4 =$ constant.

(b) Because the data are in solar units, we must return to the more general formula $L \propto R^2/\lambda^4$. For Arcturus, we have $\lambda = 1.36\lambda_\odot = 680$ nm (red), while for Orionis, $\lambda = 0.447\lambda_\odot = 223$ nm (blue).

8.5 (a) Life-time is given by $t = E/L$. We take $E = 0.1$ Mc^2 and $L \propto M^a$. Hence, the star's life-time on the main sequence is $t \propto M^{1-a}$.

(b) On the main sequence, we assume the relations $R \propto M^b$, $L \propto M^a$ and $L \propto R^2 T^4$. Together, they give $T \propto M^{(a-2b)/4}$.

8.6 (a) We add electrons on both sides of $4\,p \rightarrow {}^4\text{He} + 2\,e^+ + 2\,\nu$, so that the initial system is just equivalent to four hydrogen atoms, while the final system is just a helium-4 atom, with the two electron–positron pairs canceling out to give gamma rays (energy). Hence, the mass deficit of the reaction is just the mass difference between four hydrogen atoms and one helium-4 atom: $(4 \times 1.007825 - 4.0026) \times 931.5 = 26.73$ MeV.

(b) 14 MeV.

(c) 0.862 MeV.

8.7 Consider a column of matter extending from the center to the surface and ask yourself what is the weight per unit area of material on top. In a rough calculation, we take the *average* gravitational field felt by a particle in the column as $g = GM/R^2$. The mass per unit area is $\mu = M/R^2$. Hence, $P = \mu g = GM^2/R^4$.

8.8 Since $\Delta x \propto n^{-1/3}$ and $\Delta p_x = h/\Delta x = hn^{1/3}$, we have $v_x = p_x/m$ (non-relativistic). Therefore, $P_e = nv_x p_x = h^2 n^{5/3}/m$. The electron number density n is $n = Z\rho/Am_p$, where ρ is the mass density given by $\gamma(3M/4\pi R^3)$, the factor γ is a pure number. So that $P_e \propto M^{5/3}/R^5$. Equating this to $P_g \propto M^2/R^4$, we obtain $MR^3 = $ constant.

8.9 As above, except that now $v = c$. Hence, $P_e = hcn^{4/3} \propto M^{4/3}/R^4$. The equation $P_e = P_g$ becomes an equation for M, which yields the mass limit.

8.10 If N is the number of ions ($N = M/(Am_N)$) then the total internal energy of the white dwarf is $U = 3NkT/2$. The cooling rate is $-dU/dt$, which is equal to L. Hence the cooling time is, approximately, $\tau = U/L$, and exactly $\tau = 2U/5L$ (the latter by exact integration). If we take $N = 10^{56}$ (assuming carbon, $A = 12$) and $T = 10^7$ K, then $U = 10^{40}$ J and $L = 10^{23}$ W, so that we find $\tau = 10^9$ yr.

8.11 (a) 0.2×10^{-5}; 0.2×10^{-3}; 0.15; 0.75.

(b) $28g_E$; $28 \times 10^4 g_E$; $14 \times 10^{10} g_E$; $14 \times 10^{11} g_E$.

8.12 (a) Electron gas in metals: $\varepsilon_F = 0.31$ eV, $T_F = 4000$ K.

(b) Electron gas in white dwarfs: $\varepsilon_F = 3.8 \times 10^5$ eV, $T_F = 10^9$ K.

(c) Neutron gas in neutron stars: $\varepsilon_F = 2 \times 10^7$ eV, $T_F = 10^{11}$ K.

8.13 Since 1 yr $= 3 \times 10^7$ s, an accretion rate of $10^{-10} M_\odot/$yr is equivalent to 0.7×10^{30} J/s, or 10^{30} W. If all of this goes into black body radiation, then $L \geq 10^{30}$ W. We use the formula $(L/L_\odot) = (R/R_\odot)^2 (T/T_\odot)^4$ to obtain $T \geq 2000 T_\odot = 10^7$K. Wien's displacement law then gives $\lambda \leq 0.3$ nm.

8.14 The gravitational binding energy of a proton on the surface of a star of mass M and radius R is $\varepsilon = GMm_p/R = \gamma m_p c^2$, where γ is the surface potential. For a solar-mass neutron star, we know that $\gamma = 0.15$, so that the rate of conversion is 15% the accreted mass, or 1.5×10^{16} J/kg.

8.15 To measure the mass M of a black hole, one would place a satellite into orbit about the hole. Once one has measured the size and period of the satellite's orbit, one would use Kepler's law to determine M. To measure the hole's spin, one would place two satellites in orbit, one circling in the same direction of rotation of the hole, the other in the opposite direction. By comparing the two orbital periods of the satellites, one can determine the hole's angular momentum. Finally, one would send a spacecraft, carrying test charge and instruments sensitive to electric fields, passing near the black hole, and would then measure its charge.

8.16 For $M = M_\odot$, $r = 215$ km; for $M = 100M_\odot$, $r = 10^3$ km; and for $M = 10^9 M_\odot$, $r = 2 \times 10^5$ km.

8.17 (a) Setting $J = M^2$ in the relation $M^2 = M_{\mathrm{irr}}^2 + J^2/(4M_{\mathrm{irr}}^2)$, we get $M = \sqrt{2}M_{\mathrm{irr}}$.

(b) The initial mass is M_{irr}, the final mass is $\sqrt{2}M_{\text{irr}}$. Hence, the mass increase is $\delta M = (\sqrt{2} - 1)M_{\text{irr}}$. The rotational energy is $\delta M = 0.29M$.

8.18 (a) With $E = 3NkT/2$, where $N = 10^{57}$, we have $S_\odot = 2 \times 10^{57}k$.

(b) Since $L_{\text{P}}^2 = 2.6 \times 10^{-70}$ m^2 and $A = 4\pi r_{\text{g}}^2 = 1.13 \times 10^8$ m^2, we have the entropy $S_{\text{BH}} = kA/4L_{\text{P}}^2 = 10^{77}k$, which is 10^{20} times larger than the entropy of the sun.

Chapter 9. Elementary Particles and Forces

9.1 List of particles:

Name	mc^2 in MeV	spin (J)	charge (e)
photon γ	0	1	0
gluon g	0	1	0
electron e^-	0.51	$\frac{1}{2}$	-1
muon μ^-	105.66	$\frac{1}{2}$	-1
electron-neutrino ν_e	$< 3 \times 10^{-6}$	$\frac{1}{2}$	0
muon-neutrino ν_μ	< 0.19	$\frac{1}{2}$	0
tau-neutrino ν_τ	< 18.2	$\frac{1}{2}$	0
u-quark u	1.5 to 4.5	$\frac{1}{2}$	$+\frac{2}{3}$
d-quark d	5 to 8.5	$\frac{1}{2}$	$-\frac{1}{3}$
s-quark s	80 to 155	$\frac{1}{2}$	$-\frac{1}{3}$
pion $\pi^\pm$	139.57	0	± 1
pion π^0	134.98	0	0
η meson	547.30	0	0
$f_0(600)$ meson	400 to 1200	0	0
ρ meson	771.1	1	$\pm 1, 0$
ω meson	782.57	1	0
η' mesons	957.78	0	0
$f_0(980)$ meson	980	0	0
a_0 meson	984.7	0	0
$K^\pm$ meson	493.68	0	± 1
$K^0, \overline{K^0}$ mesons	497.67	0	0
$K^*(892)$ mesons	~ 892	1	$\pm 1, 0, 0$
proton p	938.27	$\frac{1}{2}$	$+1$
neutron n	939.57	$\frac{1}{2}$	0

9.2 Using $\hbar c = 197$ MeV fm, and the rest energy mc^2 from the table in Problem 1, we obtain the ranges for the exchange of ρ, ω, η, η' to be respectively 0.26 fm, 0.25 fm, 0.36 fm, and 0.21 fm.

9.4 From the table in Problem 1, the energy needed to create a $\pi^+\pi^-$ pair is $2 \times 139.57 = 279.14$ MeV. This translates into a distance $r = 279.14$ MeV$/\sigma = \hbar c/573$ MeV $= 0.34$ fm.

9.5 Verification of relation $Q = I_3 + q/6$.

	I_3	q	Q	$I_3 + q/6$
u	$+\dfrac{1}{2}$	$+1$	$+\dfrac{2}{3}$	$+\dfrac{2}{3}$
d	$-\dfrac{1}{2}$	$+1$	$-\dfrac{1}{3}$	$-\dfrac{1}{3}$
$\bar{u}$	$-\dfrac{1}{2}$	-1	$-\dfrac{2}{3}$	$-\dfrac{2}{3}$
$\bar{d}$	$+\dfrac{1}{2}$	-1	$+\dfrac{1}{3}$	$+\dfrac{1}{3}$

9.5 Since the central value may not be the correct value, the fit cannot be exact, so the parameters λ, A, ρ, η cannot be determined uniquely. From Footnote 3, the magnitude of the matrix elements of V are given by (we have assumed λ to be small and positive):

$$|V| = \begin{pmatrix} 1 - \dfrac{1}{2}\lambda^2 & \lambda & |A|\lambda^3\sqrt{\rho^2 + \eta^2} \\ \lambda & 1 - \dfrac{1}{2}\lambda^2 & |A|\lambda^2 \\ |A|\lambda^3\sqrt{(1-\rho)^2 + \eta^2} & |A|\lambda^2 & 1 \end{pmatrix}.$$

The central values on the webpage are

$$|V| = \begin{pmatrix} 0.9749 & 0.223 & 0.0037 \\ 0.223 & 0.974 & 0.041 \\ 0.009 & 0.041 & 0.9992 \end{pmatrix}$$

λ is fixed from $|V_{us}|$ to be 0.223. With that, $|A|$ can be fixed from $|V_{cb}|$ to be $0.041/(0.223)^2 = 0.824$. The remaining parameters can be determined from $|V_{ub}|$ and $|V_{td}|$ to satisfy $\rho^2 + \eta^2 = (0.0037/|A|\lambda^3)^2 = 0.164$, and $(1-\rho)^2 + \eta^2 = (0.009/|A|\lambda^3)^2 = 0.969$. This gives $|\rho| = 0.086$ and $|\eta| = 0.396$.

Chapter 10. Cosmology

10.1 Measurements of astrophysical parameters:

1. $v = zc = 3 \times 10^4$ km/s.

2. $s = v/H_0 = (3 \times 10^4/72)$ Mpc $= 420$ Mpc $= 4.2 \times 10^8$ pc $= 1.5 \times 10^9$ light-years.

3. (a) Recall that a difference of 5 magnitudes corresponds to a brightness ratio of 100, and that

   (b) The apparent brightness is $(10 \text{ pc/s})^2$ times the intrinsic brightness.

   (c) If y is the difference between the apparent magnitude and the absolute magnitude, then the apparent brightness is $(100)^{-y/5}$ times the intrinsic brightness.

   (d) Equating (b) and (c), and taking the logarithm to base 10 on both sides, we get $-2y/5 = 2 \times \log(10 \text{ pc/s})$. Hence, $y = 5 \times \log(s/10 \text{ pc}) = 5 \times \log(4.2 \times 10^7) = 38.1$, so the apparent magnitude of the galaxy is $38.1 - 20 = 18.1$.

10.2 Let $\rho_0 = \rho_{01} + \rho_{02}$ be the present energy density of universe, where ρ_{01}, ρ_{02} are, respectively, the matter and dark energy components. Similar notations without the subscript 0 will be used to denote the densities at any time. The equations of state for these two components are, respectively, $P_1 = 0$ and $P_2 = -0.8\rho_2$.

Recall from Sec. 10.3.5 that the equation of state $P = -w\rho$ leads to the dependence $\rho \propto a(t)^{3(w-1)}$. Hence, $\rho_1 = \rho_{01}[a(t_0)/a(t)]^3 = \rho_{01}a(t)^{-3}$, and $\rho_2 = \rho_{02}[a(t_0)/a(t)]^{0.6} = \rho_{02}a(t)^{-0.6}$. The ratio of the two densities at any time is therefore $\rho_2/\rho_1 = (\rho_{02}/\rho_{01})a(t)^{2.4} = 2a(t)^{2.4}$.

1. The effective force on a galaxy is proportional to $\rho + 3P = \rho_1 + \rho_2 + 3(P_1 + P_2) = \rho_1 + (1 - 3w)\rho_2 = \rho_1 - 1.4\rho_2$. Deceleration turns into acceleration when this force is zero. This occurs when $\rho_2/\rho_1 = 1/1.4 = 2a(t)^{2.4}$, or $a(t) = (1/2.8)^{1/2.4} = 0.65$.

2. $z = 1/a(t) - 1 = 0.54$

10.3 Recall from Footnote 13 that entropy conservation implies $g(aT)^3$ to be a constant. Using a subscript 0 to represent the present time, we have $a(t_0) = 1$, $T_0 = 2.728$ K, and $g_0 = 1.68$ (see Footnote 14). At the beginning of the Big Bang, $g = g_U$, $T = 10^{24}$ eV/$k_B = 1.16 \times 10^{28}$ K. Hence, $a = (T_0/T)(g_0/g_U)^{1/3} = 2.8 \times 10^{-28}/g_U^{1/3}$. Let a' denote the scale factor at the beginning to inflation. Hence, $a' = 10^{-25}a = 2.8 \times 10^{-53}$. Taking the present size of the universe to be $r = 14 \times 10^9$ light-years $= 14 \times 10^9 \times 0.946 \times 10^{16}$ m $= 1.3 \times 10^{26}$ m, the size before inflation is therefore $s' = a'r = 3.6 \times 10^{-27}/g_U^{1/3}$ m.

10.4 Evolution of the universe:

1. $a(t) = a(t_0)(T_0/T) = 3/(0.2 \times 10^6 \times 1.16 \times 10^4) = 1.29 \times 10^{-9}$.

2. $z = a(t_0)/a(t) - 1 = 7.7 \times 10^8$.

3. Using Eq. (10.6) with the coefficient equal to 1.32, the time it takes is $t \simeq 1.32 \times (1/0.2)^2 = 33$ s. This estimate is accurate to approximately 1 s because we have used coefficient 1.32 throughout. The time it takes to

reach a temperature of 100 MeV is quite negligible. Assuming e^+e^- annihilation to occur around 1 MeV, the time it takes to reach that can be calculated from Eq. (10.6) using the coefficient 0.74, which is 0.74 s. Since this is short compared to 33 s, we can ignore the correction in this epoch.

10.5 Wien's law:

1. The photon density at $T_0 = 2.728$ K is $(n_\gamma)_0 = 414/(\text{cm})^3$ (see Sec. 10.5.2). At $T = 27°C = 300$ K, the photon density is $n_\gamma = (n_\gamma)_0(T/T_0)^3 = 414 \times (300/2.728)^3 = 5.5 \times 10^8/(\text{cm})^3$.

2. According to Wien's law, the peak wavelength occurs at $\lambda_m = (2\pi/4.967) \times (\hbar c/k_B T) = 1.265(\hbar c/k_B T)$. Since $k_B = 1$ eV/$(1.16 \times 10^4$ K$)$ and $\hbar c = 1.97 \times 10^{-7}$ eV m, $\hbar c/k_B = 2.29 \times 10^{-3}$ m K. With $T = 300$ K, we get $\lambda_m = 1.265 \times 2.29 \times 10^{-3}/300 = 0.96 \times 10^{-5}$ m $= 9.6$ μm.

3. It is dark because the photons are in the infrared, beyond the visible range.

10.6 Elements in the universe:

1. Hydrogen.
2. ^{4}He and D are produced in Big Bang nucleosynthesis.
3. ^{4}He.

10.7 Doppler shift:

1. If $\delta\lambda/\lambda = (\lambda' - \lambda)/\lambda$ is the percentage shift of wavelength, then according to the Doppler shift formula (Sec. 10.1.1), $\delta\lambda/\lambda = v/c = 371/(3 \times 10^5) = 1.24 \times 10^{-3}$. In particular, this is true for the peak wavelength $\lambda = \lambda_m$ of the black-body radiation.

2. According to Wien's formula, $\lambda_m \propto 1/T$, hence $-\delta T/T = \delta\lambda_m/\lambda_m = 1.24 \times 10^{-3}$.

Glossary

absolute zero The lowest possible temperature. It is by definition the zero of temperature on the Kelvin scale, or about -273.15 degrees on the Celsius scale. At this temperature, thermal random motions completely disappear, but not quantum fluctuations.

asymptotic freedom A principle in quantum chromodynamics which says that the gluon-exchange forces acting between quarks become weaker as the quarks move in very close, as they often do in very high-energy collisions, so that the quarks become *free* of the forces at *asymptotically* high energies.

attractor A stable end-state toward which a system with appropriate dynamical properties will ultimately approach.

beta-decay (β-decay) Transmutation of a neutron into a proton, accompanied by the production of an electron and an electron-antineutrino; also, similar transmutation of a neutron-rich (radioactive) atomic nucleus. It is the most familiar manifestation of the weak interaction.

baryons Heavy particles of half-integral spins made up of three quarks, such as the proton and other, higher-mass particles. They can interact through any of the four known fundamental forces.

beauty One of the quark flavors; same as *bottom*.

bifurcation Splitting of an equilibrium state in two, progressively, as a control parameter is varied; hence, the bifurcation diagram.

Big Bang The singularity at the beginning of the universe.

black hole A region of space-time, centered around a gravitational singularity, that cannot be seen or otherwise observed by distant observers, because everything, light or matter, once inside, cannot escape to the outside. Physicists think that black holes exist in three forms: stellar, supermassive and primordial.

Bose–Einstein condensate (BEC) Below a critical temperature that depends on the number density, a system with fixed number of indistinguishable Bose particles has a finite fraction of it occupying the lowest-energy single-particle state. This macroscopically occupied state is called the BEC. Thus, we have the BEC of ^{4}He, ^{87}Rb, etc. The BEC shows quantum coherence effects like interference, and is associated with superfluidity — zero viscosity.

bosonic stimulation It refers to the phenomenon of a Bose particle being scattered preferentially to a state which already has a higher population of the bosons of the same kind. This follows from Bose statistics. It is involved in the kinetics of Bose–Einstein condensation.

bosons Particles having an integral-valued intrinsic spin (measured in units of $\hbar$). Photons, gravitons, gluons, W and Z — all particles associated with the transmission of forces — are bosons. Composite objects made up of even number of fermions (particles possessing half-integral spin) — *e.g.*, mesons, ^{4}He, ^{16}O — also behave as bosons. Bosons obey *Bose–Einstein statistics*, which favors the occupation of the same state by many particles. They also exhibit a phenomenon called *Bose condensation* observed at a temperature near the degeneracy temperature.

bottom-up nanotechnology Attempts to create structures by connecting molecules. See *chemical reduction, electrochemical processes, thermal decomposition* and *self-assembly*.

Cepheid variable A type of star with variable luminosity due to pulsations (regular oscillating changes in size). The luminosity can be deduced from the period of pulsation, making these stars celestial standard candles, useful for the determination of distance.

chaos Aperiodic and apparently random behavior of a dynamical system despite governing laws being deterministic; hence, deterministic chaos.

charm One of the quark flavors.

chemical reduction Method by which crystals of pure metal or metal alloys can be produced by dissolving an inorganic salt of the desired metal in an appropriate solvent and allowing the solution to react with a reducing agent.

color A quantum number for quarks and gluons.

conduction band Set of closely spaced single-electron energy levels in a metal that are partially occupied by current-carrying electrons. In an insulator or a semiconductor, it is the lowest normally empty band to which electrons may be excited from lower levels to carry electric current.

conservation Principle according to which the total amount of some physical quantity in an isolated system always remains the same, whatever internal changes may occur in the system.

Cooper pair Stable complex of two fermions having equal and opposite momenta and opposite spins in a degenerate-fermion system; such a pair behaves as a condensed boson if an effective attractive force exists between the two fermions in the pair.

cosmic rays High-energy particles (mostly protons, electrons and helium nuclei) present throughout our galaxy. Their origin is not well known. They can be detected above the earth's atmosphere. Cosmic rays are the object of study in *cosmic-ray astronomy*.

cosmic string String-like object with a high-energy density that could have been formed at the beginning of the universe. It can serve as a seed to accrete matter into galaxies.

cosmological constant A constant that can be incorporated into Einstein's gravitational theory (general theory of relativity). It represents the energy density of a matterless universe. This number is observationally known to be either zero or very small, though present theories of elementary particle physics almost invariably predict a large value.

cosmological principle This principle states that the universe looks the same to every observer anywhere in the universe if local matter fluctuation is averaged out. It is the modern version of what Copernicus advocated some 500 years ago.

critical phenomena Singular thermodynamic behavior of a system close to a second-order phase transition. Example: Divergence of the magnetic susceptibility of a magnet at the *critical point* (Curie temperature).

degeneracy A state of matter characterized by extreme compactness or density in which the quantum effects of indistinguishability (Bose or Fermi statistics) of identical particles become important. The *degeneracy* temperature is the temperature below which such effects become observable. A system of identical particles is said to be *degenerate* when they occupy the lowest possible one-particle levels consistent with their respective statistics. For bosons, this means the same one-particle state, and for fermions, all one-particle states up to a certain energy (the Fermi level), while all states with higher energies are empty. The *pressure* of degenerate matter is dominated by quantum effects and is largely independent of temperature.

dielectric material A material in which an electric field can be sustained with a minimum of energy dissipation.

Doppler cooling Slowing down of an atom moving against a beam of (laser) light whose frequency is tuned slightly below that for resonant atomic absorption, *i.e.*, red-detuned. Due to the Doppler-shift effect, the red-detuned light photon is blue-shifted into atomic resonance and absorbed, giving in the process a momentum kick directed so as to retard the atom. The subsequent spontaneous re-emission is, however, undirected (isotropic). This results in a net retarding force slowing down the atom, and hence, the Doppler cooling.

Doppler shift The change in wavelength of waves as seen by an observer when the source and observer are in relative motion. When they are moving apart the wavelength increases and, conversely, when they approach the wavelength decreases. With sound, the Doppler effect causes a shift in pitch, and with light, a shift in color of the source. Named for Christian Doppler.

electrochemical processes Processes in which an electric field is used to initiate a chemical reaction. For example, layers of nickel can be grown on a silver substrate by applying an electrode potential to the electrolyte Ni^{2+}.

electron gas A collection of electrons which interact with each other weakly enough to be considered as practically free, subject only to the exclusion principle.

energy band A continuous range of energies in a solid in which there are allowed quantum states for electrons. Energy bands may be occupied or empty, and are separated from one another by *gaps*, which are energy regions where no states for electrons exist.

entropy A thermodynamic property which gives a measure of the degree of disorder of a macroscopic system.

event horizon The boundary of a black hole. Light or any signal emitted from inside cannot escape to the outside. Only light emitted outside can escape to infinity.

exclusion principle *See* Pauli's exclusion principle.

fermion Particles having a half-integral intrinsic spin (measured in units of $\hbar$). Quarks and leptons — all particles associated with matter and antimatter — are fermions. Composite systems of odd numbers of fermions, such as baryons, ^{3}He, ^{13}C and so on, also behave as fermions. Fermions obey *Fermi statistics*, *i.e.*, the Pauli exclusion principle, which forbids two identical fermions to occupy the same one-particle state in a given system.

Feynman diagram A pictorial representation of the quantum field theoretical predictions of particle and of many-body dynamics.

field A physical quantity that extends over space, to be contrasted with a particle, which is localized at a point in space. Examples are electric field, magnetic field and temperature distribution in a spatial region.

flavor The quality that distinguishes the different varieties of quarks — up, down, strange, charmed, top (or truth) and bottom (or beauty). This property is also used to distinguish the various types of leptons (electron, muon, tau).

fractal geometry Geometry of highly irregularly shaped objects having fractional dimensions (*e.g.*, jagged coast-lines, sponges, foams). Studied first by Benoît Mandelbrot.

fundamental interactions Forces acting between the basic constituents of matter, to be contrasted with *effective* interactions, which are simplified forms of interactions between composite particles. The four known fundamental interactions are: the gravitational force, the electromagnetic force, the weak force and the strong force.

gamma rays A very energetic electromagnetic radiation, with wavelengths shorter than 10^{-11} m, or energies greater than 1 million electron-volts.

gas-phase condensation A precursor material is vaporized by some high-energy process (*e.g.*, electric arc discharge, heating, laser ablation, irradiation by electrons) and the vapor is then condensed in a controlled inert-gas environment. Many metal, ceramic nanoparticles, fullerenes and tubulites are grown in this manner. In a related process, called physical vapor synthesis, the precursor is heated by a plasma in an open atmosphere and the vapor is cooled by collisions with the atoms of a reactive gas.

gauge symmetry Invariance of the form of the dynamical equations with respect to certain (space-time dependent) changes of various fields.

gauge theories Theories of fundamental interactions based on gauge symmetry. In such a theory a spin-1 *gauge field(s)* is always present, whose longitudinal polarization is decoupled by the gauge symmetry. Particles having the same gauge quantum numbers always couple the same way to the gauge field(s).

general relativity Theory of gravitation developed by Albert Einstein. It gives the necessary equations to determine the geometry of spacetime in terms of a distribution of matter and energy.

globular cluster A gravitationally bound system of (typically 10^6) stars with a well-defined spheroidal or ellipsoidal shape. These stars are among the oldest known objects in our galaxy and are characterized by very low abundances of heavy elements.

gluon Massless boson of spin one and agent for the strong interaction of quarks.

graviton Massless boson with a spin value of two, associated with the gravitational force.

hadron Any strongly interacting particle, made up of quarks and antiquarks. Mesons and baryons are all hadrons.

Hawking's radiation Radiation emitted by a black hole when a particle–antiparticle pair, produced near the event horizon, splits into two, the antiparticle falling through the event horizon and the particle escaping to infinity; discovered by Stephen Hawking.

Heisenberg's uncertainty principle A general proposition in quantum mechanics that not all quantities in a quantum system may have simultaneously definite values. The uncertainty relations are mathematical formulas (due to Werner Heisenberg) that describe this irreducible level of uncertainty of certain pairs of dynamical variables when they are observed in identical situations.

Hertzsprung–Russell diagram The plot of stellar brightness versus surface temperature for stars (named in honor of Ejnar Hertzsprung and Henry Russell). Ordinary stars lie in a band called the *main sequence*; other types of stars are found in specific places on the diagram.

Higgs field A spin-zero field, some of whose components have non-zero values in the vacuum at a low enough temperature. This *vacuum condensate* is instrumental in giving non-zero masses to the gauge bosons (fields) and fermions.

Higgs boson Particle associated with a Higgs field which is not needed to provide the longitudinal polarization degree of freedom for a massive gauge boson; named after Peter Higgs.

holography Photography by reconstruction of light waves. The two parts of a laser beam, split into two by a half-silvered mirror, are directed at a photographic plate, one directly, the other after being scattered off the subject. The photographic plate records the interference pattern caused by the recombination of the two beams. When the *hologram*, as this photographic record is called, is illuminated by light from a laser identical in characteristics to the laser used for the recording, a three-dimensional image of the original object is reconstructed.

Hubble's law The observation, first made by E. Hubble, that distant galaxies are receding from us with a velocity proportional to their distance.

inflationary universe A theory of the very early universe constructed to overcome certain difficulties of the classical Big Bang theory. According to this theory, the universe underwent a huge expansion in its very early life, a period during which much of the matter of the universe and the energy for the explosion in the classical Big Bang was produced.

interference A characteristic property of waves whereby a new wave is formed when two waves overlap. When the two interfering waves are in step, an amplified wave motion is produced (constructive interference); otherwise, an attenuated wave motion is obtained (destructive interference).

interstellar medium The gaseous and dusty matter present in regions between a galaxy's stars.

laser Acronym for Light Amplification by Stimulated Emission of Radiation. It is a device composed of an optically active medium, a pumping mechanism to inject energy into the medium and a resonant cavity to amplify the radiation. It can generate an intense directional beam of coherent light.

lepton Generic name for a class of fermions (electron, muon, tauon and neutrinos) that respond to weak and electromagnetic forces, but not to the strong force.

light shift (also called lamp shift) The shift of energy of an atomic level due to its local interaction with light. The shift depends on the intensity and the polarization of the light and can, therefore, be made to vary in space, *e.g.*, periodically for a standing-wave of light. This is made use of in Sisyphus cooling.

lithography Production process in micro-electronics in which a pre-designed pattern is transferred from a template to a target material by exposing a selectively resistive overlay to energetic photons, electrons, ions or atoms. The patterned structure is then coated with a metal, infused with a dopant or subjected to further etching in order to fashion useful devices.

magnetic monopole A particle with a non-zero *magnetic* charge. All the known macroscopic and microscopic magnetic objects are *magnetic dipoles*, with equal and opposite magnetic charges which total to zero. A magnetic monopole has never been found, but is postulated to exist in certain theories.

magneto-optical trap (MOT) A combination of laser Doppler cooling and electromagnetic trapping for neutral atoms consisting of three pairs of mutually perpendicular, appropriately polarized counterpropagating laser beams and a static magnetic-field configuration that provides a scattering force which depends on the position (for trapping) and velocity (for cooling) of the atoms. MOT is the workhorse for most BEC research work.

maser Acronym for Microwave Amplification by Stimulated Emission of Radiation, the immediate predecessor of the laser and operating along the same principles in the microwave band.

Meissner effect Exclusion of magnetic flux from the bulk of a superconductor on cooling through the transition temperature. Hence, a superconductor is a perfect diamagnet.

meson Bosonic particle composed of a quark and an antiquark. Examples are the π-meson and the K-meson.

microwave background radiation Thermal radiation with a temperature of about 3 K uniformly distributed in the universe. The radiation is believed to be the cooled remnant of the hot radiation that was in thermal equilibrium with matter during the very early phases of the existence of the universe.

muon One of the three known kinds of charged leptons, a heavier analogue of the electron.

nanoscience Science dealing with extremely small objects (atomic-sized structures and devices, finely-structured materials, etc.).

nanotechnology Creation of functional materials, devices, and systems through control of matter at the scale of 1–100 nanometers, and the exploitation of novel properties and phenomena at the same scale.

neutrino Electrically neutral lepton. Its mass is either small or vanishing. There are three known varieties, one associated with each of the charged leptons (electron, muon and tau); their presence is always indicative of the action of the weak force.

neutron star A dense compact star consisting predominantly of neutrons and supported against gravity by degenerate-neutron pressure.

nonlinear optics The study of optical properties of matter under intense radiation fields. It is found that in such circumstances, the optical response of matter is nonlinear, *i.e.*, strongly sensitive to the applied field, and manifests itself in some unusual forms, such as a nonlinear induced polarization, frequency mixing, harmonics generation and so on.

nuclear fission Division of a heavy-mass atomic nucleus into two nuclear fragments of nearly equal masses and possibly lighter particles, accompanied by a release of energy. This process is the operational basis of nuclear fission reactors and atomic bombs.

nuclear fusion Combination of two very light nuclei into a nucleus lighter than iron. This process is the main source of energy of stars, and also forms the operational basis of controlled fusion reactors and thermonuclear weapons. Also called nuclear burning.

nucleosynthesis Synthesis of chemical elements in the universe. Very light elements are synthesized during the hot early phases after the Big Bang. Other elements, not heavier than iron, are produced by nuclear fusion in the central regions of stars. Heavier elements are produced mainly during supernova explosions.

optical chaos An uncontrollable and unpredictable response exhibited by many nonlinear optical media under an applied intense field.

optical molasses A set of three pairs of counter-propagating laser beams along three mutually perpendicular directions, red-detuned so as to slow down, through the Doppler cooling effect, an atom moving in an arbitrary direction. The retarding force is proportional and directed opposite to the atom's velocity, making the system act as a viscous fluid — hence, the optical molasses.

order parameter A physical quantity that is a measure of the order of a thermodynamic state (or phase). It can be scalar, vector, or tensor. Thus, magnetization is the order parameter for a ferromagnet.

parity Left–right symmetry with respect to mirror reflection.

Pauli's exclusion principle A general rule of quantum theory (found by Wolfgang Pauli) that forbids two fermions of the same kind to occupy the same quantum one-particle state in a given system. At any space-time point, at most one such particle can carry a given complete set of quantum numbers.

phase The phase of a wave is a measure of the state of the wave's motion. When a wavelength of a wave has passed at a given point in space, we may say that the wave motion has completed a cycle, or that it has changed its phase by $360°$. Two waves are said to be in phase if their peaks coincide; otherwise, their phase difference indicates how far apart their peaks are. In a completely different usage, the thermodynamic state of a macroscopic system is also called a *phase*. Thus, ice, water and steam represent different phases of H_2O.

phase transition Change of thermodynamic state as some parameter, *e.g.*, temperature is varied. *First-order* transitions, like freezing/melting, are discontinuous and involve latent heat; *second-order* transitions, like magnetic transitions, are continuous and have no latent heat.

photon Massless spin-one boson and quantum of the electromagnetic force.

Planck constant The quantum of action, a fundamental physical constant symbolized by h or $\hbar = h/2\pi$ and named after Max Planck. It corresponds to the scale at which quantum effects become important.

planetary nebula An expanding cloud of gas and matter around a hot star at a late evolutionary stage.

polarization (of a dielectric) Electric response of molecules in a dielectric material; it is also a measure of this effect, *i.e.*, the electric dipole moment per unit volume. Dielectric polarization may arise from the distortion of the electron distribution about the nuclei in an electric field, or from a change in dipole moment resulting from the stretching of electrical bonds between unlike atoms in molecules.

polarization (of waves) Phenomenon observed when a transverse wave is *polarized*, that is, when the displacement of vibrations is completely predictable. A transverse wave is said to be *unpolarized* when the vibrations in the plane perpendicular to the direction of propagation appear to be oriented in all directions with equal probability; no preferred pattern of orientation can be observed over a long time period.

proton–proton chain The sequence of nuclear fusion reactions that transform four protons into an α-particle in the core of ordinary stars.

pulsar A rotating neutron star emitting electromagnetic radiation, which astronomers detect in regular pulses.

quantum chromodynamics (QCD) The theory of strong interactions between quarks, in which the forces are mediated by gluons and the states characterized by color charges.

quantum electrodynamics (QED) The quantum theory of electromagnetic interactions between charged particles.

quark Fundamental particle of all hadrons. There are six known flavors of quarks, which combine in twos (a quark and an antiquark) to form mesons, and in threes to form baryons.

quasar Extremely bright source of radiation, thought to be a spinning black hole with an accretion disk from which radiation is emitted.

radiowaves Electromagnetic radiation of wavelengths 10^{-3}–10^4 m. *Radio astronomy* is the branch of astronomy devoted to the radio observations of celestial objects.

renormalization A procedure to rewrite a fundamental physical law in an effective form appropriate to a certain energy scale μ in terms of a number of measurable parameters.

renormalization group A mathematical relationship describing the fact that the arbitrary choice of μ should not affect the physical outcome of a measurable quantity.

RR Lyrae stars A class of variable stars that can be used to determine astronomical distances.

scale invariance Symmetry demanding invariance with respect to a change of scale (of length, energy, etc.). Thus, a system at its *critical point* is scale invariant, whence all microscopic length scales, such as lattice spacing, become irrelevant.

self-assembly A process in which atoms, molecules or components arrange themselves into functioning entities without human intervention. Once launched, it is driven either by the laws of thermodynamics toward an energetically stable form or by built-in sets of instructions toward a pre-designed form.

Sisyphus cooling An ingenious mechanism devised for laser cooling based on repeated cycles of absorption and re-emission so arranged as to make the atom always move up-the-potential-hill, and hence continuously slow down. Here in each cycle of absorption and re-emission the atom switches between two ground-state sub-levels which are light-shifted periodically in space in antiphase, such that the crest of one coincides with the trough of the other.

spacetime The four-dimensional space in which the three spatial coordinates and the time coordinate are treated on an equal footing, as required by the special and general theories of relativity.

special relativity The currently accepted description of space, time and motion formulated by Einstein in 1905. A key result of the theory is no material body or physical signal can exceed the speed of light.

spin Intrinsic angular momentum of a particle or system of particles. It is characteristic of the particle and independent of its motion. As a quantum entity, it can take on integral or half-integral values in units of $\hbar$. According to the Theorem of Spin and Statistics, particles with integral spins are bosons, and particles with half-integral spins are fermions.

spontaneous emission Emission of a photon by an excited molecule or atom that is independent of external radiation.

spontaneous symmetry breaking The symmetry of the state becoming lower than that of the governing law. This takes place at a phase transition.

state function A mathematical object that contains the most complete physical information about the physical state of a quantum system.

stimulated emission Emission of a photon by an excited atom or molecule under the influence of an external field.

strange attractor A region of phase space (*attractor*) to which the phase trajectories are attracted but on which they wander aperiodically. It has fractal geometry and describes deterministic chaos.

strangeness A quantum number measuring the number of strange quarks inside a hadron.

string theory A theory which holds that the elementary particles are different manifestations of the excited modes of a tiny string, of the order of 10^{-35} m in size, and that all the four fundamental forces are produced by the joining and breaking up of the strings.

superconductivity A low temperature phenomenon of zero electrical resistivity and perfect diamagnetism shown by many materials below their characteristic critical temperatures.

superfluidity Total loss of viscosity exhibited by helium close to the absolute zero of temperature, which enables it to flow through the finest capillaries.

symmetry Invariance of an object with respect to certain operations like mirror reflection. The object may be an equation expressing physical law.

supernova The explosion of a massive star toward the end of its evolution, which leaves behind a supernova remnant composed of a gaseous expanding nebula and, in most cases, a central collapsed object which evolves eventually into a compact object.

tau lepton Negatively charged lepton, similar to but heavier than both the electron and the muon.

thermal convection Transfer of heat by moving masses of matter in gases and liquids.

thermal decomposition Process in which a metal-bearing precursor, when submitted to a high temperature and refluxing in a high-boiling point solvent, decomposes and combines with other atoms in the solvent to form crystallites.

top-down nanotechnology Makes existing structures or devices smaller. See *lithography, gas-phase condensation.*

truth One of the quark flavors; same as *top.*

tunnel effect A quantum property which allows particles to pass through regions which are energetically forbidden by classical physics.

turbulent flow Fluid flow in which the velocity at any point of the fluid changes constantly in magnitude and direction, to be contrasted with a *laminar flow*, which is characterized by a regular space and time dependence.

uncertainty relations *See* Heisenberg's uncertainty principle.

universality The assertion that the critical behavior of a system (*i.e.*, behavior close to second-order phase transition) is independent of the behavior of the microscopic details and depends only on symmetry and dimensionality.

valence band The highest completely filled band of energy levels in a solid.

virtual particle A quantum particle that exists only for very short durations and, because of Heisenberg's uncertainty relations, need not satisfy the usual relations between mass, energy and momentum, in contrast to the more familiar, long lived, *real* particles. Virtual particles may appear singly when exchanged between other particles, or in particle–antiparticle pairs when spontaneously created in vacuum.

weak interaction One of the four fundamental interactions. The weak force is responsible for β-decay and any interaction involving neutrinos or antineutrinos.

white dwarf A dense compact star with mass less than 1.4 solar mass, supported against gravity by degenerate-electron pressure.

X-rays Electromagnetic radiation with wavelengths around 10^{-9} m, or energies around 1 keV. *X-ray astronomy* is the study of celestial X-ray emitters, which include stars, supernovae and active galaxies.

X-ray binary system Double-star system that emits X-rays. Such systems consist of a normal star and neutron star or black hole that accretes matter from its companion star.

zero-point motion Quantum vibrational motion that is always present in a system, even at absolute zero temperature. The energy due to these quantum fluctuations is a minimum, irreducible amount of energy called the zero-point energy. Its existence may be regarded as a consequence of Heisenberg's uncertainty principle.

Index